Manufacturing Processes For Advanced Composites

Manufacturing Processes For Advanced Composites

F. C. Campbell

UK	Elsevier Ltd, The Boulevard, Langford Lane, Kidlington, Oxford OX5 1GB, UK
USA	Elsevier Inc, 360 Park Avenue South, New York, NY 10010-1710, USA
JAPAN	Elsevier Japan, Tsunashima Building Annex, 3-20-12 Yushima, Bunkyo-ku, Tokyo 113, Japan

Reprinted 2004

British Library Cataloguing in Publication Data
Campbell, Flake C.
Manufacturing processes for advanced composites
1.Composite materials 2.Manufacturing processes
I.Title
620.1'18

ISBN 978-1-85617-415-2

Library of Congress Cataloging-in-Publication Data
Campbell, Flake.
Manufacturing processes for advanced composites / Flake Campbell, Jr.
p. cm.
Includes bibliographical references and index.
ISBN 1-85617-415-8 (hardcover)
1. Polymeric composites. 2. Fibrous composites. 3. Manufacturing processes. I. Title.
TA418.9.C6C265 2003
668.9–dc22 2003049415

Published by
Elsevier Advanced Technology
The Boulevard, Langford Lane, Kidlington, Oxford OX5 1GB, UK
Tel: +44(0) 1865 843000
Fax: +44(0) 1865 843971

Transferred to Digital Printing 2007

Contents

Chapter 9 Liquid Molding

Chapter 10 Thermoplastic Composites

Chapter 11 Commercial Processes

Preface

This book deals is intended for anyone wishing to learn more about the materials and manufacturing processes used to fabricate and assemble advanced composites. Although advanced composites can mean many different types of fibers in either polymer, metal or ceramic matrices, this book deals with the three main fibers (glass, aramid and carbon) in polymeric matrices.

The book (Chapter 1) starts with an overview of fibers, matrices and product forms. Then, a brief introduction to the various fabrication processes is covered along with the advantages and disadvantages of composites. The first chapter wraps up with some of the applications for advanced composites. Chapter 2 examines the reinforcements and prepregging in more detail, while Chapter 3 covers the main thermosetting resin systems, including polyesters, vinyl esters, epoxies, bismaleimides, polyimides and phenolics. The principles of resin toughening are also presented along with an introduction to the physiochemical tests that are used to characterize resin and cured laminates.

Chapters 4 through 7 form a natural progression with Chapter 4 covering the basics of cure tools followed by ply collation in Chapter 5. Important ply collation methods include manual lay-up, flat ply collation, automated tape laying, filament winding and fiber placement. Vacuum bagging in preparation for cure is also discussed in Chapter 5. Chapter 6 discusses the cure process for both addition and condensation curing thermosets. The importance of both lay-up and cure variables are discussed including hydrostatic resin pressure, chemical composition, resin and prepreg, debulking and caul plates. Residual cure stresses are also covered followed by exotherm, in-process cure monitoring and cure modeling. In Chapter 7, a case study is presented on the interaction of chemical composition and processing on laminate quality.

Adhesive bonding and integrally cocured structures are introduced in Chapter 8. The basics of adhesive bonding are covered along with the advantages and disadvantages. The important of joint design, surface preparation and bonding procedures are discussed along with honeycomb bonded assemblies, foam bonded assemblies and integrally cocured assemblies.

Chapter 9 on liquid molding covers preforming technology (weaving, knitting, stitching and braiding) followed by the major liquid molding processes, namely resin transfer molding (RTM), resin film infusion (RFI) and vacuum assisted resin transfer molding (VARTM).

Thermoplastic composites are discussed in Chapter 10 including the major matrix materials and product forms. Consolidation is then covered followed by the different methods of thermoforming thermoplastics. Finally, the joining processes that are unique to thermoplastic composites are discussed.

Some of the important commercial composite processes are presented in Chapter 11 with an emphasis on lay-up, compression molding, injection molding, structural reaction injection molding and pultrusion.

Chapter 12 covers the processes that are unique to assembling composite structures. In this chapter, the emphasis is on mechanical joining including the hole preparation procedures and fasteners used for composite assembly. Sealing and painting of composite structure is also briefly discussed.

The final chapter (Chapter 13) covers two topics: nondestructive inspection and repair. NDI methods covered include visual, ultrasonics, radiographic and thermographic inspection methods. The part of repair includes fill repairs, injection repairs, bolted repairs and bonded repairs.

It should be pointed out that this book deals solely with the materials and processes used to manufacture advanced composites. It does not address the mechanics of lamina, laminates, adhesive bonding or bolted joints. If it were used as text, it would probably be more appropriate as a second course in composite materials.

The author would like to thank the following colleagues for reviewing the chapters in this book: Gary Bond, Ray Bohlmann, John Griffith, Mike Karal, Dan King, Bob Kisch, Doug McCarville, Mike Paleen and Bob Rapp. Any and all mistakes remain the responsibility of the author.

F.C. Campbell
St. Louis, Missouri
December 2003

The Author

Flake C. Campbell

Flake Campbell is currently a Senior Technical Fellow in the field of Manufacturing Technology within The Boeing Company's Phantom Works R&D organization. He currently conducts R&D programs in both advanced composites and metallic structures. His 34-year career at Boeing has been split about equally between engineering and manufacturing. He has worked in the engineering laboratories, manufacturing R&D, composites engineering on three production aircraft programs, and in production operations. Prior to being named a Senior Technical Fellow, he was Director of Manufacturing Process Improvement for the St. Louis operations for five years and was Director of Advanced Manufacturing Technology for nine years. He holds B.S. and M.S. degrees in Metallurgical Engineering and a Masters of Business Administration.

Chapter 1

Introduction to Composite Materials and Processes: Unique Materials that Require Unique Processes

A composite material can be defined as a combination of two or more materials that results in better properties than when the individual components are used alone. As opposed to metal alloys, each material retains its separate chemical, physical and mechanical properties. The two constituents are normally a fiber and a matrix. Typical fibers include glass, aramid and carbon, which may be continuous or discontinuous. Matrices can be polymers, metals or ceramics. This book will deal with continuous and discontinuous fibers embedded in polymer matrices, with an emphasis on continuous-fiber high-performance structural composites. Examples of continuous reinforcements include unidirectional, woven cloth and helical winding, while discontinuous reinforcements include chopped fibers and random mat (Fig. 1).

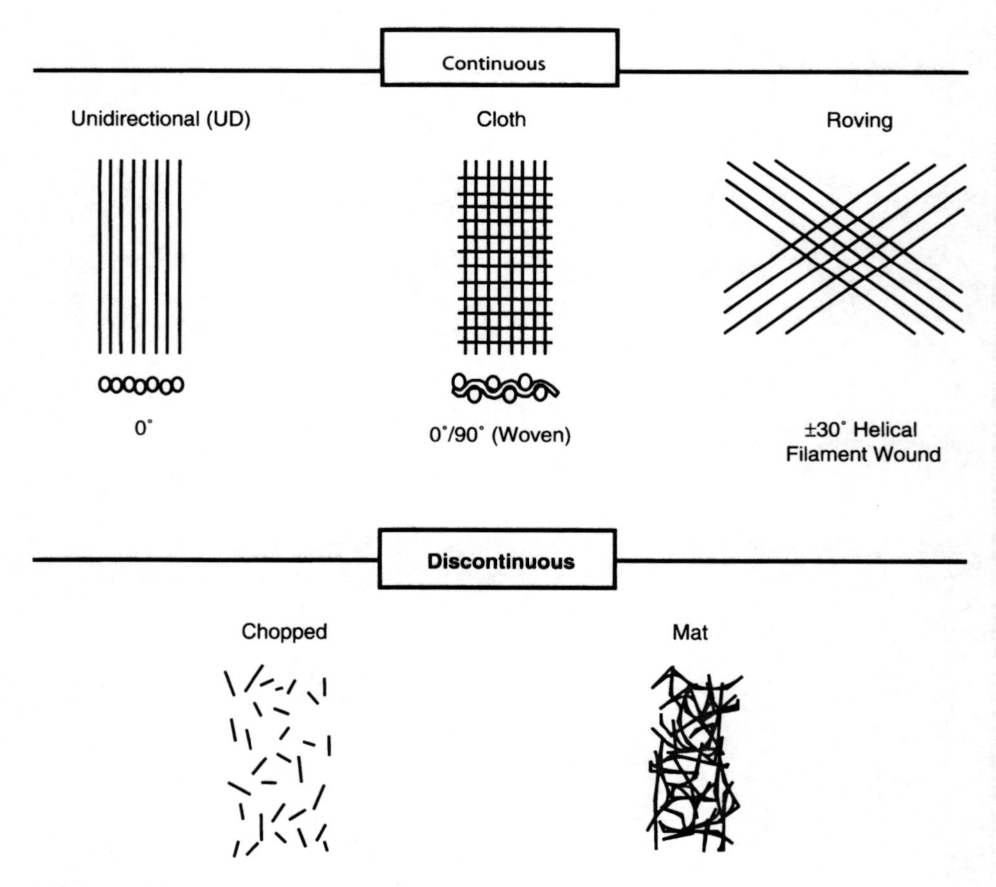

Fig. 1. Reinforcement Options

1.1 Laminates

Continuous-fiber composites are laminated materials (Fig. 2) in which the individual layers, plies or laminae are oriented in directions that enhance the strength in the primary load direction. Unidirectional (0°) laminates are extremely strong and stiff in the 0° direction; however, they are also very weak in the 90° direction because the load must be carried by the much weaker polymeric matrix. While a high-strength fiber can have a tensile strength of 500 ksi or more, a typical polymeric matrix normally has a tensile strength of only 5-10 ksi (Fig. 3). The longitudinal tension and compression loads are carried by the fibers, while the matrix distributes the loads between the fibers in tension and stabilizes and prevents the fibers from buckling in compression. The matrix is also the primary load carrier for interlaminar shear (i.e., shear between the layers) and transverse (90°) tension. The relative roles of the fiber and the matrix in determining the mechanical properties are summarized in Table 1.

Since the fiber orientation directly impacts the mechanical properties, it would seem logical to orient as many layers as possible in the main load-carrying direction. While this approach may work for some structures, it is usually necessary to balance the load-carrying capability in a number of

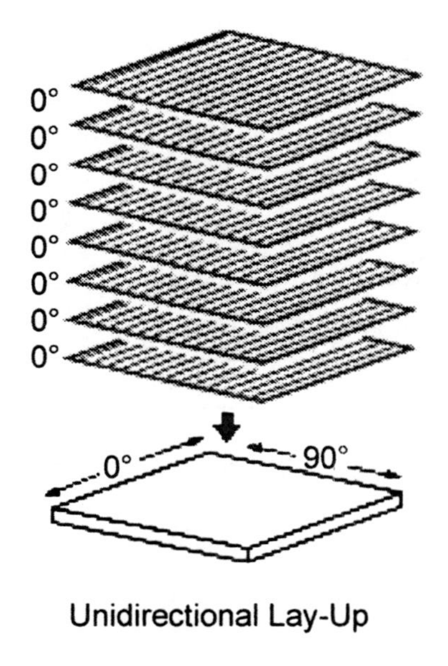

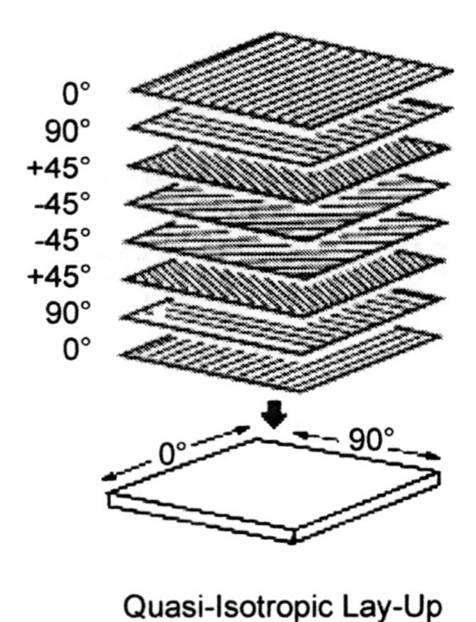

Reference 1: with permission

Fig. 2. Quasi-Isotropic Laminate Lay-Up

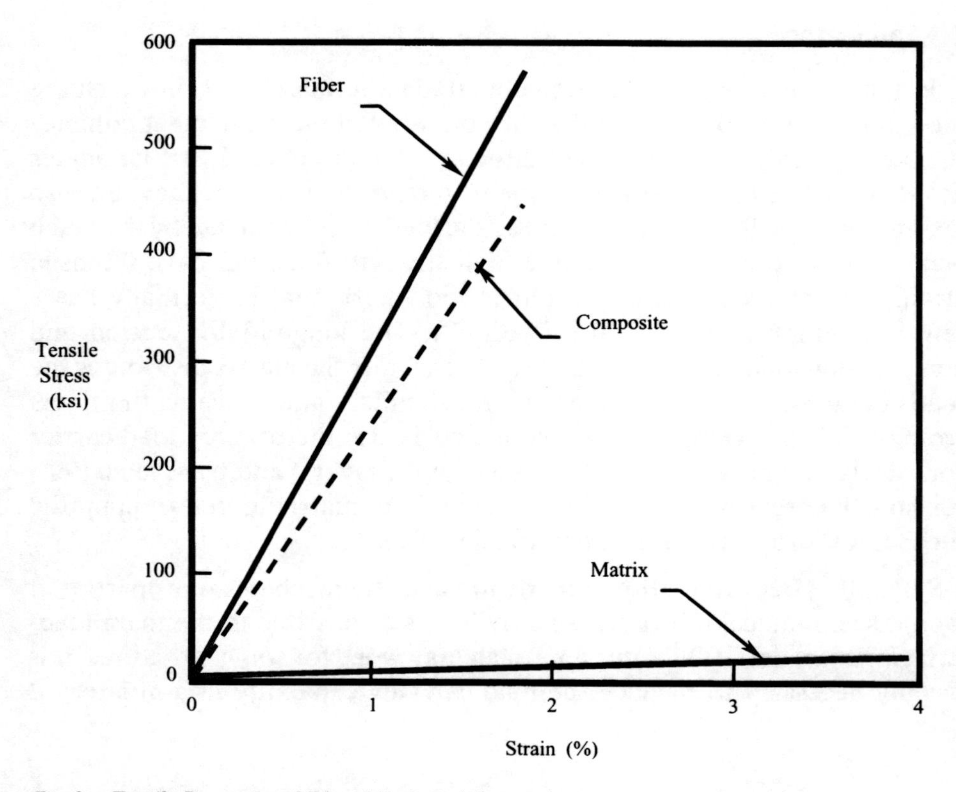

Fig. 3. Tensile Properties of Fiber, Matrix and Composite

Table 1.1 Effect of Fiber and Matrix on Mechanical Properties

Mechanical Property	*Dominating Composite Constituent*	
	Fiber	*Matrix*
Unidirectional		
0° Tension	✓	
0° Compression	✓	✓
Shear		✓
90° Tension		✓
Laminate		
Tension	✓	
Compression	✓	✓
In-Plane Shear	✓	
Interlaminar Shear		✓

different directions, such as the 0°, +45°, –45° and 90° directions. Fig. 4 shows a photomicrograph of a cross-plied continuous carbon fiber reinforcement in an epoxy resin matrix. A balanced laminate with equal numbers of plies in the 0°, +45°, –45° and 90° directions is called a quasi-isotropic laminate, since it carries equal loads in all four directions. Fig. 5 provides a graphical presentation of the preferred laminate orientations. These are preferred orientations because they are fairly balanced laminates that carry loads in multiple directions.

1.2 Fibers

The primary role of the fibers is to provide strength and stiffness. However, as a class, high-strength fibers are brittle; posses linear stress–strain

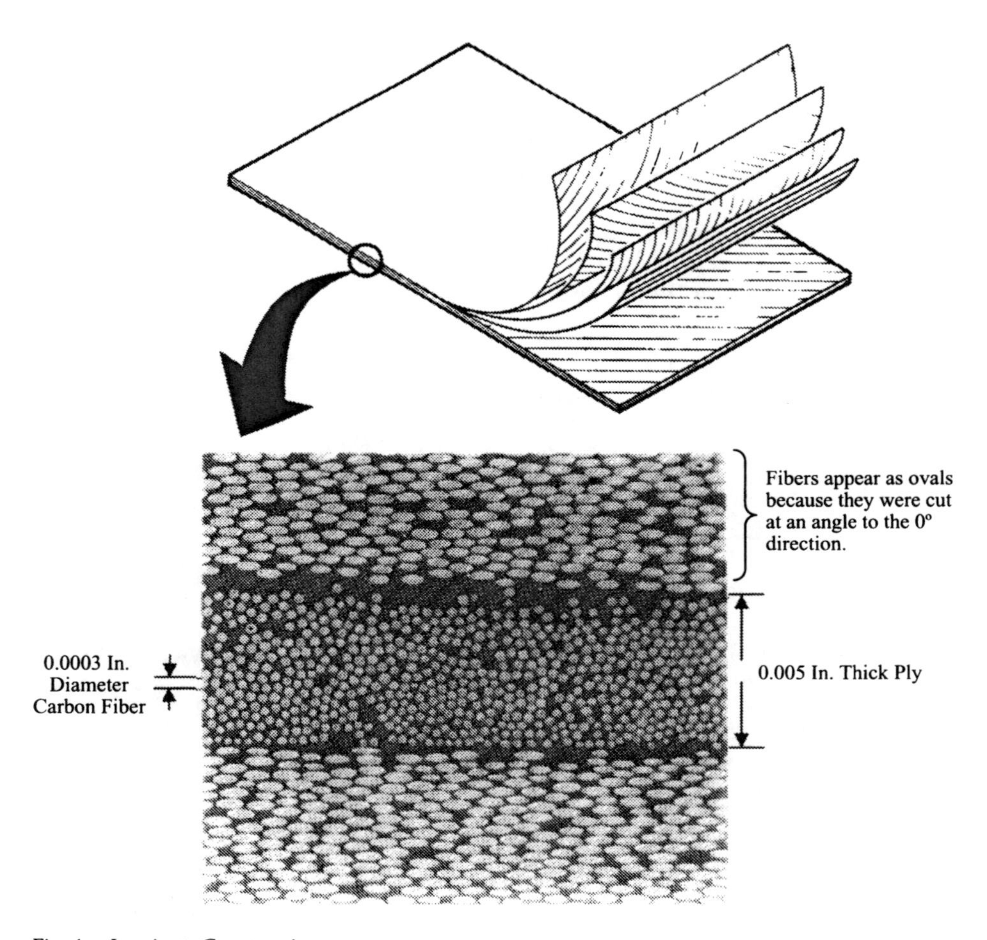

Fig. 4. Laminate Construction

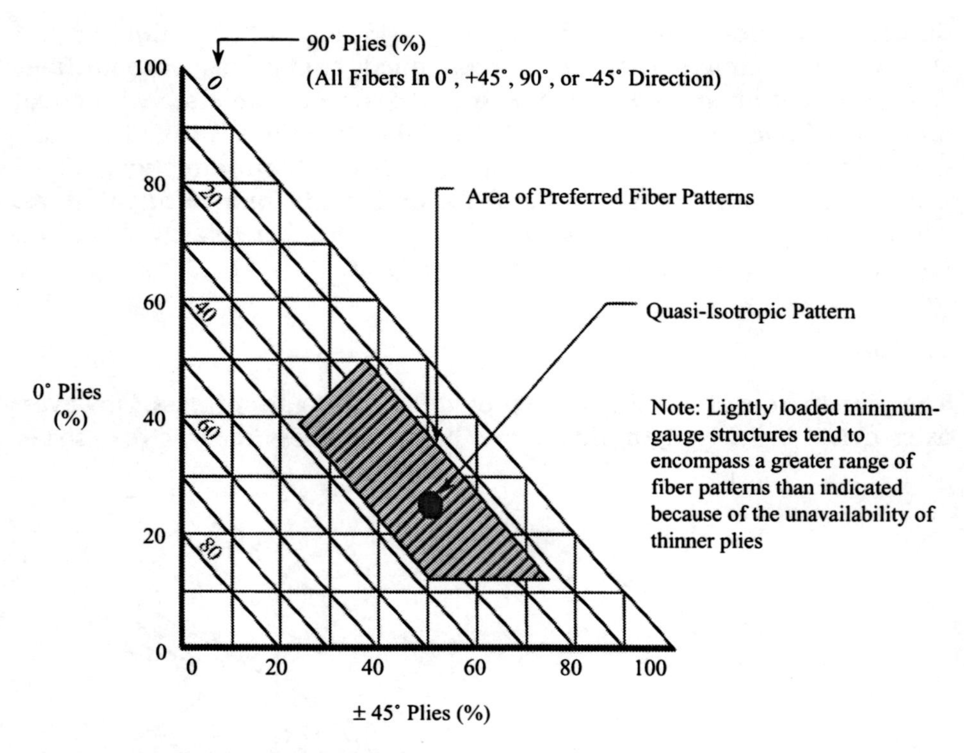

Fig. 5. Preferred Laminate Orientations

Table 1.2 Properties of Typical High Strength Fibers

Fiber	*Density lb/in³*	*Tensile Strength (ksi)*	*Elastic Modulus (msi)*	*Strain to Failure (%)*	*Diameter (Mils)*	*Thermal Expansion Coefficient 10-6 in/in/ F*
E-glass	0.090	500	11.0	4.8	0.36	2.8
S-glass	0.092	650	12.6	5.6	0.36	1.3
Quartz	0.079	490	10.0	5.0	0.35	1.0
Aramid (Kelvar 49)	0.052	550	19.0	2.8	0.47	–1.1
Spectra 1000	0.035	450	25.0	0.7	1.00	–1.0
Carbon (AS4)	0.065	530	33.0	1.5	0.32	–0.2
Carbon (IM-7)	0.064	730	41.0	1.8	0.20	–0.2
Graphite (P-100)	0.078	350	107	0.3	0.43	–0.3
Boron	0.093	520	58.0	0.9	4.00	2.5

behavior with little or no evidence of yielding; have a low strain to failure (1-2% for carbon); and exhibit larger variations in strength than metals. Table 2 presents a summary of the major composite reinforcing fibers.

Glass fibers are the most widely used reinforcement due to their good balance of mechanical properties and low cost. E-glass, or "electrical" glass, is the most common glass fiber and is used extensively in commercial composite products. E-glass is a low-cost, high-density, low-modulus fiber that has good corrosion resistance and good handling characteristics. S-2 glass, or "structural" glass, was developed in response to the need for a higher-strength fiber for filament-wound pressure vessels and solid rocket motor casings. Its density value, performance level and cost lie between those of E-glass and carbon. Quartz fiber is used in many electrical applications due to its low dielectric constant; however, it is very expensive.

Aramid fiber (e.g., Kevlar) is an extremely tough organic fiber with low density, and exhibits excellent damage tolerance. Although it has a high tensile strength, it performs poorly under compression. It is also sensitive to ultraviolet light and its use should be limited to long-term service at temperatures less than 350 °F.

Another organic fiber is made from ultrahigh molecular weight polyethylene (UHMWPE; e.g., Spectra). It has a low density with excellent radar transparency and a low dielectric constant. Due to its low density, it exhibits very high specific strength and modulus at room temperature. However, being a polyethylene its use is limited to within 290 °F. Like aramid, Spectra has excellent impact resistance; however, poor adhesion to the matrix is a problem, although plasma treatments can be employed to improve the adhesion .

Carbon fiber contains the best combination of properties but is also more expensive than either glass or aramid. It has a low density, a low coefficient of thermal expansion (CTE) and is conductive . It is structurally very efficient and exhibits excellent fatigue resistance. It is also brittle (strain-to-failure less than 2 %) and exhibits low impact resistance. Being conductive, it causes galvanic corrosion if placed in direct contact with aluminum. Carbon fiber is available in a wide range of strength (300-1000 ksi) and stiffness (modulus 30-145 msi). With this wide range of properties, carbon fibers (Table 3) are frequently classified as: (1) high-strength, (2) intermediate-modulus or (3) high-modulus fibers.

The terms carbon and graphite are often used to describe the same material. However, carbon fibers contain ~95% carbon and are carbonized at 1800-2700 °F, while graphite fibers contain ~99% carbon and are first carbonized and then graphitized at temperatures between 3600 °F and 5500 °F. In general, the graphitization process results in a fiber with a higher modulus. Carbon and graphite fibers are made from rayon, polyacrylonitrile (PAN) or petroleum-based pitch. PAN-based fibers produce the best combination of properties. Rayon was developed as a precursor prior to PAN but is rarely used today, due to its higher cost and lower yield. Petroleum-based pitch fibers were also developed as a lower-

Table 1.3 Properties of PAN Based Carbon Fibers

		Aerospace		
Property	*Commercial High Strength*	*High Strength*	*Intermediate Modulus*	*High modulus*
Tensile Modulus (Msi)	33	32–35	40–43	50–65
Tensile Strength (Ksi)	550	500–700	600–900	600–800
Elongation at Failure (%)	1.6	1.5–2.2	1.3–2.0	0.7–1.0
Electrical Resistivity (μΩ-cm)	1650	1650	1450	900
Thermal Conductivity (Btu/ft-h-°F)	11.6	11.6	11.6	29–46
Coefficient of Thermal Expansion Axial Direction (10^{-6} K)	–0.4	–0.4	–0.55	–0.75
Density (Ib/in.^{3})	0.065	0.065	0.065	0.069
Carbon Content (%)	95	95	95	+99
Filament Diameter (μm)	6–8	6–8	5–6	5–8

Reference 2

cost alternative to PAN but are mainly used to produce high- and ultra-high-modulus graphite fibers. Both carbon and graphite fibers are produced as untwisted bundles called tows. Common tow sizes are 1k, 3k, 6k, 12k and 24k, where k = 1000 fibers. Immediately after fabrication, carbon and graphite fibers are normally surface treated to improve their adhesion to the polymeric matrix. Sizings, often epoxies without a curing agent, are frequently applied as thin films (1 % or less) to improve handleability and protect the fibers during weaving or other handling operations.

Several other fibers are occasionally used for polymeric composites. Boron fiber was the original high-performance fiber before carbon was developed. It is a large-diameter fiber that is made by pulling a fine tungsten wire through a long slender reactor where it is chemically vapor-deposited with boron. Since it is made one fiber at a time, rather than thousands of fibers at a time, it is very expensive. Due to its large diameter and high modulus, it exhibits outstanding compression properties. Among the disadvantages, it does not conform well to complicated shapes and is very difficult to machine. High-temperature ceramic fibers, such as silicon carbide (Nicalon), aluminum oxide and alumina boria silica (Nextel), are frequently used in ceramic-based composites but rarely in polymeric composites.

The following factors should be considered when choosing between glass, aramid and carbon fibers:

- *Tensile strength* – If tensile strength is the primary design parameter, E-glass may be the best selection because of its low cost.

- *Tensile modulus* – When designing for tensile modulus, carbon has a distinct advantage over both glass and aramid.
- *Compression strength* – If compression strength is the primary requirement, carbon has a distinct advantage over glass and aramid. Due to its poor compression strength, aramid should be avoided.
- *Compression modulus* – Carbon fibers are the best choice, with E-glass having the least desirable properties.
- *Density* – Aramid fibers have the lowest density, followed by carbon and then S-2 and E-glass.
- *CTE* – Aramid and carbon fibers have a CTE that is slightly negative, while S-2 and E-glass are positive.
- *Impact strength* – Aramid fibers have excellent impact resistance, while carbon is brittle and should be avoided. It should be noted that the matrix also has a significant influence on impact strength.
- *Environmental resistance* – Matrix selection has the biggest impact on composite environmental resistance. However, (1) aramid fibers are degraded by ultraviolet light and the long-term service temperature should be kept below 350 °F; (2) carbon fibers are subject to oxidation at temperatures exceeding 700 °F although long-term 1000 h thermal oxidation stability tests in polyimides have shown strength decreases in the 500-600 °F range; and (3) glass sizings tend to be hydrophilic and absorb moisture.
- *Cost* – E-glass is the least expensive fiber, while carbon is the most expensive. The smaller the tow size, the more expensive the carbon fiber. Larger tow sizes help reduce labor costs because more material is deposited with each ply. However, large tow sizes in woven cloth can increase the chances of voids and matrix microcracking due to larger resin pockets.

1.3 Matrices

The matrix holds the fibers in their proper position, protects the fibers from abrasion, transfers loads between fibers, and provides interlaminar shear strength. A properly chosen matrix also provides resistance to heat, chemicals and moisture; it has a high strain-to-failure; and it cures at as low a temperature as possible and yet has a long pot or out-time life and is not toxic. The most prevalent thermoset resins used for composite matrices (Table 4) are polyesters, vinyl esters, epoxies, bismaleimides, polyimides and phenolics.

Matrices for polymeric composites can be either thermosets or thermoplastics. Thermoset resins usually consist of a resin (e.g., epoxy) and a compatible curing agent. When the two are initially mixed they form a low-viscosity liquid that cures as a result of either internally generated (exothermic) or externally applied heat. The curing reaction forms a series

Table 1.4 Relative Characteristics of Composite Resin Matrices

Polyesters	Used extensively in commercial applications. Relatively inexpensive with processing flexibility. Used for continuous and discontinuous composites.
Vinyl Esters	Similar to polyesters but are tougher and have better moisture resistance.
Epoxies	High performance matrix systems for primarily continuous fiber composites. Can be used at temperatures up to 250-275F. Better high temperature performance than polyesters and vinyl esters.
Bismaleimides	High temperature resin matrices for use in the temperature range of 275-350F with epoxy-like processing. Requires elevated temperature post cure.
Polyimides	Very high temperature resin systems for use at 550-600F. Very difficult to process.
Phenolics	High temperature resin systems with good smoke and fire resistance. Used extensively for aircraft interiors. Can be difficult to process.

of cross-links between the molecular chains so that one large molecular network is formed, resulting in an intractable solid that cannot be reprocessed on reheating. On the other hand, thermoplastics start as fully reacted high-viscosity materials that do not cross-link on heating. On heating to a high enough temperature, they either soften or melt, so they can be reprocessed a number of times. Since thermosets dominate the advanced composite marketplace, a discussion of thermoplastic composites will be deferred until Chapters 10 (Thermoplastic Composites) and 11 (Overview of Commercial Processes).

The first consideration in selecting a resin system is the service temperature required for the part. The glass transition temperature, T_g, is a good indicator of the temperature capability of the matrix. For a polymeric material T_g is the temperature at which it changes from a rigid glassy solid into a softer, semi-flexible material. At this point the polymer structure is still intact but the cross-links are no longer locked in position. A resin should never be used above its T_g unless the service life is very short (e.g., a missile body). A good rule of thumb is to select a resin in which the T_g is 50 °F higher than the maximum service temperature. Since most polymeric resins absorb moisture, which lowers the T_g, it is not unusual to require that the T_g be as much as 100 °F higher than the service temperature. Fig. 6 shows the effects of temperature and moisture on the hot-wet compression strength of a glass/epoxy system. It should be noted that different resins absorb moisture at different rates and the saturation levels can be different; therefore, the specific resin candidate must be evaluated for environmental performance. Most thermoset resins are fairly resistant to solvents and chemicals.

In general, the higher the temperature performance required, the more brittle and less damage tolerant the matrix. Toughened thermoset resins are available but are more expensive and their T_g's are typically lower. High-temperature resins are also more costly and more difficult to process.

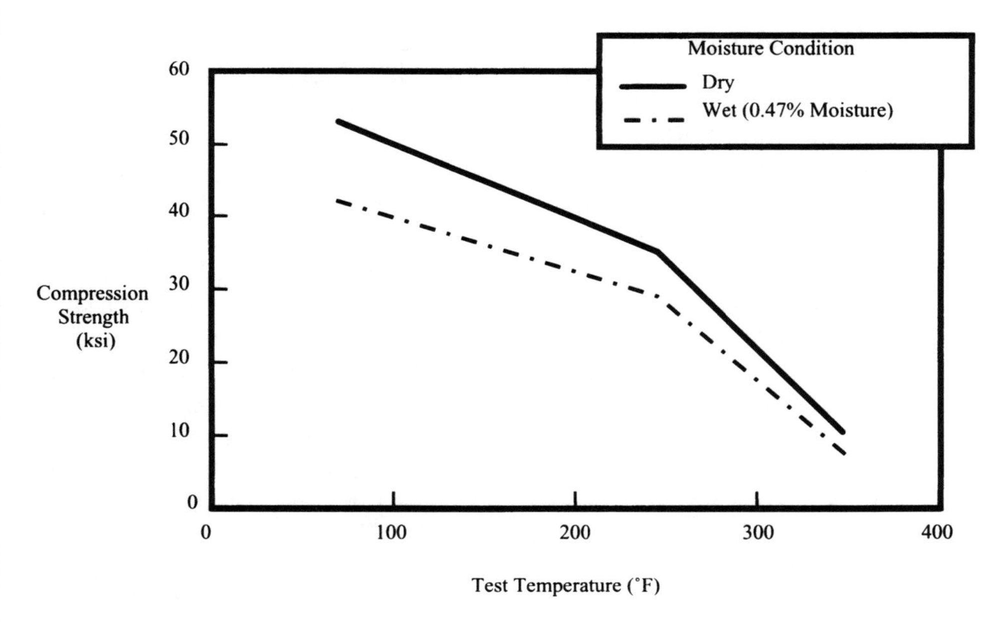

Fig. 6. Hot-Wet Compression Strength of Glass/Epoxy

Temperature performance is difficult to quantify because it is dependent on time at temperature, but it is important to thoroughly understand the environment in which the matrix is expected to preform.

Although the fiber selection usually dominates the mechanical properties of the composite, the matrix selection can also influence performance. Some resins wet out and adhere to fibers better than others, forming a chemical and/or mechanical bond that affects the fiber-to-matrix load transfer capability. The matrix can also microcrack during cure or in-service. Resin-rich pockets and brittle resin systems are susceptible to microcracking, especially when the processing temperatures are high and the use temperatures are low (e.g., –65 ºF), since this condition creates a very large difference in thermal expansion between the fibers and the matrix. Again, toughened resins help in preventing microcracking but often at the expense of elevated temperature performance.

The selection of a matrix material can profoundly affect the processing conditions. The following factors should be considered when selecting a resin matrix:

- *Pot-life or working life –* This is the time period that a matrix has when the handling characteristics remain suitable for the intended use. Typically, pot-life refers to neat resins (unreinforced) and working life refers to prepregs (reinforced). A long pot-life is desired for processes that use neat resin, such as wet filament winding, resin transfer molding and pultrusion. A short pot-life requires frequent resin bath changes and increased scrapped material. A short pot-life can also negatively affect the quality of the part in a wet process by decreasing fiber wet-out.

- *Shelf life* – This is the length of time a matrix material can be stored for under certain environmental conditions while meeting all performance and handling requirements. Thermoset prepreg materials are generally stored in a freezer and have a 6-12 month shelf life before re-certification is required. Thermoset materials, whose part A (resin) and part B (curing agent) are supplied in separate containers, generally have longer (~2 years) shelf lives at room temperature. Although not as reactive as a prepreg, viscosity and chemical changes occur over time. Refrigeration slows down the process and extends the shelf life.
- *Viscosity* – The viscosity of an uncured resin can be described as its resistance to flow. It is measured in terms of flow, using water as the standard, which has a viscosity of 1 cP (centipoises). Viscosity requirements depend on the process but, typically, the lower the viscosity the easier it is to process and the better the wetability of the matrix to fiber. As a resin is heated, the viscosity initially drops and then rises as the chemical reactions proceed until it sets up or gels. For wet processing of thermosets, typically viscosities of less than 1000 cP are preferred. A thermoset is typically considered gelled when it reaches a viscosity of 100 000 cP.
- *Cure time* – For thermoset resins, the cure time is the time it takes for the cross-linking reactions to take place. Typically, higher-T_g resins require longer cure times. Epoxies generally have cure times of 2-6 h at elevated temperatures. A post-cure may not be required for some epoxies, polyesters and vinyl esters; therefore, elimination of post-cure requirements should be evaluated as a way to decrease processing costs. Higher-T_g resins, such as bismaleimides and polyimides, require longer cure cycles and post-cures. Post-curing further develops higher-temperature mechanical properties and improves the T_g of the matrix for some epoxies, bismaleimides and polyimides. Very short cure times are desired for some processes, such as compression molding and pultrusion. Cure temperatures can range from 250 °F to 350 °F for epoxies. Bismaleimide cure temperatures typically range from 350 °F to 475 °F (including post-cure). Polyimide cure temperatures range from 600 °F to 700 °F.

1.4 Product Forms

There are a multitude of material product forms used in composite structures, some of which are illustrated in Fig. 7. The fibers can be continuous or discontinuous. They can be oriented or disoriented

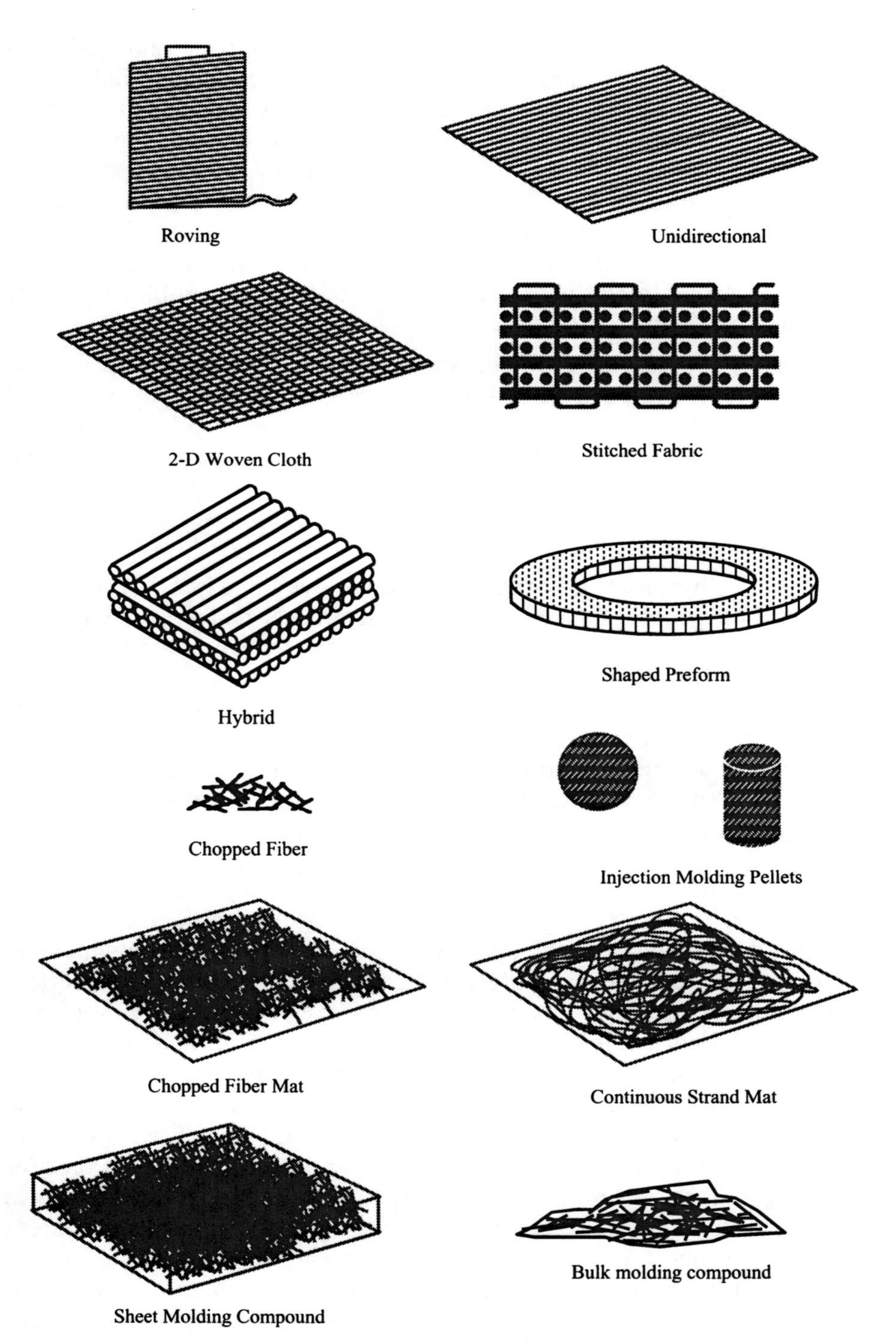

Fig. 7. Product Forms Used in Composites

(random). They can be furnished as dry fibers or pre-impregnated with resin (prepreg). Not all fiber or matrix combinations are available in a particular material form, since the market drives availability. In general, the more operations required by the supplier, the higher the cost. For example, prepreg cloth is more expensive than dry woven cloth. While complex dry preforms may be expensive, they can translate into lower fabrication costs by reducing or eliminating hand lay-up costs. If structural efficiency and weight are important design parameters, then continuous reinforced product forms are normally used because discontinuous fibers yield lower mechanical properties.

Rovings, tows and yarns are collections of continuous fiber. This is the basic material form that can be chopped, woven, stitched or prepregged into other product forms. It is the least expensive product form and available in all fiber types. Rovings and tows are supplied with no twist, while yearns have a slight twist to improve their handleability. Some processes, such as wet filament winding and pultrusion, use rovings as their primary product form.

Continuous thermoset prepreg materials are available in many fiber and matrix combinations. A prepreg is a fiber form that has a predetermined amount of uncured resin impregnated on the fiber by the material supplier. Prepreg rovings and tapes are normally used in automated processes, such as filament winding and automated tape laying, while unidirectional tape and prepreg fabrics are used for hand lay-up. Unidirectional prepreg tapes offer improved structural properties over woven prepregs due to the absence of fiber crimp and the ability to more easily tailor the designs. However, woven prepregs offer increased drapeablility. With the exception of predominantly unidirectional designs, unidirectional tapes require placement of more individual plies during lay-up. For example, with cloth, for every 0° ply in the lay-up, a 90° reinforcement is also included. With unidirectional tape, a 0° ply and a separate 90° ply must be placed onto the tool. Prepregs are supplied with either a net resin (prepreg resin content = final-part resin content) or excess resin (prepreg resin content > final-part resin content). The excess-resin approach relies on the matrix flowing through the plies, removing entrapped air, while the extra resin is removed by impregnating bleeder plies on top of the lay-up. The amount of bleeder used in the lay-up dictates the final fiber and resin content. Accurate calculations of the number and areal weight of bleeder plies for a specific prepreg are required to ensure proper final physical properties. Since the net resin approach contains the final resin content weight in the fabric, no resin removal is necessary. This is an advantage because the fiber and resin volumes can easily be controlled; however, void contents may be higher because excess resin does not flow through the part, removing the entrapped air. Thermoset prepreg properties include volatile content, resin content, resin flow, gel time, tack, drape, shelf life and out-time. Careful testing, evaluation and control of these characteristics are necessary to ensure the prepreg materials handling characteristics are optimal and final-part structural performance is obtained.

Woven fabric is the most common continuous dry material form. A woven fabric consists of interlaced warp and fill yarns. The warp is the 0° direction as the fabric comes off the roll while the fill, or weft, is the 90° fiber. Typically, woven fabrics are more drapeable than stitched materials; however, the specific weave pattern affects their drapeablility characteristics. The weave pattern also affects the handleability and structural properties of the woven fabric. Many weave patterns are available. All weaves have their advantages and disadvantages, and consideration of the part configuration is necessary during fabric selection. Most fibers are available in woven fabric form. However, it can be very difficult to weave some high-modulus fibers due to their inherent brittleness. Advantages of woven fabric include drapeablility, ability to achieve high fiber volumes, structural efficiency and market availability. A disadvantage of woven fabric is the crimp that is introduced to the warp or fill fiber during weaving. Finishes or sizings are typically put on the fibers to aid in the weaving process and minimize fiber damage. It is important to ensure that the finish is compatible with the matrix selection when specifying a fabric

A stitched fabric consists of unidirectional fibers oriented in specified directions that are then stitched together to form a fabric. A common stitched design includes 0°, +45°, 90° and –45° plies in one multidirectional fabric. Advantages include: (1) the ability to incorporate off-axis orientations as the fabric is removed from the roll. Off-axis cutting is not needed for a multidirectional stitched fabric that reduces scrap rates (up to 25%) when compared to conventional woven materials; (2) labor costs are also reduced when using multi-ply stitched materials because less plies are required to cut or handle during fabrication of a part; and (3) ply orientation remains intact during handling due to the z-axis stitch threads. Disadvantages are: (1) availability of specific stitched ply set designs. Typically, a special order is required due to the tailoring requested by customer such as fiber selection, fiber volume, and stitching requirements. Not as many companies stitch as weave; and (2) drapeablility characteristics are reduced (however, this can be an advantage for parts with large simple curvature). Careful selection of the stitching thread is necessary to ensure compatibility with the matrix and process temperatures.

Hybrids are material forms that make use of two or more fiber types. Common hybrids include glass/carbon, glass/aramid and aramid/carbon fibers. Hybrids are used to take advantage of properties or features of each reinforcement type. In a sense, a hybrid is a "trade-off" reinforcement that allows increased design flexibility. Hybrids can be interply (two alternating layers), intraply (present in one layer), or in selected areas. Glass/carbon interply hybrids can be used to prevent galvanic corrosion of aluminum by the carbon. Hybridization in selected areas is usually done to locally strengthen or stiffen a part. Carbon/aramid hybrids have low thermal stresses compared to other hybrids due to their similar CTEs, increased modulus and compressive strength over an all-aramid design, and increased toughness over an all-carbon design. Carbon/E-glass hybrids have better properties compared to an all-E-glass design, but have lower cost than an all-carbon design. The CTE of each fiber type in a hybrid needs careful evaluation to ensure that high internal stresses are not introduced to the laminate during cure, especially at higher cure temperatures.

A preform is a pre-shaped fibrous reinforcement that has been formed into shape, on a mandrel or in a tool, before placing into a mold. The shape of the preform closely resembles the final-part configuration. A simple multi-ply stitched fabric is not a preform unless it is shaped to near its final configuration. The preform is the most expensive dry, continuous, oriented fiber form; however, using preforms can reduce fabrication labor. A preform can be made using rovings, chopped, woven, stitched, or unidirectional material forms. These reinforcements are formed and held in place by stitching, binders or tactifiers, braiding or three-dimensional weaving.

Advantages of preforms include reduced labor costs, minimal material scrap, reduced fiber fraying of woven or stitched materials, improved damage tolerance for three-dimensional stitched or woven preforms; and the desired fiber orientations are locked in place. Disadvantages include high preform costs, fiber wetability concerns for complex shapes, tackifier or binder compatibility with the matrix; and limited flexibility if design changes are required. A common defect is the preform being out of tolerance that makes placement in the tool difficult or trimming required.

The use of preforms is not appropriate for all applications. All issues need to be evaluated carefully before a preform is baselined. If the component is not a very complex configuration, the money saved in labor reduction may not offset the cost of the preform. Each application should be evaluated individually to determine if the preform approach offers cost and quality advantages.

Chopped fiber is made by mechanically chopping rovings, yarns, or tows into short lengths, typically 1/4–2 inches long. The minimum lengths of chopped fibers are very important to ensure maximum reinforcement efficiency. Milled fibers, typically 1/32–1/4 inches long, have low aspect ratios (length/diameter) that provides minimal strength and, therefore, should not be considered for structural applications. The strength of the composite will be greatly improved with increased fiber length. Stiffness properties are much less affected by fiber length. Lengths of chopped fiber greater than 1–2 inches do not add any structural advantage because the full efficiency of the random fiber has been achieved. Chopped glass fibers are often embedded in thermoplastic or thermoset resins in the form of pellets for injection molding.

Chopped fibers or entangled continuous strands are combined with a binder to form mat or veil materials. A veil is a thin mat used to improve the surface finish of a molded composite. This material form is used extensively for automotive, industrial, recreational and marine applications, where high-quality class-A exterior surface finishes are required. These are very inexpensive product forms, with E-glass being the most common fiber.

Other fibers may be available in mat form but are generally expensive due to specialty nature of the product. Presently, carbon mats are uncommon due to low demand. Some carbon veils, for EMI or lightning protection, are available but are expensive. Historically, the aerospace industry has not been interested in this product form due to the lack of structural efficiency and the associated weight penalty for its discontinuous reinforcement. On the other hand, the automotive industry, which widely uses mats, would not pay the cost penalty for the carbon fiber material.

Sheet molding compounds (SMCs) consist of flat sheets of chopped, randomly oriented fibers, typically 1–2 inches in length, with a B-stage matrix, usually consisting of E-glass fiber in either a polyester or vinyl ester resin. The material is available either as rolls or pre-cut sheets.

Bulk molding compounds (BMCs) are also short randomly oriented pre-impregnated materials; however, the fibers are only 1/8–1-1/4 inches long, and the reinforcement percentage is lower. As a result, the mechanical properties for BMC composites are lower than for SMC composites. BMCs are available in doughlike bulk form or extruded into logs for easier handling. Vinyl esters, polyesters and phenolics are the most common matrices used for BMCs. Sometimes, for convenience in terminology, chopped prepreg, used for compression molding, is referred to as BMC.

Table 5 presents a summary of some of the product forms and their processes, while Table 6 provides a relative comparison of using dry fiber/neat resin and prepreg

1.5 Overview of Fabrication Processes

Prepreg lay-up is a process in which individual layers of prepreg are laid up on a tool and then cured, as shown in Fig. 8. The layers are laid up in the required directions and to the correct thickness. A thin nylon vacuum bag is then placed over the lay-up and the air is evacuated to draw out the air between the plies. The bagged part is placed in an oven or an autoclave (a

Table 1.5 Typical Material Product Forms vs. Process

Material Form \ Process	Pultrusion	RTM	Compression Molding	Filament Winding	Hand Lay-Up	Auto Tape Laying
Discontinuous:						
Sheet Molding Compound			●			
Bulk Molding Compound			●			
Random Continuous:						
Swirl Mat/Neat Resin	●	●	●	●	●	
Oriented Continuous:						
Undirectional Tape			●		●	●
Woven Prepreg			●		●	
Woven Fabric/Neat Resin	●	●		●	●	
Stitched Material/Neal Resin		●				
Prepreg Roving				●		
Roving/Neat Resin	●			●	●	
Preform/Neat Resin		●				

Table 1.6 Relative Comparison of Dry Fiber / Neat Resin and Prepreg

	Dry Fiber / Neat Resin	Prepreg
Cost	Lower	Higher
Shelf Life	Better	Worse
Storage	Better	Worse
Material Handling		
• Drapeability	Better	Worse
• Tack	Worse	Better
Resin Control	Worse	Better
Fiber Volume Control	Worse	Better
Part Quality	Worse	Better

heated pressure vessel) and cured under the specified time, temperature and pressure. If oven curing is used, the maximum pressure that can be obtained is atmospheric (14.7 psia or less). An autoclave (Fig. 9) works on the principle of differential gas pressure. The vacuum bag is evacuated to remove the air and the autoclave supplies gas pressure to the part. The vessel contains a heating system with a blower to circulate the hot gas. An autoclave offers the advantage that much higher pressures (e.g., 100 psig) can be used resulting in better compaction, higher fiber volume percentages, and less voids and porosity. Presses can also be used for this

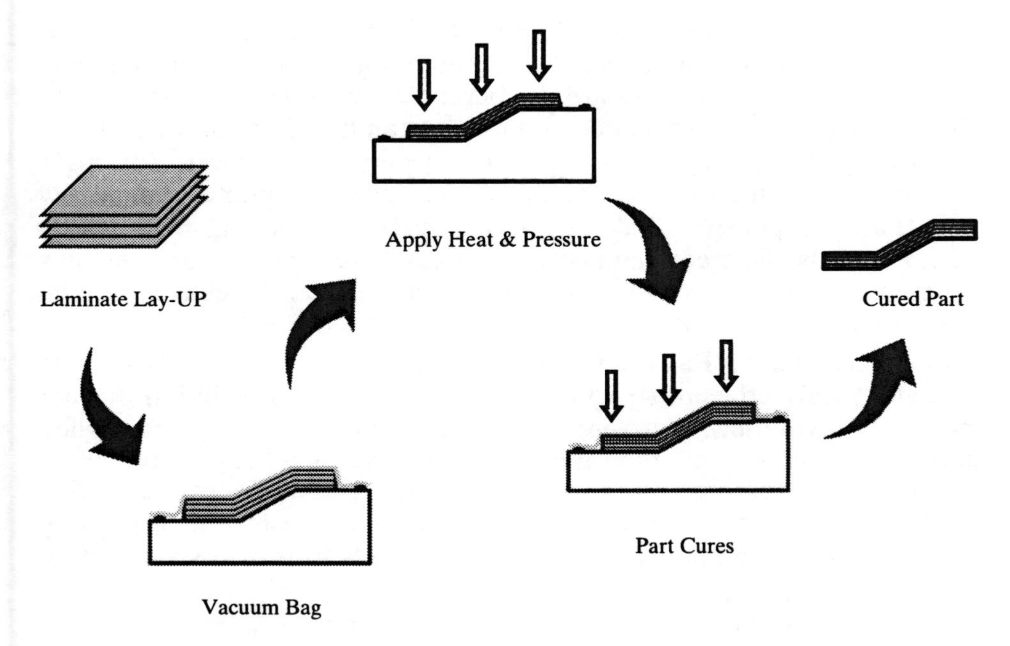

Fig. 8. Prepreg Lay-Up Process

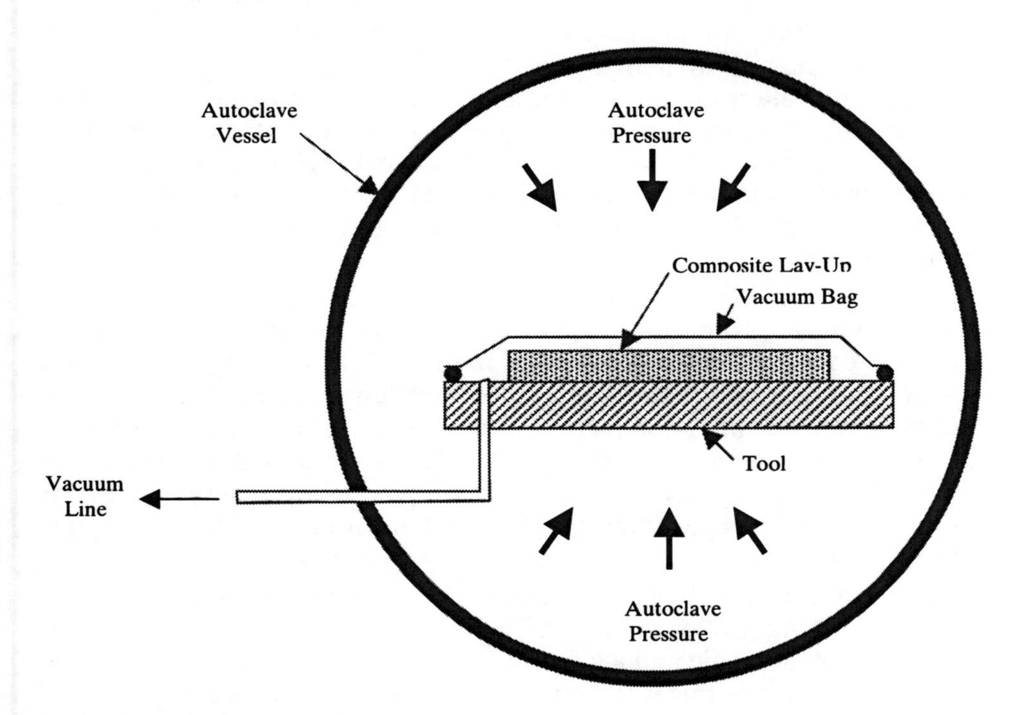

Fig. 9. Principle of Autoclave Curing

process but they have several disadvantages: (1) the size of the part is limited by the press platen size; (2) the platens may produce high- and low-pressure spots if the platens are not parallel; and (3) complex shapes are difficult to produce. Automated ply cutting, manual collation or lay-up and autoclave curing (Fig. 10) is the most widely used process for high-performance composites in the aerospace industry. While manual ply collation is expensive, this process is capable of making high-quality complex parts. Since cost has become a major driver, there is a tremendous amount of research being conducted to find more cost-effective material product forms and processes.

Filament winding (Fig. 11) is a process that has been used for many years to build highly efficient structures that are bodies of revolution or near bodies of revolution. Wet winding, in which dry fiber rovings are pulled through a resin bath prior to winding on the mandrel, is the most prevalent process, but prepregged roving or tows can also be filament wound. Curing is usually conducted in an oven with or without a vacuum bag. Hoop windings are often applied over separator sheets to provide compaction pressure during cure.

Wet lay-up, as shown in Fig. 12, is often used to build large structures such as yacht hulls. It can be a cost-effective process when the quantities required are small. Dry reinforcement, usually woven cloth or mat material, is hand-laid up one ply at a time. During or before lay-up, each ply is impregnated with a low viscosity resin. After the ply is placed on the lay-up, hand rollers are used to remove excess resin and air and compact the plies. After lay-up, cure can be done at room or elevated temperatures. Frequently, cure is conducted without a vacuum bag, but vacuum pressure helps to improve laminate quality. Since cure is usually conducted at room or low temperatures, very inexpensive tooling (e.g., wood) can be used to minimize cost.

Spray-up (Fig. 13) is a more cost-effective process than wet lay-up, but the mechanical properties are much lower due to the use of randomly oriented chopped fibers. Normally continuous glass roving is fed into a special gun that chops the fibers into short lengths and simultaneously mixes them with either a polyester or vinyl ester resin that is then sprayed onto the tool. Manual compaction with rollers is again used to compact the lay-up. Vacuum bag cures can be used to improve part quality but are not normally used. Since the fibers are short and the orientation is random, this process is not used to make structural load-bearing parts.

The term liquid molding covers a fairly extensive set of processes. In resin transfer molding (RTM), shown in Fig. 14, a dry preform or lay-up is placed in a matched metal die and a low-viscosity resin is injected under pressure to fill the die. Since this is a matched-die process, it is capable of holding very tight dimensional tolerances. The die can contain internal heaters or

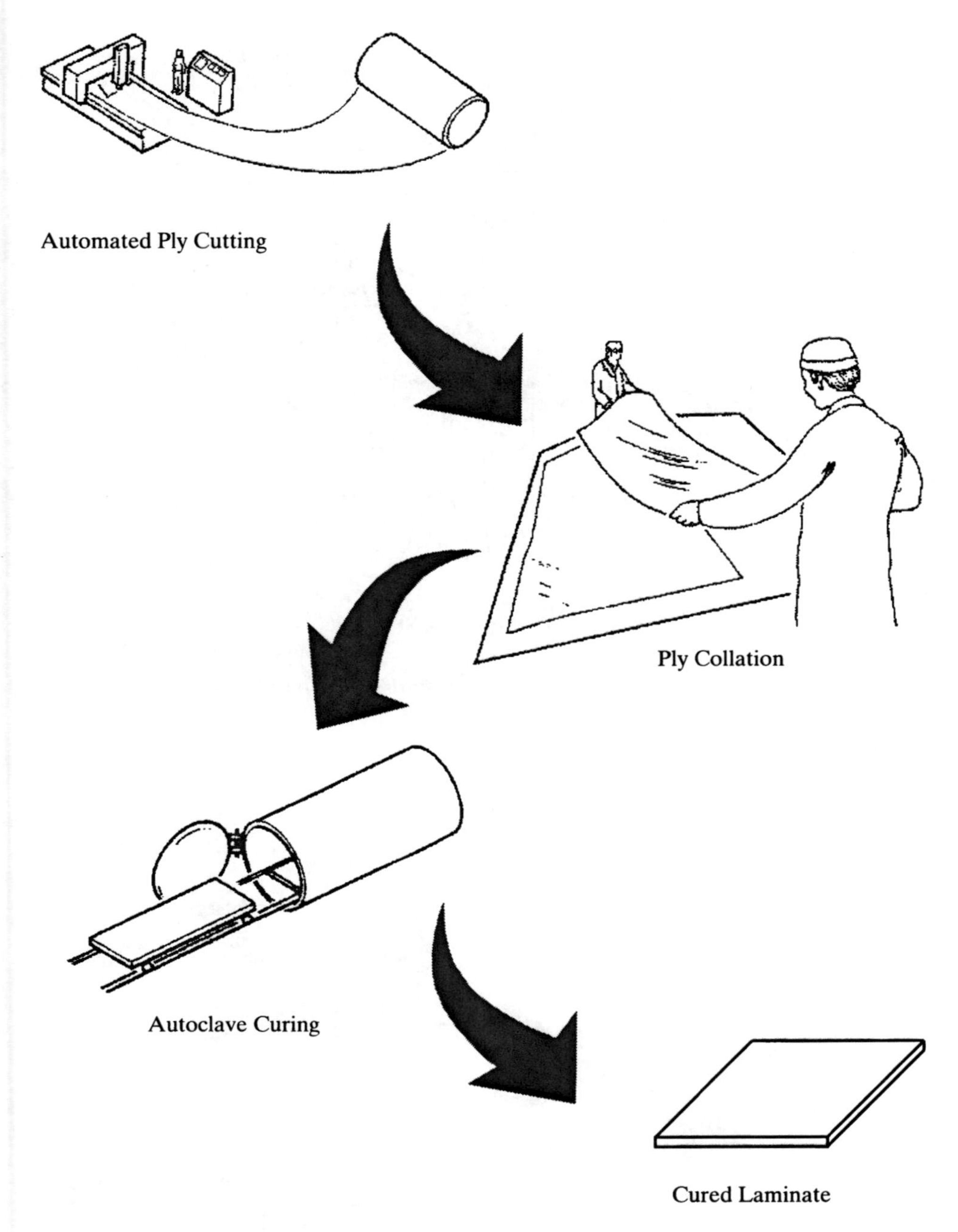

Fig. 10. Traditional Lay-Up and Autoclave Cure

can be placed in a heated platen press for cure. Other variations of this process include vacuum-assisted RTM (VARTM), in which a single-sided

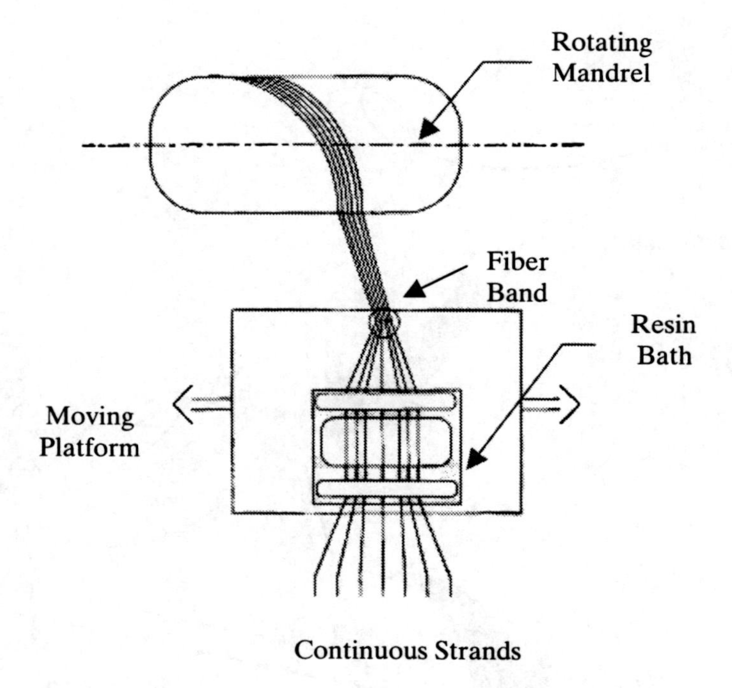

Fig. 11. Filament Winding

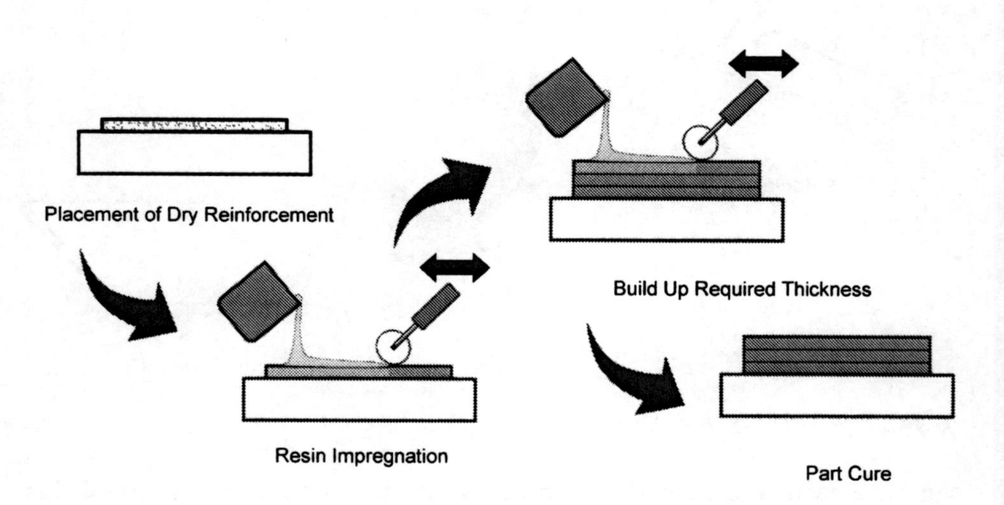

Fig. 12. Wet Lay-Up Process

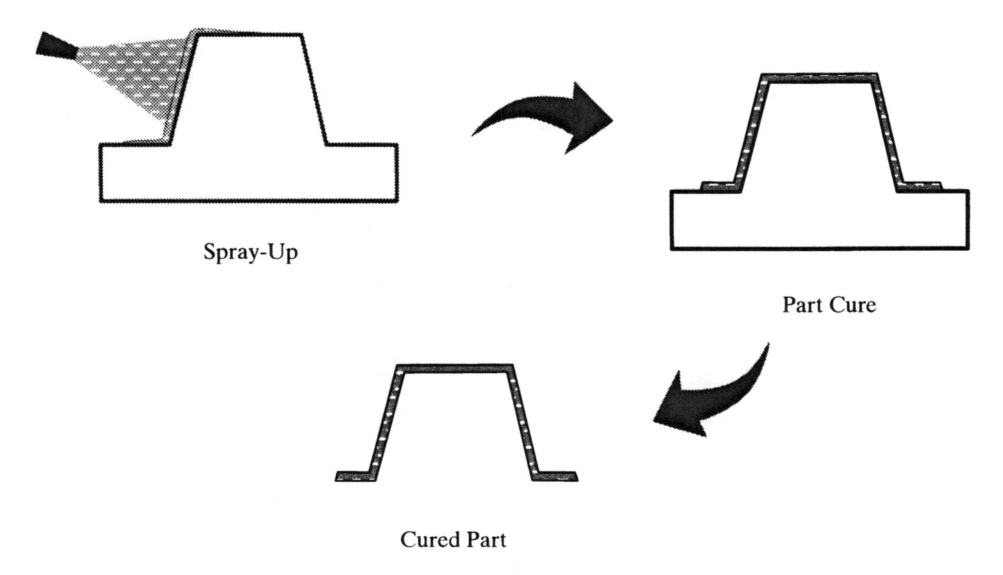

Fig. 13. Spray-Up Process

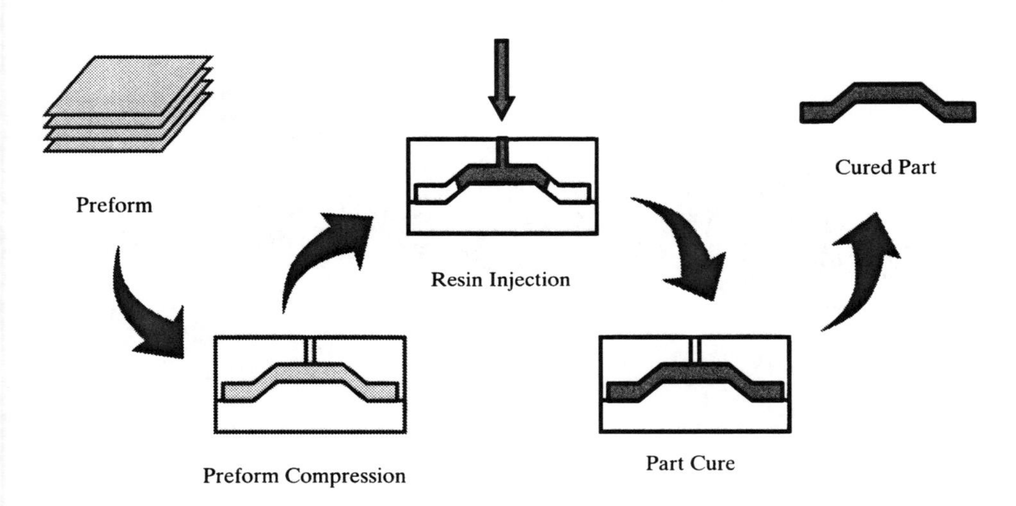

Fig. 14. Resin Transfer Molding

tool is used along with a vacuum bag. Instead of injecting the resin under pressure, a vacuum pulls the resin through a flow medium that helps impregnate the preform.

Compression molding (Fig. 15) is another matched-die process that uses either discontinuous, randomly oriented SMC or BMC. A charge of predetermined weight is placed between the two dies and then heat and

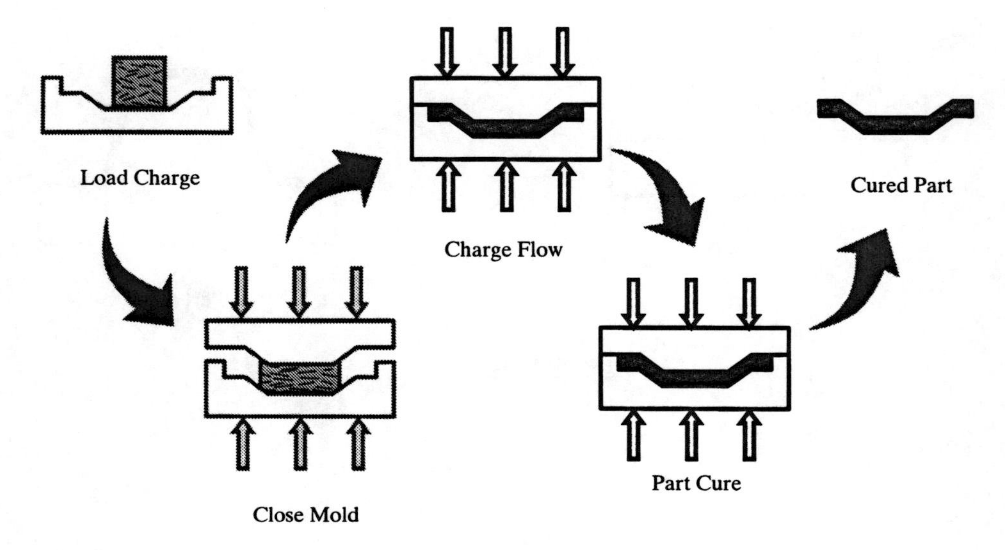

Fig. 15. Compression Molding

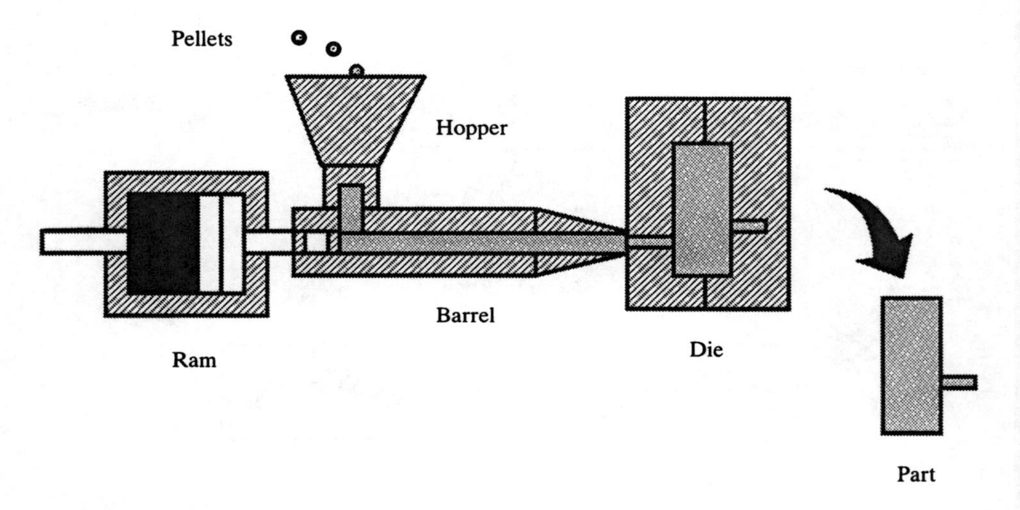

Fig. 16. Injection Molding

pressure are applied. The molding compound flows to fill the die and then rapidly cures in 1–5 min, depending on the type of polyester or vinyl ester used. Thermoplastic composites, usually consisting of glass fiber and polypropylene, are also compression molded for the automotive industry.

Injection molding (Fig. 16) is a high-volume process capable of making small- to medium-size parts. The reinforcement is usually chopped glass fibers with either a thermoplastic or thermoset resin, although the majority of applications use thermoplastics because they process faster and have higher toughness. In the injection-molding process, pellets containing embedded fibers or chopped fibers and resin are fed into a hopper. They are heated to their melting temperature and then injected under high pressure into a matched metal die. After the thermoplastic part cools or the thermoset part cures, they are ejected and the next cycle is started.

Pultrusion, illustrated in Fig. 17, is a rather specialized composite fabrication process that is capable of making long constant-thickness parts. Dry E-glass rovings are normally pulled through a wet resin bath and are then preformed to the desired shape before entering a heated die. Mats and veils are frequently incorporated into the part. Cure occurs inside the die and the cured part is pulled to the desired length and cut off. Quick curing polyesters and vinyl esters are the predominant resin systems.

There are many trade-offs that must be considered when selecting a material and process. Some considerations as summarized in Table 7.

1.6 Advantages and Disadvantages of Composite Materials

The advantages of composites are many, including lighter weight, the ability to tailor the lay-up for optimum strength and stiffness, improved

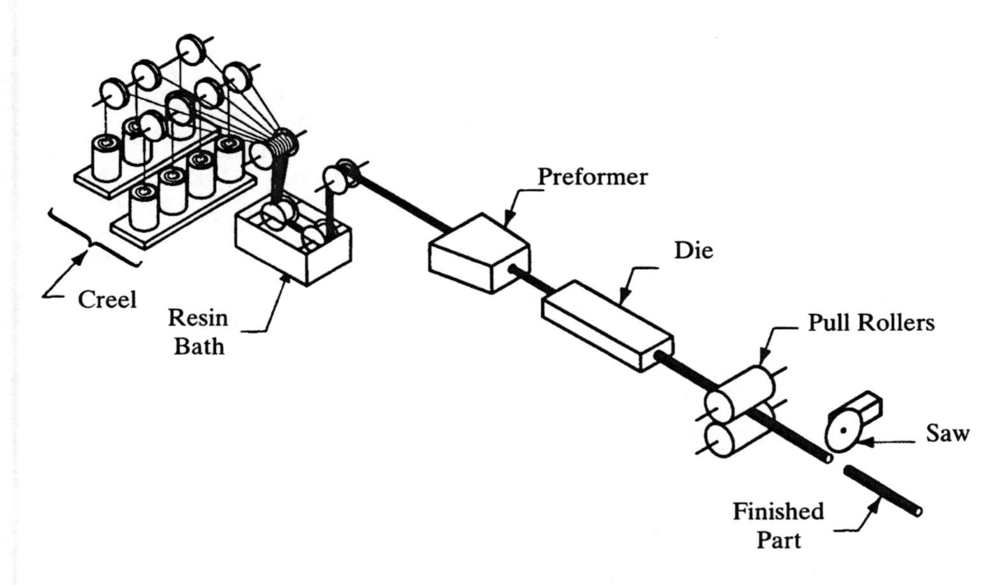

Fig. 17. Pultrusion

fatigue life, corrosion resistance, and with good design practice, reduced assembly costs due to fewer detail parts and fasteners.

The specific strength (strength/density) and specific modulus (modulus/density) of high-strength fibers, especially carbon, are higher than those of other comparable aerospace metallic alloys (Fig. 18). This translates into greater weight savings resulting in improved performance, greater payloads, longer range and fuel savings. The overall structural efficiencies of carbon/epoxy, Ti-6Al-4V and 7075-T6 aluminum are compared in Fig. 19.

Table 1.7 Common Trade-Offs When Selecting Composite Materials

Design Decision	*Common Tradeoffs*	*Typically Lowest Cost*	*Typically Highest Performance*
Fiber Type	Cost, Strength, Stiffness, Density (Weight), Impact Strength, Electrical Conductivity, Environmental Stability, Corrosion, Thermal Expansion	E-Glass	Carbon
Tow Size (If Carbon Is Slected)	Cost, Fiber Volume, Improved Fiber Wet-out, Structural Efficiency (Minimize Ply Thickness), Surface Finish	12K Tow	3K Tow
Fiber Modulus (If Carbon Is Slected)	Cost Stiffness, Weight, Brittleness	Lowest Modulus Carbon	Highest Modulus Carbon
Fiber Form (Continuous vs. Discontinuous)	Cost, Strength, Stiffness, Weight, Fiber Volume, Design Complexity	Random/ Discontinuous	Oriented And Continuous
Matrix	Cost, Service Temperature, Compressive Strength, Interlaminar Shear, Environmental Performance (Fluid Resistance, UV Stability, Moisture Absorption), Damage Tolerance, Shelf Life, Processability, Thermal Expansion	Vinyl Ester and Polyester	High Temperature - Polyimide[a] Low-ModerateTemperature - Epoxy Toughness - Toughened Epoxy
Composite Material Forms	Cost (Material And Labor), Process Compatibility, Fiber Volume Control, Material Handling, Fiber Wet-out, Material Scrap	Base Form-Neat Resin/Rovings	Prepreg[b]

[a] Depends on how "Highest Performance" is defined – high temperature, toughness, and superior mechanical properties.

[b] Material form is not driven by performance, but typically defined by the manufacturing process.

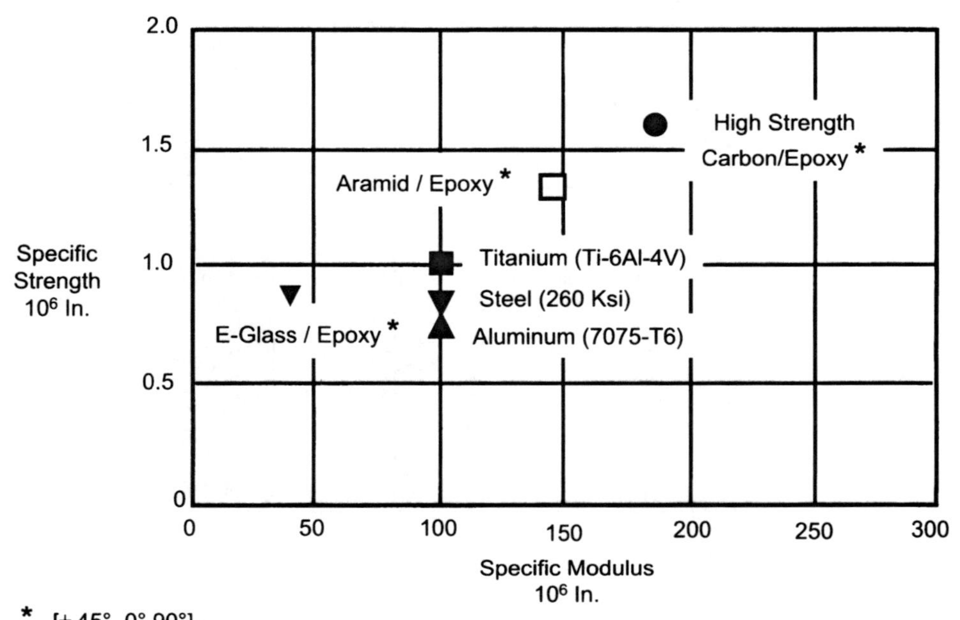

Fig. 18. Specific Property Comparaison

The chief engineer of aircraft structures for the US Navy once told the author that he liked composites[1]. Corrosion of aluminum alloys is a major cost and constant maintenance problem for both commercial and military aircraft. The corrosion resistance of composites can result in major savings in supportability costs. While carbon fiber composites cause galvanic corrosion of aluminum if the fibers are placed in direct contact with the metal surface, bonding a glass fabric electrical insulation layer on all interfaces that contact aluminum eliminates this problem. The fatigue resistance of composites compared to high-strength metals is shown in Fig. 20. As long as reasonable strain levels are used during design, fatigue of carbon fiber composites should not be a problem.

Assembly costs usually account for about 50 % of the cost of an airframe. Composites offer the opportunity to significantly reduce the amount of assembly labor and fasteners. Detail parts can be combined into a single cured assembly during either initial cure or by secondary adhesive bonding.

Disadvantages of composites include: their raw material costs are high, and they usually incur high fabrication and assembly costs; they are adversely affected by both temperature and moisture; they are weak in the

[1] "they don't rot (corrode) and they don't get tired (fatigue)."

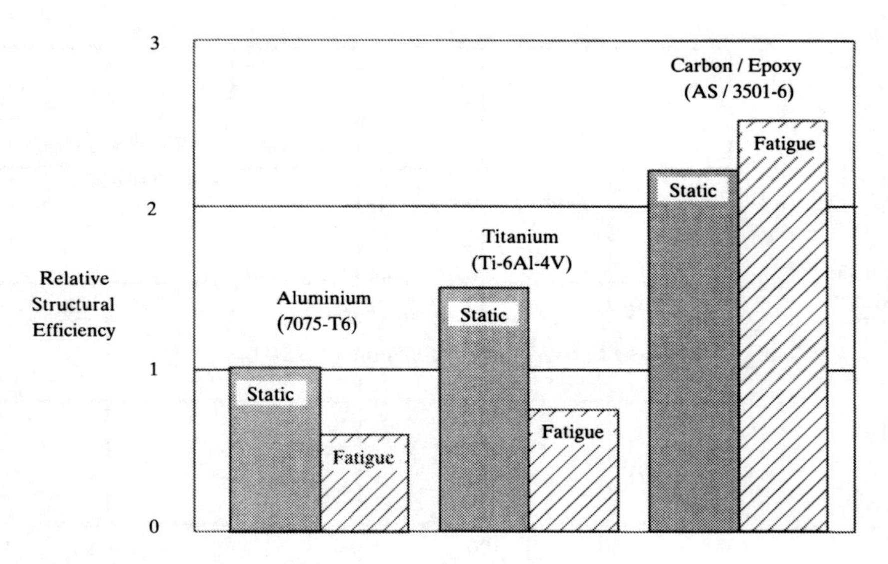

Fig. 19. Relative Structural Efficiency of Aircraft Materials

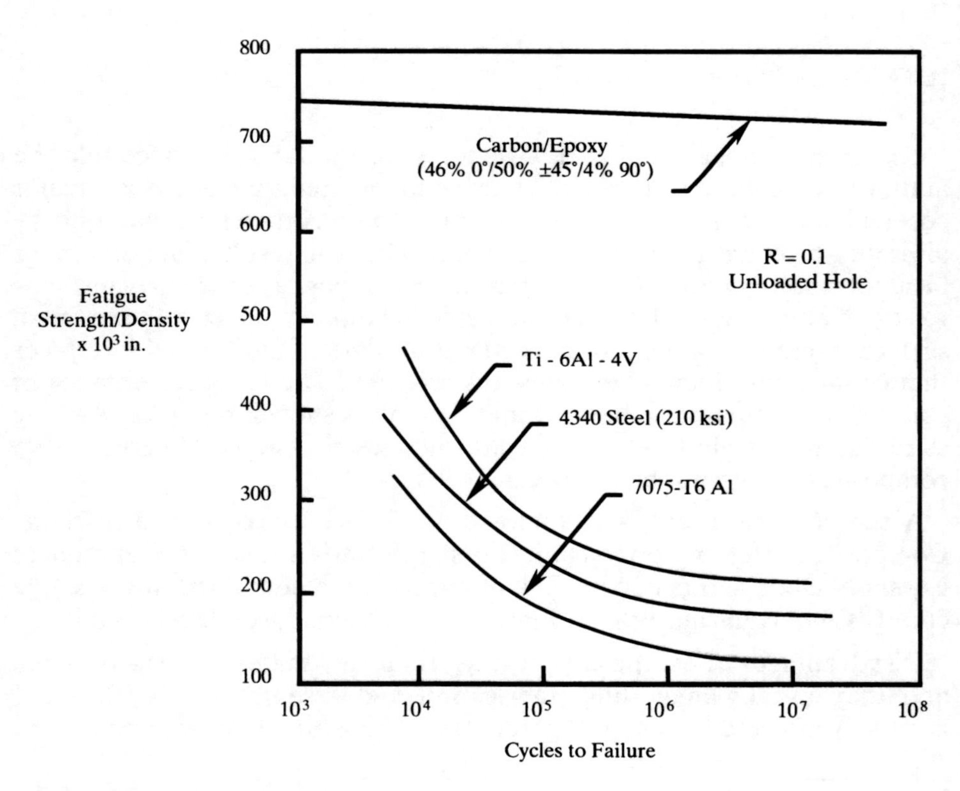

Fig. 20. Fatigue Properties of Aerospace Materials

out-of-plane direction where the matrix carries the primary load and should not be used where load paths are complex (e.g., lugs and fittings); composites are susceptible to impact damage and delaminations or ply separations can occur; and they are more difficult to repair than metallic structures.

The major cost driver in fabrication for a conventional hand-laid-up composite part is the cost of laying up or collating the plies. This cost (see Fig. 21) generally consists of 40–60% of the fabrication cost, depending on part complexity. Assembly cost is another major cost driver, accounting for about 50% of the total part cost. As previously stated, one of the potential

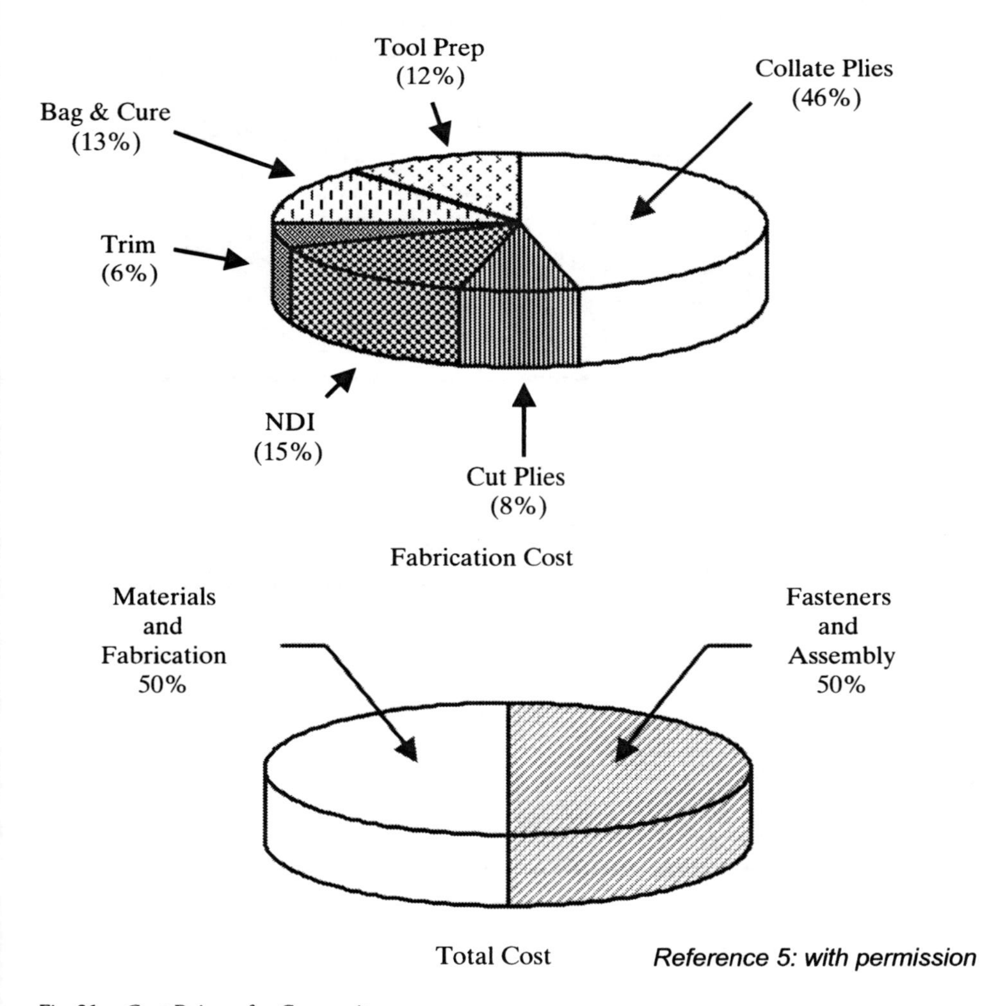

Fig. 21. Cost Drivers for Composites

advantages of composites is to cure or bond a number of detail parts together to reduce assembly costs and the number of required fasteners.

Temperature affects the mechanical properties of a composite. Typically, as the temperature increases, the matrix-dominated mechanical properties decrease. Fiber-dominated properties are somewhat affected by cold temperatures but the effects are not as severe as those of elevated temperatures on the matrix-dominated properties. As shown in Fig. 22, the design parameters for carbon/epoxy are cold–dry tension and hot–wet compression. The amount of absorbed moisture (Fig. 23) is dependent upon the matrix material and the relative humidity. Elevated temperature

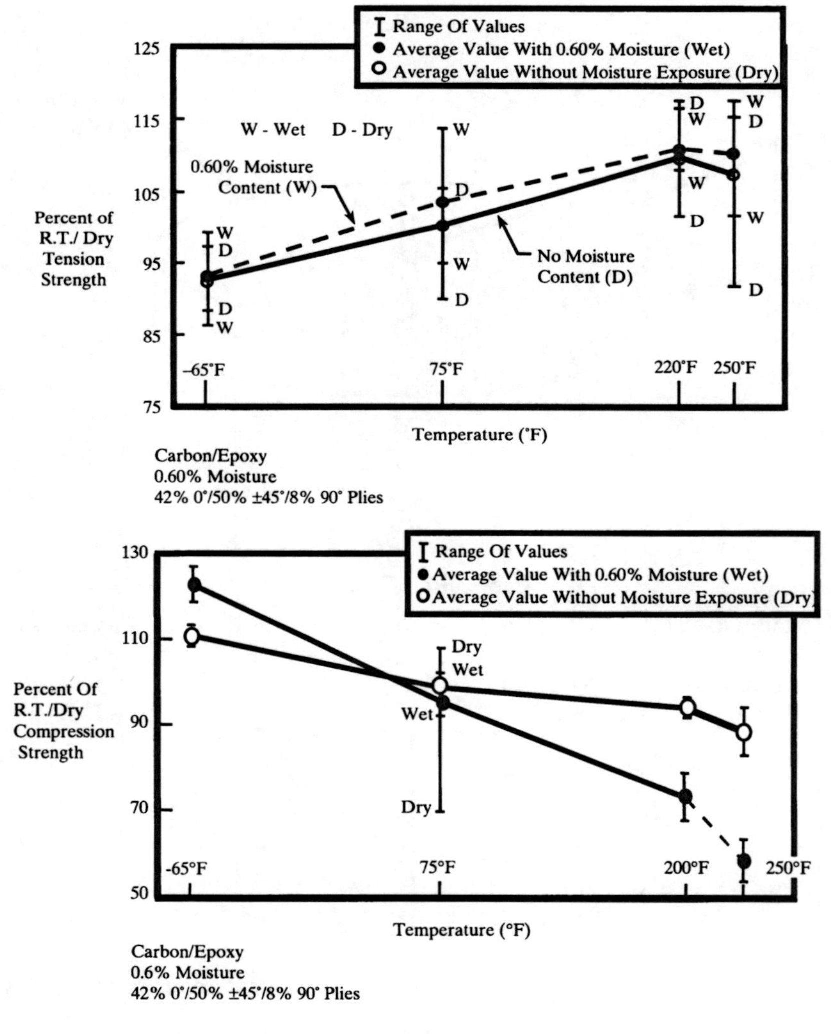

Fig. 22. Effects of Temperature and Moisture on Strength of Carbon/Epoxy

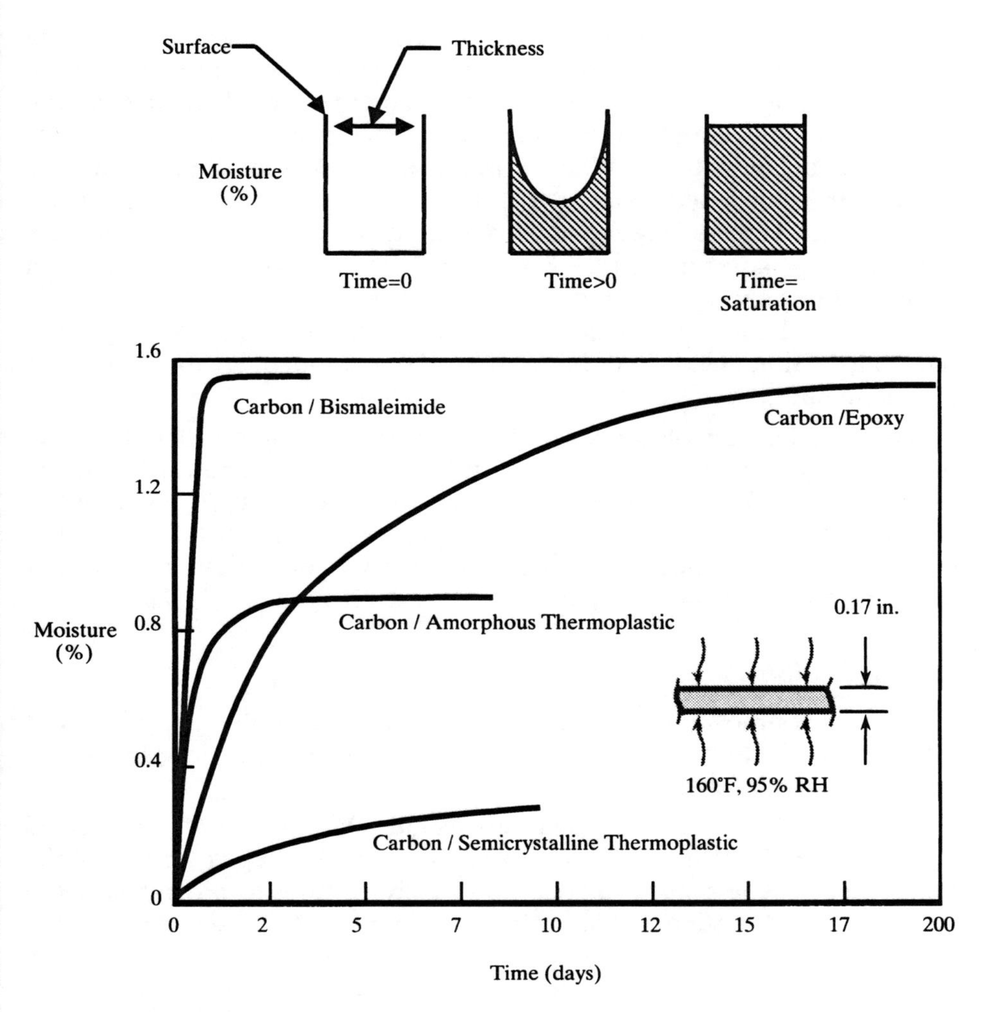

Fig. 23. Absorption of Moisture for Polymer Composites

speeds the rate of moisture absorption. Absorbed moisture reduces the matrix-dominated mechanical properties. Absorbed moisture also causes the matrix to swell. This swelling relieves locked-in thermal strains from elevated-temperature curing. These strains can be large, and large panels fixed at their edges can buckle due to the swelling strains. During freeze–thaw cycles, the absorbed moisture expands during freezing and can crack the matrix. During thermal spikes, absorbed moisture can turn to steam. When the internal steam pressure exceeds the flatwise tensile strength of the composite, the laminate delaminates.

Composites are susceptible to delaminations (ply separations) during fabrication, assembly and in-service. During fabrication, foreign materials, such as prepreg backing paper, can be inadvertently left in the lay-up. During assembly, improper part handling or incorrectly installed fasteners can cause delaminations. When in-service, low-velocity-impact damage (LVID) from dropped tools or fork lifts running into aircraft can cause damage. The damage may appear as only a small indentation on the surface but can propagate through the laminates forming a complex network of delaminations and matrix cracks as depicted in Fig. 24. Depending on the size of the delamination, it can reduce the static and fatigue strength and the compression buckling strength. If it is large enough, it can grow under fatigue loading.

Typically, damage tolerance is a resin-dominated property. The selection of a toughened resin can significantly improve the resistance to impact damage. In addition, aramid fibers are extremely tough and damage tolerant. During the design phase, it is important to recognize the potential for delaminations and use conservative enough design strains that damaged structure can be repaired.

1.7 Applications

The use of composite materials is extensive and expanding. Applications (see Fig. 25) include aerospace, automotive, marine, sporting goods and, more recently, infrastructure.

Examples of aerospace applications include the horizontal and vertical tail planes on the Boeing 777 that are made of carbon/epoxy. Some of the newer business jets have almost completely composite airframes. Both small and large commercial aircraft rely on composites to decrease weight and increase fuel performance. In military aircraft, weight is most significant for performance and payload reasons and composites often approach 20–30 % of the airframe weight. For decades, helicopters have used glass-fiber-reinforced rotor blades for improved fatigue resistance and in recent years have expanded into largely composite airframes. Composites are also used extensively in both reusable and expendable launch vehicles and satellite structures.

The major automakers are increasingly turning towards composites to help them meet performance and weight requirements, as well as the styling achievable by smooth molded components. Cost is a major driver for commercial transportation, and composites offer lower weight and lower maintenance costs. Typical materials are glass/ polyurethane made by liquid or compression molding and glass/polyester made by compression molding. Recreational vehicles have long used glass fibers, mostly for durability and weight savings over metal. The product form is

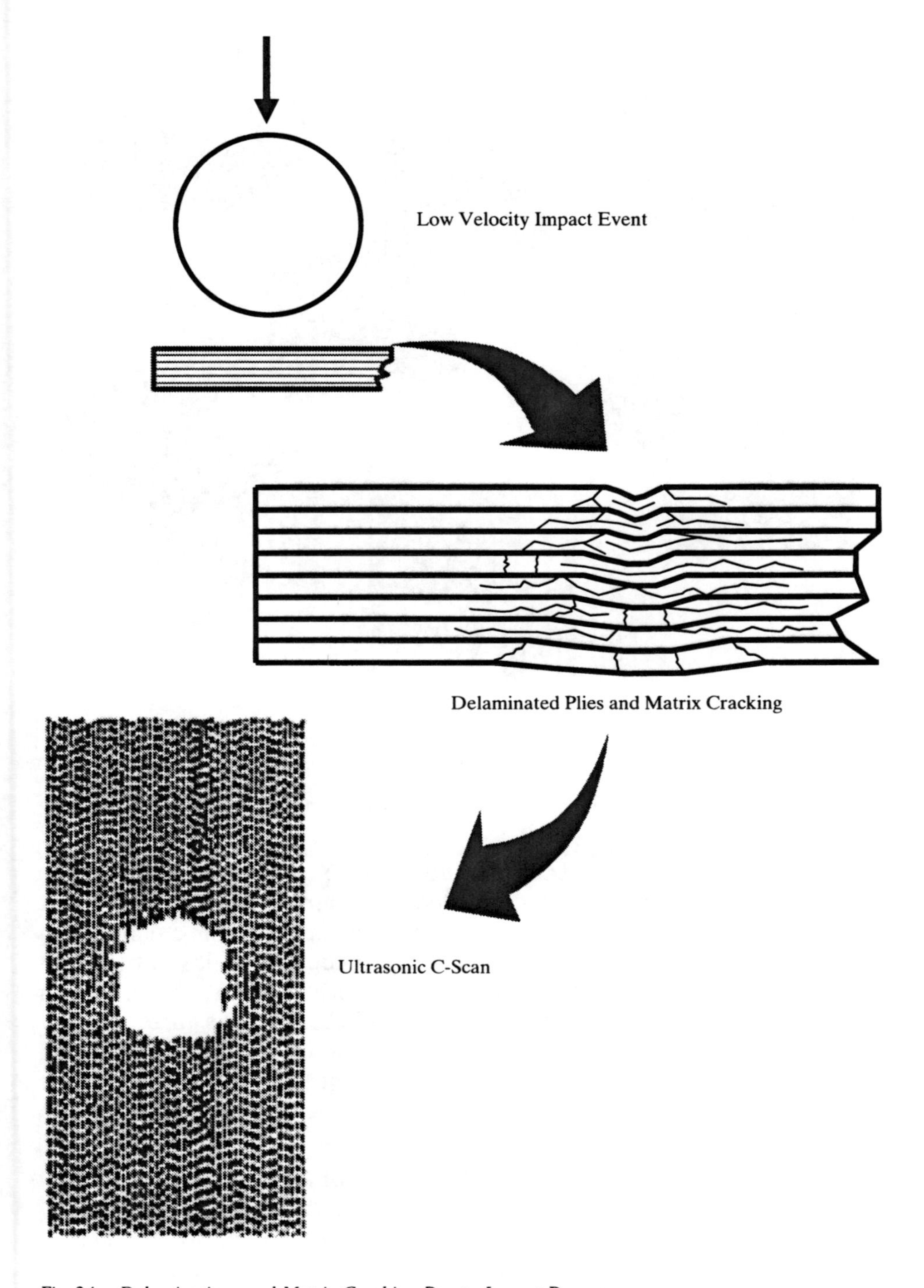

Fig. 24. Delaminations and Matrix Crasking Due to Impact Damage

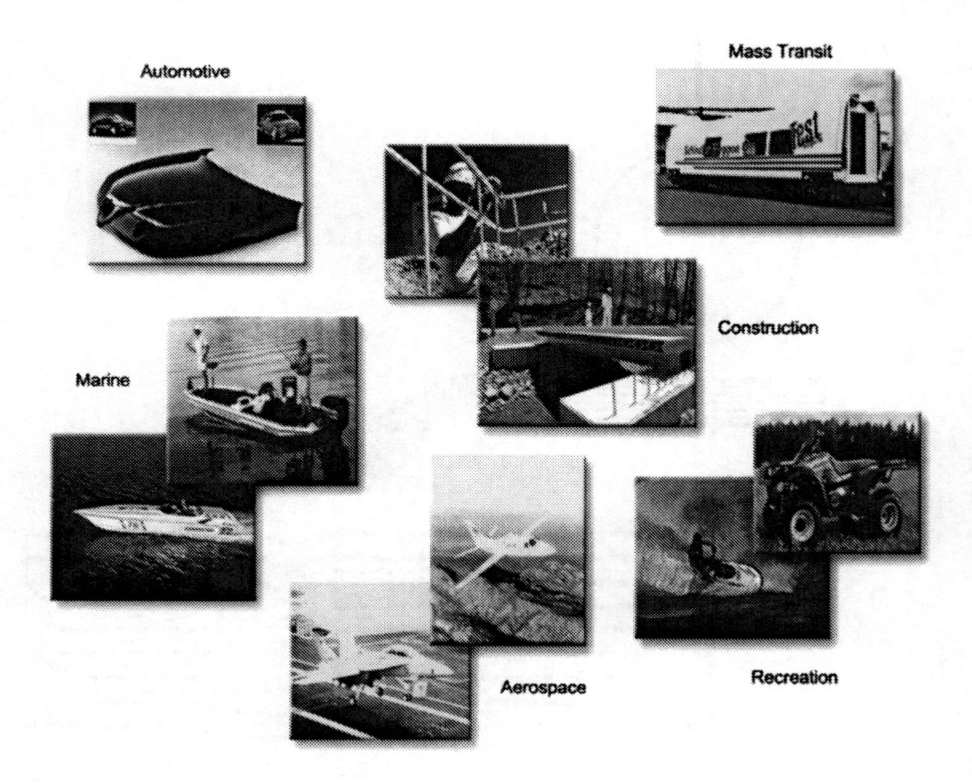

Fig. 25. Composite Applications Are Varied and Expanding

typically glass SMC, and the process is often compression or injection molding.

Maintenance is a major headache and expense for the marine industry. Composites help minimize these problems, primarily because they do not corrode like metals or rot like wood. Boat hulls ranging from small fishing boats to large racing yachts are routinely made from glass fibers and polyester or vinyl ester resins. Masts are frequently fabricated from carbon fiber composites. Glass-filament-wound SCUBA tanks are another example of composites improving the marine industry. Lighter tanks can hold more air yet require less maintenance than their metallic counterparts. Jet skis and boat trailers often contain glass composites to help minimize weight and reduce corrosion.

Tennis racquets have been made for years out of glass, and many golf-club shafts are made of carbon. Processes can include compression molding for tennis racquets and tape wrapping or filament winding for golf shafts. Composites also make possible lighter, better, stronger skis and surfboards. Snowboards are another example of a composite application that takes a beating yet keeps on performing. They are typically made by using a sandwich construction (composite skins with a honeycomb core) for maximum specific stiffness.

Using composites to improve the infrastructure of our roads and bridges is a relatively new but exciting application. Many of the world's roads and bridges are badly corroded and require constant maintenance or replacement.

Composites offer much longer life with less maintenance requirements due to their corrosion resistance. Typical materials/processes include wet lay-up repairs and corrosion-resistant pultruded products. Pultruded fiberglass rebar strengthens concrete, and glass fibers are used in some shingling materials. With the number of mature tall trees dwindling, the use of composites for electrical towers and light poles is greatly increasing. Typically, these are pultruded or filament-wound glass. The blades for large windmills are normally made from composites to improve the efficiency of electrical energy generation.

1.8 Summary

Advanced composites consist of either continuous or discontinuous fibers embedded in a matrix. Common fibers include glass, aramid and carbon. Glass fibers are used extensively in commercial applications because of their good balance of properties and low cost. Aramid fibers, being organic, have low densities and outstanding toughness. Carbon fibers have the best combination of strength and stiffness but are also the most expensive of the three.

The matrix holds the fibers in proper position and protects them from abrasion. The matrix transfers loads between the fibers and supports them in compression. Matrices commonly used for composites include polyesters, vinyl esters, bismaleimides, phenolics and polyimides, with polyesters and vinyl esters dominating the commercial markets and epoxies dominating the aerospace markets. The matrix usually determines the upper temperature up to which the composite can be used. The glass transition temperature (T_g), an important property of the matrix, is the temperature at which the matrix transforms from a glassy solid to a softer, more rubber-like material. A matrix should never be used above its glass transition temperature unless the service life is extremely short.

There are a multitude of product forms available. Rovings, tows and yarns are the basic building blocks of almost all composite product forms. Prepregs, in which the rovings, tows or yarns are impregnated with a predetermined amount of resin, are the primary product forms for high-performance structural applications. They are available as unidirectional tape or woven broadgoods. To take advantage of certain features of different fibers, hybrids such as aramid and carbon (greater toughness) or glass and carbon (lower cost) are occasionally used. Dry woven cloth, stitched fabrics and preforms that reduce lay-up costs are available for liquid-molding processes. Where high-strength properties are not required, short fibers, primarily glass, are used for compression molding (SMCs) and injection molding.

Composite fabrication processes include:

- Prepreg lay-up is expensive but produces the highest-quality parts. Prepreg parts are usually cured in an autoclave to maximize part quality.
- Filament winding is a process capable of making parts that are bodies of revolution or near bodies of revolution. Filament winding can be conducted by using either prepreg roving or wet winding.
- Wet lay-up, along with low-temperature curing resins, is capable of making very large structures. It is an excellent process for low-volume large parts. Wet laid-up parts can be cured with or without a vacuum bag.
- Spray-up is another process capable of making very large part size. Due to the random discontinuous nature of the fibers, the mechanical properties are lower than parts made by continuous fiber prepreg or wet lay-up.
- Liquid-molding processes use a dry preform that is placed in a matched-die tool, liquid resin is injected and the part is cured in the mold. The resin can be injected under pressure or pulled through the preform with a vacuum.
- Compression molding is another matched-die process that uses either SMCs or BMCs that is loaded as a predetermined charge. Heat and pressure form and cure the part to the required shape
- Injection molding is a high-volume process capable of making millions of parts per year. A thermoplastic or thermoset resin reinforced with short fibers is injected into a precision mold under high pressures, where it either cools (thermoplastic) or cures (thermoset).
- Pultrusion is a continuous process for making structural shapes of constant cross-section. Rovings are normally pulled through a resin bath and into a heated die where the part cures as it travels through the die. It is then pulled to the desired length and cut with a mechanical saw.

Advanced composites are a diversified and growing industry due to their distinct advantages over competing metallics, including lighter weight, higher performance and corrosion resistance. They are used in aerospace, automotive, marine, sporting goods, and more recently in infrastructure applications. *The major disadvantage of composites is their high cost*. The proper selection of materials (fiber and matrix), product forms and processes have a major impact on the cost of the finished part. The remainder of this book will address these processes and cost drivers in some detail.

Acknowledgments

The majority of the information contained in this chapter is from a training course taught by Ray Bohlmann, Mike Renieri, Gary Renieri and Russ Miller entitled "Advanced Materials and Design for Integrated Topside Structures" given to Thales in The Netherlands during 15–9 April 2002.

References

[1] "Prepreg Technology", Hexcel Composites, January 1997.
[2] Walsh, P.J., "Carbon Fibers", in *ASM Handbook,* Volume 21 Composites, ASM International, 2001, p. 38.
[3] Mantel, S.C., Cohen, D. "Filament Winding," in *Processing of Composites,* Hanser, 2000.
[4] Groover, M.P., *Fundamentals of Modern Manufacturing–Materials, Processes, and Systems,* Prentice-Hall, Inc.
[5] Taylor, A., "RTM Material Developments for Improved Processability and Performance", *SAMPE Journal* **36** (4), July-August 2000, pp. 17-24.

Chapter 2

Fibers and Reinforcements: The String That Provides the Strength

Reinforcements for composite materials can be particles, whiskers or fibers. Particles have no preferred orientation and provide minimal improvements in mechanical properties. They are frequently used as fillers to reduce the cost of the material. Whiskers are single crystals, which are extremely strong but are difficult to disperse uniformly in the matrix. In comparison to fibers, they are small in both length and diameter. Fibers have a very long axis compared to particles and whiskers. They are usually circular or nearly circular and are significantly stronger in the longer direction, because they are normally made either by drawing or pulling during the manufacturing process, which orients the molecules so that tension loads on the fibers pull more against the molecular chains themselves than against a mere entanglement of chains. Due to the strength and stiffness of fibers, they are the predominantly used reinforcements for advanced composites. Fibers may be continuous or discontinuous, depending on the application and manufacturing process.

2.1 Fiber Terminology

Before examining the various types of fibers used for composite reinforcements, the major terminology used for fiber technology will be reviewed[1]. Fibers are produced and sold in many forms:

- *Fiber* – a general term for a material that has a long axis, which is many times greater than its diameter. The term "aspect ratio" (fiber length divided by its diameter, l/d), is frequently used to describe short fiber lengths. For fibers, aspect ratios are normally greater than 100.
- *Filament* – the smallest unit of a fibrous material. For spun fibers, this is the unit formed by a single hole in the spinning process. It is synonymous with fiber.
- *End* – a term used primarily for glass fibers, which refers to a group of filaments in long parallel lengths.
- *Strand* – another term associated with glass fibers that refers to a bundle or group of untwisted filaments. Continuous strand rovings provide good overall processing characteristics through fast wet–out (penetration of resin into the strand), and even tension and abrasion resistance during processing. They can be cut cleanly and dispersed evenly throughout the resin matrix during molding.
- *Tow* – similar to a strand for glass fiber, tow is used for carbon and graphite fibers to describe the number of untwisted filaments produced at one time. Tow size is usually expressed as *X*k. For example, a 12k tow contains 12,000 filaments.
- *Roving* – a number of strands or tows collected into a parallel bundle without twisting. Rovings can be chopped into short fiber segments for sheet molding compound (SMC), bulk molding compound (BMC) or injection molding.

- *Yarn* – a number of strands or tows collected into a parallel bundle with twist. Twisting improves the handleability and makes processes such as weaving easier, but the twist also reduces the strength.
- *Band*– the thickness or width of several rovings, yarns or tows as it is applied to a mandrel or tool. A common term used in filament winding.
- *Tape* – a composite product form in which a large number of parallel filaments (e.g., tows) are held together with an organic matrix material (e.g., epoxy). The length of the tape, in the direction of the fibers, is much greater than the width, and the width is much greater than the thickness. Typical tape product forms are several hundred feet long, 6-60 inches wide, and 0.005-0.010 inch thick.
- *Woven cloth* – another composite product form made by weaving yarns or tows in various patterns to provide reinforcement in two directions, usually 0° and 90°. Typical two–dimensional (2D) woven cloth is several hundred feet long, 24-60 inches wide and 0.010-0.015 inch thick.

Additional fiber and textile terminology will be introduced as different fiber types, processes and product forms are discussed.

2.2 Glass Fibers

Due to their low cost, high tensile strength, high impact resistance and good chemical resistance, glass fibers are used extensively in commercial composite applications. However, their properties cannot match those of carbon fibers for high–performance composite applications. They possess a relatively low modulus and have inferior fatigue properties compared to carbon fibers. The three most common glass fibers used in composites are: E–glass, S–2 glass and quartz. E–glass is the most common and least expensive, providing a good combination of tensile strength (500 ksi) and modulus (11.0 msi). S–2 glass, having a tensile strength of 650 ksi and a modulus of 12.6 msi is more expensive, but is 40% stronger than E–glass and retains a greater percentage of its strength at elevated temperatures. Quartz fiber is a rather expensive, ultra–pure silica glass with a low dielectric fiber and is used primarily in demanding electrical applications. The physical and mechanical properties of some of the commercially important fibers are given in Table 1.

Glass is an amorphous material consisting of a silica (SiO_2) backbone with various oxide components to give specific compositions and properties. Glass fibers are made from silica sand, limestone, boric acid and minor amounts of other ingredients such as clay, coal, and fluorspar. In the glass manufacturing process, the ingredients .are dry mixed and melted in a high–temperature, refractory furnace. As shown in Fig.1, there are two processes that are used to make high–strength glass fibers[2]. In the marble

Table 1.1 Comparative Properties of High Strength Fibers

Type of Fiber	*Tensile Strength (Ksi)*	*Tensile Modulus (Msi)*	*Elongation at Failure (%)*	*Density (gm/cm³)*	*Coefficient of Thermal Expansion (10⁻⁶°C)*	*Fiber Diameter (μm)*
Glass						
E-Glass	500	11.0	4.7	2.58	4.9-6.0	5-20
S-2 Glass	650	12.6	5.6	2.48	2.9	5-10
Quartz	490	10.0	5.0	2.15	0.5	9
Organic						
Kelvar 29	525	12.0	4.0	1.44	-2.0	12
Kevlar 49	550	19.0	2.8	1.44	-2.0	12
Kevlar 149	500	27.0	2.0	1.47	-2.0	12
Spectra 1000	450	25.0	0.7	0.97	-----	27
PAN Based Carbon						
Standard Modulus	500-700	32-35	1.5-2.2	1.80	-0.4	6-8
Intermediate Modulus	600-900	40-43	1.3-2.0	1.80	-0.6	5-6
High Modulus	600-800	50-65	0.7-1.0	1.90	-0.75	5-8
Pitch Based Carbon						
Low Modulus	200-450	25-35	0.9	1.9	-----	11
High Modulus	275-400	55-90	0.5	2.0	-0.9	11
Ultra High Modulus	350	100-140	0.3	2.2	-1.6	10

Note: Representative only. For specific properties, contact the fiber manufacturer.

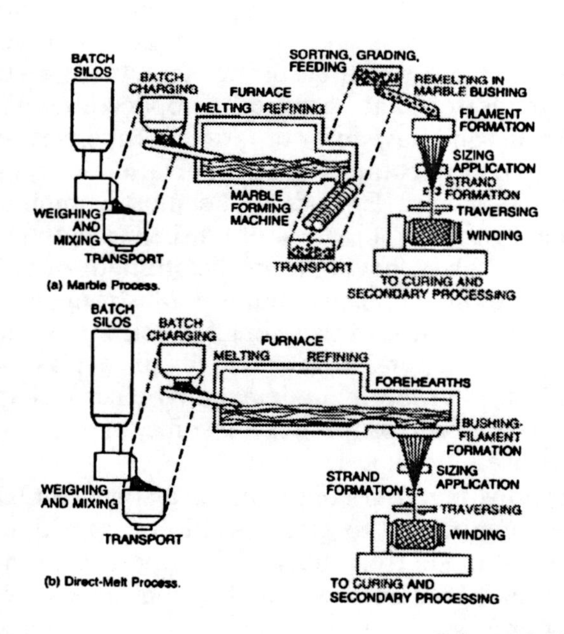

Reference 2: with permission

Fig. 1. Manufacturing Processes for Fiberglass Fibers

process, the glass ingredients are shaped into marbles, sorted by quality and then remelted into fiber strands. Alternatively, the molten glass is introduced directly to form fiber strands. After heating to approximately 2,300°F, the molten glass flows through small bushing holes at speeds approaching 180 ft s^{-1} to form filaments that are immediately quenched with water or an air spray to yield an amorphous structure. The diameter is controlled by hole size, draw speed, temperature, viscosity of melting and cooling rate. In typical glass fiber terminology, a number of individual strands (or ends) are usually incorporated into a roving to provide a convenient form for subsequent processing. Rovings are preferred for most reinforcements because they possess higher mechanical properties than twisted yarns. Rovings are wound onto individual spools (Fig. 2) containing 20–50 lb of fiber. If the material is to be used for weaving, it is usually twisted into a yarn to provide extra strength during the weaving operations. The strands are specified by their yield (yards per pound) or "denier" (the mass (in grams) per 9,000 meters of fiber). Another textile terminology frequently encountered is the term "tex," which is the mass (in grams) per 1,000 m of fiber.

Fig. 2. Glass Roving

Virgin glass filaments are susceptible to mechanical abrasion, which degrades the strength; therefore, a sizing is applied immediately after manufacturing, to prevent scratches from forming on the surface during spooling and mechanical damage from weaving, braiding and other textile processes. Sizings are extremely thin coatings that account for only about 1–2 % by weight. The sizing, usually a starch and a lubricant, is removed either by solvents or by heat scouring , after all mechanical operations are complete. After the sizing is removed, it is replaced with a surface finish, which greatly improves the fiber–to–matrix bond. For example, organosilane coupling agents have one end group that is compatible with the silane structure of the glass and another end group that is compatible with the organic matrix. Coupling agents are critical to the performance of the glass–reinforced composites. Improvements of over 100% have been demonstrated in composite tensile, flexural and compression strength. The coupling agent also helps to protect the glass fiber from attack by water. Some sizings also function as coupling agents and, therefore, remain on the fiber throughout the manufacturing process. There are many different types of sizings/finishes available, and it is important that the sizing/finish selected for an application is compatible with both the fiber and the matrix.

Anti–static agents and lubricants can also be used to improve the handling and processing characteristics such as hardness or softness. If the glass fiber is going to be chopped into short lengths for fiber spraying, hardness is a desirable property since it improves the chopping ability. On the other hand, softness is a desirable property if the fiber is going to be used in a lay–up operation where drapability and forming are important. The maximum use temperature used for glass fibers ranges from 930°F for E–glass up to 1,920°F for quartz.

2.3 Aramid Fibers

Aramid fibers are truly organic fibers with stiffness and strength intermediate between that of glass and carbon. Dupont's Kevlar fiber is the most frequently used fiber. These aromatic polyamides are part of the nylon family. Kevlar is made from reacting paraphenylene diamine with terephthaloyl chloride in an organic solvent to form polyparaphenylene terephthalamide (aramid). This is a condensation reaction that is followed by extrusion, stretching and drawing. The polymer is washed and then dissolved in sulfuric acid. At this point, the polymer is a partially oriented liquid crystal. The polymer solution is then extruded through small holes (spinnerets). The fibers are oriented in solution and during spinning and again as they pass through the spinneret. The fibers are then washed, dried and wound up.

Aramid fibers are actually thermoplastics that have a glass transition temperature (T_g) higher than their degradation temperature. They have highly oriented molecular chains in the fiber direction that are held together by strong covalent bonds resulting in high longitudinal tensile strength. However, the chains in the transverse direction are held together by weaker hydrogen bonds resulting in low transverse strength. Unlike glass and carbon or graphite fibers, Kevlar fibers are not surface treated because no acceptable surface treatment for aramid fibers has been developed to date. Sizings are used when the fiber is going to be woven, made into rope or for ballistic applications. Aramid fibers have a combination of good tensile strength and modulus; they are lightweight and have excellent toughness and outstanding ballistic and impact resistance. However, due to the lack of adhesion to the matrix, they exhibit relatively poor transverse tension, longitudinal compression and interlaminar shear strengths. Like carbon and graphite, aramid fibers also exhibit a negative coefficient of thermal expansion.

The three most prevalent aramid fibers are Kevlar 29 (low modulus), Kevlar 49 (intermediate modulus) and Kevlar 149 (high modulus). These differences among the various grades of Kevlar are due to changes in process conditions that enhance crystallinity in the high– and ultra–high modulus fibers. Kevlar 29 has high toughness; Kevlar 49 has a higher modulus; and Kevlar 149 has ultrahigh modulus. Normal bundle size ranges from 134 to 10,000 filaments per bundle. Due to its organic nature, the maximum temperature used for aramid is limited to about 350°F. As in the case of carbon, they are available in tows or yarns of various weights that can be converted to woven cloth or chopped fiber mat. However, being an extremely tough fiber, they are more difficult to cut, which causes some handling problems. Due to their extreme toughness, they are often used for ballistic protection. Aramid fibers are resistant to flames and to most solvents except strong acids and bases; however, aramid fibers are hygroscopic and absorb moisture.

2.4 Ultrahigh Molecular Weight Polyethylene (UHMPE) Fibers

High–modulus polyethylene fibers can be made by solid–state drawing of high–density polyethylene. These fibers are limited to temperatures of 290°F or lower. They have good moisture resistance, high impact resistance and attractive electrical properties (i.e., low dielectric constant and low loss tangent). At a density of 0.97 g cm^{-3}, they are even lighter than aramid fibers. They do not form a strong bond to the matrix, which results in poor transverse tension and compression strengths. Being a thermoplastic fiber, they are subject to creep under continuous loading.

2.5 Carbon and Graphite Fibers

Carbon and graphite fibers are the most prevalent fiber form in high–performance composite structures. Carbon and graphite fibers can be produced with a wide range of properties; however, they generally exhibit superior tensile and compressive strength, possess high moduli, have excellent fatigue characteristics and do not corrode. Although the terms are often used interchangeably, graphite fibers are (1) subjected to heat treatments above 3,000°F, (2) possess 3D ordering of their atoms, (3) have carbon contents greater than 99% and (4) have elastic moduli (E) greater than 50 msi. Carbon fibers have lower carbon contents (93–95%) and are heat treated at lower temperatures.

Carbon and graphite fibers can be made from rayon, pitch or polyacrylonitrile (PAN) precursors. Although PAN fibers are more expensive than rayon fibers, they are extensively used for structural carbon fibers because their carbon yield is almost double that of rayon fibers. The pitch process produces fibers that have lower strength than those produced from PAN, but it can produce ultrahigh modulus fibers (50–145 msi).

Fibers made from PAN (Fig. 3) are stretched, heat set, carbonized, graphitized and then surface treated. The stretching process helps to orient the molecules and some tension is maintained throughout the entire process. Heat setting or oxidation at 390–570°F cross–links the PAN and stabilizes the structure, which prevents melting during the carbonization process. The heat setting process is conducted in an atmosphere of air and

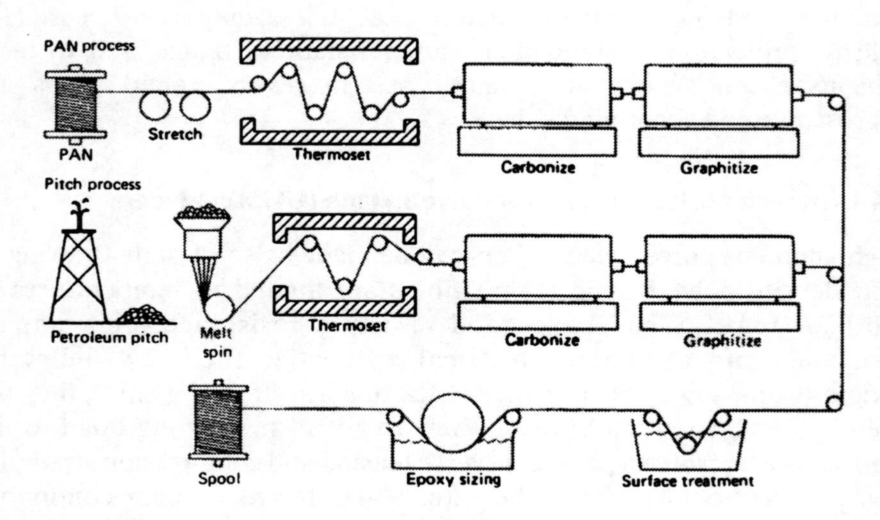

Fig. 3. Manufacturing Processes for Carbon and Graphite Fibers

converts the thermoplastic PAN into a non–plastic cyclic or ladder compound, capable of withstanding the high temperatures present during carbonization.[3] Carbonization converts the PAN to carbon and is conducted in a nitrogen atmosphere at 1,800–2,700°F. During carbonization, the fibers shrink in diameter and lose approximately 50% of their weight. If a true graphite fiber is desired, then the fiber is graphitized at temperatures between 3,600°F and 5,500°F, which produces a more crystalline structure and a higher elastic modulus. The final step, surface treatment, is conducted to improve the fiber–to–matrix bond and is usually done in an electrolytic alkaline bath. The surface treatment attaches carboxyl, carbonyl and hydroxyl groups to the fiber surface, which can bond with the polymeric matrix. If the fiber is going to be woven, sizes (usually epoxy) are applied to the fiber to protect the fiber surface from mechanical abrasion.

Pitch–based carbon and graphite fibers are made by heating coal–tar pitch for up to 40 h at 800°F, forming a high–viscosity liquid with a high degree of molecular order known as a mesophase. Orientation is responsible for the ease of consolidation of pitch into carbon. The mesophase is then spun through a small orifice, aligning the molecules along the fiber axis. Pitch–based fibers are then processed following the same basic steps as used in PAN–based fiber manufacturing, namely, stretching, carbonization, graphitization and surface treatment. Pitch–based graphite fibers are characterized as having a higher modulus and lower strength than carbon fibers manufactured by the PAN process. Pitch–based high–modulus graphite fibers having a modulus between 50 and 145 msi are often used in space structures requiring high rigidity. The higher temperatures used in the graphitization process for graphite fibers result in closer orientation of the graphite cystallites towards the fiber axis. The better the alignment of the crystallites, the higher the modulus of the fiber. However, high crystallinity also makes the fiber weak in shear, which results in lower compressive strength. Therefore, high crystalline graphite fibers do not exhibit balanced tensile and compressive mechanical properties. In addition to their high modulus and low thermal expansion, pitch–based graphite fibers have high values of thermal conductivity; e.g., values of 900-1,000 W $m^{-1}K^{-1}$ as compared to values of only 10-20 W $m^{-1}K^{-1}$ for PAN–based carbon fibers. These high thermal conductivities are used to remove and dissipate heat in space–based structures. High–modulus graphite fibers can also be manufactured using the PAN process; however, the highest modulus attainable is around 85 msi.[4]

The strength of carbon and graphite fibers depends on the type of precursor used, the processing conditions during manufacturing, such as fiber tension and temperatures, and the presence of flaws and defects. Flaws in the carbon fiber microstructure include internal pits and inclusions, external gouges, scratches, and stuck filament residues, as well as undesirable characteristics such as striations and flutes. These flaws can have a considerable impact on the fiber tensile strength, but little effect, if any, on modulus, conductivity or thermal expansion.[5] Both carbon and graphite fibers usually have a slightly negative coefficient of thermal expansion that becomes more negative as the modulus E increases. One consequence of using high– and ultrahigh–modulus carbon fibers is the increased possibility of matrix microcracking during processing or environmental exposure due to the larger mismatch in the coefficients of thermal expansion between the fibers and the matrix.

Carbon fibers with a wide range of strength and moduli are available from a number of producers. PAN–based carbon fibers having strengths ranging from 500 to 1,000 ksi and moduli ranging from 30 to 45 msi with elongations of up to 2% are commercially available. Standard–modulus PAN fibers have good properties and lower cost, while higher–modulus PAN fibers cost more because high processing tempera-tures are required. Heating the fibers to 1,800°F yields PAN fibers containing 94% carbon and 6% nitrogen, while heating to 2,300°F removes the nitrogen and raises the carbon content to around 99.7%. Higher processing temperatures increase the tensile modulus by refining the crystalline structure and the 3D nature of the structure. The diameter of carbon fibers usually ranges from 0.3 to 0.4 mil. Carbon fibers are provided in untwisted bundles of fibers called "tows." Tow sizes can range from as small as 1,000 fibers per tow up to >200,000 fibers per tow. A typical designation of "12k tow" indicates that the tow contains 12,000 fibers. Normally, as the tow size decreases, the strength and cost increases. The small 1k tow size is normally not used unless the property advantages outweigh the cost disadvantages. For aerospace structures, normal tow sizes are 3k, 6k and 12k, with 3k and 6k being the most prevalent for woven cloth and 12k for unidirectional tape. It should be noted that there are very large tow sizes (>200k), which are primarily used for commercial applications, and are normally broken down after manufacturing into smaller tow sizes (e.g., 48k) for subsequent handling and processing. The costs of carbon fibers are dependent on the manufacturing process, the type of precursor used, the final mechanical properties and the tow size. Costs can vary anywhere from below $10 a pound for large tow commercial fibers to several hundreds of dollars per pound for small– tow ultrahigh–modulus pitch–based fibers. The maximum use temperature for carbon and graphite fibers in an oxidizing atmosphere is 930°F.

The ideal engineering material would have high strength, high stiffness, high toughness and low weight. Carbon fibers combined with polymer matrices meet these criteria more closely than any other material. Carbon fibers are elastic to failure at normal temperatures, creep resistant and not susceptible to failure, chemically inert, except in strong oxidizing environments or in contact with certain molten metals, and have excellent damping characteristics. Some disadvantages of carbon fibers are: they are brittle and have low impact resistance; they have low strains to failure; their compressive strengths are less than their tensile strengths; and they are relatively expensive compared to glass fibers.

2.6 Woven Fabrics

Two–dimensional woven products (Fig. 4) are usually offered as 0° and 90° constructions. However, bias weaves (45°, 45°) can be made by twisting the basic 0° and 90° constructions. Weaves are made on a loom by interlacing two orthogonal (mutually perpendicular) sets of yarns (warp and fill). The warp direction is parallel to the length of the roll, while the fill, weft, or woof direction is perpendicular to the length of the roll. Textile looms (Fig. 5) produce woven cloth by separation of the warp yarns and insertion of the fill

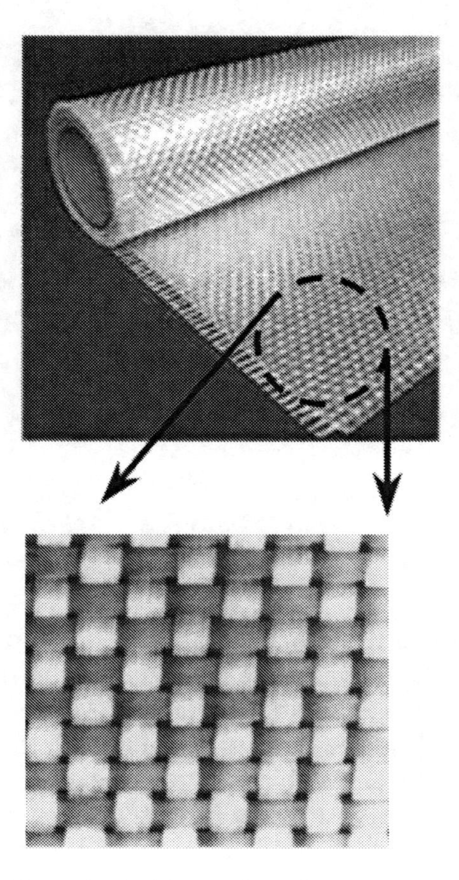

Fig. 4. Typical Woven Glass Cloth

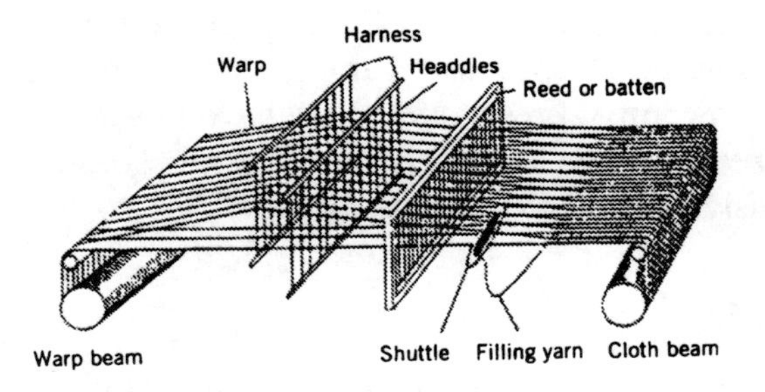

Reference 6: with permission

Fig. 5. Schematic of Basic Weave Process

Fig. 6. Examples of Hybrid Weaves

yarns.[6] The individualwarp yarnsare manipulated by the harnesses that, in turn, move the headdles. The fill yarns are carried by a shuttle system that moves back and forth creating an interlacing fabric.[7] Most weaves contain similar number of fibers and use the same material in both the warp and fill directions. However, hybrid weaves (Fig. 6) such as carbon and glass and weaves dominated by warp yarns are frequently used. These hybrid weaves can be used to obtain specific properties, such as mixing carbon witharamid to take advantage of the toughness of aramid, or to reduce costs, such as

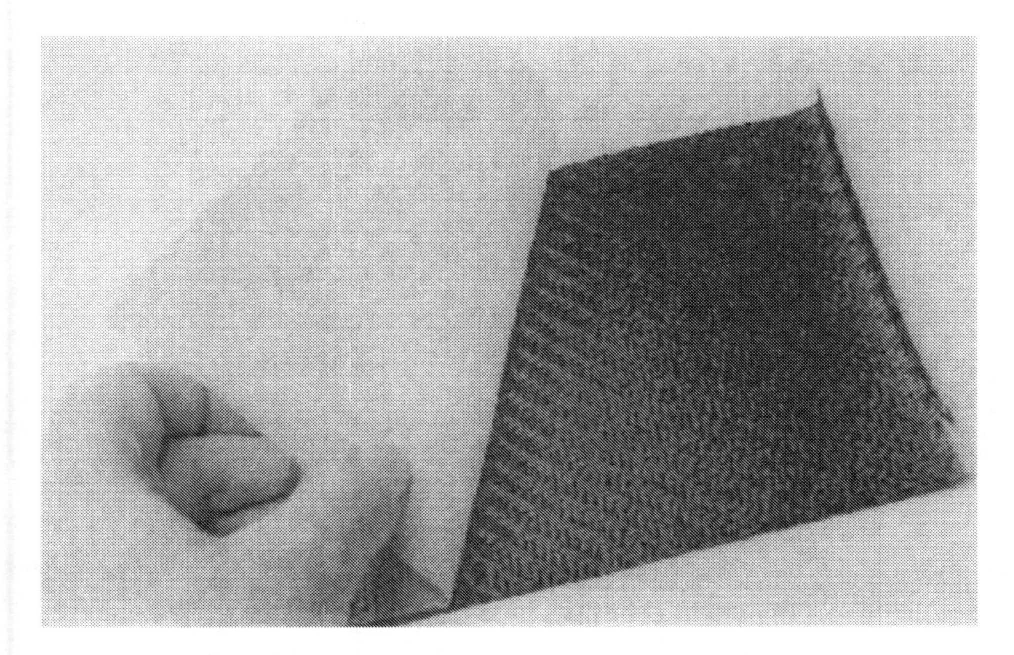

Fig. 7. Woven Carbon Cloth Prepreg With Backing Paper

mixing glass with carbon fibers. Woven broadgoods may be purchased either as a dry preform or pre–impregnated with a B–staged resin (Fig. 7). In most applications, multiple layers of 2D weaves are laminated together. As with tape laminates, the layers are oriented to tailor strength and stiffness.

Weaves may be classified by the pattern of interlacing as shown in Fig. 8. The simplest pattern is the plain weave. In the plain weave, every warp and fill yarn goes alternatively over and then under successive warp and fill yarns, respectively. Plain weaves have the most interlaces per unit area than any other type of weave and, therefore, the tightest basic fabric design and are the most resistant to in–plane shear movement. Therefore, plain weaves resist distortion during handling, but may be difficult to form on complex contours. They are also more difficult to wet–out during impregnation. Another disadvantage of the plain weave is the frequent exchanges of position from top to bottom made by each yarn. This waviness or yarn crimp reduces the strength and stiffness of the composite.

The basket weave is a variation of plain weave in which two (or more) warp and two (or more) fill yarns are woven together. An arrangement of two warps crossing two fills is designated as a 2×2 basket, but the arrangement of fibers need not be symmetrical; i.e., it is possible to have 8×2, 5×4 and other variations. The basket weave has less crimps than the plain weave and is, therefore, somewhat stronger. Plain weaves are

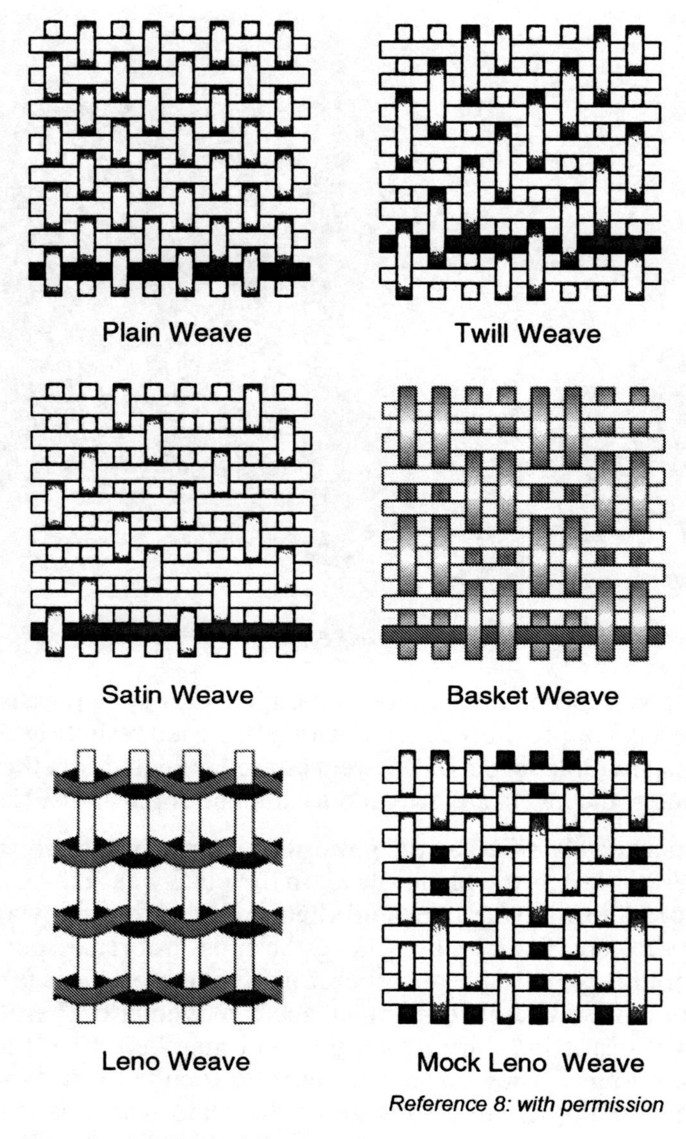

Fig. 8. Common 2-D Weave Styles

often used for less curved parts while five– and eight–harness satin weaves are used for more highly contoured parts.

The satin weaves are characterized by a minimum of interlacing and, therefore, have less resistance to in–plane shear movement and have the best drapability. In the four–harness satin weave, the warp yarns skip over three fill yarns and then under one fill yarn. In the five–harness satin weave, the warp yarns skip over four fill yarns and then under one fill yarn. Finally, in the eight–harness satin weave, the warp yarns skip over seven fill yarns and then under one fill yarn. Due to less fiber crimp, satin weave fabrics are of a higher strength than plain weave fabrics. They also provide smooth surface finishes at a minimum per ply thickness. The eight–harness satin weave has the best drapability of this group; however, five–harness weaves normally use a 6k carbon tow that is less expensive than the 3k carbon tow used for eight–harness satin weaves. The trend in the industry has been to move towards greater use of the less expensive five–harness satin weaves.

Twill weaves are occasionally used because they have better drapablity than the plain weaves and are known for their extremely good wet–out during impregnation. In this weave, one or more warps alternately weave over and under two or more fills in a regular repeating manner, producing the visual effect of a straight or broken "rib" to the fabric.

The leno and mock leno weaves are rarely used for structural composites. The leno weave is also a form of the plain weave in which the adjacent warp fibers are twisted around consecutive fill fibers to form a spiral pair, effectively locking each fill in place. This construction produces an extremely open fabric with a low fiber content. The leno weave is frequently used to tie the edges of dry fabric together (Fig. 9) so it will not unravel during handling. The mock leno, also a version of the plain weave, has occasional warp fibers at regular intervals, but several fibers apart, which deviate from the alternate under–over interlacing and instead interlace every two or more fibers. This happens with similar frequency in the fill direction, and the overall result is a fabric with increased thickness, a rougher surface and more porosity. Woven fabrics often contain tracer yarns. For example, yellow aramid tracers are often woven on two–inch

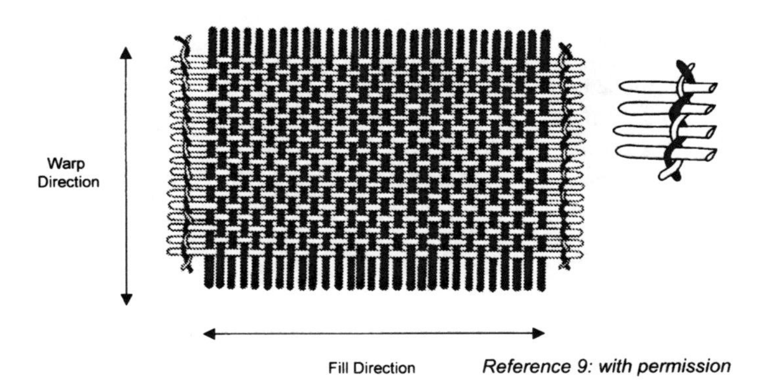

Fig. 9. Plain Weave Cloth with Leno Selvage

spacings in carbon cloth along the warp direction to help fabricators identify warp and fill direction during composite part lay–up.

The selection of a weave involves manufacturing considerations as well as final mechanical properties. The weave type affects dimensional stability and the conformability (or drape) of the fabric over complex surfaces. Satin weaves, for example, exhibit good conformability. Unfortunately, good conformability and resistance to shear are mutually exclusive. Thus, while woven fabrics are frequently the material of choice for complex geometries, the designer must be aware that specified material directions may be impossible to maintain on compound contours and other complex shapes; i.e., initially orthogonal yarns may not remain orthogonal in the finished product. Several of the commonly used weave styles for high–performance composites are shown in Table 2.

2.7 Reinforced Mats

Reinforced mats, shown in Fig. 10, are made of either chopped strands or continuous strands laid down in a swirl pattern. Mats are generally held together by resinous binders. These are used for medium–strength parts with uniform cross–sections. Both chopped and continuous strand reinforcing mats are available in weights varying from 0.75 to 4.5 oz ft^{-2} and in various widths. Surfacing mats or veils are thin, lightweight materials

Table 1.2 Common Weave Styles Used in High Performance Composites

Weave	*Type Of Fiber*	*Construction Yarns/in. Warp X Fill*	*Fiber Areal Weight (g/m^2)*	*Approximate Cured Per Ply Thickness[a] (in.)*
Style 120	E–Glass	60 × 58	107	.005
Style 7781	E–Glass	57 × 54	303	.010
Style 120	Kevlar 49	34 × 34	61	.004
Style 285	Kevlar 49	17 × 17	170	.010
8-Harness satin	3K Carbon	24 × 23	370	.014
5-Harness satin	6K Carbon	11 × 11	370	.014
5-Harness satin	1K Carbon	24 × 24	125	.005
Plain	3K Carbon	11 × 11	193	.007

a. Actual cured per ply thickness depends on resin system, resin content and processing conditions.

Fig. 10. Glass Reinforced Mat

used in conjunction with reinforcing mats and fabrics to provide good surface finish. They are effective in blocking out the fiber pattern of the underlying mat or fabric. Combination mats, consisting of one ply of woven roving chemically bonded to chopped strand mat, are available from several processors of glass reinforcements. These products form a drapable reinforcement that combines the bidirectional fiber orientation of woven roving with the multidirectional fiber orientation of chopped strand mat. This saves time in hand lay–up, since two layers can be placed in the mold in a single operation. Other combinations are available for surface finish improvement as well as for multilayer reinforcement.

2.8 Chopped Fibers

Chopped fibers (Fig. 11) produce higher strength in parts produced by compression and injection molding. Chopped fibers are usually available in various lengths from 0.125 to 2 inches, although shorter milled fibers and longer fibers are available. They are blended with resins and other additives to prepare molding compounds for compression or injection moldings, encapsulation and other processes. Chopped glass reinforcement is available with many surface treatments to ensure optimum compatibility with most thermosetting and thermoplastic resin systems. The shorter chopped reinforcements are best suited for blending with

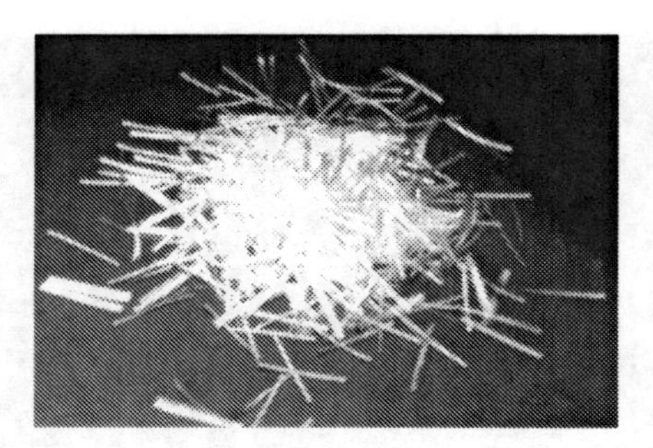

Fig. 11. Chopped Glass

thermoplastic resin systems for injection molding. Longer chopped reinforcements are blended with thermosetting resins for compression and transfer molding. Milled fibers combine reinforcing properties with processing ease in encapsulation or injection molding. Milled fibers are 1/32 to 1/8 inches long fibrous glass fibers. They are used for reinforcement of thermoplastic parts where strength requirements are low to moderate and for reinforcing fillers and adhesives.

2.9 Pregreg Manufacturing[10]

There are several stages a resin goes through during the manufacturing process. Resins are normally made by batch manufacturing in which the ingredients are placed in mixers (Fig. 12) and slowly heated to the A–stage condition or initial mixed state, where the resin has a very low viscosity that allows flow and impregnation of the fibers. Since the resins and curing agents used in composite matrices can be quite reactive, careful temperature control during mixing is critical to prevent the occurrence of an exothermic reaction that could result in a fire. Some components may require premixing before adding to the main mix. After mixing, the resin is usually placed in plastic bags and frozen until it is needed for prepregging or shipping as a resin for processes such as wet filament winding, liquid molding or pultrusion.

Prepreg is the most prevalent product form used in advanced composite manufacturing. It usually consists of a single layer of fibers embedded in a B–staged resin, as shown in Fig. 13. A large roll of 48 inches wide carbon/epoxy tape is shown in Fig. 14. Some fabricators use prepreg containing multiple layers if their usage warrants the additional costs. During prepregging, the resin advances to a B–stage condition in which it is a semi–solid at room temperature, which remelts and flows during the cure cycle. The B–staged resin normally contains some tack or stickiness to allow it to adhere to itself and tooling details during the lay–up operation. Being a resin in a state of continual advancement (i.e., reaction), the degree of

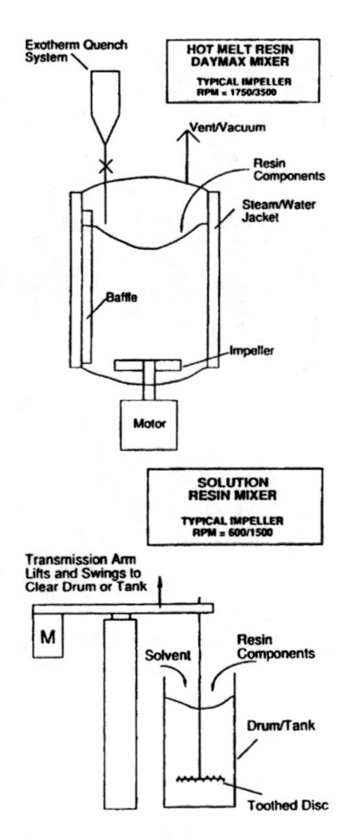

Fig. 12. Resin Mixing Schematics

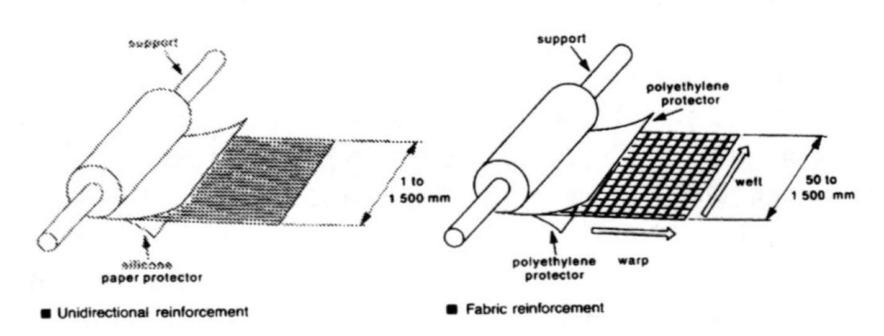

Fig. 13. Composite Tape and Broadgoods

advancement and the resultant tack and flow behavior will change unless it is kept refrigerated when not in use.

Several variables define a prepreg, namely, the fiber type, the fiber form (e.g., unidirectional or woven), the resin type, the fiber areal weight

The Boeing Company

Fig. 14. 48 Inch Wide Carbon/Epoxy Broadgoods

(FAW), the prepreg resin content (RC) and the cured per ply thickness (CPT). Fiber areal weight is simply the weight of the fiber in a given area and is usually specified in g m^{-2}. Prepreg resin content specifies the percentage of resin by weight in the prepreg. This is not necessarily the resin content of the cured part. Some resins are prepregged with excess resin (e.g., 42%) that will be bled off during cure to yield a final cured resin content of 28–30% by weight. Other resins, called net resin content prepregs, will be prepregged to almost the same resin content as the final cured part; therefore, no bleeding of excess resin is required for this product form. Finally, cured per ply thickness is specified as the thickness of each ply in inches. It should be noted that the final cured per ply thickness is dependent on part configuration, especially part thickness and the processing conditions employed by the user.

Prepreg is usually supplied as either roving (or towpreg), unidirectional tape or woven cloth. Prepreg rovings are bundles of fibers that are used primarily for filament winding or fiber placement. As the name implies, a single bundle of fibers is impregnated with the resin during prepregging. The cross–section of both product forms is a flat rectangle with a width of between 0.10 and 0.25 inches. The material is supplied in lengths, up to 20,000 ft, on a single spool. Unidirectional tape prepreg is the combination of multiple tows aligned in parallel that are impregnated with resin. Typical fiber areal weights range from as low as 30 g m^{-2} to as high as 300 g m^{-2}, with typical values being 95, 145 and 190 g m^{-2} that corresponds to cured per ply thicknesses of 0.0035, 0.005 and 0.0075 inches, respectively. Width ranges anywhere from 6 to 60 inches. Automated tape laying machines usually use 6 or 12 inches wide material, while the wider 60 inches broad goods are machine–cut into ply shapes and used for hand lay–up. Fabric prepregs consist of a woven fabric impregnated with a resin. Since fabric prepregs are primarily used for hand lay–up, the material is usually supplied as wide rolls, again up to 60 inches wide, to minimize the number of splices required in a part. Fabric prepregs usually have higher fiber areal weights than unidirectional tape and thicker cured per ply thickness (e.g., 0.014 inches per ply).

Prepregging can be accomplished by (1) hot–melt impregnation, (2) resin filming or (3) solvent impregnation. In the original hot–melt process (Fig. 15), the fibers were fed from a creel, collimated and impregnated with

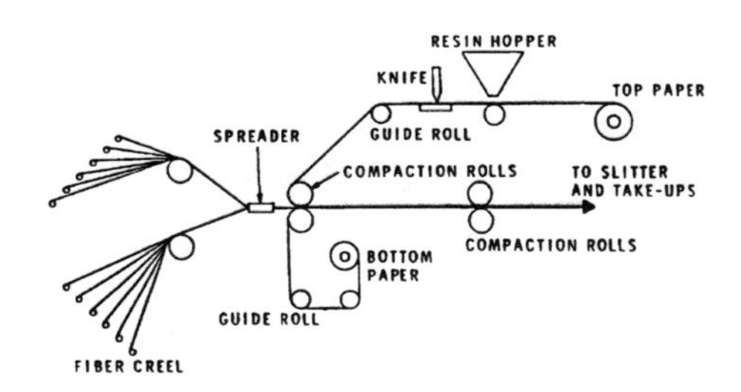

Fig. 15. Hot Melt Resin Impregnation Process

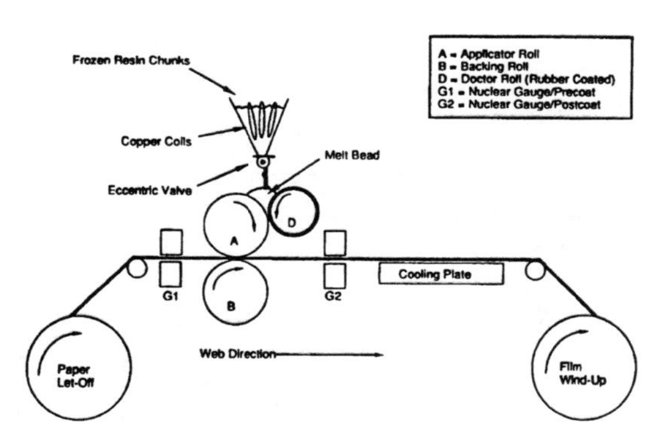

Reference 10: with permission

Fig. 16. Resin Filming Process

the melted resin and then immediately cooled prior to spooling on the roll. The newer resin filming process takes place in two different operations. First the resin is filmed to a controlled thickness on backing paper, as shown in Fig. 16. Then the spooled film can be either taken directly to the prepregging operation or frozen for future use. The majority of prepregs are currently made by the filming technique, which allows better control of resin content and fiber areal weight. Typical filming weights are in the range of 20-80 g m^{-2} with speeds up to 40 ft min^{-1}. When the resin film is ready for impregnation, it is conducted on a separate machine (Fig. 17) in which the fiber web is protected on both surfaces with a backing paper. Impregnation is achieved by the application of heat and nip roll pressure as the fibers, resin film and upper and lower backing papers are pulled through the line. After the material passes through the second set of nip rollers, it is immediately chilled to raise the resin viscosity and produce the semi–solid prepreg. At the exit, the upper paper sheet is removed and discarded, the edges are trimmed straight with slitter blades and the

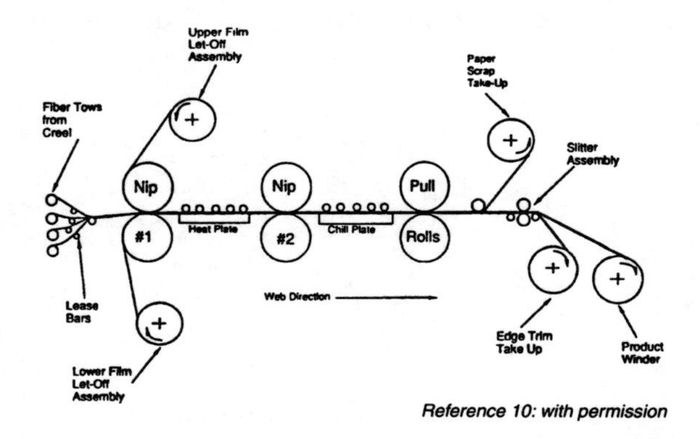

Fig. 17. Hot Melt Tape From Resin Film

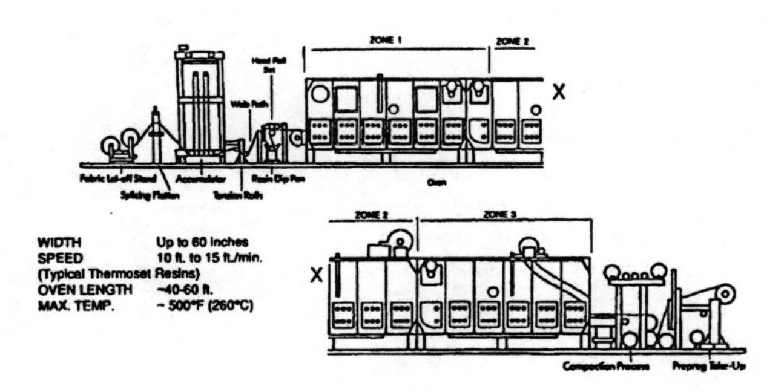

Fig. 18. Solution Impregnation Line

finished prepreg is rolled up on the spool. This process runs at about 8-20 ft min^{-1} in widths up to 60 inches.

The third method, solvent impregnation (Fig. 18) is used almost exclusively for towpreg, woven fabrics or high-temperature resins (e.g., polyimides), which are not amenable to hot-melt prepregging and must be dissolved in solvents. A disadvantage of this process is that the residual solvent may remain in the prepreg and cause a volatile problem during cure; therefore, the recent trend has been to use hot-melt or resin-film methods for both unidirectional and fabric prepregs. The solution process is operated with a treater line. The fabric web is drawn off a reel into a dip tank containing the resin solution (acetone is a common solvent for epoxies) and then pulled through a controlled set of nip rollers to set the resin content. Typical speeds are 10-15 ft min^{-1} for 60 inches wide material. The web moves down a hot-air oven that serves to both evaporate the bulk of the solvent and advance the resin for tack control. At the end of the oven, the material is spooled up with a layer of plastic film applied to one side to provide a separator. Some products require compaction at this stage to close the weave and others may be distorted during the passage through the treater line and require reworking in a tenter, which pulls and straightens the fabric to restore alignment.

2.10 Summary

Due to their low cost, high tensile strength, high impact resistance and good chemical resistance, glass fibers are used extensively in commercial composite applications. However, their properties cannot match those of carbon fibers for high–performance composite applications. They posses a relatively low modulus and have inferior fatigue properties compared to carbon fibers. The three most common glass fibers used in composites are E–glass, S–2 glass and quartz.

Aramid fibers have a combination of good tensile strength and modulus, light weight, excellent toughness with an outstanding ballistic and impact resistance. However, due to the lack of adhesion to the matrix, they exhibit relatively poor transverse tension, longitudinal compression and interlaminar shear strengths.

Carbon and graphite fibers are the most prevalent fiber forms in high–performance composite structures. Carbon and graphite fibers can be produced with a wide range of properties; however, they generally exhibit superior tensile and compressive strength, possess high moduli, have excellent fatigue characteristics and do not corrode. Carbon and graphite fibers can be made from rayon, pitch or PAN precursors, although rayon is rarely used because of its low yield and high cost. PAN-based carbon fibers having strengths ranging from 500 to 1,000 ksi and moduli ranging from 30 to 45 msi with elongations of up to 2% are commercially available.

Pitch-based high-modulus graphite fibers having a modulus between 50 and 145 msi are often used in space structures requiring high rigidity.

Two–dimensional woven products are usually offered as 0°, 90° constructions. Weaves are made on a loom by interlacing two orthogonal (mutually perpendicular) sets of yarns (warp and fill). The warp direction is parallel to the length of the roll, while the fill, weft, or woof direction is perpendicular to the length of the roll. Weaves may be classified by the pattern of interlacing including plain weaves, basket weaves, satin weaves, twill weaves, leno and mock leno weaves. Reinforced mats (chopped fibers or swirled fibers) and chopped fibers are also available for parts requiring lower mechanical properties.

Prepregging is a process in which a predetermined amount of an advanced resin is put on the fibers by the supplier. It can be accomplished by (1) hot–melt impregnation, (2) resin filming or (3) solvent impregnation. Hot–melt impregnation has been replaced to a large extent with the resin filming process because of the better quality and tighter controls obtainable with resin filming. Solvent impregnation is used for materials that are not amenable to hot melt or resin filming; however, the presence of residual solvent can cause processing problems in the form of voids and porosity during cure.

Reinforcing fibers can be supplied in dry unimpregnated forms or as prepregs ready for lay–up. Ply lay–up or collation of prepregs will be discussed in Chapter 5 on Ply Collation, while dry preform fabrication techniques for liquid molding will covered in Chapter 9 (Liquid Molding).

References

[1] Price T.L., Dalley G., McCullough P.C., Choquette L., "*Handbook: Manufacturing Advanced Composite Components for Airframes*", Report DOT/FAA/AR–96/75, Office of Aviation Research, April 1997.

[2] Strong A.B., *Fundamentals of Composite Manufacturing: Materials, Methods, and Applications*, SME, 1989.

[3] Tsai Jin–Shy, "Carbonizing Furnace Effects on Carbon Fiber Properties", *SAMPE Journal*, May/June 1994.

[4] "Pitch Fibers Take The Heat (Out)", High Performance Composites, September/December 2001.

[5] Schultz David A., "Advances in UHM Carbon Fibers", *SAMPE Journal,* March/April 1987.

[6] Backer S., "Textiles: Structures and Processes", in *The Encyclopedia of Materials Science and Engineering*, Pergamon Press, 1986.

[7] Gutowski T.G., "Cost, Automation, and Design", in *Advanced Composites Manufacturing*, Wiley, 1987.

[8] SP Systems "Guide to Composites"

[9] "Fabrics and Preforms", *ASM Handbook, Volume 21, Composites*, ASM International, 2001.

[10] Smith C., Gray M., "ICI Fiberite Impregnated Materials and Processes – An Overview", unpublished white paper.

[11] Hexcel Product Literature, "Prepreg Technology", 1997.

Chapter 3

Thermoset Resins: The Glue That Holds the Strings Together

The role of the matrix is to bind the fibers together in an orderly array and protect them from the environment. The matrix transfers loads to the fibers, which is critical in compression loading by preventing premature failure due to fiber microbuckling. The matrix also provides the composite with toughness, damage tolerance, and impact and abrasion resistance. The properties of the matrix also determine the maximum usage temperature, resistance to moisture and fluids, and thermal and oxidative stability.

Polymeric matrices for advanced composites are classified as either thermosets or thermoplastics. Thermosets are low-molecular-weight, low-viscosity monomers ($\approx$2,000 cP) that are converted during curing into three-dimensional (3D) cross-linked structures that are infusible and insoluble. Cross-linking (Fig. 1) results from chemical reactions that are driven by heat generated either by the chemical reactions themselves (i.e., exothermic heat of reaction) or by externally supplied heat. As curing progresses, the reactions accelerate and the available volume within the molecular arrangement decreases, resulting in less mobility of the molecules and an increase in viscosity. After the resin gels and forms a rubbery solid, it cannot be remelted. Further heating causes additional cross-linking until the resin is fully cured. This progression through cure is shown in Fig. 2. Since cure is a thermally driven event requiring chemical reactions, thermosets are characterized as having rather long processing times. In contrast, thermoplastics (Fig. 3) are not chemically cross-linked with heat and, therefore, do not require long cure cycles. They are high-molecular-weight polymers that can be melted, consolidated and then cooled. They may also be subsequently reheated for forming or joining operations; however, due to their inherently high viscosity and high melting points, high temperatures and pressures are normally required for processing. This chapter will concentrate on thermoset systems, while thermoplastics will be covered in Chapter 11 on Thermoplastic Composites.

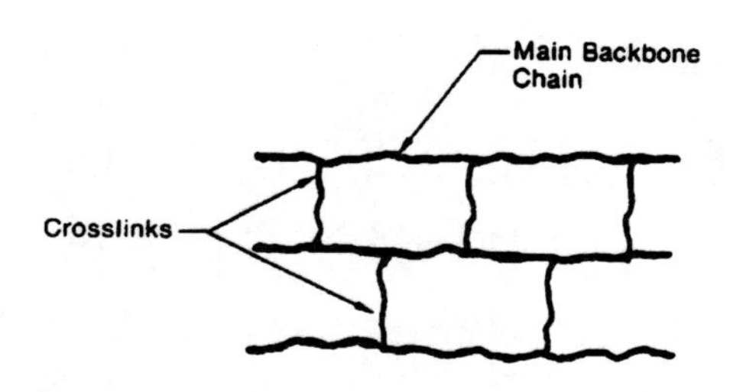

Fig. 1. Thermoset Crosslinked Structure

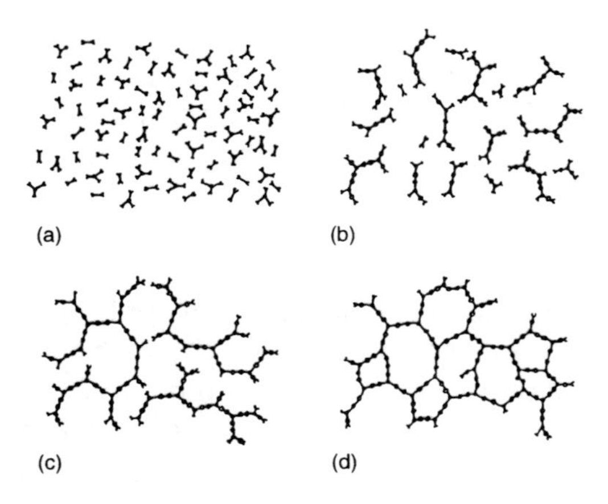

(a) Polymer and curing agent prior to reaction
(b) Curing initiated with size of molecules increasing
(c) Gelation with full network formed
(d) Full cured and crosslinked

Reference 1: with permission

Fig. 2. Stages of Cure for Thermoset Resin

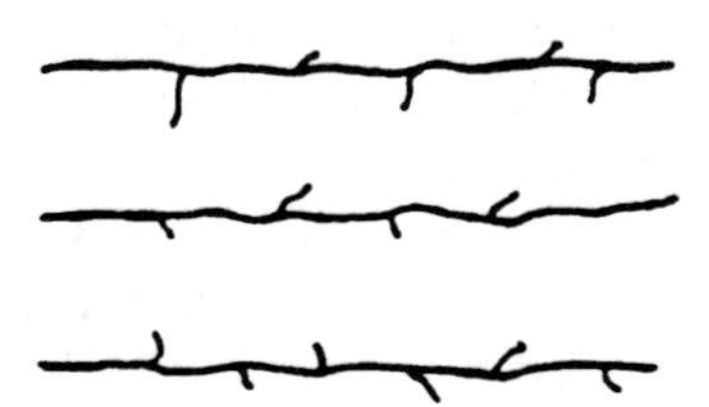

Fig. 3. Thermoplastic Structure

3.1 Thermosets

Thermoset composite matrices include polyesters, vinyl esters, epoxies, bismaleimides, cyanate esters, polyimides and phenolics. Epoxies currently are the dominant resins used for low and moderate temperatures (up to 275 °F). Bismaleimides are used primarily in the temperature range of 275-350 °F. For very high temperature applications (up to 550-600 °F), polyimides are usually the material of choice. Polyesters and vinyl esters, which can be used at approximately the same temperatures as epoxies, are used quite extensively for commercial applications but are rarely used for high-performance composite matrices because of their lower properties. Cyanate esters are a relatively new class of resins that were designed to compete with both epoxies and bismaleimides and offer some advantages in lower moisture absorption and attractive electrical properties but at a significantly higher price. Phenolics are high-temperature systems that offer outstanding smoke and fire resistance and are frequently used for aircraft interior components. Due to their high char yield, they can also be used as ablators and as precursors for carbon-carbon (C — C) components.

3.2 Polyester Resins

Polyesters are used extensively in commercial applications but are limited in use for high-performance composites. Although of a lower cost than epoxies, they generally have lower temperature capability, lower mechanical properties, inferior weathering resistance and exhibit more shrinkage during cure. Polyesters are cured by addition reactions in which unsaturated carbon-carbon double bonds (C = C) are the locations where cross-linking occurs. A typical polyester consists of at least three ingredients: a polyester, a cross-linking agent such as styrene and an initiator, usually a peroxide, such as methyl ethyl ketone peroxide (MEKP) or benzoyl peroxide (BPO). Styrene acts as the cross-linking agent and also lowers the viscosity to improve processability. Styrene is not the only curing agent (cross-linker). Others include vinyl toluene, chlorostyrene (which imparts flame retardance), methyl methacrylate (improved weatherability) and diallyl phtalate, which has a low viscosity and is often used for prepregs. The properties of the resultant polyester are strongly dependent on the cross-linking or curing agent used. One of the main advantages of polyesters is that they can be formulated to cure at either room temperature or elevated temperatures, allowing great versatility in their processing.

The basic chemical structure of a typical polyester (Fig. 4) contains ester groups and an unsaturated or double-bond reactive groups (C = C). Polyesters are usually viscous liquids consisting of a solution of polyester in a monomer, usually styrene. Styrene up to the amount of 50% reduces

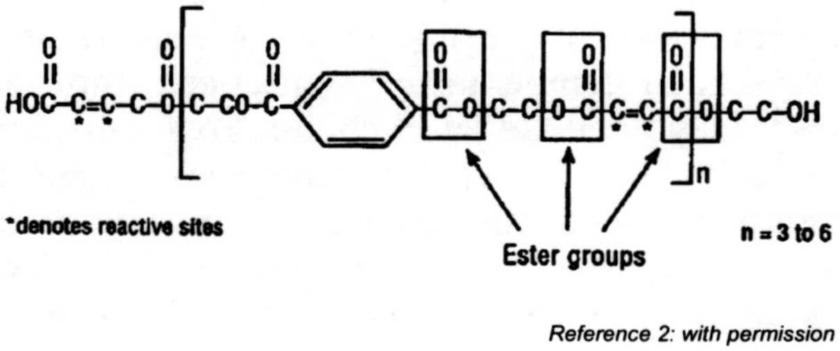

Reference 2: with permission

Fig. 4. Typical Polyester Chemical Structure

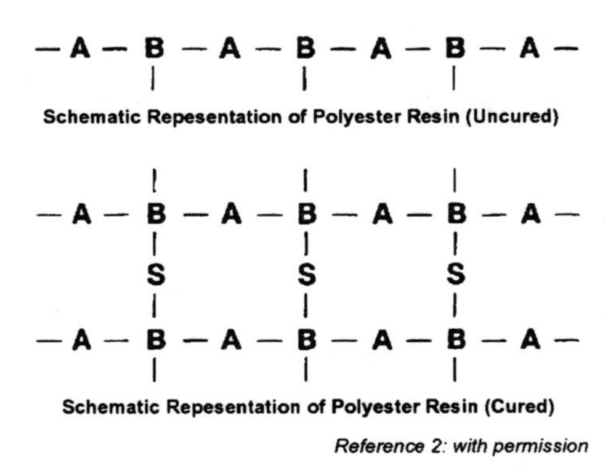

Fig. 5. Polyester Crosslinking Mechanism

the viscosity of the solution, thereby improving processablility and also reacts with the polyester chain to form a rigid cross-linked structure. The formation of the cross-linked polyester network with styrene is shown schematically in Fig. 5. Since polyesters have a limited pot life and set or gel at room temperature over a long period of time, small quantities of inhibitors, such as hydroquinone can be added to slow the reaction rate and extend the out-time. A solution of polyester and styrene alone polymerizes too slowly for practical purposes; therefore, small amounts of accelerators or catalysts are always added to speed up the reaction. Catalysts are added to the resin just prior to use, to initiate the polymerization reaction. The catalyst does not actually take part in the chemical reaction, but simply activates the process. Accelerators, such as cobalt napthenate, diethyl aniline and dimethyl aniline can also be added to speed up the reaction. There are a wide variety of monomers and curing agents available that yield a wide range of physical and mechanical properties. For example, the benzene ring improves rigidity and thermal stability.

Vinyl esters (Fig. 6) are very similar to polyesters but have only reactive groups at the end of the molecular chain. Since this results in lower cross-link densities, vinyl esters are normally tougher than the more highly cross-linked polyesters. In addition, since the ester group is susceptible to hydrolysis by water and vinyl esters have fewer ester groups than polyesters, they are more resistant to degradation from water and moisture.

3.3 Epoxy Resins

Epoxies are the most common matrix material for high-performance composites and adhesives. They have an excellent combination of strength, adhesion, low shrinkage and processing versatility. Commercial epoxy matrices and adhesives can be as simple as one epoxy and one curing agent;

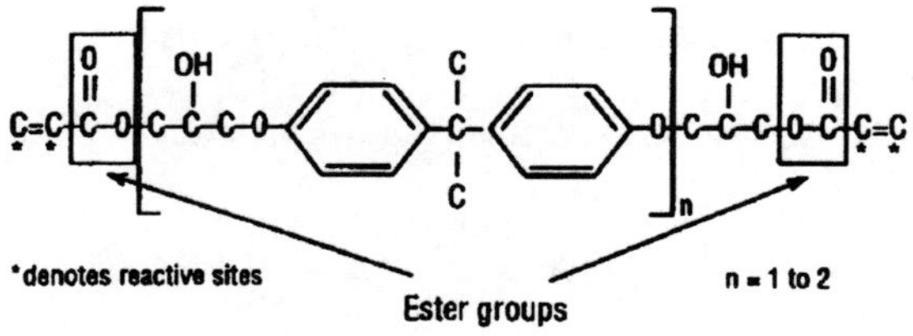

*denotes reactive sites

Ester groups

n = 1 to 2

Reference 2: with permission

Fig. 6. Vinyl Ester Structure

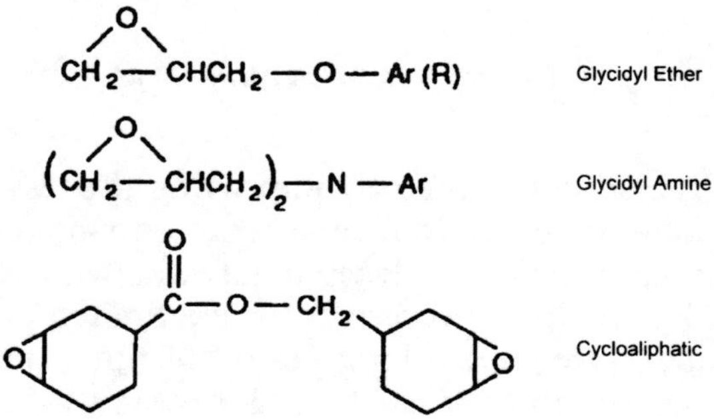

CH_2— $CHCH_2$ — O — Ar (R) Glycidyl Ether

$(CH_2$— $CHCH_2)_2$— N — Ar Glycidyl Amine

Cycloaliphatic

Reference 3: with permission

Fig. 7. Types of Epoxy Resins

however, most contain a major epoxy, one to three minor epoxies and one or two curing agents. The minor epoxies are added to provide viscosity control, improve high-temperature properties, lower moisture absorption or improve toughness. There are two major epoxies used in the aerospace industry: (i) diglycidyl ether of Bisphenol A (DGEBA), which is used extensively in filament winding, pultrusion and some adhesives; and (ii) tetraglycidyl methylene dianiline (TGMDA), also known as tetraglycidyl-4,4"-diaminodiphenylmethane (TGGDM), which is the major epoxy used for a large number of the commercial composite matrix systems.

The epoxy group or oxirane ring is the site of cross-linking.

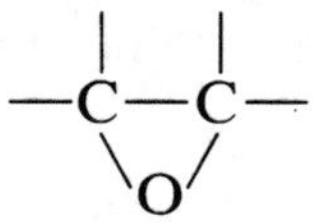

This three-member epoxy group is usually present as either a glycidyl ether, a glycidyl amine or as part of an aliphatic ring (Fig.7).[3] The cure of epoxy resins is based on the oxirane ring opening and cross-linking with the curing agent. Glycidyl ethers and amines are normally used for composites and cycloaliphatics are used extensively in electrical applications or as a minor epoxy in composite matrix systems. It should be emphasized that the properties of a cured epoxy are strongly dependent on the specific curing agent used. Like polyesters, they may be formulated to cure at either room temperature or elevated temperatures.

DGEBA is the most widely used of the epoxy types (Fig. 8). Since it is available as a liquid at several viscosities, it is frequently used for filament winding and pultrusion. DGEBA is a difunctional epoxy (i.e., two epoxy end-groups that can react) that can be either a liquid or a solid. If the repeat unit (n) is between 0.1 and 0.2, it is a liquid with a viscosity in the range of 6,000 to 16,000 cP. As n approaches 2, it becomes a solid. For n values greater than 2, it is not useful as a matrix material because the cross-link density becomes too low. The effect of the repeat unit on viscosity is shown in Table 1.

Glycidyl amines contain higher functionality (i.e., three or four reactive epoxy end-groups) because the resins are based on aromatic amines. The most important glycidyl amine is TGMDA, also known as TGGDM. This resin, shown in Fig. 9, is the base resin used in the majority of the commercial epoxy matrix systems. Its high functionality (four) provides highly cross-linked structures that exhibit high strength, rigidity and elevated temperature resistance. It is available in a variety of viscosities

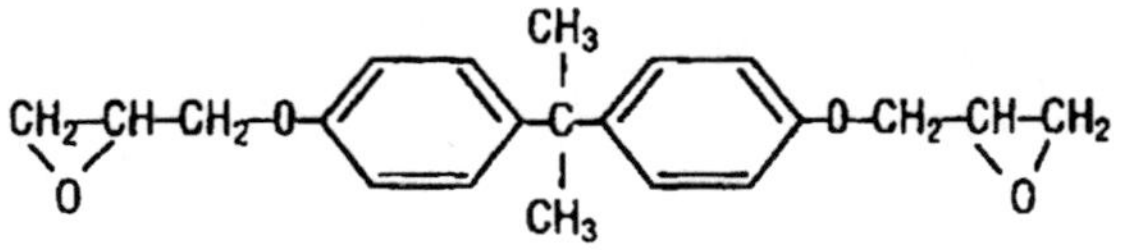

Reference 2: with permission

Fig. 8. Diglycidyl Ether of bisphenol A

Table 3.1 Effect of Repeat Unit on WPE and Viscosity for DGEBA Resins

Repeat Unit n	*Weight per Epoxide*	*Viscosity or Melting Point*
0	170-178	4-6000 cps
0.07	180-190	7-10000 cps
0.14	190-200	10-16000 cps
2.3	450-550	65-80 °C
4.8	850-1000	95-105 °C
9.4	1500-2500	115-130 °C
11.5	1800-4000	140-155 °C
30	4000-6000	115-165 °C

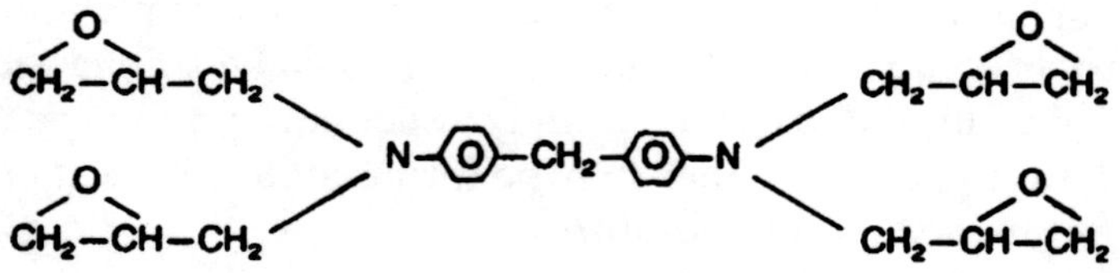

$(CH_2(O)CH{-}CH_2)_2N{-}C_6H_4{-}CH_2{-}C_6H_4{-}N(CH_2{-}CH(O)CH_2)_2$

Reference 4: with permission

Fig. 9. Tetraglycidyl Methylene Dianiline (TGMDA)

and is sold commercially as MY-720 (viscosity of 8,600-18,000 cP) and MY-721 (viscosity of 3,000-6,000 cP).[5] In some adhesive systems, where toughness (i.e., peel strength) is an important property, suppliers mix difunctional DBEBA with tetrafunctional TGMDA to help provide more flexibility in the cured adhesive. Several other less frequently used major epoxies are shown in Fig. 10. The novolacs are usually high-viscosity liquids or semi-solids that are often mixed with other epoxies to improve processability.

Minor epoxies are frequently added to improve processability (viscosity), improve elevated-temperature performance or improve other properties of the cured resin system. Typical minor epoxies include amine-based phenols, novalacs, cycloaliphatics and others. The composition[6] of a widely used epoxy matrix system is:

Component	*Total Weight Percent*
Tetraglycidyl Methylenedianiline (TGMDA)	56.5
Alicyclic Diepoxy Carboxylate	9.0
Epoxy Cresol Novalac	8.5
4,4' Diaminodiphenyl Sulfone (DDS)	25.0
Boron Trifluoride Amine Complex (BF_3)	1.1

TGMDA is the major epoxy, while alicyclic diepoxy carboxylate and epoxy cresol novolac are the two minor epoxies. The curing agents are DDS and BF_3.

Diluents are sometimes added to epoxy resin systems to reduce viscosity, improve shelf and pot life, lower the exotherm and reduce shrinkage. They are normally used in only small amounts (3-5%) because higher concentrations degrade the mechanical and thermal properties of the cured system. Typical diluents include butyl glycidyl ether, cresyl glycidyl ether, phenyl glycidyl ether and aliphatic alcohol glycidyl ethers.

There are a wide variety of curing agents that can be used with epoxies, but the most common for adhesives and composite matrices include aliphatic and aromatic amines, anhydrides and catalytic curing agents.[3]

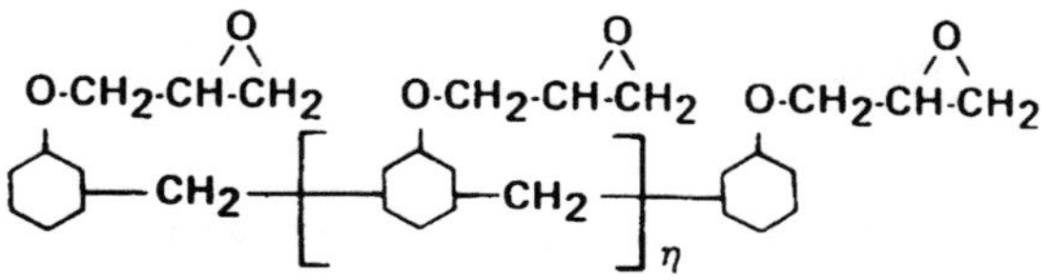

Polyglycidyl Ether of Phenol-Formalhyde Novolac

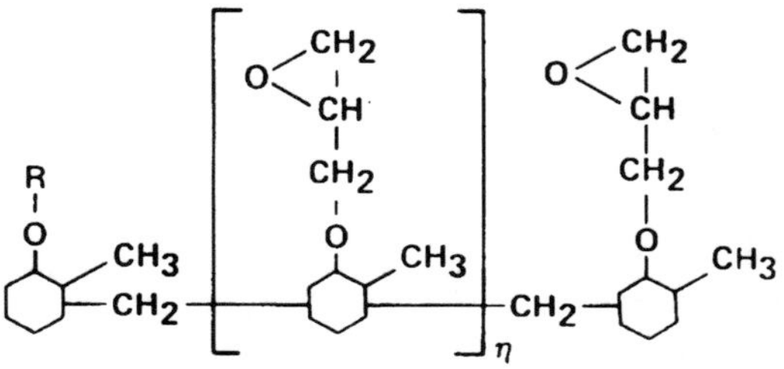

Polygicidyl Ether of 0-Cresol Formaldehyde Novolac

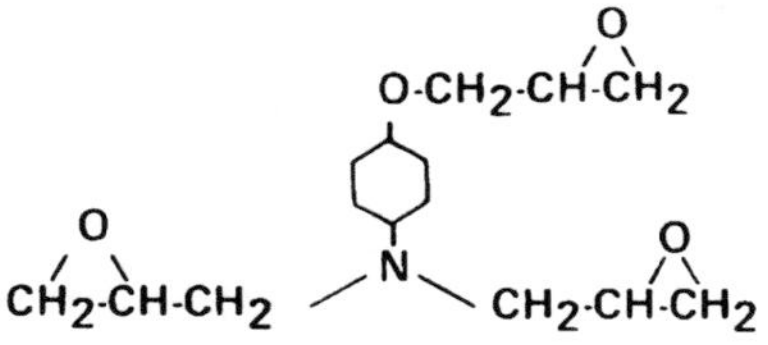

Triglycidyl P-Aminophenol

Fig. 10. Major Epoxy Resins

Aliphatic amines are very reactive, producing enough exothermic heat given off by the reaction to cure at room temperature or slightly elevated temperature. In fact, if they are mixed in large mass, the exotherm can be large enough to cause a fire. However, since these are room-temperature curing systems, their elevated-temperature properties are lower than the elevated-temperature cured aromatic amine systems. These systems form the basis for many room-temperature curing adhesive systems. Their elevated-temperature performance can sometimes be improved by initially curing at room temperature followed by a second cure cycle at elevated temperatures. Typical aliphatic amine curing agents are shown in Fig. 11.

Aromatic amines require elevated temperatures, usually 250 °F to 350 °F to obtain full cure. These systems are widely used for curing matrix resins, filament winding resins and high-temperature adhesives. Aromatic amines produce structures with greater strength, lower shrinkage and better temperature capability but with less toughness than aliphatic amines.

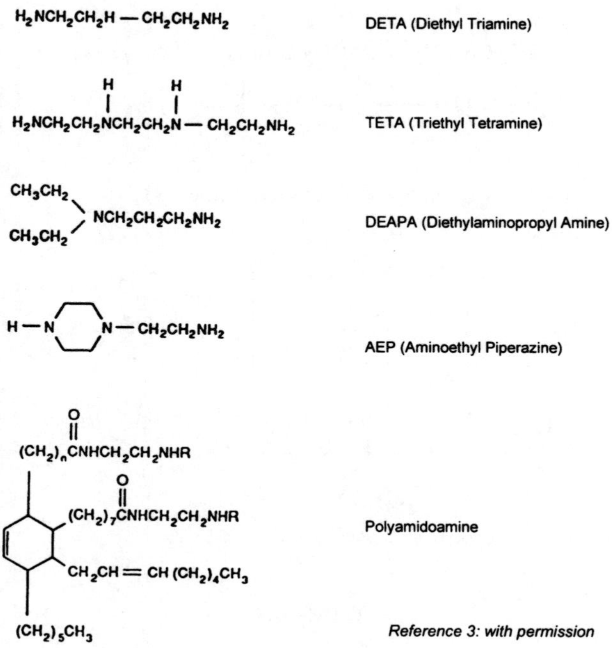

$H_2NCH_2CH_2H-CH_2CH_2NH_2$ — DETA (Diethyl Triamine)

$H_2NCH_2CH_2NHCH_2CH_2NH-CH_2CH_2NH_2$ — TETA (Triethyl Tetramine)

$(CH_3CH_2)_2NCH_2CH_2CH_2NH_2$ — DEAPA (Diethylaminopropyl Amine)

H—N(piperazine)N—$CH_2CH_2NH_2$ — AEP (Aminoethyl Piperazine)

$(CH_2)_nCONHCH_2CH_2NHR$, $(CH_2)_7CONHCH_2CH_2NHR$, $CH_2CH=CH(CH_2)_4CH_3$, $(CH_2)_5CH_3$ — Polyamidoamine

Reference 3: with permission

Fig. 11. Aliphatic Amine Curing Agents

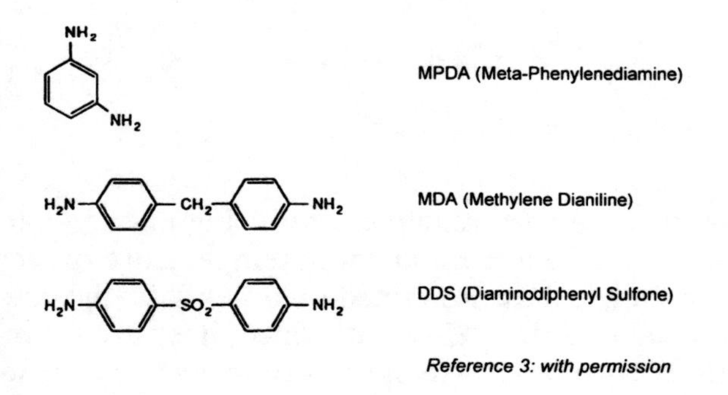

Fig. 12. Aromatic Amine Curing Agents

Aromatic amine curing agents are usually solids at room temperature that must be melted although some low-melting-point eutectic liquids are available. The three predominant aromatic amines are shown in Fig. 12.[3] DDS (diaminodiphenyl sulfone) is by far the most common curing agent used in composite matrices and in the majority of high-temperature epoxy adhesive systems. It should be noted that methylene dianiline, the curing agent used in the high-temperature polyimide PMR-15 and occasionally used with epoxies is a suspected carcinogen that can either be inhaled or absorbed through the skin.

Anhydride curing agents require high temperatures and long times to achieve full cure. They are characterized by their long pot lives and low exotherms.[3] They yield good high-temperature properties, good chemical resistance and have good electrical properties. They can be blended with epoxies to reduce viscosity. Anhydride curing agents generally require the addition of a catalyst to proceed at a rapid rate. One anhydride group reacts with one epoxy group during cure. Anhydride curing agents are susceptible to moisture pickup, which can inhibit the cure reaction. The predominant anhydride curing agents are shown in Fig. 13, with nadic methyl anhydride (NMA) being the most frequently used.

Catalytic curing agents, such as BF_3, promote epoxy-to-epoxy or epoxy-to-hydroxyl reactions.[3] They do not serve as cross-linking agents. They produce very tightly cross-linked structures and are characterized by their long shelf lives. A typical catalytic curing agent is boron trifluoride mono ethyl amine (BF_3:MEA), which is called a Lewis acid. It is normally used

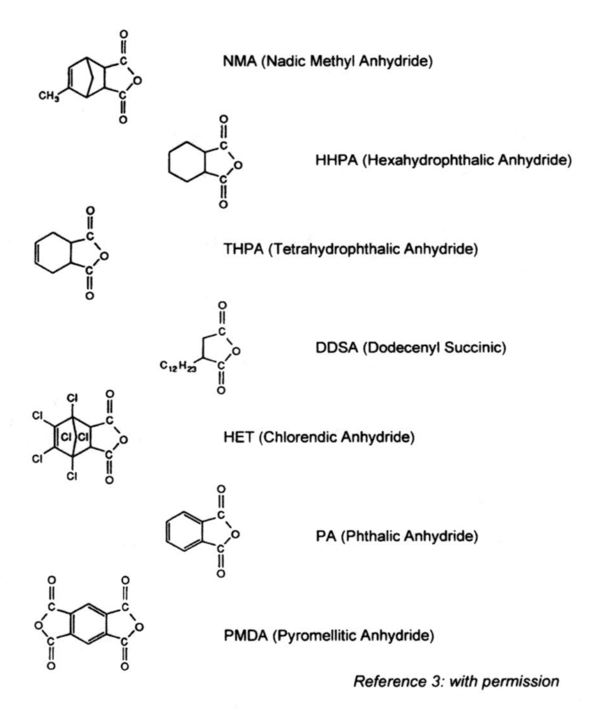

Fig. 13. Anhydride Curing Agents

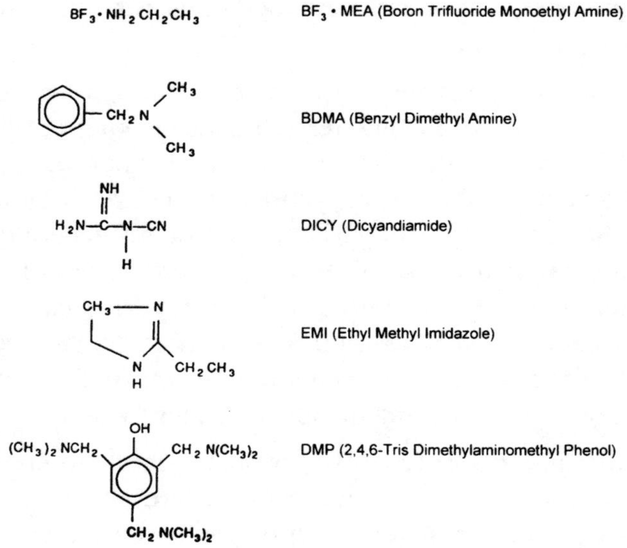

$BF_3 \cdot NH_2CH_2CH_3$ — $BF_3 \cdot$ MEA (Boron Trifluoride Monoethyl Amine)

BDMA (Benzyl Dimethyl Amine)

DICY (Dicyandiamide)

EMI (Ethyl Methyl Imidazole)

DMP (2,4,6-Tris Dimethylaminomethyl Phenol)

Reference 3: with permission

Fig. 14. Catalytic and Miscellaneous Curing Agents

along with another curing agent (e.g., DDS) in small percentages (1-5 phr) to reduce the flow and improve the processablity of the composite matrix. It is a latent curing agent with a long pot life that requires 200 °F or higher to initiate curing. However, once curing is initiated, it proceeds at a very rapid rate. Another class of catalytic curing agents are Lewis bases, which are often used as accelerators for anhydride curing agents. Dicyanidiamide (dicy) is also a very important latent curing agent used in both prepregs and adhesives. It is a solid powder that must be thoroughly mixed into the resin to provide uniform cures. The major catalytic and latent curing agents are shown in Fig. 14.

The curing of an epoxy resin system consists of low-molecular-weight resins and curing agents, reacting under heat at room temperature or elevated temperature to yield high cross-linked structures. The important points to remember are:

- Commercial matrix resins and adhesives are usually a blend of two or more epoxies combined with one or two curing agents. The major epoxy in the majority of epoxy matrix systems and high-temperature adhesives is TGMDA. Frequently, two, or sometimes three, minor epoxies are added to control viscosity or the influence the final cured properties, such as modulus or toughness. The major curing agent used in matrix resins and a many adhesives is DDS. Catalytic curing agents, primarily BF_3, can be added to reduce flow and accelerate the curing process. Epoxy resins for both composite matrices and adhesives are truly engineered systems to yield the best combination of processablity and final properties

- Higher curing temperatures and long curing times give the highest T_g's. When combined with high functionality (e.g., four reactive end-groups), the highest possible cross-link densities are achieved, which yields strong and stiff but somewhat brittle structures. The resin is frequently toughened by various means, but this frequently results in lower usage temperatures. The use of flexibilizing units (either the epoxy or curing agent) gives a high elongation and impact strength at the expense of T_g, tensile and compression strength and modulus. However, recent advances in epoxy chemistry and formulation have allowed much tougher resin systems with acceptable elevated-temperature performance.
- Epoxy matrices and adhesives, in fact almost all thermoset resins, absorb moisture from the atmosphere, which degrades their elevated-temperature matrix-dependent properties (Fig. 15). However, the moisture problem is well understood and can be accounted for in the structural design process.

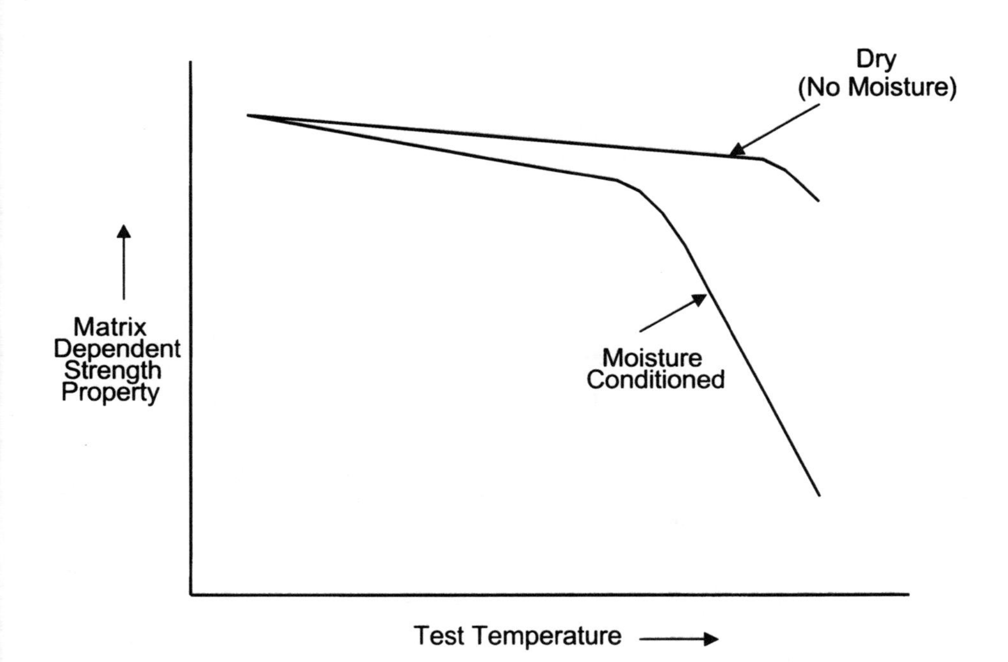

Fig. 15. Effect of Moisture Absorption on Hot-Wet Matrix Dependent Mechanical Properties

3.4 Bismaleimide Resins

Bismaleimides were developed to bridge the temperature gap between epoxies and polyimides with dry T_g's in the range of 430-600 °F and usage temperatures of 300-480 °F. They process similar to epoxies, by curing through addition reactions at 350 °F. To obtain their high-temperature properties, they are given free-standing post-cures at 450-475 °F to complete the polymerization reactions. BMI composites can be processed by autoclave curing, filament winding and resin transfer molding. The tack and drape of most BMIs are quite good due to the liquid component of the reactants. Since BMIs process at the same temperature (i.e., 350 °F) and pressures (e.g., 100 psig) as epoxies, conventional nylon bagging films, bleeder and breather materials and other expendables can be used. In contrast, the higher-temperature traditional polyimide materials usually require higher-temperature cures (600-700 °F) and usually higher pressures (e.g., 200 psig or higher) resulting in more expensive and difficult tooling and bagging materials.

BMI chemistry is quite varied, with many potential paths to producing matrix materials. Both bismaleimides and polyimides contain the imide group shown in Fig. 16. Bismaleimide monomers are synthesized by reactions with a primary diamine with maleic anhydride (Fig. 17). The most prevalent BMI base monomer for matrices and adhesives is 4,4'-bismaleimidedodiphenylmethane.

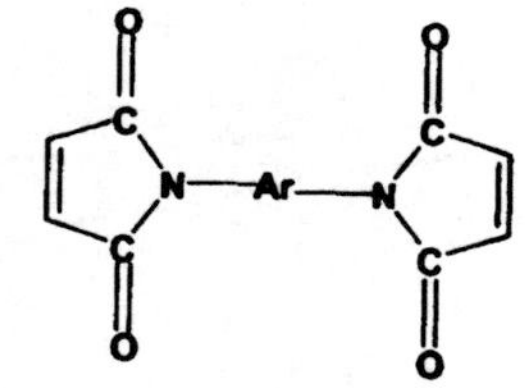

O C N—Ar—N C O C O C O

Reference 7: with permission

Fig. 16. Bismaleimide Chemical Structure

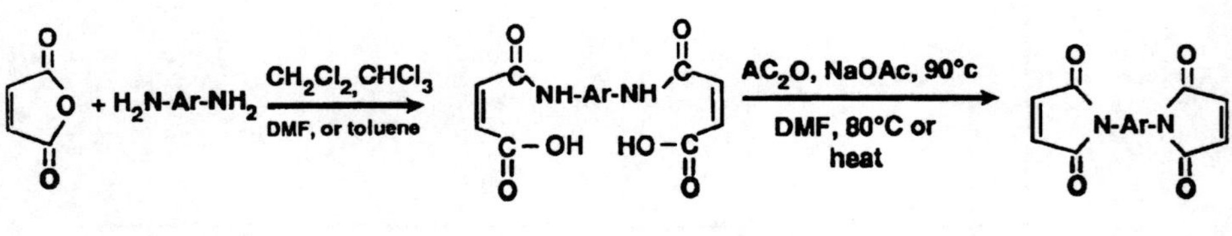

O O O + H_2N-Ar-NH_2 CH_2Cl_2, $CHCl_3$ DMF, or toluene → O C NH-Ar-NH C O C-OH HO-C O O AC_2O, NaOAc, 90°c DMF, 80°C or heat → O N-Ar-N O O O

Reference 8: with permission

Fig. 17. Synthesis of BMI Resin

The wide range of possible chemical reactions involved in curing BMIs are shown in Fig. 18. Homopolymerization (Fig. 18a) under heat is simple, but results in extremely brittle structures. A more practical approach (Fig. 18b) is copolymerization with a diamine, which produces a chain of more extended and flexible polymer. Copolymerization with an olefinic compound is another route to toughening, as shown for divinylbenzene (Fig. 18c), bispropenylphenoxy (Fig. 18d) and diallylphenyl or diallyphenoxy compounds (Fig. 18e). All three of these approaches have been successfully used in commercial products such as V378A, Matrimid 5292 and Compimide 796. BMIs have also been reacted with epoxies (Fig. 18f) in attempts to provide epoxy-like processing with BMI high-temperature performance. The copolymerization with 0,0'-discyanobisphenol A (Fig. 18g) has been used to produce some resins known as BT resins, where the B stands for BMI and the T stands for triazine, a type of cyanate ester. The BCB (benzocyclobutene)-BMI copolymers (Fig. 18h) exhibit high T_g's, good thermal-oxidative behavior, high toughness and processing ease, but the BDB monomer is expensive and commercially unavailable. The ATT (addition type thermoplastics), in which certain acetylene-terminated materials (Fig. 18j) undergo copolymerization with BMIs, allows the use of addition curing low-molecular-weight and low-viscosity starting materials. In addition, due to their linear backbone they are inherently tough but also possess good thermal-oxidative stability as a result of a fully aromatic structure.

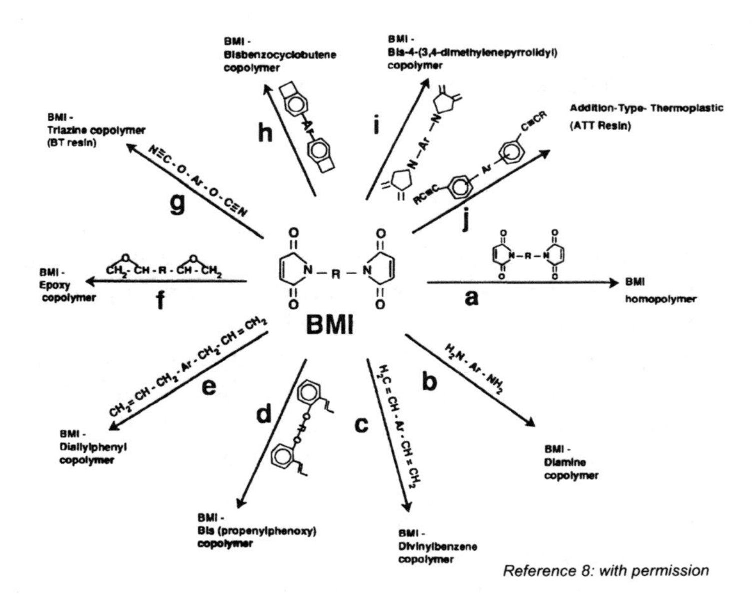

Fig. 18. Chemistry of BMI Resins

Commercial BMIs are usually one of five forms:[8] (1) BMIs or mixtures of different BMIs, (2) blends of BMIs and BMI-diamines, (3) BMI and olefinic monomers and/or oligomelic blends, (4) BMI and epoxy blends and (5) BMI and 0,0'-dicyanobisphenol A mixtures.

Although early BMI materials were characterized as being hard to process (low tack and short out-times) and possessing low toughness (brittle), BMI materials being produced today have much better tack and long out-times. In addition, some BMIs (e.g., Cytec 5250-4) are as tough as most toughened epoxies. Using liquid molding processes such as RTM can also readily process bismaleimides.

One potential usage problem with bismaleimides and any polymer containing the imide end-group is a phenomenon known as "imide corrosion." This is a form of hydrolysis that results in degradation of the polymer itself. It was originally observed with carbon fiber/bismaleimide composites, galvanically coupled to aluminum in a sump environment (i.e., a mixture of salt water and jet fuel). If the aluminum corrodes, the composite (since it is electrically coupled to the aluminum through the carbon fibers) becomes the cathode. Water reduction in the presence of oxygen occurs at the cathode leading to the formation of hydroxyl ions that attack the imide-carbonyl linkage of the BMI. The mechanism is shown schematically in Fig. 19. No corrosion occurs with non-conductive fibers such as glass or aramid, nor does corrosion occur if the metal is galvanically similar to carbon, such as titanium or stainless steel. Studies[9] have shown that increases in temperature and bare exposed carbon edges accelerate the deterioration. Electrically isolating the carbon from the aluminum prevents the problem. This can be accomplished by curing a layer of fiberglass on the composite faying surface and then sealing the edges with a polysulfide sealant.

3.5 Cyanate Ester Resins

Cyanate esters are often used in applications requiring low dielectric loss properties, such as antennas and radomes. They are potential substitutes for both epoxies and bismaleimides with dry T_g's ranging from 375 °F to 550 °F.[10] However, due to a rather limited market, they are expensive materials. The prepreg is also susceptible to moisture pickup that can produce CO_2 during the curing process. Their adhesion or bonding ability is inferior to that of epoxies. The cured laminates exhibit lower moisture absorption than epoxies or bismaleimides and are inherently flame resistant.

Cyanate esters are bisphenol derivatives containing a ring forming cyanate functional group. This family of thermosetting monomers and their prepolymers are esters of bisphenols and cyanic acid which cyclotrimerize to substituted triazine rings upon heating (Fig. 20).[11] During

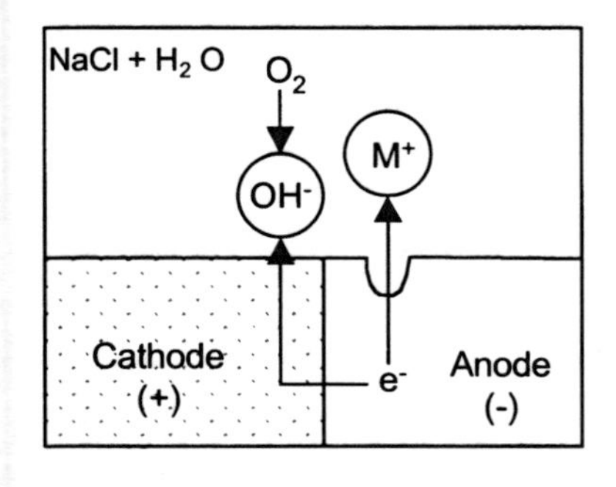

Half Reactions

Anode: $Al \rightarrow Al^{3+} + 3e^-$

Cathode: $H_2O + 1/2\ O_2 + \rightarrow 2OH^-$

Polyimide Reaction

N-R + $2OH^-$ $\xrightarrow{H_2O}$ OH^- OH^- + H_2NR

Reference 9: with permission

Fig. 19. Galvanic Corrosion Mechanism for Imide Linkage

cure, 3D networks of oxygen-linked triazine rings and bisphenol units cross-link via addition reactions. The high aromatic content of the triazine and benzene rings provides a high T_g. The single-atom oxygen linkages function somewhat like ball joints to dissipate localized stresses. Moderate cross-link densities contribute to toughness. In addition, both rubber and thermoplastic toughening mechanisms have successfully been used with cynate esters to further enhance toughness.

The attractive electrical properties, such as low dielectric constant and dissipation factor, are a result of the balanced dipoles and the absence of strong hydrogen bonding. This lack of polarity along with the symmetry of the triazine rings makes cyanate esters more resistant to water absorption than most epoxies and bismaleimides. The low moisture absorption (in the range of 0.6-2.5%) creates less outgassing, a critical factor in the dimensional stability required for space structures.

3.6 Polyimide Resins

Polyimides are high-temperature matrix materials intended for usage temperatures as high as 500-600 °F. They can either be condensation or addition curing systems. The condensation systems give off water, or water and alcohol, which causes a severe volatile management problem during cure. If the volatiles are not removed prior to resin gellation they become entrapped as voids and porosity, which lowers the matrix-dependent mechanical properties. In addition, polyimides are usually formulated with high-temperature solvents, such as DMF (dimethylformamide), DMAC (dimethylactamide), NMP (N-methylpyrrolidone) or DMSO (dimethylsufoxide), which must also be removed either prior to or during the curing cycle.

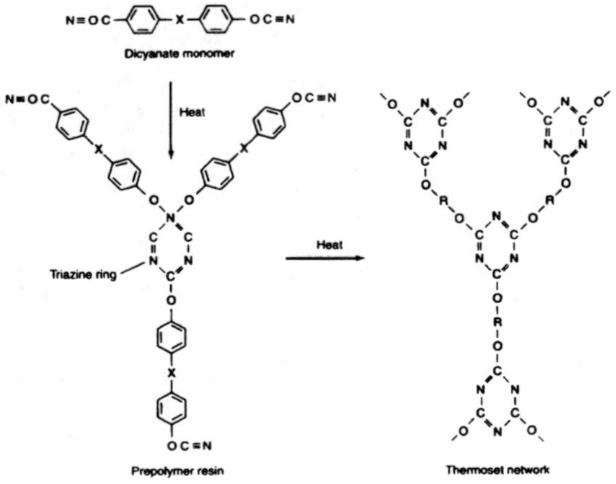

Reference 11: with permission

Fig. 20. Cure of Cyanate Resins

Polyimides can be thermoplastics or thermosets. For example, the Avmid K polymers, which have T_g's in the range of 430-480 °F, are linear thermoplastic polyimides.[12] They can be thermoformed into simple shapes above their Tg's for a few times and then higher temperatures and pressures are required for further forming. The Torlons polyamideimides have lower T_g's (470-525 °F) than regular polyimides but much better toughness. The amide groups give flexibility, elongation and good tensile strength. Ultem, a polyetherimide, is another thermoplastic polyimide that has good thermal resistance, mechanical properties and good processability but a lower T_g (420 °F) and poor resistance to solvents. Polyimides can also produce thermoset structures, Skybond is an example of a condensation curing thermoset material.

Polyimides are much more difficult to process than epoxies or bismaleimides. They require high processing temperatures (e.g., 600-700 °F), long cycles and higher pressures (e.g., 200 psig). Volatiles and voids are always potential problems when processing polyimides. Even the addition curing systems can exhibit volatile problems as the low-molecular-weight monomers are usually dissolved in solvents during manufacturing.

PMR-15 is the best known of the addition curing polyimide systems. PMR stands for Polymeric Monomer Reactants and the –15 stands for a molecular weight of 15,000. In PMR-15, three types of monomers (Fig. 21) are mixed together along with a solvent, usually methyl or ethyl alcohol. However, one of the monomers, 4,4'-methylene dianaline is a suspected carcinogen that can be absorbed through the skin or inhaled if used as a spray. There have been several non-MDA polyimides formulated and sold;

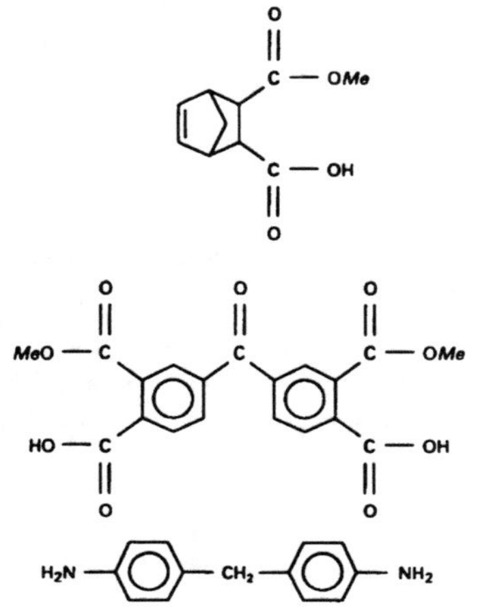

Monomethyl Ester of
5-Norbornene-2-3-Diacarboxylic Acid

Dimethyl Ester of
3,3',4',4-Benzophenone-Tetracarboxylic Acid

4,4'- Methylene Dianailine

Reference 13: with permission

Fig. 21. Components of PMR-15

however, their elevated-temperature performance is not as good as the original formulation. Even though PMR-15 is classified as an addition curing reaction, it undergoes condensation reactions early in the curing cycle during the imidization stage. This creates a volatile management problem and the impregnation solvent must be removed before the resin gels, otherwise the solvent results in voids and porosity. PMR-15 has a usage temperature of 550-600 °F for 1,000-10,000 h, depending on the specific usage temperature. The main disadvantages of PMR-15 are poor tack and drape, inadequate resin flow for fabricating thick and complicated structures, the tendency toward microcracking and the health and safety concerns with MDA.

Considerable effort has been expended over the last 25 years on the development of high-temperature polymers that have good thermal-oxidative behavior in the 500-600 °F range, yet are easy to process. Much of this effort has either been led or funded through NASA, most recently to support the High Speed Civil Transport (HSCT) program conducted during the mid-1990s. The goal was to develop a resin system capable of withstanding 350 °F for 60,000 h.[14] After screening the available materials, the most promising resin developed was a resin called PETI-5, a phenylethynyl terminated imide. A matrix resin, an adhesive, a resin transfer molding (RTM) grade and a resin film infusion (RFI) grade were developed during the program. As with other high-temperature resin systems that use high-boiling-point solvents for manufacturing, NMP in this case, management of volatiles during curing is a major consideration. However, successful demonstration parts were fabricated using all of these product forms.[15]

3.7 Phenolic Resins

Phenolics are normally very brittle and exhibit large shrinkage during cure. Their primary use is for aircraft interior structures because of their low flammability and smoke production. They are also used for high-temperature heat shields due to their excellent ablative resistance and as the starting material for C — C composites because of their high char yield during graphitization.

Phenolics are made by a condensation reaction with phenol and formaldehyde that gives off water as a byproduct.[16] A typical phenolic reaction is shown in Fig. 22. Phenolics are usually classified as either resoles or novolacs. If the phenol and formaldehyde reaction is carried out with an excess of formaldehyde and a base catalyst, the result is a low-molecular-weight liquid resole. If the reaction is carried out with an excess of phenol and an acid catalyst, the result is a solid novolac. Resoles are normally used for phenolic prepregs. Like epoxies and bismaleimides, recent efforts have resulted in phenolics with improved toughness.[17] In addition, phenolics have been developed that do not give off water during cure,[17] always a major processing concern.

Phenolics are one route to the production of high-temperature resistant C — C composites. The phenolic is charred or pyrolized to produce a carbon matrix. Since the charring processes produces a porous structure due to the vaporization of the part of the phenolic, the process has to be repeated several times. Before each subsequent pyrolization, the porous structure is impregnated with either pitch, phenolic resin or directly with carbon by chemical vapor deposition. This is a slow process that must be done with great care to prevent delaminations and severe matrix cracking. Quite frequently, 3D reinforcements will be used to resist ply delaminations. C — C composites can also be made directly using

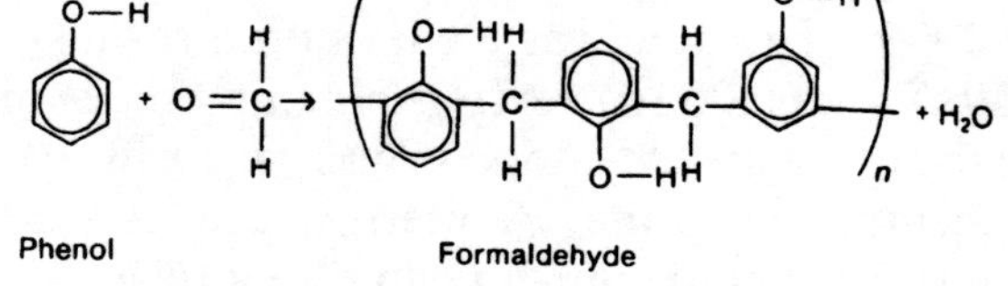

Phenol Formaldehyde

Reference 16: with permission

Fig. 22. Typical Phenolic Condensation Reaction

chemical vapor deposition. A carbon perform is impregnated using methane gas. Since there is a tendency for the deposited carbon to seal off internal voids and porosity, intermittent machining operations are required to remove the surface layer to allow the carbon deposit access to the internal structure.

3.8 Toughening Approaches

The toughness limitation of thermoset matrix is a direct result of its rigid, highly cross-linked, glassy polymer structures that form during cure. These rigid structures have both advantages and disadvantages. The main advantages are high-temperature capability and the ability of the rigid matrix to stabilize the reinforcing fibers during compression loading. The greatest disadvantage is their susceptibility to delaminations when impacted. Of particular concern is low-velocity impacts causing internal delaminations that cannot be detected during visual walk-around inspections.

Over the last 20 years, extensive effort has been expended to develop resin systems that are tougher and less susceptible to impact damage. Two candidates emerged in the mid-1980s: damage tolerant thermoplastic composites and toughened thermoset composites. Although radically different in chemical structure, their resultant properties are somewhat similar. Both have improved resistance to low-velocity impact damage, resulting in greater load-carrying capability after being impacted. Likewise, compared to the stiff thermoset systems, both have a somewhat lower resin modulus and, therefore, exhibit somewhat lower compression strengths. Although certainly not true in every case, the tougher systems generally have less heat resistance than the rigid glassy thermosets.

Due to their inhomogeneous nature, where the properties vary in many directions, toughness is more complicated for composites than for homogeneous metallic materials. In composite structures, in-plane loading is controlled primarily by the reinforcing fibers, while out-of-plane loading is dominated by the properties of the resin matrix. Therefore, composite structures are intentionally designed so that the load paths are stiff and are primarily in-plane. However, out-of-plane loading can occur. Out-of-plane loads develop during in-plane compressive buckling but, more importantly, they are induced by a variety of design features. Five of the more common design details that can cause out-of-plane loading are shown in Fig. 23. Even under normal in-plane loading, interlaminar shear and normal stresses develop at these locations. These indirect loads can either act alone or in combination with direct out-of-plane loads, such as fuel or air pressures. If the out-of-plane loads become large enough, interply delaminations can form and propagate. Fortunately, current design criteria are conservative enough that even if a small delamination is present, either from a manufacturing defect or service abuse, it does not normally propagate.

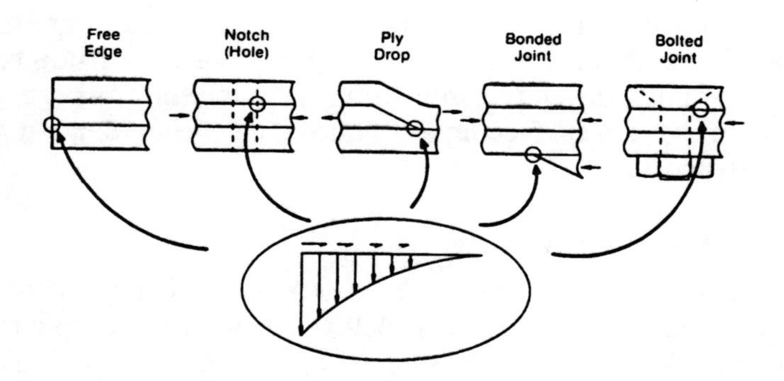

Fig. 23. Design Details That Cause Out-of-Plane Loads

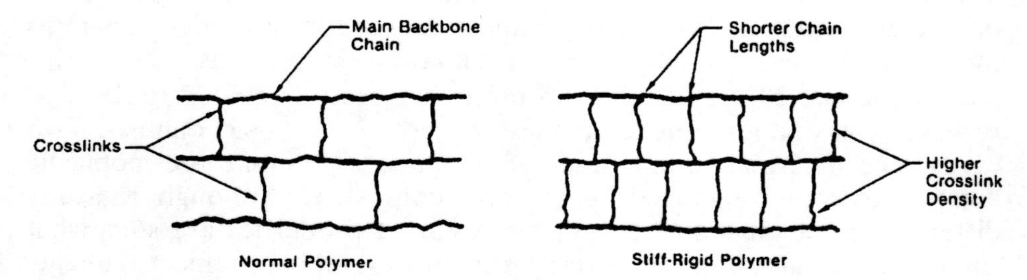

Fig. 24. Effect of Crosslink Density on Rigidity

Cross-links are the chemical ties between two or more polymer chains that give the cured polymer its strength, rigidity and thermal resistance. As shown conceptually in Fig. 24, the higher the cross-link density (i.e., number of cross-links per unit volume) and the shorter the polymer chain lengths between the cross-links, the more constrained is the chain motion and thus the stiffer and more thermally resistant is the molecular structure. Further stiffening can be imparted by the use of stiffener backbones in the main chains. However, this rigidity results in brittleness, low strain-to-failures and poor impact and post-impact properties. Highly cross-linked rigid structures usually exhibit good thermal stability due to their ability to restrict the relative motion of the polymer chains by the chemical bonds holding them together. Since the cross-link bonds are primary covalent bonds, they retain a large portion of their strength as the temperature is increased. When the cross-link density is high and the bond lengths are short, acceptable properties are maintained up to near the T_g of the resin. Above the glass transition temperature, the rigid solid polymer converts to a softer rubber-like material as the main backbone chains themselves soften. Therefore, highly cross-linked polymers possess moderate to high strength and stiffness and excellent temperature resistance. However, being rigid and glassy structures, they are somewhat brittle and susceptible to impact damage.

The molecular structure of a thermosetting polymer determines how it processes and its resultant properties. Examples of properties that are a function of molecular structure include glass transition temperature, moisture absorption, strength, modulus, elongation and toughness. By altering the molecular structure, it is possible to alter these performance properties. The molecular structure is controlled by its backbone structure (i.e., the main polymer chains) and its network structure (i.e., the number and types of cross-links). The chemical structure of the main monomer defines the backbone structure and, to some extent, the network structure. The network structure is also influenced by the type of curing agent or the hardener used in the curing reaction. Resin formulators have expended a significant amount of effort to develop new molecular structures that result in polymers with improved toughness.

The improved post-impact compression strength of the toughened systems is a direct result of the amount of damage inflicted during the impact event. As shown in Fig. 25, the toughened systems (Hexcel's IM-7/8551-7 in this example) experience much less internal damage when impacted. Therefore, when the specimens are subsequently loaded to failure, the toughened systems have a larger cross-sectional area of undamaged material to help support the compression load.

To impart greater toughness to the cross-linked thermoset polymers, a number of different approaches have evolved. Single approaches as well as a combination of approaches are used to further enhance toughness. Four toughening approaches will be discussed: (1) network alteration, (2) rubber elastomer second-phase toughening, (3) thermoplastic elastomer toughening and (4) interlayering.

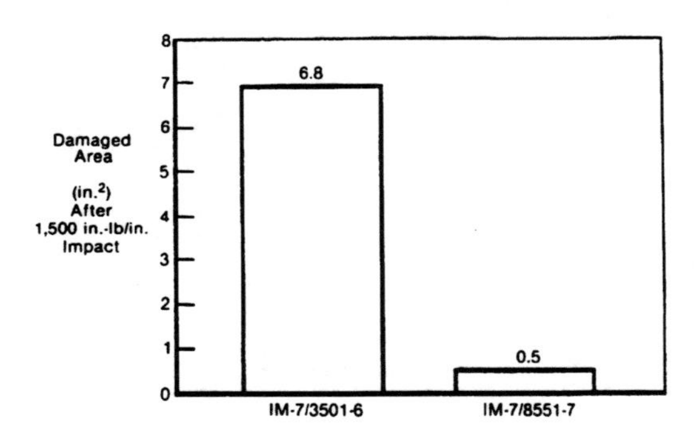

Fig. 25. Damage Comparison During Low Velocity Impact

Network Alteration – Since the brittleness of thermosetting polymers is a direct consequence of their high cross-link densities, one method of toughening a thermoset polymer is to lower the cross-link density. This approach, when taken to the extreme case where there are no more cross-links, results in a thermoplastic polymer. Since amorphous thermoplastic polymers are inherently tough, the more the cross-link density is reduced the tougher the resulting polymer. However, the decrease in cross-link density is also accompanied by a decrease in desirable properties such as glass transition temperature and resin modulus.

There are two well-known methods for reducing the cross-link density of thermoset polymers. The first is to alter the main monomer backbone chain by increasing the molecular weight between cross-links with long chain monomers as shown schematically in Fig. 26. The resulting decrease in glass transition temperature can be somewhat offset by constructing a long-chain monomer which is very rigid and contains bulky side groups. The decreased mobility of the polymer chains will somewhat compensate for the loss in glass transition temperature due to the lower cross-link density. A second method is to lower the monomer functionality. Most highly cross-linked thermosetting polymers have a functionality of four, which means there are four reactive end-groups that can react and cross-link during cure. As shown in Fig. 27, if a portion of the polymer mix contains a monomer with a functionality of two, there are less available sites for cross-linking during cure and the toughness will, therefore, be improved due to the lower cross-link density. However, the heat resistance is again affected as evidenced by the lower glass transition temperatures of difunctional monomers.

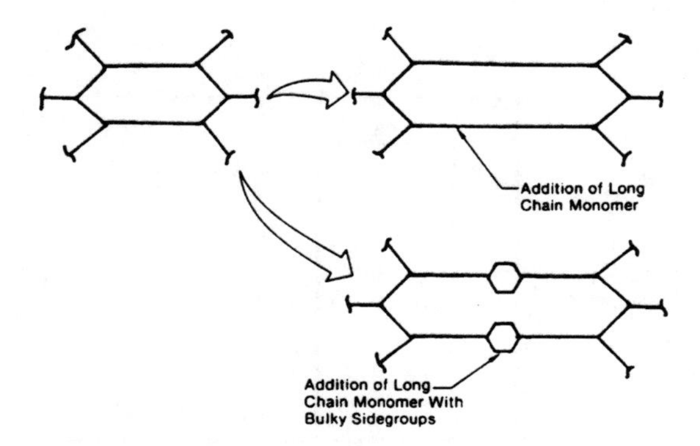

Fig. 26. Reducing Crosslink Density with Long Chain Monomers

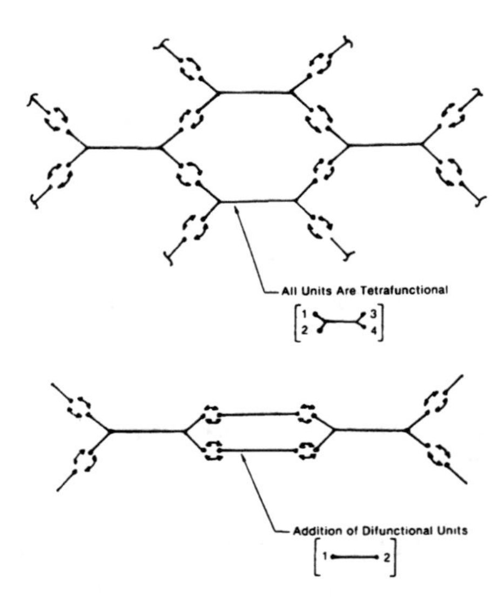

Fig. 27. Effect of Monomer Functionality on Crosslink Density

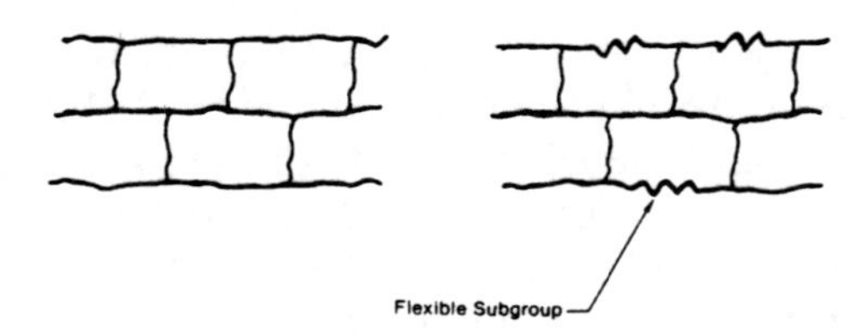

Fig. 28. Toughening with Flexible Subgroups

Another method frequently used to improve toughness is to incorporate flexible subgroups into the main chain backbone of either the resin or the curing agent. Although depicted as "springs" in Fig. 28, in actuality, more flexible aliphatic segments are used in preference to the more rigid and bulky aromatic groups that contain the large benzene ring. As with the other approaches described above, there is a trade-off in glass transition temperature. This can be minimized by using some stiff segments along with the flexible ones.

Rubber Elastomer Second-Phase Toughening – When a crack initiates in a brittle glassy solid, it requires very little energy to propagate. In fiber-reinforced composites, the fibers will prevent in-plane crack growth. However, if the crack is interlaminar (i.e., between the plies), the fibers are of no help in preventing crack propagation. One way of reducing crack propagation is to use second-phase elastomers. Discrete rubber particles

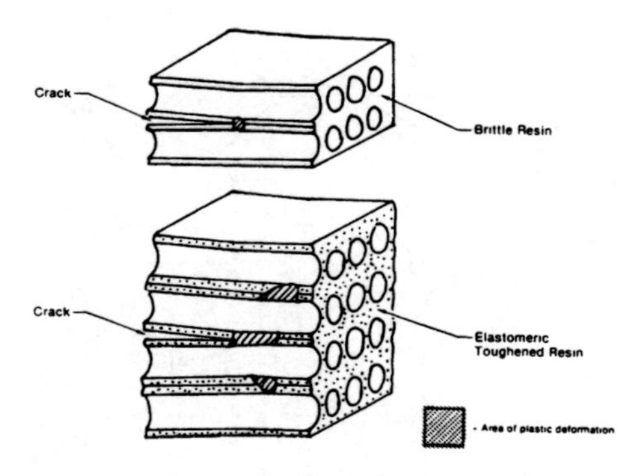

Fig. 29. Interlaminar Crack Propagation in Composite Laminates

help to blunt crack growth by promoting greater plastic flow at the crack tip as shown in Fig. 29. The size of the elastomeric domains is a critical factor in determining the microdeformation processes that control toughening. Rubber particles, usually round, having domain diameters of 100-1,000 Å initiate shear yielding, while larger domains (10,000-20,000 Å) are generally believed to lead to crazing. Since matrix crazing is not possible in highly cross-linked systems due to their low tensile elongations, very small domains are used to enhance the shear deformation processes. However, if the cross-link density permits the use of larger domain sizes, bimodal distributions (i.e., a mixture of large and small domains) of the elastomer will result in both crazing and shear. Since these mechanisms complement each other, the toughening effect can be nearly doubled.

A finely dispersed phase of elastomer-rich domains can be obtained by using either preformed rubber particles or a rubber elastomer system that is initially soluble in the liquid resin, but then the phase separates (i.e., precipitates) during cure. Performed particles are advantageous because they can be used as additives or fillers. Also, their domain size is easier to control than when a phase separation process is used. Unfortunately, commercially available preformed rubber particles are suitable only for toughening some of the lower-temperature thermosets. When elastomer domains are to be formed during cure, it is necessary to use a rubber with suitable solubility. The rubber must initially be soluble in the resin and then separate into the desired small domains during the cure process. An elastomer that is too soluble will remain in solution too far into cure and significant quantities will be trapped in solution when the resin gels. The trapped elastomer will then act as a plasticizer and subsequently lower the glass transition temperature and hot-wet properties. On the other hand, if the elastomer is not soluble enough, a stable solution with the resin will not be possible and a fine dispersion of particles will not be obtained. If just the right amount of compatibility exists, then the elastomer will initially remain in solution and once the resin begins to cure, small particles will uniformly precipitate into the resin matrix. Early and complete phase separation is necessary to maintain good hot-wet matrix performance.

In epoxy-based composites and adhesives, reactive liquid polymers, such as CTBN (carboxyl terminated butadiene-acrylonitrile) rubbers, are often used to provide the desired solution and compatibility characteristics. CTBN rubbers, which contain functional carboxyl groups, are usually blended and prereacted with one of the epoxy monomers to provide the desired solubility. In addition, the carboxyl groups react with some of the epoxy groups to form light cross-links that increase the cohesive strength of the elastomeric domains. Good bonding between the elastomer domains and the continuous resin phase is important. If the bond is poor, the elastomer can debond from the resin during cooling and form voids.

The elastomer itself must have good rubbery characteristics. Specifically, the elastomer-rich domain must have a T_g lower than about –100 °F, so that when a crack travels rapidly through the material, the domains still act as elastomers. If the T_g is too high (i.e., too close to room temperature), then at high deformation rates it will behave like a glass and the desired toughening effect will not be achieved. Another requirement is that the elastomer must be thermally and thermally-oxidatively stable. If unstable rubbers are used, they are likely to cross-link or degrade when oxidized, which would embrittle the elastomeric domains.

Elastomeric second-phase toughening is not always effective. If the resin is too highly cross-linked, then the resin lacks the ability to deform locally. In the absence of some localized deformation, the addition of an elastomeric second phase is largely ineffective. As the cross-link density is lowered, the benefits of second-phase elastomeric toughening rapidly increase. In epoxy resins, it has been shown that the relatively small gains in toughness obtained by modification of the monomer stiffness or length, or by altering the network structure can be greatly amplified by the use of second-phase toughening.

Thermoplastic Elastomeric Toughening – A number of important commercially available toughened thermoset composite systems are based on thermoplastic toughening.[18] Thermoplastic toughening can exhibit four distinct morphologies in the cured composite: (1) homogeneous (single phase), (2) particulate (thermoplastic particles in a continuous thermoset matrix), (3) co-continuous (both the thermoplastic and thermoset are continuous) and (4) phase inverted (the thermoplastic is continuous and the thermoset is discontinuous). It has been shown that the co-continuous morphology results in the greatest toughness improvements.[18] Polyetherimide (PEI), polyethersulfone (PES) and polysulfone (PS) have all been evaluated.[19]

In a co-continuous structure, the thermoplastic increases the toughness, while the cross-linked thermoset helps to retain a high glass transition temperature and hot-wet performance. A thermoplastic with a high T_g helps to maintain the final T_g, since the T_g is at least the average of the T_g's of the separate components. The selection of the optimum thermoplastic component depends on its compatibility, heat resistance and thermal stability. Chemical compatibility is required so that the thermoplastic will remain in solution (i.e., not phase separate) until the thermoset network has gelled. In addition, the viscosity of the ungelled resin must be controlled so that it remains low enough to allow fiber wetting and ply bonding during cure. Finally, the thermoplastic must have good heat resistance and be thermally stable to maintain hot-wet performance.

Interlayering – A "mechanical" or "engineered" approach to toughness can be achieved by incorporating a thin layer of a tough ductile resin between the individual prepreg plies. As shown in Fig. 30, this layer is usually thin (0.001 inches or less) and must remain fairly discrete during the composite curing operation. The rationale behind this approach is that the stiff brittle matrix supports the carbon fibers in compression and thus helps to maintain the hot-wet performance, while the tough interlayers provide the desired increase in resistance to low-velocity impact. Since the tough interlayer has a high strain-to-failure, it helps to reduce the interlaminar shear and normal forces that can induce delaminations. The original development work was conducted using tough film adhesives that were simply placed between the prepreg plies during lay-up.

More recently, discrete thermoplastic toughening particles have been added to the surfaces during the prepregging operation. The particles are larger than the fibers and, therefore, remain on the interlayers during prepregging and cure. The interlayering approach can be used with either

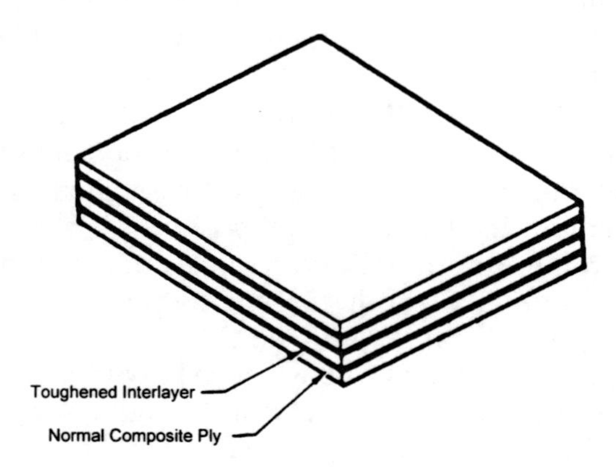

Fig. 30. Toughening by Interlayering

brittle or toughened resin systems. Although a toughened resin system used in combination with interlayers provides better impact resistance, the hot-wet compression strength decreases somewhat as compared to when using a stiff brittle matrix by itself.

3.9 Physiochemical Testing and Quality Control

Composites are unlike metals in that their final properties are dependent upon chemical reactions. Since the fabricator cures the part, methods must be used to control fabrication, and physical, chemical and thermal testing must be performed on the starting materials rather than just the finished product. The quality of a composite part can be assured by either post-fabrication testing of "tagalong" specimens fabricated with the production hardware, using the same materials or by being certain that the proper raw materials were used and processed correctly. In the early phases of implementing a new material into production, usually both approaches are used. The supplier runs a battery of physical, chemical and mechanical property tests before shipping the material. When the fabricator receives the material, these tests are often repeated. During part fabrication, the same material that is used in the part will also be used to fabricate tagalong (or process control) specimens. After the part is fabricated, the part is normally subjected to nondestructive testing and the process control specimens are destructively tested. As more confidence is gained in the material and processing, it is normal practice to back off on this rather expensive battery of testing. Quite frequently, the material supplier is given the responsibility for all of the testing of the raw material and they supply certification sheets to the user. As the fabricator gains more experience and confidence, it is normal practice to scale back the number of in-process control specimens fabricated with the production part.

Physiochemical testing is a series of chemical, rheological and thermal tests used to characterize both uncured resins and cured composites. These tests are useful in evaluating new resins, developing processing parameters and for quality assurance of incoming materials. The major factors that must be considered to ensure the uniformity of a resin include the type of ingredients, the purity of the ingredients, the concentration of the individual ingredients, as well as the homogeneity and advancement of the resin mix.

3.10 Chemical Testing

Chromatographic methods can be used to ensure that the correct ingredients are present in the resin in the prescribed ratios. These methods accomplish separation of the resin by the interaction of soluble resin components with a flowing liquid or mobile phase and a solid stationary phase. Frequently, they are used in conjunction with spectrographic

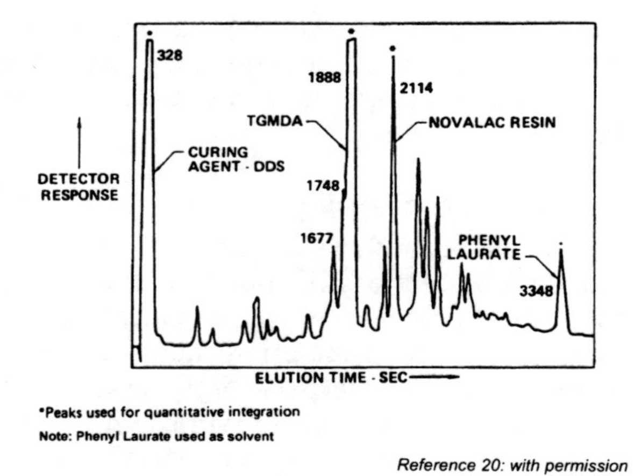

Reference 20: with permission

Fig. 31. HPLC Curve for 3501-6 Epoxy Resin

methods to identify and quantify the specific components in the resin. High-performance liquid chromatography (HPLC) uses a mobile liquid phase and a solid stationary phase. A resin sample is dissolved in a suitable organic solvent, injected into the chromatograph and swept through a column packed with fine solid particles. A detection device senses the presence of each molecular fraction. The detector monitors the concentrations of the separated components and its signal response, recorded as a function of the time after injection and provides a "fingerprint" of the resin's chemical composition. If the components are known and sufficiently well resolved and if standards are available, then quantitative information can be obtained for the sample. A chromatograph of the epoxy resin system 3501-6 is shown in Fig. 31. Peak locations are associated with the chemical structure of the individual components and the areas under the peaks are proportional to the amounts of each component.[20] A subset of HPLC is gel permeation chromatography (GPC), in which the components are separated based on the molecular size. In GPC, the components are separated by their permeation into a porous packing gel.

Infrared (IR) spectroscopy is based on molecular vibrations in which a continuous beam of electromagnetic radiation is passed through or reflected off the surface of the sample, which may be a solid, liquid or a gas. Individual molecular bonds and bond groupings vibrate at characteristic frequencies and selectively absorb IR radiation at matching frequencies. The amount of radiation absorbed or passed through unchanged depends on the chemical composition of the sample and the resultant curve is known as an IR spectrum, as shown in Fig. 32 for 3501-6 epoxy resin.[21] IR

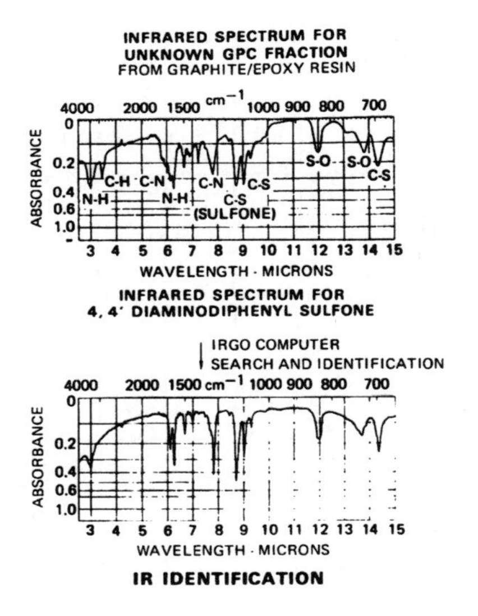

Reference 21: with permission

Fig. 32. IR Identification of DDS Curing Agent

spectroscopy can also be used to identify the specific ingredients by comparing the sample with a computerized spectrum search. IR spectroscopy is sensitive to changes in the dipole moments of vibrating groups of molecules, which yields useful information about the identity of the resin components. It also provides a "fingerprint" of the resin composition and may be used with gases, liquids or solids. Fourier transform IR spectroscopy (FTIR) is a computer-assisted method in which Fourier transformation of the IR spectra is used to enhance the signal-to-noise ratio and provide improved spectra for interpretation. Computerized libraries of spectra for common materials exist for direct comparison and identification of the resin ingredients. For metallic catalysts such as BF_3, atomic absorption (AA) is a useful test.

The equivalent weight or weight percent epoxide (WPE) is a measure of the number of active sites available for cross-linking. Knowing the equivalent weight allows one to calculate the amount of curing agent required to fully cure the resin. Equivalent weight is usually determined by titration that utilizes a known concentration of one substance to react with and detect the amount of reaction species in the unknown sample.

3.11 Rheological Testing

Rheology is the study of deformation and flow. It is frequently used to determine the flow (viscosity changes) of uncured resins. The viscosity of an uncured resin is usually determined by a parallel plate rheometer, in

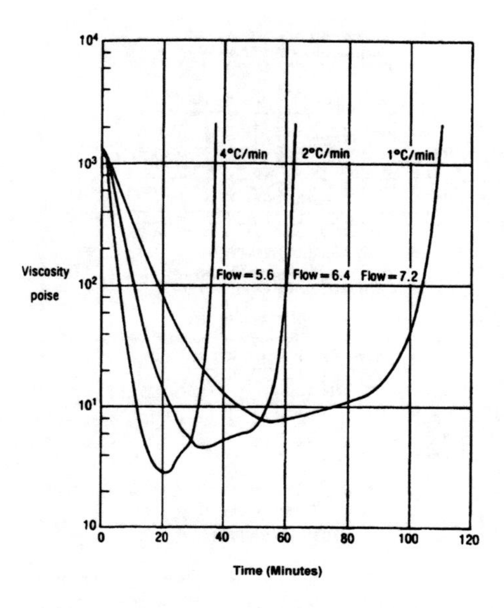

Fig. 33. Viscosity Curves at Three Different Heating Rates

which a small sample is placed between two oscillating parallel plates and then heated. During curing, the resin is converted from a liquid (not cross-linked) or semi-solid into a rigid cross-linked solid. As shown in the viscosity curves in Fig. 33, as heat is applied, the semi-solid resin melts and flows. During this point in the cure there are two competing forces: (1) as the temperature increases, the mobility of the molecules increases and the viscosity falls; and (2) as the temperature increases, the molecules start to react and grow in size, leading to an increase in viscosity. As further heating occurs, the cross-linking reactions cause the viscosity to rapidly rise and eventually the resin gels and transforms from a liquid to a rubbery state. Gellation is the point in the curing process where the reacting molecules have become so large that no further flow can take place. At gellation, the resin is normally about 58-62% cross-linked.[22] The gel point represents the point at which the viscous liquid becomes an elastic gel and the beginning of an infinite cross-linked network. The behavior of the flow affects the way in which the resin can be processed and gellation marks the end of flow. Gellation is often arbitrarily taken as the point where the resin reaches a viscosity equal to 1,000 P. The resin is then held at the cure temperature, usually 250-350 °F, until the cross-linking reactions are completed and the resin is fully cured. Further heating above 480 °F results in resin degradation for a typical epoxy resin.

3.12 Thermal Analysis

Thermal analysis is the general term given to a group of analytical techniques that measure the properties of a material as it is heated or cooled. Techniques such as differential scanning calorimetery (DSC), thermal mechanical analysis (TMA) and thermogravimetric analysis (TGA) are used to determine the degree of cure, the rates of cure, heat levels in the reactions, melting points of thermoplastics and thermal stability.

DSC is the widest utilized method for obtaining the degree of cure and reaction rate. DSC is based on the measurement of the differential voltage, converted into heat flow in a calorimeter, necessary to obtain thermal equilibrium between the resin sample and an inert reference. A small sample of resin is placed in a sealed packet and then inserted into the calorimeter heating chamber along with a known standard of material. The temperature is increased and the amount of heat given off (exothermic) or taken in (endothermic) is compared to a known standard. DSC can be run in either an isothermal or a dynamic heating mode. In the isothermal mode, it measures the thermal changes as a function of time at a constant temperature. In the dynamic mode, it measures the thermal changes as a function of temperature at a predetermined heating rate. A typical dynamic DSC scan is shown in Fig. 34.[23] Critical points on the curve are:

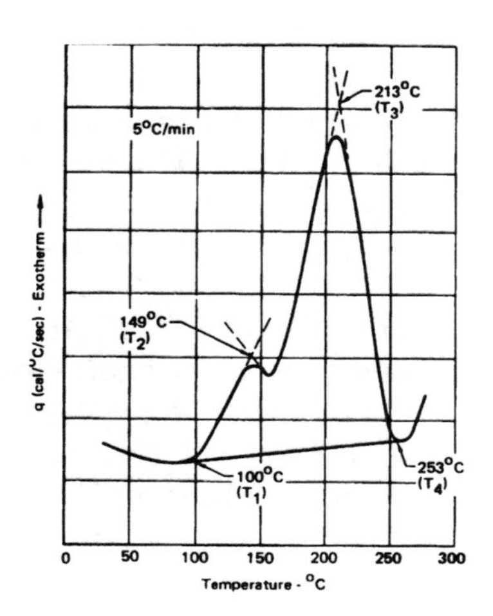

T1 (Ti)- Onset Temperature or Beginning of Polymerization
T2 (Tm) - Minor Exotherm Peak due to BF_3 Catalyst
T3 (Texo)- Peak Exotherm Peak
T4 (Tf)- Final Cure

Reference 23: with permission

Fig. 34. Dynamic DSC Scan for 3501-6 Epoxy Resin

- T_i, the initiation or onset temperature of the reaction indicating the beginning of polymerization;
- T_m, a minor exotherm peak associated with the presence of an accelerator, such as BF_3 in 3501-6 epoxy;
- T_{exo}, the major exotherm peak temperature;
- T_f, the final temperature, indicating the end of heat generation and the completion of cure.

As the resin cures, the T_g rises until it becomes fixed. The extent of cure at any given time can be defined as the degree of cure, α. Since cure of thermoset resins is an exothermic process, α can be related to the heat released during the cure reaction:

$$\alpha = \Delta H_t / \Delta H_R$$

where ΔH_t is the heat of reaction at time t and ΔH_R is the total heat of reaction.

For a 100% cure, the degree of cure $\alpha = 1$. A higher cure temperature results in faster curing. As previously stated, DSC can be conducted isothermally or dynamically. There are two methods commonly used for isothermal DSC measurements.[24] In the first, the resin sample is placed in a previously heated calorimeter or an unheated calorimeter and the temperature is raised as quickly as possible to the cure temperature. In the second method, the resin sample is cured for various times until no additional curing can be detected and then the sample is scanned at a heating rate of 2-20°C min^{-1} to measure the residual heat of reaction (ΔH_{res}). The degree of cure can be calculated from:

$$\alpha = (\Delta H_R - \Delta H_{res}) / \Delta H_R$$

TGA measures the weight gain or loss of a material as a function of temperature as it is heated isothermally or dynamically. The TGA unit is a sensitive microbalance with a precision-controlled furnace. As the material is heated, various changes in weight are recorded, either as a function of temperature (isothermal) or time (dynamic). It can be used to measure the amount of moisture (which will evolve at 212 °F), the total volatiles in the sample and the temperature and rate of thermal decomposition. TGA is often used in conjunction with mass spectroscopy (MS) so that as the gases evolve from the sample they can be collected and analyzed for their composition. One of the most important uses of TGA is in determining the thermal degradation profile of a composite as it is heated to higher and higher temperatures.

3.13 Glass Transition Temperature

The cured glass transition temperature (T_g) of a polymeric material is the temperature at which it changes from a rigid glassy solid into a softer, semi-flexible material. At this point the polymer structure is still intact but the cross-links are no longer locked in position. The T_g determines the maximum use-temperature for a composite or an adhesive, above which the material exhibits significantly reduced mechanical properties. Since most thermoset polymers absorb moisture that severely depresses the T_g, the actual use temperature should be about 50 °F lower than the wet or saturated T_g. The cured glass transition temperature can be determined by several methods, such as TMA, DSC and DMA, all of which give different results because each measures a different property of the resin. Schematic outputs from these three methods are shown in Fig. 35.

TMA or thermal mechanical analysis measures the thermal expansion of a sample as it is heated. The slope of the thermal expansion vs. temperature curve changes at the T_g due to changes in the coefficient of thermal expansion as it passes through the T_g. Materials below their T_g expand linearly at a lower rate than above their T_g, where since the material is more mobile and, as a result, the material expansion rate increases dramatically. Cured thermosets usually exhibit two linear regions. The first is associated with the glassy state and is followed by a change at the T_g to a second linear region of higher slope associated with the rubbery state.

DSC can also be used to measure the T_g since the glass transition is an exothermic reaction. Although DSC is an accepted method of determining T_g, it is sometimes difficult to accurately determine the T_g from the curve, especially for high cross-linked resin systems. An advantage of DSC is that a small sample size (25 mg) can be used to determine T_g.

DMA or dynamic mechanical analysis measures the ability of a material to store and dissipate mechanical energy upon deformation. Therefore, DMA is a mechanical measure of the T_g in which a cured sample is loaded under a sinusoidal stress and the displacement is measured. The T_g is the point where the storage modulus (G' or E') or stiffness drops by several orders of magnitude. The loss modulus (G'' or E'') is usually associated with the viscosity or toughness of the material. The specimen may be loaded in either torsion or flexure. In a typical DMA test, a small bar of cured composite is either twisted (torsion) or flexed (tension) while it is being heated. Two properties are normally reported: the torsional (G') or bending (E') stiffness and the loss tangent (tan δ). As the material nears its T_g, it loses stiffness and drops rapidly with increasing temperature, as shown in Fig. 35. The loss tangent is a measure of the energy stored in the polymer's structure as a result of its viscoeleastic behavior. This stored energy rises to a sharp peak and then drops rapidly. The peak in the tan δ curve is often reported as the glass transition temperature but this gives artificially high values. A more realistic T_g is the interception of the two tangent lines on the storage modulus curve, since it represents the point at which the composite starts exhibiting a significant loss in mechanical stiffness. The complex modulus (G^*), the storage modulus (G'), the loss

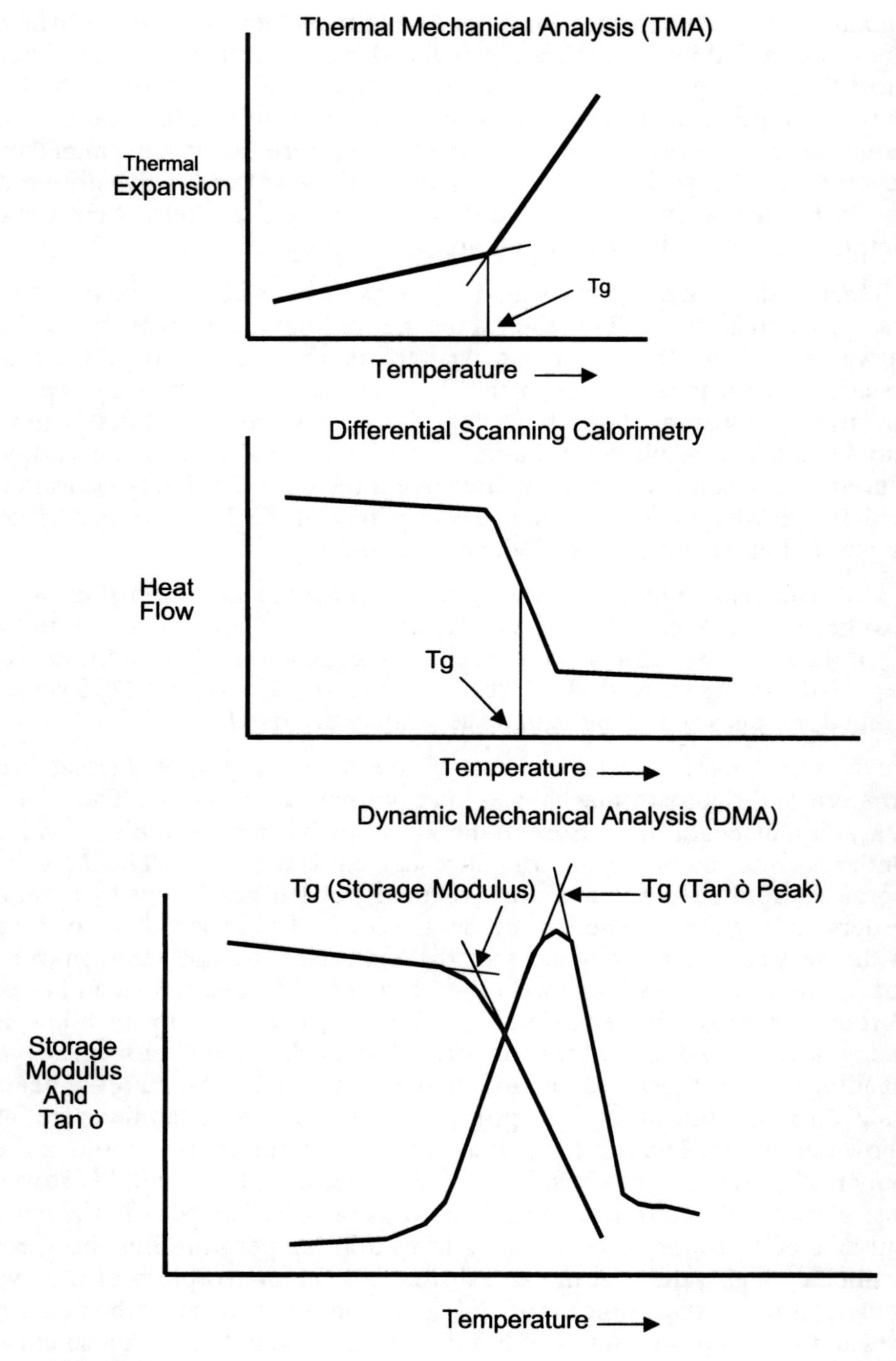

Fig. 35. Schematics of Tg Curves

modulus (G'') and the loss tangent ($\tan \delta$) are related by the following relations:

$$G^* = G' + G''$$

$$\tan \delta = G''/G'$$

3.14 Summary

The function of the matrix is to bind the fibers together in an orderly array and protect them from the environment. The matrix transfers loads to the fibers and is critical in compression loading in preventing premature failure due to fiber microbuckling. The matrix also provides the composite with toughness, damage tolerance, impact and abrasion resistance. The properties of the matrix also determine the maximum usage temperature, resistance to moisture and fluids, and thermal and oxidative stability.

Polymeric matrices for advanced composites are usually thermosetting resins. Thermosets are low-molecular-weight, low-viscosity monomers that are converted during curing into infusible and insoluble 3D cross-linked structures. Cross-linking results from chemical reactions that are driven by heat generated by either the chemical reactions themselves (i.e., exothermic heat of reaction) or by externally supplied heat. As curing progresses, the reactions accelerate and the available volume within the molecular arrangement decreases, resulting in less mobility of the molecules and an increase in viscosity. After the resin gels and forms a rubbery solid, it cannot be remelted. Further heating causes additional cross-linking until the resin is fully cured. Since curing is a thermally driven event requiring chemical reactions, thermosets are characterized as having rather long processing times.

Although the fibers provide the main strengthening mechanism for advanced composite structures, the matrix resins and their properties generally control the manufacturing and processing. Thermoset composite matrices include polyesters, vinyl esters, epoxies, bismaleimides, cyanate esters, polyimides and phenolics. Epoxies currently are the dominant resins used for low and moderate temperatures (up to 275 °F). bismaleimides are used primarily in the temperature range of 275-350 °F. For very high temperature applications (up to 550-600 °F), polyimides are usually the materials of choice. Polyesters and vinyl esters, which can be used at approximately the same temperatures as epoxies, are used quite extensively for commercial applications but are rarely used for high-performance composite matrices because of their lower properties. Cyanate esters are a relatively new class of resins that were designed to compete with both epoxies and bismaleimides and offer some advantages (lower moisture absorption and attractive electrical properties) but at a significantly higher price. Phenolics are high-temperature systems that offer outstanding smoke and fire resistance and are frequently used for aircraft interior components.

To impart greater toughness to the cross-linked thermoset polymers, a number of different approaches have evolved. Some are used by themselves, while others are often combined to further enhance toughness. The four main toughening approaches are: (1) network alteration, (2) rubber elastomer second-phase toughening, (3) thermoplastic elastomer toughening and (4) interlayering.

Physiochemical testing is used to characterize both uncured resins and cured composites. These chemical, rheological and thermal tests are useful in evaluating new resins, developing processing parameters and for quality assurance of incoming materials.

In general, the specific resin chemistries and their formulation practices are closely held trade secrets of the material suppliers. However, they are amenable to working with their customers to select the right material system for the required application and in helping develop the composite characterization methods and processing cycles. It is important to seek their help early and often when implementing any new material system. Since uncured resin systems can contain materials that are potentially hazardous to health and safety of workers, it is again important to closely coordinate with the material supplier and examine the MSDS (material safety data sheet) for any new material.

References

[1] Prime R.B., Chapter 5 in *Thermal Characterization of Polymeric Materials*, E.A. Turi Editor, Academic Press, 1981.

[2] SP Systems "Guide to Composites".

[3] Smith W., Chapter 3 "Resin Systems" in *Processing and Fabrication Technology*, Vol. 3, Delaware Composites Design Encyclopedia, Technomic Publishing Company, Inc., 1990, p. 15-86.

[4] Strong A.B., *Fundamentals of Composite Manufacturing: Materials, Methods, and Applications*, Society of Manufacturing Engineers, 1989.

[5] Vantico Inc. Data Sheets for MY-720 and MY-721.

[6] Carpenter J.F., "Processing Science For AS/3501-6 Carbon/Epoxy Composites", Technical Report N00019-81-C-0184, Naval Air Systems Command, 1983.

[7] Stenzenberger H.,"Bismaleimide Resins" in *ASM Handbook 21 Composites*, ASM International Inc., 2001, p. 97.

[8] Prater R.H., "Thermosetting Polyimides: A Review", SAMPE Journal, Vol. 30, No. 5, September/October 1994, p. 29-38.

[9] Rommel M.I., Postyn, A.S., Dyer, T.A., "Accelerating Factors in Galvanically Induced Polyimide Degradation", SAMPE Journal, Vol. 30, No. 2 March/April 1994, p. 10-15.

[10] McConnell V.P., "Tough Promises from Cyanate Esters", Advanced Composites, May/June 1992, p.28-37.

[11] Robitaile S.,"Cyanate Ester Resins", in *ASM Handbook 21 Composites*, ASM International Inc., 2001, p. 126-131.

[12] Scola D.A., "Polyimide Resins", in *ASM Handbook 21 Composites*, ASM International, Inc., 2001, p. 107.

[13] Jang B.Z., "Fibers and Matrix Resins" in *Advanced Polymer Composites: Principles and Applications*, p. 24.
[14] Hergenrother P.M., "Development of Composites, Adhesives and Sealants for High-Speed Commercial Airplanes", SAMPE Journal, Vol. 36, No. 1, January/February 2000, p. 30-41.
[15] Criss J.M., Arendt C.P., Connell J.W., Smith J.G., Hergenrother P.M., "Resin Transfer Molding and Resin Infusion Fabrication of High-Temperature Composites", SAMPE Journal, Vol. 36, No. 3, May/June 2000, p. 31-41.
[16] Harrington H.J., "Phenolics" in *ASM Vol. 3 Engineered Materials Handbook: Engineering Plastics*, ASM International Inc., 1988, p.242-245.
[17] Bottcher A., Pilato L.A., "Phenolic Resins for FRP Systems", SAMPE Journal, Vol. 33, No. 3, May/June 1997, p.35-40.
[18] Almen G.R., Byrens R.M., MacKenzie P.D., Maskell R.K., McGrail P.T., Sefton M.S., "977- A Family of New Toughened Epoxy Matrices", 34th International SAMPE Symposium, 8-11 May 1989, 259-270.
[19] Park J.W., Kim S.C., "Phase Separation and Morphology Development During Cure of Toughened Thermosets" in *Processing of Composites*, Hanser, 2000, p. 108-136.
[20] Sewell T.A., "Quality Assurance of Graphite/Epoxy by High-Performance Liquid Chromatogrphy in *Composite Materials: Quality Assurance and Processing*, ASTM STP 797, 1983, p.3-14.
[21] Carpenter J.F., "Assessment of Composite Starting Materials: Physiochemical Quality Control of Prepregs", AIAA/ASME Symposium on Aircraft Composites: The Emerging Methodology for Structural Assurance, San Diego, CA, 24-25 March 1977.
[22] Haung M.L., Williams J.G., "Macromolecules" Vol. 27, 1994, p. 7423-7428.
[23] Carpenter J.F., "Physiochemical Testing of Altered Composition 3501-6 Epoxy Resin", 24th National SAMPE Symposium, San Francisco, CA, 8-9 May 1979.
[24] Veronica M.A. Calado, Advani S.G., "Thermoset Resin Cure Kinetics and Rheology", in *Processing of Composites*, Hanser, 2001, p. 32-107.

Chapter 4

Cure Tooling: You Can Pay Me Now ...or Pay Me Later

Tooling for composite fabrication is a major up-front non-recurring cost. It is not unusual for a large bond tool to cost as much as $500,000–$1,000,000. Unfortunately, if the tooling is not designed and fabricated correctly, it can become a recurring headache, requiring continual maintenance and modifications, and, in the worst scenario, replacement. This chapter will cover some of the basics of bond tools for composites, primarily for autoclave curing. As other composite manufacturing processes are introduced in later chapters, additional tooling information specific to that process will be discussed, particularly in Chapter 8, since the fabrication of unitized structure is tooling intensive.

Tooling for composite structures is a complex discipline in its own right, largely built on years of experience. It should be pointed out that there is no single correct way to tool a part. There are usually several different approaches that will work with the final decision based largely on experience of what has worked in the past and what did not work. The purpose of the bond tool is to transfer the autoclave heat and pressure during cure, to yield a dimensionally accurate part and there are a number of alternatives that will usually work. Although this chapter serves as an introduction to tooling, the interested reader is referred to Ref. 1 for a much more comprehensive coverage. A shorter review of some of the key principles of Ref. 1 can be found in Refs. 2–4. Refs. 5 and 6 also contain useful sections on composite cure tooling.

4.1 General Considerations

There are many requirements a tool designer must consider before selecting a tooling material and fabrication process for a given application. Some of these requirements are listed in Fig. 1; however, the number of parts to be made on the tool and the part configuration are often the overriding factors in the selection process. It would not make good economic sense to build an inexpensive prototype tool that would only last

- Stable at Use Temperature (Usually 350°F)
- Withstand Loads of 100 psi
- Smooth Finish in Part Area
- Acceptable to Parting Agent
- Have Expansion Factor Compensation
- Wear Resistant to Scraping
- Resistant to Solvent Cleaning
- Machinable or Capable of Lamination
- Locate and Support All Components
- Capable of Producing Production Article within Tolerance and Process Specification
- May Require Vacuum Integrity
- Uniform Heat-Up Rate
- Light Weight
- Compatible to Shop Equipment

Fig. 1. Potential Requirements for Composite Cure Tooling

for several parts when the application calls for a long production run, or vice versa. Part configuration or complexity will also drive the tooling decision process. For example, while welded steel tools are often used for large wing skins, it would not be cost effective to use steel for a highly contoured fuselage section due to the high fabrication cost and complexity.

One of the first choices that must be made is which side of the part should be tooled, i.e., the inside or outside surface as shown in Fig. 2. Tooling a skin to the outside surface or outer moldline (OML) surface provides the opportunity to produce a part with an extremely smooth outside surface finish. However, if the part is going to be assembled to substructure, for example, with mechanical fasteners, tooling to the inside or inner moldline (IML) surface will provide better fit with fewer gaps and less shimming required. An example is the wing skin shown in Fig. 3, which was tooled to the IML surface to ensure the best possible fit to the substructure during assembly. A caul plate is used on the bag side during cure to provide an acceptable OML aerodynamic surface finish. As shown in Fig. 4, this part contains severe thickness transitions on the IML surface. If the part were of constant thickness or contained little thickness variation, it might have then made more sense to tool it to the OML surface. Ease of part fabrication is another concern. Returning to the channel section shown in Fig. 2, it would certainly be easier to collate or lay-up the plies on the male tool shown than down inside the cavity of a female tool.

Selection of the material used to make the tool is another important consideration. Several of the key properties of various tooling materials are given in Table 1. Normally, reinforced polymers can be used for low to intermediate temperatures, metals for low to high temperatures, and monolithic graphite or ceramics for very high temperatures. Traditionally, tools for autoclave curing have generally been made of either steel or

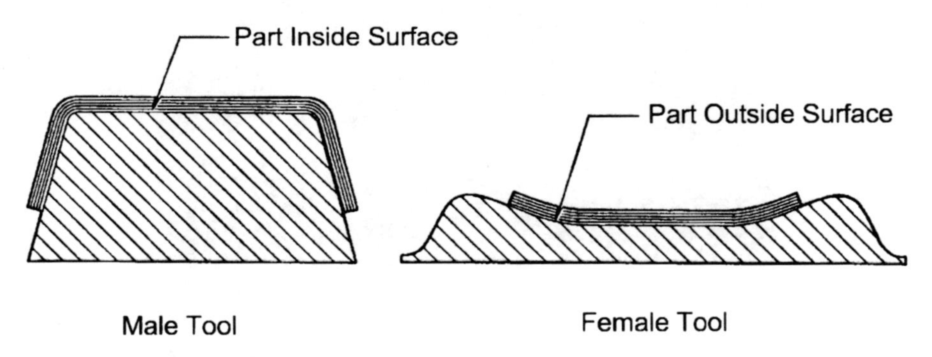

Fig. 2. Inside Surface and Outside Surface Tooling

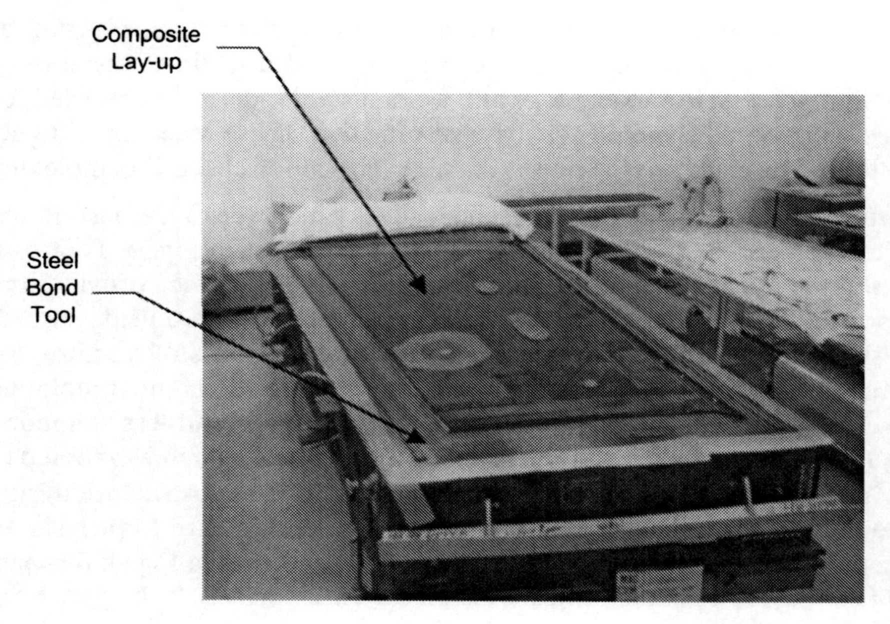

Fig. 3. F/A-18 Inner Wing Skin Tooled to Inside Surface

Fig. 4. Inside Surface of F/A-18 Inner Wing Skin Showing Thickness Changes

Table 4.1 Properties of Typical Tooling Materials

Material	*Max Service Temp. (°F)*	*CTE $X10^{-6}$/°F*	*Density lbs./in.^{3}*	*Thermal Conductivity Btu./h × ft. × °F*
Steel	1000	6.3–7.3	0.29	30
Aluminium	500	12.5–13.5	0.10	104–116
Electroformed Nickel	550	7.4–7.5	0.32	42–45
Invar/Nilo	1000	0.8–2.9	0.29	6–9
Carbon/Epoxy 350 °F	350	2.0-5.0	0.058	2–3.5
Carbon/Epoxy RT/350 °F	350	2.0–5.0	0.058	2–3.5
Glass/Epoxy 350 °F	350	8.0–11.0	0.067	1.8–2.5
Glass/Epoxy RT/350 °F	350	8.0–11.0	0.067	1.8–2.5
Monolithic Graphite	800	1.0–2.0	0.060	13–18
Mass Cast Ceramic	1650	0.40–0.45	0.093	0.5
Silicone	550	45–200	0.046	0.1
Isobutyl Rubber	350	≈90	0.040	0.1
Fluoroelastomer	450	≈80–90	0.065	0.1

Note/ For reference only. Check with material supplier for exact values.

aluminum. Electroformed nickel became popular in the early 1980s followed by the introduction of carbon/epoxy and carbon/bismaleimide composite tools in the mid-1980s. Finally, in the early 1990s, a series of low expansion iron–nickel alloys was introduced under the trade names Invar and Nilo.

Steel has the attributes of being a fairly cheap material with exceptional durability. It is readily castable and weldable. It has been known to withstand over 1,500 autoclave cure cycles and still be capable of making good parts. However, steel is heavy, has a higher coefficient of thermal expansion (CTE) than the carbon/epoxy parts usually built on it, and for large massive tooling it can experience slow heat-up rates in an autoclave. When a steel tool fails in-service, it is usually due to a cracked weldment.

Aluminum, on the other hand, is much lighter and has a much higher coefficient of thermal conductivity. It is also much easier to machine than steel but is more difficult to produce pressure-tight castings and weldments. The two biggest drawbacks of aluminum are: being a soft material, it is rather susceptible to scratches, nicks and dents, and it has a very high coefficient of thermal expansion. Aluminum tools are quite frequently hard anodized to improve the durability. Hard anodize coatings do tend to spall and flake-off as the aluminum temper ages during multiple thermal cycles. Due to its lightweight and ease of machinability, aluminum is often used for what are called "form block" tools. As shown in Fig. 5, a number of aluminum form block tools can be placed on a large flat aluminum

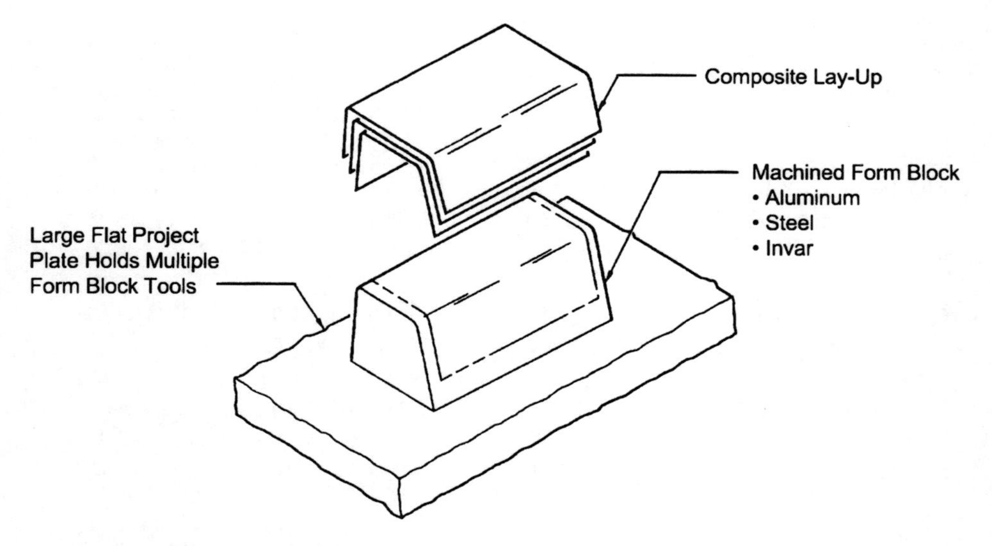

Fig. 5. Typical Form Block Tooling

project plate, and then the plate with all of the parts is covered with a single vacuum bag for cure, a considerable cost savings compared to bagging each individual part. Another application for aluminum tools is matched-die tooling, where all surfaces are tooled, as shown for the spar in Fig. 6.

Electroformed nickel has the advantages that it can be made into complex contours and does not require a thick faceplate. When backed with an open tubular-type substructure, this type of tool experiences excellent heat-up rates in an autoclave. However, to make an electroformed nickel tool requires that a mandrel be fabricated to the exact contour of the final tool.

Carbon/epoxy or glass/epoxy tools (Fig. 7) also require a master or mandrel for lay-up, during tool fabrication. A distinct advantage of carbon/epoxy tools is that their CTE can be tailored to match that of the carbon/epoxy parts they build. In addition, composite tools are relatively light and exhibit good heat-up rates during autoclave curing, and a single master can be used to fabricate duplicate tools. On the downside, there has been a lot of negative experience with composite tools that are subject to 350 °F autoclave cure cycles. The matrix has a tendency to crack and, with repeated thermal cycles, develop leaks. Another consideration when specifying composite tools is that the surface is somewhat soft and easily scratched. If plies are cut directly on the tool, it is necessary to place a metal shim between the ply and the tool to prevent scouring of the tool surface. In general, this is a bad practice since it increases the chance of the shim being left in the lay-up through cure, thus producing a possible repair or

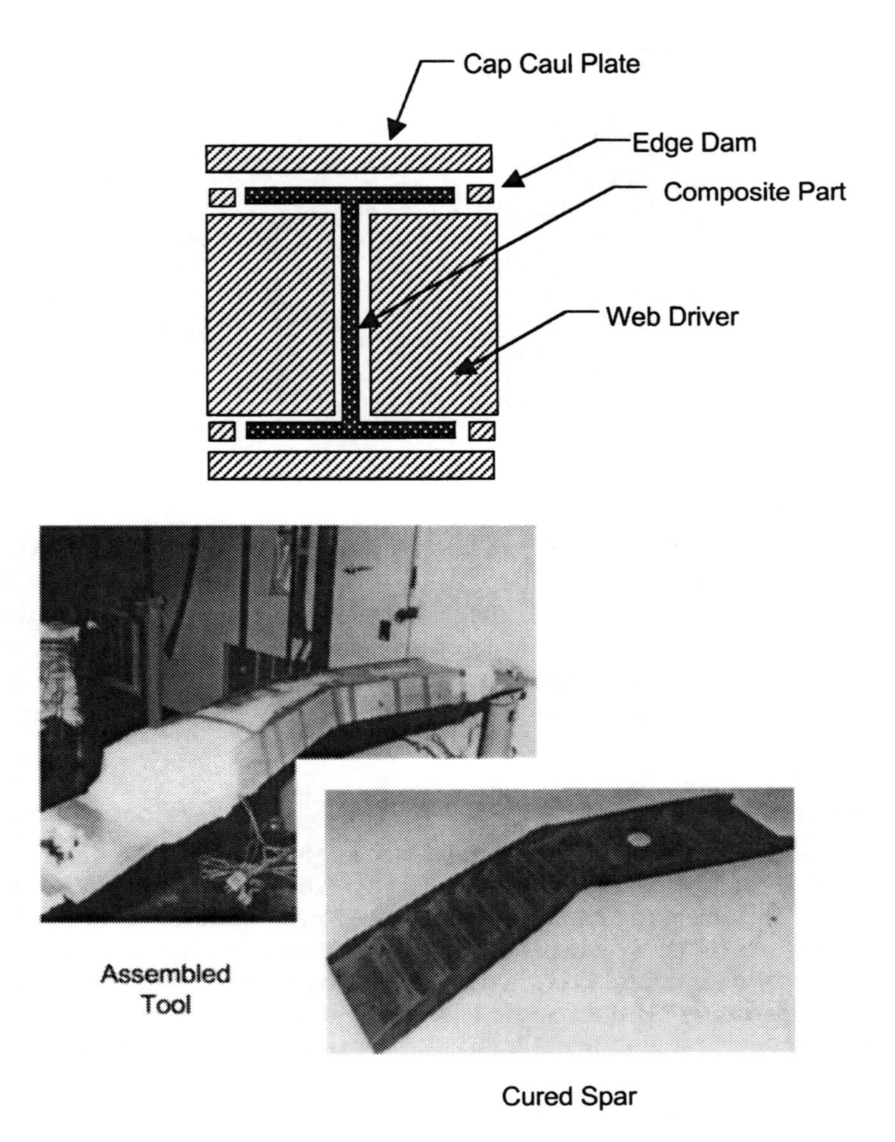

Fig. 6. Example of Matched Die Tool

scrap condition. An additional consideration is that composite tools will absorb moisture if not in continual use. It may be necessary to slowly dry tools in an oven after prolonged storage to allow the moisture to diffuse out. A moisture-saturated tool placed directly in an autoclave and heated to 350 °F could very easily develop blisters and internal delaminations due to the absorbed moisture.

Fig. 7. Large Carbon/Epoxy Tool

Invar and the Nilo series of alloys were introduced in the early 1990s as the answer for composite tooling. Being low-expansion alloys, they very closely match the CTE of the carbon/epoxy parts. Their biggest disadvantages are cost and slow heat-up rates. The material itself is very expensive and it is more difficult to work with than even steel. It can be cast, machined and welded. It is used for premium tooling applications such as the wing skin shown in Fig. 8[7].

4.2 Thermal Management

Since many common tooling materials, such as aluminum and steel, expand at greater rates than the carbon/epoxy part being cure on them, it is necessary to correct their size or compensate for the differences in thermal expansion. As the tool heats up during cure, it grows or expands more than the composite laminate. During cool-down, the tool contracts more than the cured laminate. If not handled correctly, both of these conditions can cause problems ranging from incorrect part size to cracked and damaged laminates.

Thermal expansion is normally handled by shrinking the tool at room temperature using the calculation method shown in Fig. 9. For example, an aluminum tool producing a part 120.0 inches in length might actually be made as 119.7 inches long, assuming it will be cured at 350 °F.

Another correction required for tooling of parts with geometric complexity is spring-in. When sheet metal is formed at room temperature, it normally springs back or opens up after forming. In order to correct the spring-back, sheet metal parts are over-formed. The opposite phenomenon occurs in composite parts. They tend to spring-in or close up during the cure process. Therefore, it is necessary to compensate angled parts as shown in Fig. 10. The degree of compensation required is somewhat

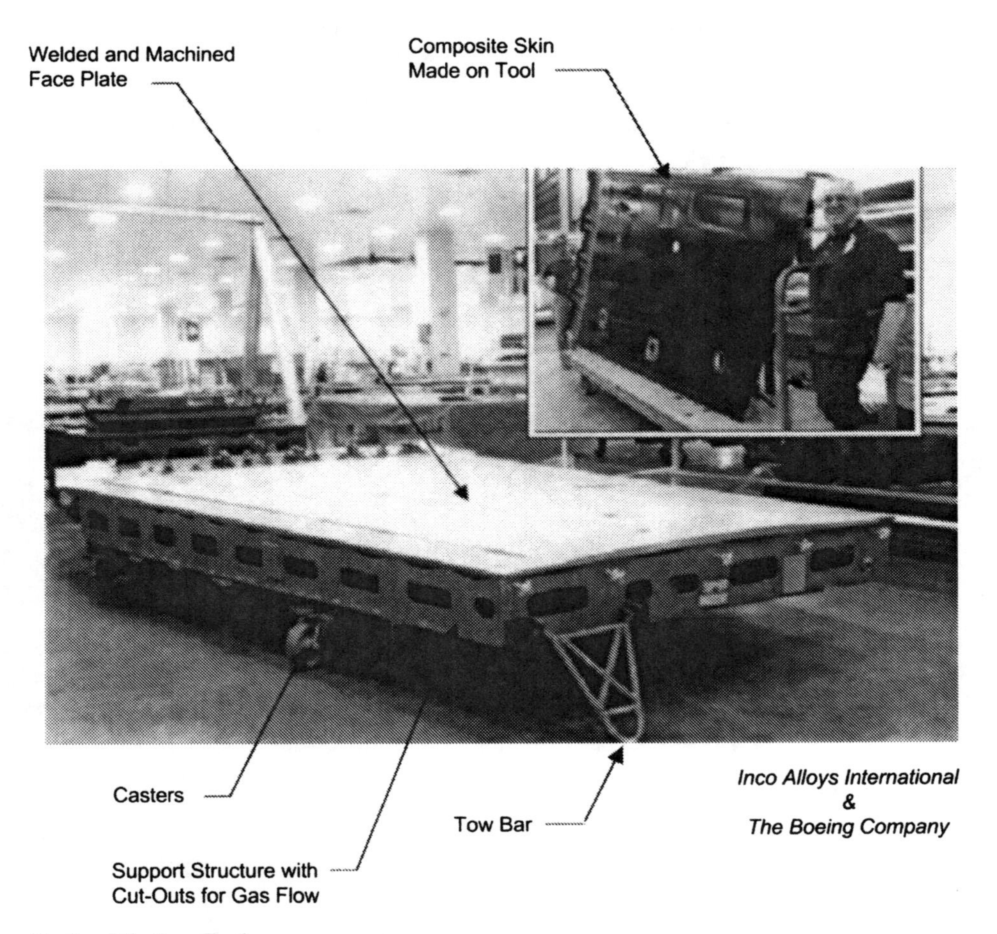

Fig. 8. Nilo Cure Tool

dependent on the actual lay-up orientation and thickness of the laminate. A great deal of progress has been made in calculating the degree of spring-in using finite element analysis, but it still usually requires some experimental data for the particular material system, cure conditions, orientation and thickness to establish tool design guidelines. A more comprehensive explanation of spring-in is given in Chapter 6 on Curing.

Cool-down from cure can also cause problems because now the tool shrinks or contracts at a faster rate than the part. For a tooling material with a large CTE such as aluminum, the tool can actually bind the part causing ply cracking or delaminations as shown in Fig. 11. For skins and other parts, Teflon shear pins (Fig. 12) are often used to prevent damage. It is possible to hard-pin a tooling detail at one or possibly two locations on the bond tool, but the detail must be allowed to freely contract separately

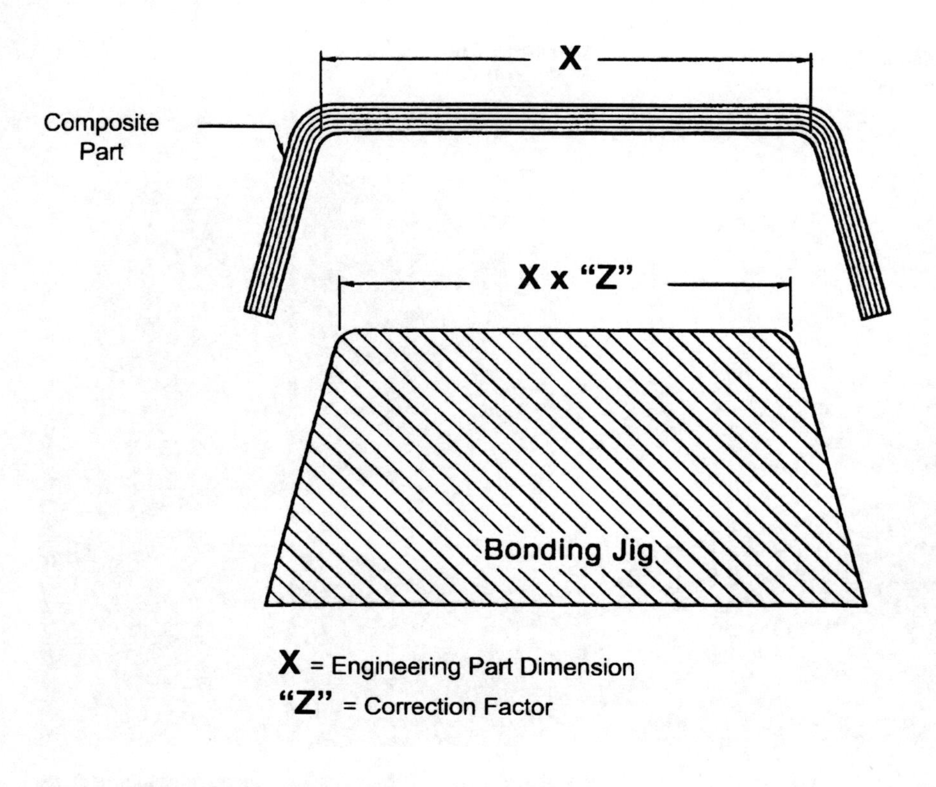

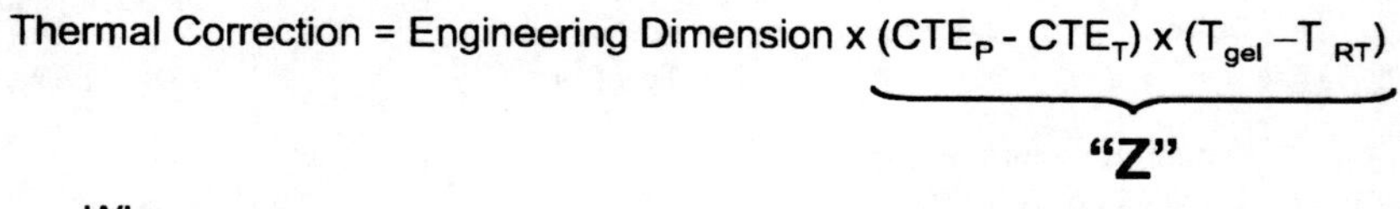

Where:

CTE_P = Coefficient of Thermal Expansion of Part
CTE_T = Coefficient of Thermal Expansion of Tool
T_{gel} = Temperature of Resin Gellation
T_{RT} = Room Temperature

Fig. 9. Thermal Expansion Correction Factors for Tooling

from the bond tool on cooling. Another example is shown in Fig. 13, where the draft has been incorporated in a tool pocket to allow the part to be pushed out from the pocket during cool-down avoiding the possibility of ply cracking. Molded-in rabbets for doors and leading and trailing edges also require special tooling details, as shown in Fig. 14.

Unless the tool is a form block tool that can be placed on a project plate, the tool generally requires a substructure to support the faceplate. An example of one type of substructure is shown in Fig. 15. The design of the substructure is important, because it can affect the gas flow to the tool

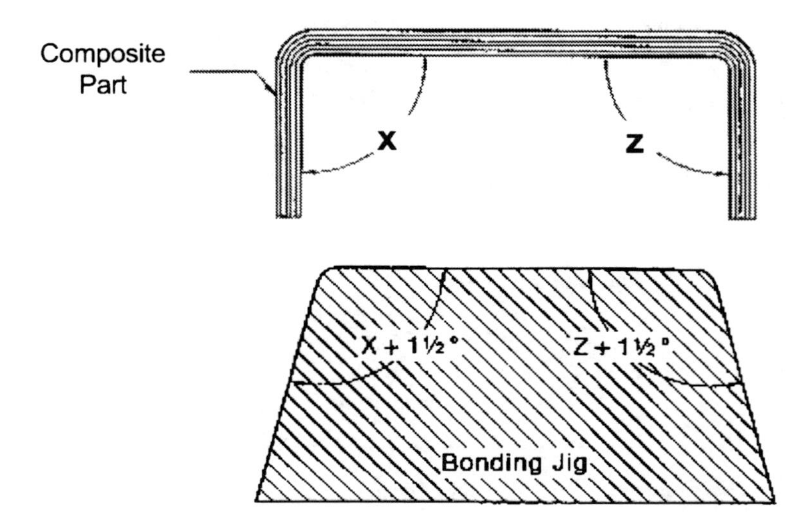

Fig. 10. Spring–In Correction factors

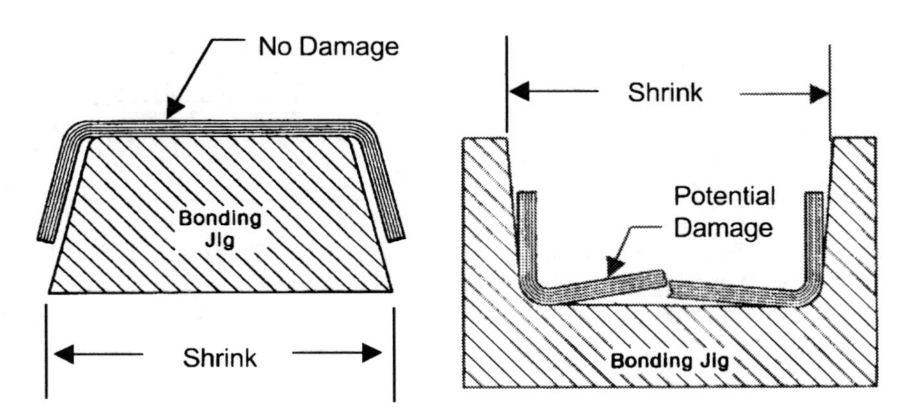

Fig. 11. Potential Effects of Tool Shrinkage on Part Quality

during autoclave curing. In general, the more "open" the substructure, the better the gas flow and the faster the heat-up rate. In one study,[8] three different tool designs were compared for their heat-up rates during autoclave processing. The designs, shown in Fig. 16, consisted of a steel faceplate and a welded egg crate substructure; an NC machined aluminum form block tool placed on an aluminum project plate; and an electroformed nickel tool with an open tubular substructure. The steel and aluminum tools were used to produce the same part configuration, a wing panel,

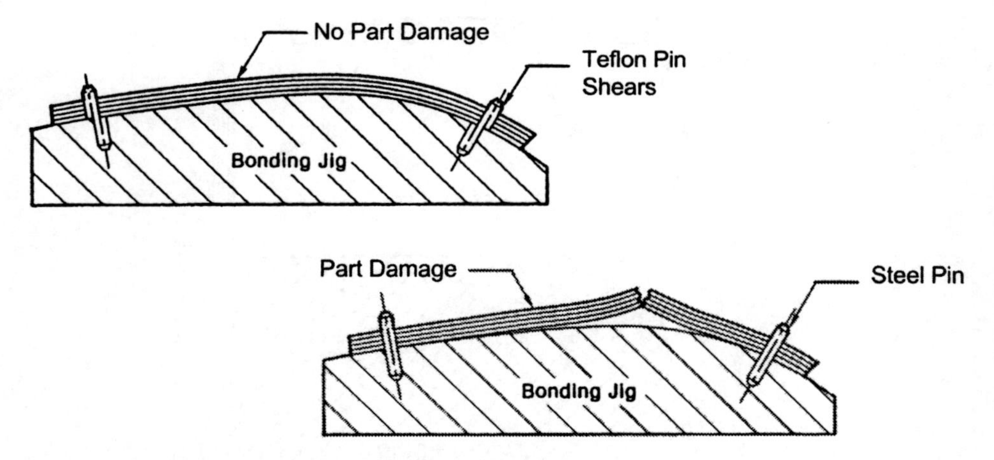

Fig. 12. Shear Pins Used to Eliminate Tool Shrinkage Damage

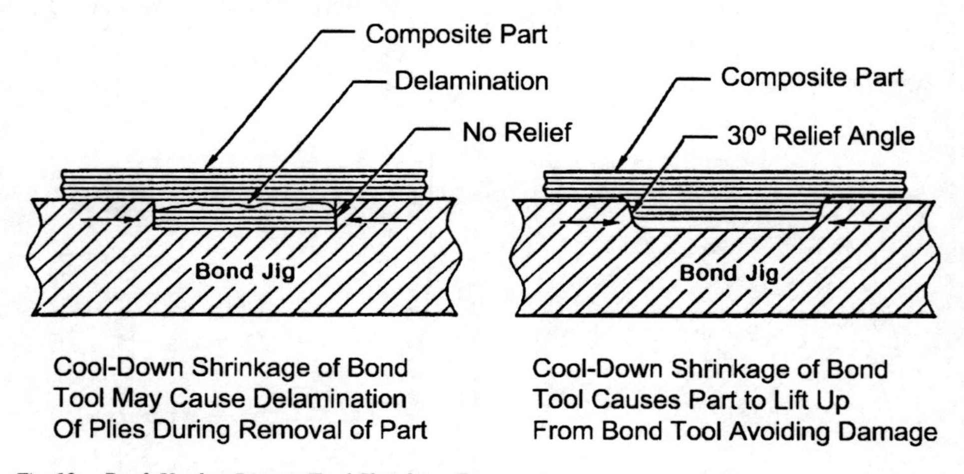

Fig. 13. Draft Used to Prevent Tool Shrinkage Damage

whereas the electroformed nickel tool was used to produce a highly contoured fuselage sidewall panel containing cocured hat stiffeners on the inside surface. The faceplate on the steel tool was 0.45–0.55 inch thick. The welded steel egg crate support contained circular cutouts to improve autoclave gas flow. This tool was identified as tool "A." The NC machined aluminum project plate tool was identified as tool "B." Although it is used to make the same aircraft part as tool "A," the design concept is entirely different. It consists of a thick aluminum slab (2.2–2.3 inches) that was NC machined to moldline contour on one surface and flat on the other surface. During autoclave cure, it is placed on a standard 1.0 inch thick aluminum

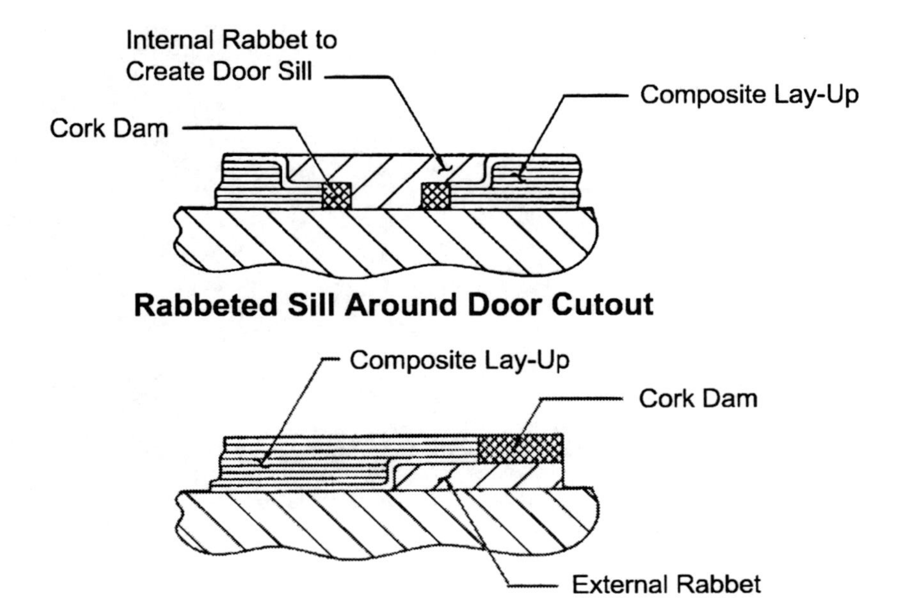

Fig. 14. Typical Tooling for Molded–In Rabbets

project plate. Normally, a number of these types of tools are placed on each project plate. The electroformed nickel bond tool, identified as tool "C," consisted of a relatively thin (0.25–0.38 inch) electroformed nickel shell supported by an open support structure. Rubber mandrels and pressure intensifiers were used at locations requiring co-cured hat stiffeners.

The nesting positions (autoclave locations) for the three batches were predetermined simply by rotating the tool position for each batch. The same production autoclave was used for all three batches. After the tools were vacuum bagged and leak checked, they were nested in the autoclave using the positions shown in Figs. 17–19. Additional thermocouples were then placed above the tool surfaces to measure the autoclave free air temperature. The autoclave was pressurized to between 85 and 100 psig for each run. The tools were then heated through a typical cure cycle profile to 350 °F while thermocouple data were recorded during heat-up and cool-down.

The results of the autoclave characterization tests revealed three significant findings:

- The aluminum project plate tool (tool "B") exhibited slower heat-up rates than either the steel (tool "A") or electroformed nickel tool (tool "C"). This is shown in Figs. 17 and 18. The slower heat-up rate was a function of the large thermal mass of the aluminum tool.

Bond Tool Substructure With Attach Points

Completed Bond Tool

The Boeing Company

Fig. 15. Typical Bond Tool Support Structure

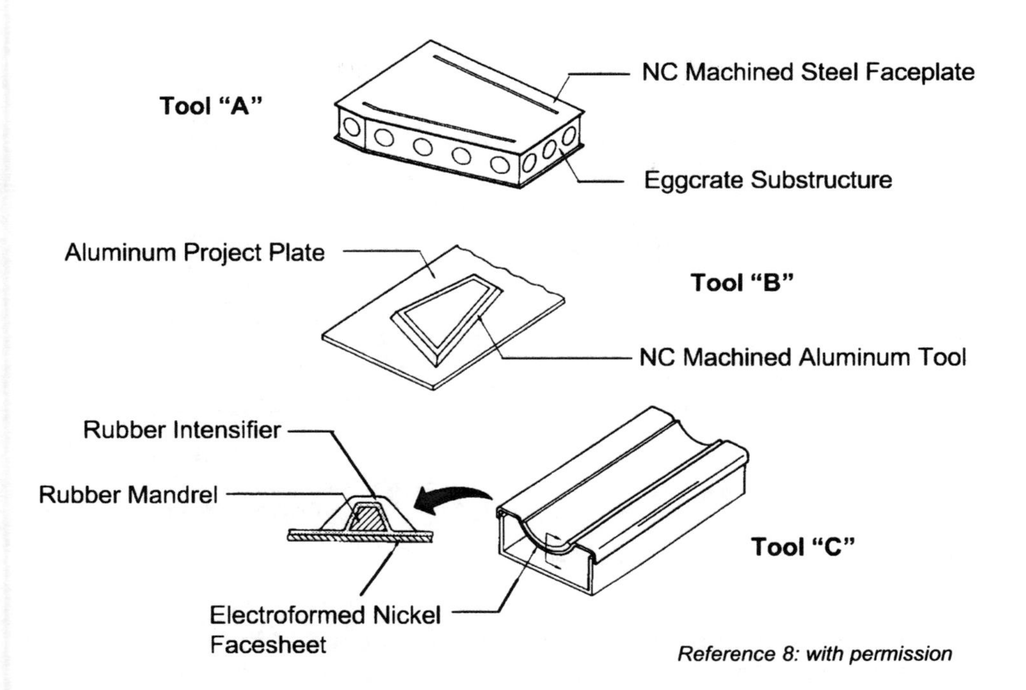

Fig. 16. Tools Used for Autoclave Heat–Up Tests

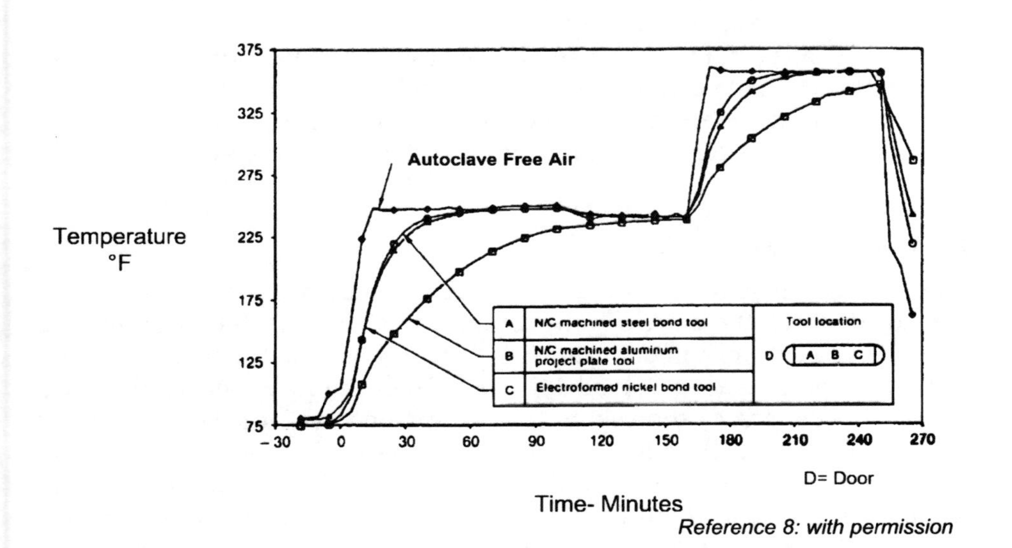

Fig. 17. First Autoclave Heat–Up Rate Test

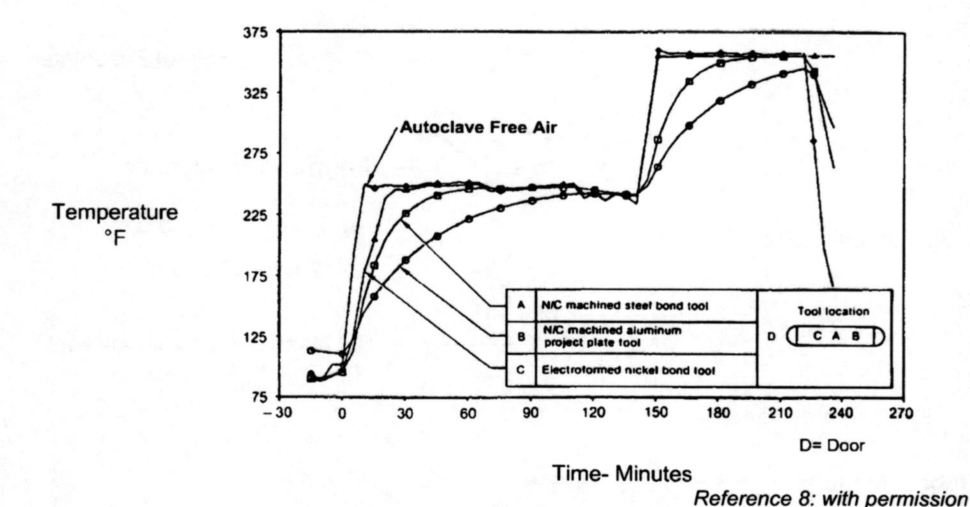

Fig. 18. Second Autoclave Heat–Up Rate Test

However, the heat-up rate for the aluminum tool improved when it was located at the front of the autoclave (Fig.19), where higher gas velocity increased the heat transfer rate from the autoclave free air to the tool.

- The electroformed nickel tool (tool "C") exhibited the fastest thermal response; however, the rubber mandrels used to support the co-cured hats created localized cold spots on the tool, due to their heat sink effect and insulative qualities.
- The heat-up rate for the steel tool (tool "A") was about the same, irrespective of nesting in the front, middle or back of the autoclave. A possible explanation is that the thick egg crate support structure acted as a flow blocker and minimized the effects of autoclave gas velocity differences.

4.3 Tool Fabrication

The fabrication sequence for a typical large steel bond tool is shown in Fig. 20. Prior to the start of fabrication, the steel faceplate sections are formed, rolled or machined to contour. The egg crate substructure is made from steel plate that is welded together. Then the faceplate sections are welded to sections of the egg crate. The final contour of the faceplate is then NC machined. Finally, if the tool is very large like the one shown, the sections are welded together. All welds are ground smooth, the faceplate is finished to give a smooth surface and the tool is vacuum leak checked. Normally, this type of tool will have integral vacuum manifolds, incorporated around the periphery of the tool with multiple lines for

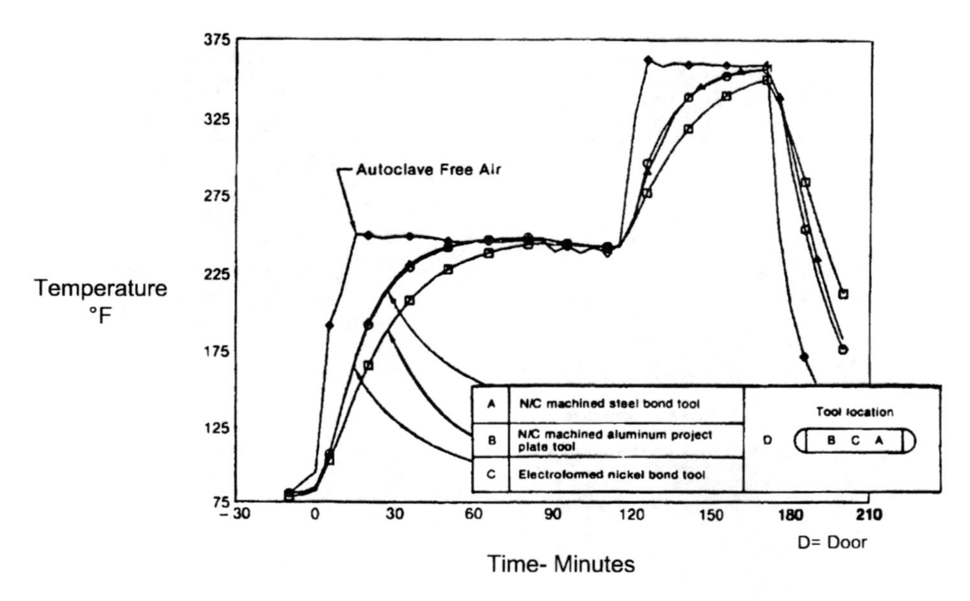

Fig. 19. Third Autoclave heat–Up Rate Up

vacuum and static readings. At a minimum, two are required: one for the dynamic vacuum and one for the static gauge. As a rule of thumb, vacuum ports should be located no further than 6–10 ft apart. To make production easier, most vacuum and static ports are equipped with quick-disconnect fittings. Thermocouples are normally welded to the backside of the tool. Handling provisions, such as casters with wheels and tow bars, are provided for movement. The fabrication approach for Invar-type tools is very similar to that of steel tools except that more cast shapes are used that are then welded to final contour. If the contour is simple, it may be possible to roll a metal faceplate to contour (Fig. 21) and attach it directly to the substructure. For tools with thin faceplates, it is a common practice to use threaded connectors (refer back to Fig. 15), so that adjustments can be made to the faceplate contour.

Electroformed nickel and composite tools require a master model to be fabricated prior to tool fabrication, i.e., it takes a tool to make a tool. Master models must have the correct shape, a smooth surface finish, be able to withstand the processing temperature for tool curing, be leak free, have a sealed surface and must be mold released prior to use. There are many methods of making master models. One of the oldest is shown in Fig. 22 in which contour boards are cut at different sections from the part model and then set up into a template setup. The template setup is then

The Boeing Company

Fig. 20. Manufacturing Sequence for Large NC Machined Steel Bond Tool

usually filled in with metal screen between the contour boards and plaster is applied and allowed to cure. After curing, the plaster is splined smooth to the tops of the boards and then sealed and mold released. If higher temperatures are required than the plaster can withstand, there are both one and two part epoxy doughs[9] that can be used for the master surface. With the advent of more powerful CAD/CAM computer packages, many masters are now directly NC machined from materials such as mass cast epoxies, foams, plasters, wood and laminated polyurethane tooling boards as shown in Fig. 23. Once the master model is complete, a plastic faced plaster (PFP) is usually made from the master. PFPs can be made of pure plaster, can have epoxy faces backed by plaster, or can have fabric reinforced faces backed by plaster. Higher-temperature intermediate or facility tools can also be fabricated off the master or PFP, depending on the processing temperature required for the final tool. The number of steps

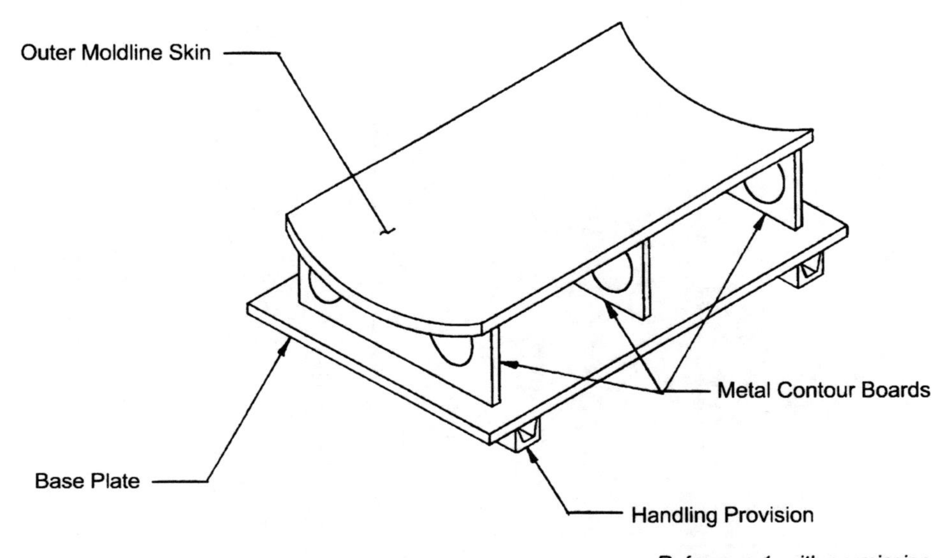

Fig. 21. Simple Contour Tool Fabricated by Rolling Faceplate

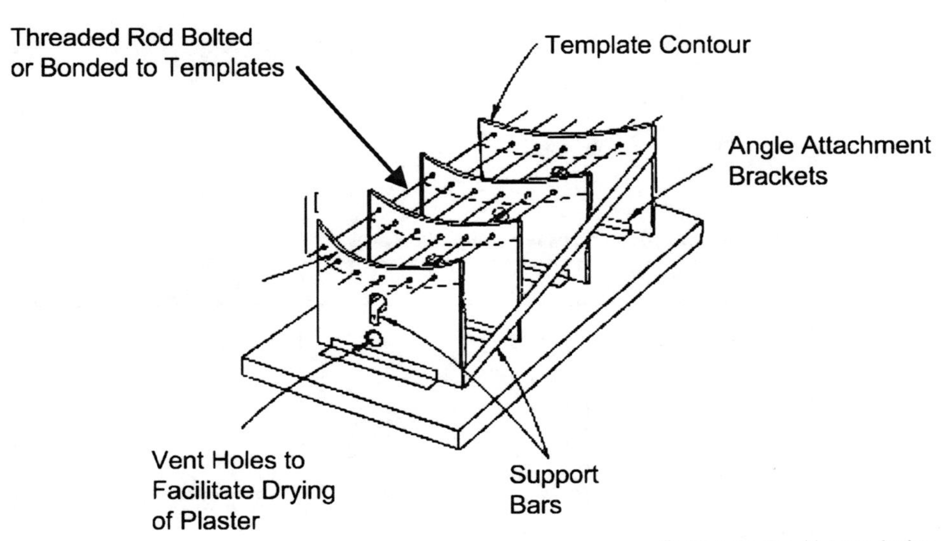

Fig. 22. Master Pattern Contour Board Assembly

involved in fabricating a composite tool can be numerous, as shown in Table 2, depending on which surface you want to work from and what material system you use. Note that you lose some accuracy each time you replicate the tool surface, primarily due to material shrinkage during cure.

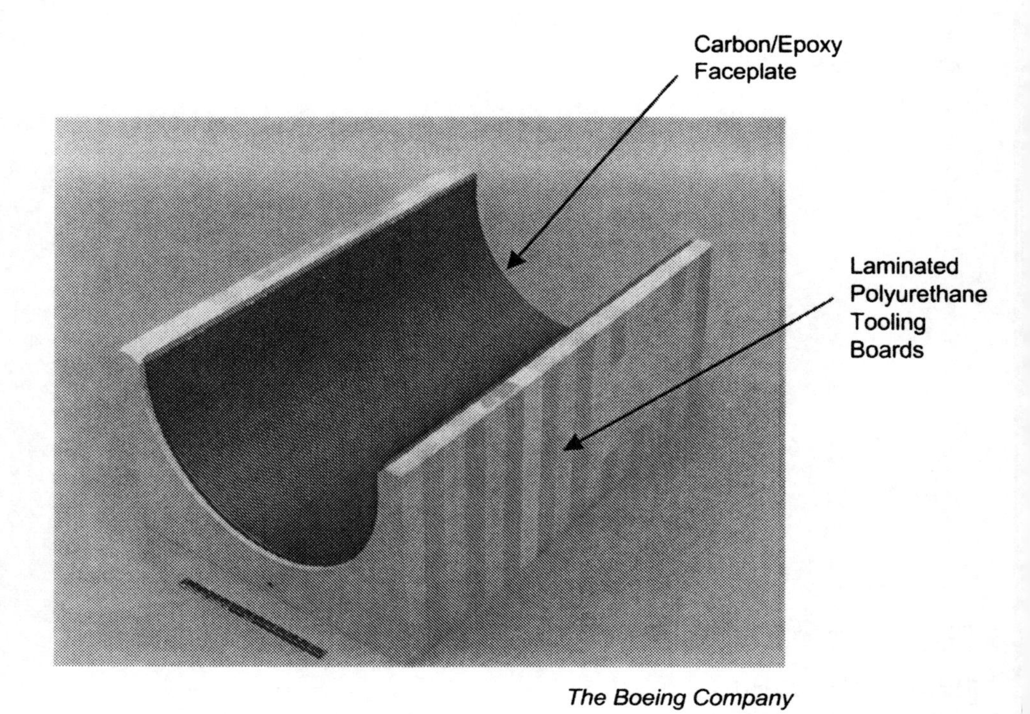

Fig. 23. Master Pattern Containing Cured Carbon/Epoxy Faceplate

Table 4.2 Comparison of Process Steps for Composite Tools

ACCURACY	NUMBER OF STEPS	MASTER PATTERN	MASTER MOLD	INTERMEDIATE OR PRODUCTION TOOL		
MOST ACCURATE	1 STEP		ROOM TEMP/HIGH TEMP PREPREG MOLD			
	2 STEPS		PFP	200°F PREPREG MOLD		
	3 STEPS		PFP	ROOM TEMP/HIGH TEMP INTERMEDIATE MOLD	200°F OR 350°F PREPREG MOLD	
	3 STEPS		SPLASH	PFP	200°F PREPREG MOLD	
LEAST ACCURATE	4 STEPS		SPLASH	PFP	ROOM TEMP/HIGH TEMP INTERMEDIATE MOLD	200°F OR 350°F PREPREG MOLD

Reference 1: with permission

The use of a PFP to fabricate an electroformed nickel tool is shown in Fig. 24. The PFP or plaster splash is used to make a laminated tool, usually glass/polyester (0.4–0.5 inch thick) that is then coated with a thin layer of silver so that it will be conductive when placed in the electroforming bath. It is important that the electroforming mandrel remain stable during the plating process. If the tool is large, the plating mandrel is frequently supported with a substructure to insure stability. The surface of the plating mandrel must be as smooth as the desired tool smoothness, since the plating procedure will reproduce the mandrel smoothness. The electroforming procedure usually takes several weeks to build up the required faceplate thickness, usually around 0.2–0.4 inch thick. The plating bath temperature is approximately 120 °F. After removal, the faceplate is attached to the substructure before the plating mandrel is removed. Electroforming is capable of making quite large tools, essentially limited by the size of the plating bath. Tools with greater than 200 ft^2 of surface area have been fabricated. Sharp corners and recesses are difficult to plate uniformly; the plating tends to build up excessively in these areas. Fittings, tooling pin locators, vacuum connectors and threaded pins for attaching the substructure can be embedded during the plating operation to provide

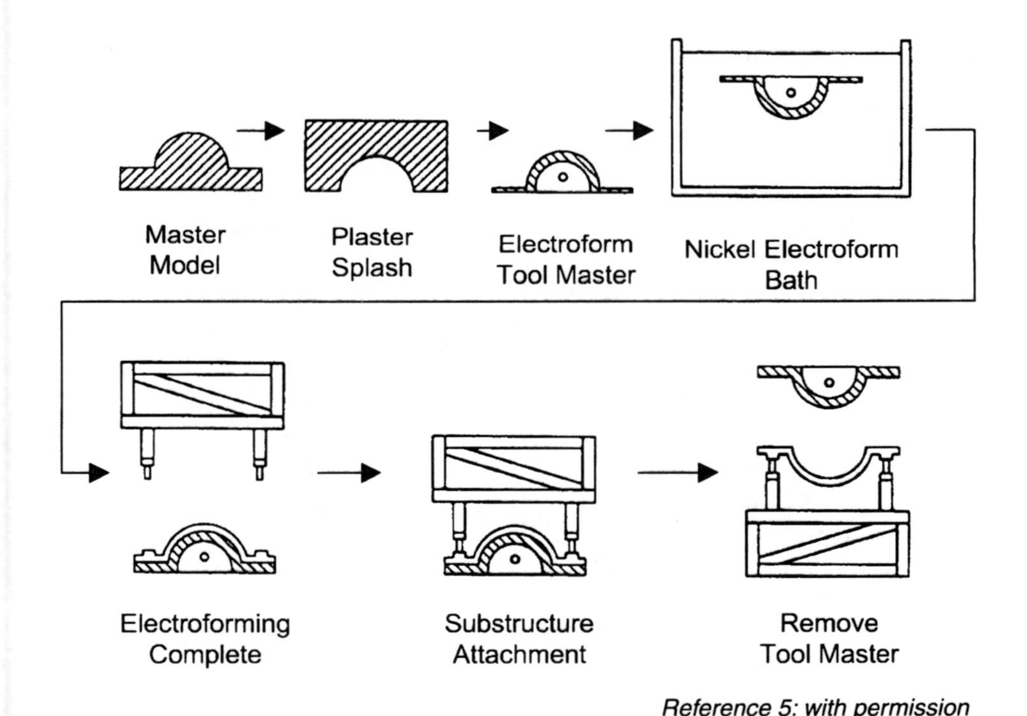

Fig. 24. Process Flow for Electroformed Nickel Tools

vacuum and pressure-tight connections. In general, nickel has better damage tolerance than aluminum or composite tools and is more scratch resistant. Electroformed nickel tools can be repaired by soldering or brazing; however, this is generally more difficult than the weld repair of a steel or Invar tool. The lead in some repair solders can react with the nickel causing joint cracking and leaks. After plating, the tool is polished to give a smooth laminating surface If a superior surface hardness is required, the tool surface can be chrome plated.

The fabrication sequence for a composite tool is shown in Fig. 25. Again a PFP is used to lay-up and cure a carbon or glass laminate. For aerospace tools, epoxies are normally used, while polyesters and vinyl esters dominate commercial applications. The predominate reinforcements are twill, plain or satin weave carbon or glass fabrics. Balanced, symmetric quasi-isotropic lay-ups are normally used in which the plies are cut into large rectangles and then butt spliced during lay-up. Finer fabrics can be used on the surface to improve surface finish, whereas heavier fabrics are used internally to reduce lay-up time. For vacuum bag cured tools, gel coats may also be useful in improving surface finish, while prepreg films or adhesive layers are often used for autoclave cured tools. Periodic vacuum

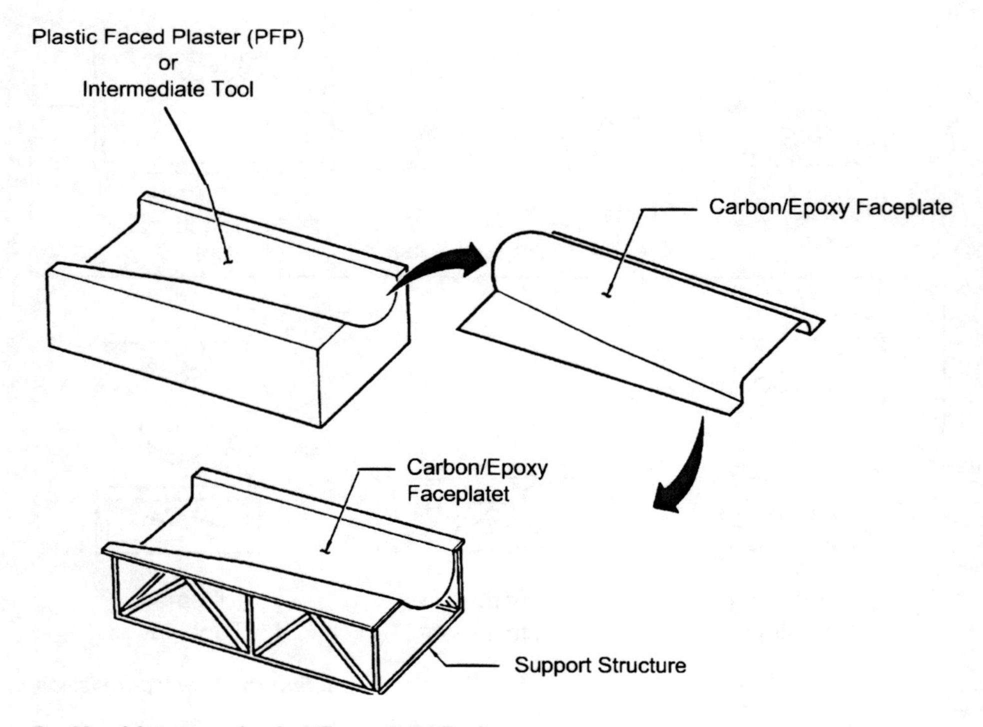

Fig. 25. fabrication of carbon/Epoxy Bond Tool

debulking to consolidate the lay-up generally improves tool quality. Several different material systems have been used for fabricating composite tools: (1) wet lay-up in which a liquid resin is used to impregnate dry cloth layer by layer that is cured at room or elevated temperature; (2) high temperature/high temperature (HT/HT) prepregs in which the tool is again laid-up layer by layer and then cured at elevated temperature (e.g., 350 °F) followed by elevated temperature post-cure (e.g., 350 °F); and more recently (3) low temperature/high temperature (LT/HT) prepreg that is laid up and cured at low temperatures (e.g., 150 °F) followed by a high-temperature (e.g., 350 °F) post-cure. If the curing is conducted at elevated temperature, it can be accomplished either in an oven or in an autoclave.

The highest-quality tools are made using either HT/HT or LT/HT prepregs that are autoclave cured to give maximum compaction and low void contents. Voids need to be avoided in tool facesheets, because they can serve as initiation points for microcracks that will eventually propagate through the thickness, resulting in potential leak paths. The biggest advantage of the LT/HT systems is that they can initially be cured on low-temperature capability PFPs and then removed from the PFP for the elevated temperature post-cure. On the other hand, HT/HT prepregs require an intermediate or facility tool with greater temperature tolerance than a PFP. To prevent distortion during the post-cure, the egg crate support structure is usually attached to the faceplate prior to post-cure. Details of a typical egg crate support structure are shown in Fig. 26. The egg crate support structure can be made from honeycomb laminates, carbon/epoxy tooling board, or constructed from carbon/epoxy prefabricated tubes. Alternatives to egg crate support systems include stiffeners molded to the backside of the tool and tubular support systems. To help protect the edges of the faceplate from handling damage, they should either be even, slightly recessed or have rolled edges. A typical faceplate is usually about 0.25 inch thick, which, in the author's opinion, is one of the major problems with leakage during service. Another potential problem with leakage is putting holes in the faceplate for vacuum connections and manifolds. Although molded-in pass-throughs are better than ones that are potted in after cure, an even better approach is to never put a hole in a composite faceplate and use through-the-vacuum bag fittings for evacuation and static readings. Thicker faceplates with no penetrations are more tolerant of both handling damage and thermal cycling. When a leak does develop in a composite tool, it can be extremely difficult to locate and repair. The leak may start at an edge or surface and propagate in a winding path down through the laminate layers and exit several feet away from where it started.

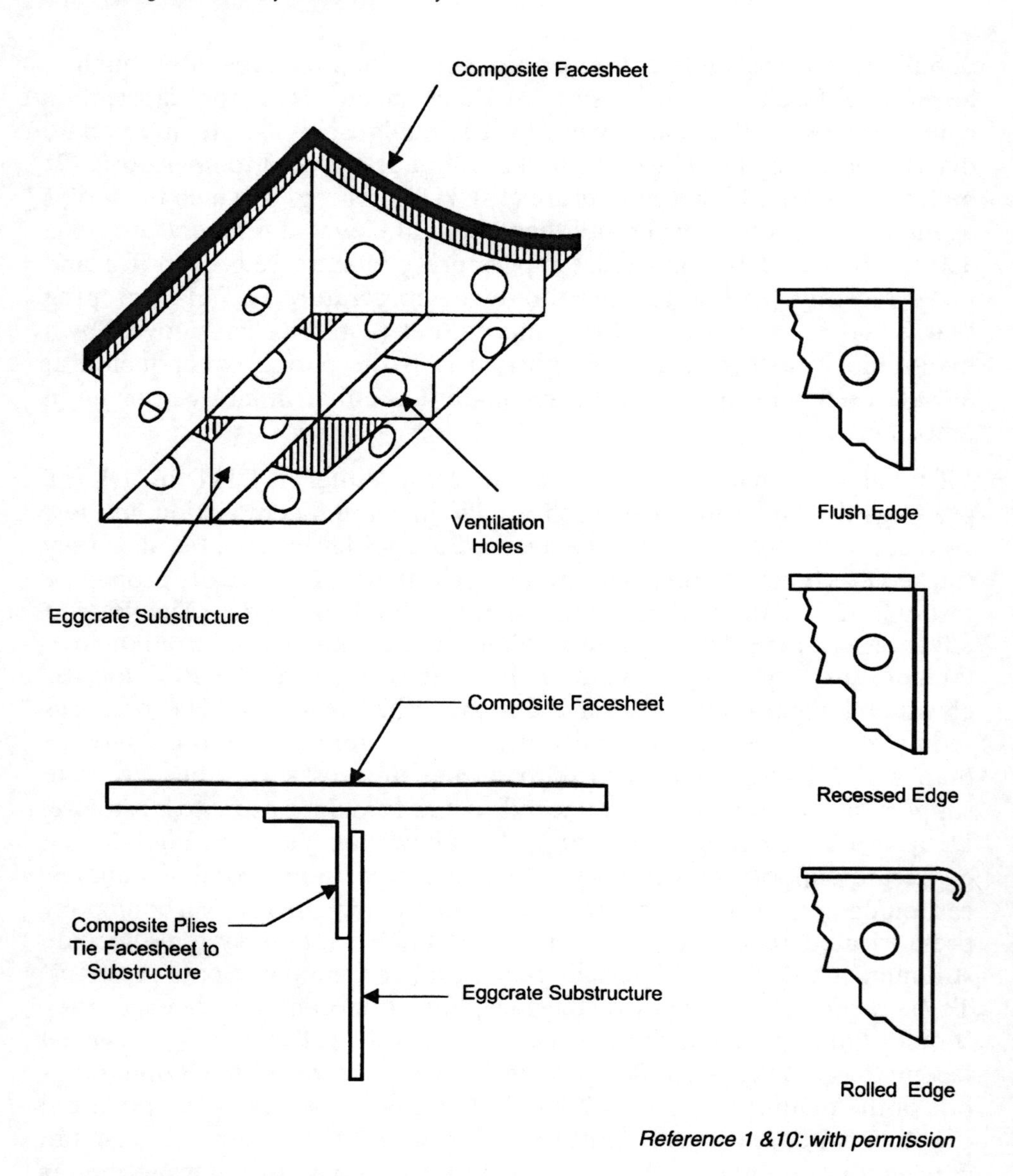

Fig. 26. Eggcrate Support for Composite Tools

If high processing temperatures are required, e.g., 500–700 °F for high-temperature polyimides or thermoplastic composites, either steel or Invar 42 tooling can be used. However, monolithic graphite and mass cast ceramics are also choices. Blocks of monolithic graphite can be bonded together and then NC machined to final contour, while mass cast ceramics and mixed, poured and cured . Both of these materials have excellent temperature capability and low CTEs. Monolithic graphite is lightweight, easy to machine (but messy), has a high thermal conductivity and can be used to temperatures up to 800 °F in air. Ceramics are also relatively light weight; however, their thermal conductivities are lower and they must be thoroughly surface sealed prior to use. Ceramic tools are also more difficult to fabricate than monolithic graphite due to their slow cures and their tendency to crack during cure. Reinforcing bars or grids are often embedded in the casting to strengthen it. The biggest drawback to both monolithic graphite and cast ceramics is that they are very brittle materials and are susceptible to breakage and handling damage.

Elastomeric tooling materials, such as silicone rubber, butyl rubber and fluoroelastomers, are frequently used for caul pads and pressure intensifiers. They are used to either intensify pressure or redistribute pressure during the cure cycle. They are frequently used in areas where it is difficult to vacuum bag and there is a danger of the bag bridging, such as radii in corners. Elastomers can be used in one of two ways (Fig. 27): (1) the fixed volume method in which the elastomer is totally contained within a hard tool and expands during heating to apply pressure to the part or (2) the variable volume method in which the elastomer is allowed to vent or

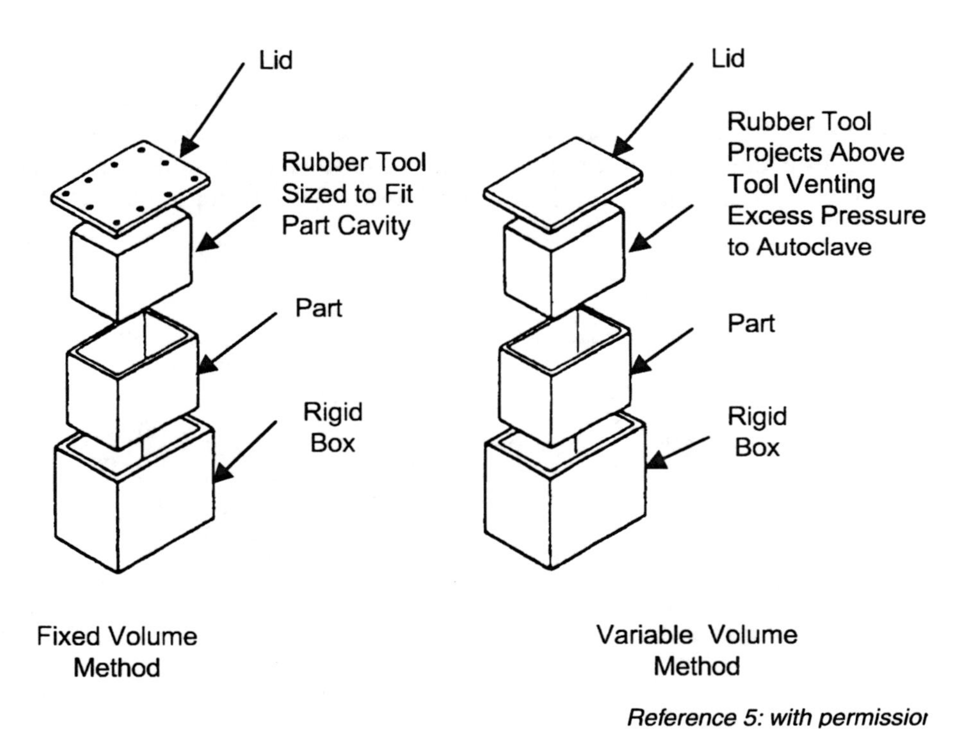

Fig. 27. Fixed Volume and Variable Volume Elastomeric Tooling Methods

release excess pressure against the autoclave pressure (usually 85–100 psig). The problem with the fixed volume method is that it requires quite precise calculations of the volume of rubber required or it can result in either too much or too little pressure. Methods for calculating the required amount of rubber can be found in Ref. 5. The fabrication sequence for a typical caul pad is shown in Fig. 28. In this sequence the caul pad is made from a calendered elastomeric sheet stock that is placed over a "dummy" part and cured on the bond tool. A dummy part is simply a first article part made on a tool without an intensifier. After cure, any surface wrinkles are filled and splined smooth. It is then used to make the caul pad. If a dummy part is unavailable, the part thickness can be built up with layers of special tooling waxes. Elastomeric tooling materials generally come as either calendered sheet stock or liquids that can be cast to shape in molds and room-temperature vulcanized (RTV) or cured. The life of elastomeric tooling details is limited, usually less than 30 autoclave cycles. Resins tend to attack some elastomers, and the continual heat cycling causes shrinkage and inbrittles them leading to cracking and tearing. Layers of carbon cloth are often embedded into elastomeric intensifiers to provide local stiffening and improve the durability. Elastomers also have low thermal conductivities and can act as heat sinks if large masses are used in a local area.

For flat or mildly contoured parts, it is common practice to use a caul sheet or caul plate on the non-tooled surface (i.e., the bag side) to provide improved surface finishes. Caul plates can be made from metal, reinforced glass or carbon or elastomers. It is important that the caul plate is flexible

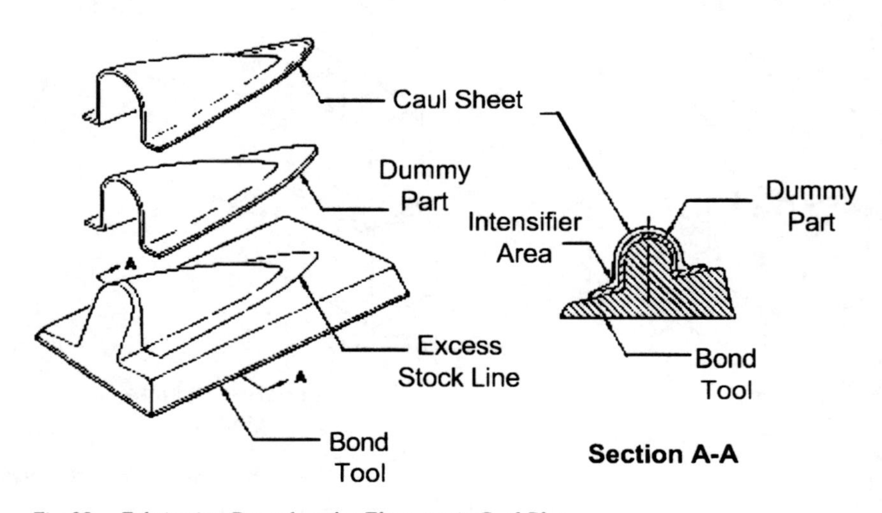

Fig. 28. Fabrication Procedure for Elastomeric Caul Plate

enough that local surface bridging does not occur. The caul plates may be perforated to allow air evacuation and, in some cases, resin bleeding.

4.4 Summary

There are many requirements a tool designer must consider before selecting a tooling material and fabrication process for a given application; however, the number of parts to be made on the tool and the part configuration are usually the overriding factors in the selection process.

Selection of the material used to make the tool is an important consideration. Normally, reinforced polymers can be used for low to intermediate temperatures, metals for low to high temperatures, and monolithic graphite or ceramics for very high temperatures. Tools for autoclave curing have traditionally been made of either steel or aluminum. Electroformed nickel became popular in the early 1980s followed by the introduction of carbon/epoxy and carbon/bismaleimide composite tools in the mid-1980s. Finally, in the early 1990s, a series of low-expansion iron–nickel alloys was introduced under the trade names Invar and Nilo.

Steel is a fairly cheap material with exceptional durability. It is readily castable and weldable. It has been known to withstand over 1,500 autoclave cure cycles and still be capable of making good parts. However, steel is heavy, has a higher CTE than the carbon/epoxy parts usually built on it and, for large massive tooling, can experience slow heat-up rates in an autoclave.

Aluminum is much lighter and has a much higher coefficient of thermal conductivity than steel. It is also much easier to machine but is more difficult to produce pressure-tight castings and weldments.

Electroformed nickel has the advantages that it can be made into complex contours and does not require a thick faceplate. When backed with an open tubular-type substructure, this type of tool experiences excellent heat-up rates in an autoclave. However, to make an electroformed nickel tool requires a mandrel be fabricated to the exact contour of the final tool.

Carbon/epoxy, carbon/bismaleimide and glass/epoxy tools also require a master or mandrel for lay-up during tool fabrication. A distinct advantage of carbon/epoxy and carbon/bismaleimide tools is that their CTEs can be tailored to match the carbon/epoxy parts they build. In addition, composite tools are relatively light, exhibit good heat-up rates during autoclave curing, and a single master can be used to fabricate duplicate tools. On the downside, there has been a lot of negative experience with composite tools that are subject to 350 °F autoclave cure cycles. The matrix has a tendency to crack and, with repeated thermal cycles, it develops leaks.

Invar and the Nilo series of alloys were introduced in the early 1990s. Being low-expansion alloys, they very closely match the CTE of the carbon/epoxy parts. Their biggest disadvantages are cost and slow heat-up rates. The material itself is very expensive and it is more difficult to work with than even steel. It can be cast, machined and welded.

Since many common tooling materials, such as aluminum and steel, expand at greater rates than the carbon/epoxy part being cure on them, it is necessary to correct their size or compensate for the differences in thermal expansion. Another correction required for tooling for parts with geometric complexity is spring-in or close up during the cure process. Therefore, it is necessary to compensate angled parts. The degree of compensation required is somewhat dependent on the actual lay-up orientation and thickness of the laminate.

Tooling is a major up-front non-recurring cost for any program. There is often the temptation to skimp on tooling to save money. This is truly false economy. A well- designed and fabricated tool is crucial to achieving part quality in production. Tool design and tool fabrication requires the same level of skill as required for the engineering design of the part. This rather short chapter did not attempt to cover all aspects of the design and fabrication process but rather gave an introduction to the complexity of the process and its potential effects on composite processing.

References

[1] Morena J.J., *Advanced Composite Mold Making*, Van Nostrand Reinhold, 1988.

[2] Morena, J.J., “Mold Engineering and Materials–Part I,” *SAMPE Journal* **31**(2), March/April 1995, pp. 35–40.

[3] Morena J.J., “Advanced Composite Mold Fabrication: Engineering, Materials, and Processes–Part II,” *SAMPE Journal* **31**(3), May/June 1995, pp. 83–87.

[4] Morena J.J., “Advanced Composite Mold Fabrication: Engineering, Materials, and Processes–Part III,” *SAMPE Journal* **31**(6), November/December 1995, pp. 24–28.

[5] *Volume 1 – Engineered Materials Handbook: Composites*, ASM International, 1987.

[6] *Volume 21 – ASM Handbook: Composites*, ASM International, 2001.

[7] “The Nilo Nickel-Iron Alloys for Composite Tooling,” Inco Alloys International, 1998.

[8] Griffith J.M., Campbell F.C., Mallow A.R., “Effect of Tool Design on Autoclave Heat-up Rates,” SME Composites in Manufacturing 7, December 1987.

[9] Black S., “Epoxy-based Pastes Provide Another Choice for Fabricating Large Parts,” High-Performance Composites, January/February 2001, pp. 20–24.

[10] Niu M.C.Y., Composite Airframe Structures: Practical Design Information and Data, Conmilit Press, Hong Kong, 1992.

Chapter 5

Ply Collation: A Major Cost Driver

Cutting and ply collation are the major cost drivers in composite part fabrication, normally accounting for 40–60% of the cost depending on part size and complexity. Ply collation or lay-up can be accomplished by hand, automated tape laying, filament winding or fiber placement. Hand or manual lay-up is generally the most labor intensive method but may be the most economical if the number of parts to be built is limited, the part size is small or the part configuration is too complex to automate. Automated tape laying (ATL) is advantageous for flat or mildly contoured skins, such as large thick wing skins. Filament winding is a high-rate process that is used primarily for bodies of revolution or near bodies of revolution. Fiber placement is a hybrid process that possesses some of the characteristics of ATL and filament winding. It was developed to allow the automated fabrication of large parts which could not be fabricated by either tape laying or filament winding.

5.1 Prepreg Control

Composite thermoset prepregs and adhesives are perishable and must be stored in a refrigerator. They will advance or age when stored at room temperature. Several things happen when the resin advances: (1) there is a noticeable loss of tack that can make the plies hard to lay-up; (2) the prepreg becomes boardy and stiff; (3) during cure, there will be less resin flow which can result in thicker than desired parts; and (4) in some systems, the resin may not properly cure if taken to extreme out-times. As the resin ages at room temperature, the curing agent slowly reacts with the base resin. The effect of out-time on resin flow is illustrated in Fig. 1, which shows that the viscosity decreases with increasing out-time at room temperature[1]. Therefore, prepregs are shipped from the supplier wrapped in plastic in refrigerated trucks or packed in boxes with dry ice. Since prepregs and adhesives have room temperature out-times or shelf lives, usually 10–30 d, it is important to immediately store them in refrigerated containers, usually 0 °F or lower on receipt from the vendor. Once a roll is removed from the refrigerator for use, the time of removal should be documented. Likewise, when it is repackaged and returned to the freezer, the time is again documented. This allows a record to be kept of the total cumulative out-time of the material. After removing a roll from the freezer, it is important to let the roll warm to room temperature before removing the protective wrapping material. If the roll is opened prematurely, moisture can condense on the material possibly causing a void or porosity problem when it is cured. Freezer life is normally 3–12 months from the date of manufacture, depending upon the particular resin system used.

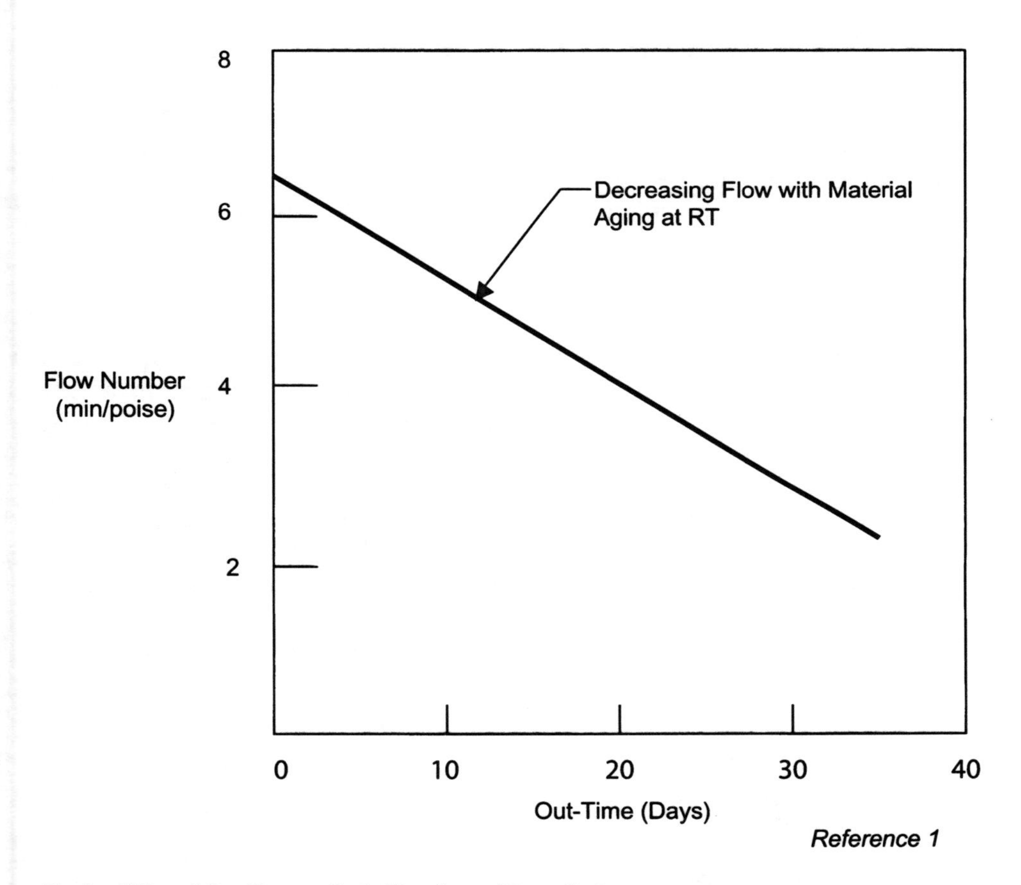

Fig. 1. Effect of Out-Time on Resin Flow for an Epoxy Resin

5.2 Lay-up Room Environment

All prepreg and adhesive films absorb moisture from the atmosphere. Absorbed moisture can cause voids and porosity when the laminate is heated for cure, i.e., the absorbed moisture comes out of solution when heated above the boiling point of water (212 °F) and forms voids and porosity. Therefore, prepreg lay-up operations should be conducted in air-conditioned and humidity controlled rooms. The lay-up rooms should also contain a slight positive pressure to prevent dust and dirt from entering the room when the doors are opened. The pressure for clean rooms is maintained by filtered air kept at a slight positive pressure by blowers.

5.3 Tool Preparation

All tools should be thoroughly cleaned and inspected prior to lay-up. Nicks and scratches on the tool surface can translate into surface defects on the cured laminate. The tool needs to be thoroughly coated with mold release to prevent resin adhesion during cure. Normally, new tools are coated with a permanent mold release. Then, prior to each use, another coat of mold release is applied to the tool surface. Many types of mold release agents have been used, such as waxes, silicone compounds and fluorocarbon compounds. Waxes and silicones have been known to evolve oils and gases during cure so they should be used with caution.

5.4 Manual Lay-up

Manual hand collation (Fig. 2) is conducted using either prepreg tape (24 in maximum width) or broadgoods (60 in maximum width). Prior to the actual lay-up, the plies are usually precut and kitted into ply packs for the part. The cutting operations are normally automated unless the number of parts to be built does not justify the cost of programming an automated ply cutter. However, if hand cutting is selected, templates to facilitate the cutting operation may have to be fabricated. If the lay-up has any contour of the plies, the contour will have to be factored into the templates. Templates for hand cutting are typically made of either steel or aluminum. Although aluminum templates are easier and less expensive to make than

The Boeing Company

Fig. 2. Typical Hand Collation of a Carbon/Epoxy/Part

steel, it is not uncommon for small particles of aluminum to be chipped off the template during the cutting operation and become embedded in the lay-up. Steel-bladed carpet knives, round blade pizza cutters and seam rippers are normally used for manual cutting operations. The operator lays the template on the prepreg in the correct orientation and cuts around the periphery. The primary disadvantages of this method are high labor costs, rather poor material utilization and the chance of orienting the template in the wrong direction on the prepreg during the cutting operation.

Automated ply cutting of broadgoods, usually 48–60 in wide material, is the most prevalent method used today. Over the past 30 years, a number of different automated cutting techniques have been used: lasers, steel rule dies, water jets, reciprocating knives and ultrasonics. Laser ply cutters were one of the first automated systems developed because they could successfully cut boron fibers, which were used quite extensively in the late 1960s and early 1970s. The primary disadvantages of laser ply cutters were that they were limited to cutting one ply thickness at a time due to the small focal point of the laser beam, and the high heat generated in the process left a cured crust around the ply edges. Steel rule die cutters, in which as many as 10–15 plies could be cut at one time, were popular for a while in the mid-1970s; however, steel rule die cutters are permanent tools that must be reworked if a ply or part configuration changes. A number of composite fabricators have used water jets, in which a high-pressure (up to 60,000 psi) stream of water is forced through a small nozzle to cut prepreg plies. A concern with this process is the potential contamination of the prepreg with water that could cause a problem during the curing operation, although no objective evidence has ever been reported that the small amount of water on the plies has caused any problems during curing.

Reciprocating knives and ultrasonically driven cutters are the most prevalent ply cutting methods currently used in the composites industry. The reciprocating knife concept originated in the garment industry. In this process, a carbide blade reciprocates up and down, similar to a saber saw, while the lateral movement is controlled by a computer- controlled head. To allow the blade to penetrate the prepreg, the bed supporting the prepreg consists of nylon bristles that allow the blade to penetrate during the cutting operation. With a reciprocating knife cutter, normally one to five plies can be cut during a single pass. A bed containing a number of cut plies is shown in Fig. 3.

The ultrasonic ply cutter operates in a similar manner; however, the mechanism is chopping rather than a cutting action. Instead of a bristle bed that allows the cutter to penetrate, a hard plastic bed is used with the ultrasonic method. A typical ultrasonic ply cutter shown in Fig. 4 can cut at speeds approaching 2,400 fpm while holding accuracies of ±0.003 inch. Prior to cutting with either a reciprocating knife or ultrasonic ply cutter, the

The Boeing Company

Fig. 3. Reciprocating Knife Plies After Cutting

The Boeing Company

Fig. 4. Large Ultrasonic Ply Cutter

bed is covered with a plastic film, and a vacuum is applied. The broadgoods are rolled out onto the bed and covered with an additional layer of release paper. Multiple layers of broadgoods may be applied at this point. Finally, an additional layer of plastic film is placed over the stack and a vacuum is applied to hold the stack flat during cutting.

One of the primary advantages of any of the automated methods (laser, water jet, reciprocating knife, and ultrasonic) is that they can be programmed offline, and nesting routines are used to maximize the material utilization. A typical nest for ply cutting is shown in Fig. 5. In

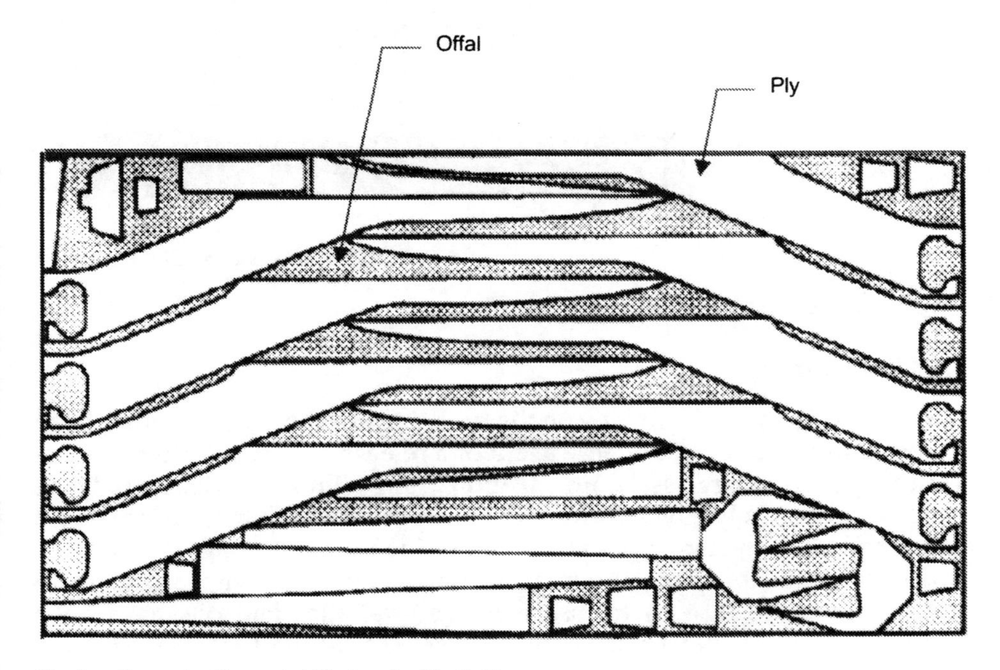

Fig. 5. Computer Generated Nesting for Ply Cutting

addition, many of these systems have automated ply labeling systems (Fig. 6) in which the ply identification label is placed directly on the release paper. A typical ply label will contain both the part number and the ply identification number. This makes the sorting and cutting operations much simpler. Modern automated ply cutting equipment is fast and produces high-quality cuts.

If the plies are going directly to lay-up, they are frequently sorted by part and ply number and placed in racks on carts that are then rolled to the lay-up area. If they are not going to be used immediately, they can be sorted, placed in sealed plastic bags and placed back in the freezer.

The Boeing Company

Fig. 6. Automated Ply Labeling Head

5.5 Ply Collation

For ply-by-ply collation directly on the tool, the tool should have either been coated with a liquid mold release agent or a release film, such as non-porous Teflon-coated fiberglass cloth, Teflon film (fluoroethyleneproplyene (FEP)), Tedlar (polyvinyl fluoride (PVF)) or silicone rubber. Caution should be exercised when using silicone rubber inside a vacuum bag. It should be baked at least 100 °F higher than the maximum cure temperature for a sufficient time (e.g., 4 h) to insure that all volatiles and oils have been removed. Some lay-ups also require a peel ply on the tool surface if the surface is going to be painted or adhesively bonded after cure. Peel plies are normally nylon, polyester or fiberglass fabrics. Some are coated with release agents and some are not. It is important to thoroughly characterize any peel ply material that is bonded to a composite surface, particularly if that surface is going to be structurally adhesively bonded in a subsequent operation.

The plies are placed on the tool in the location and orientation as specified by the engineering drawing or shop work order. Prior to placing a ply onto the lay-up, the operator should make sure that all of the release paper is removed and that there are no foreign objects on the surface. Large Mylar (clear polyester film) templates are often used to define ply location and orientation. However, these are quite bulky and difficult to use and are rapidly been displaced by laser projection units. These units, shown schematically in Fig. 7, use low-intensity laser beams to project the

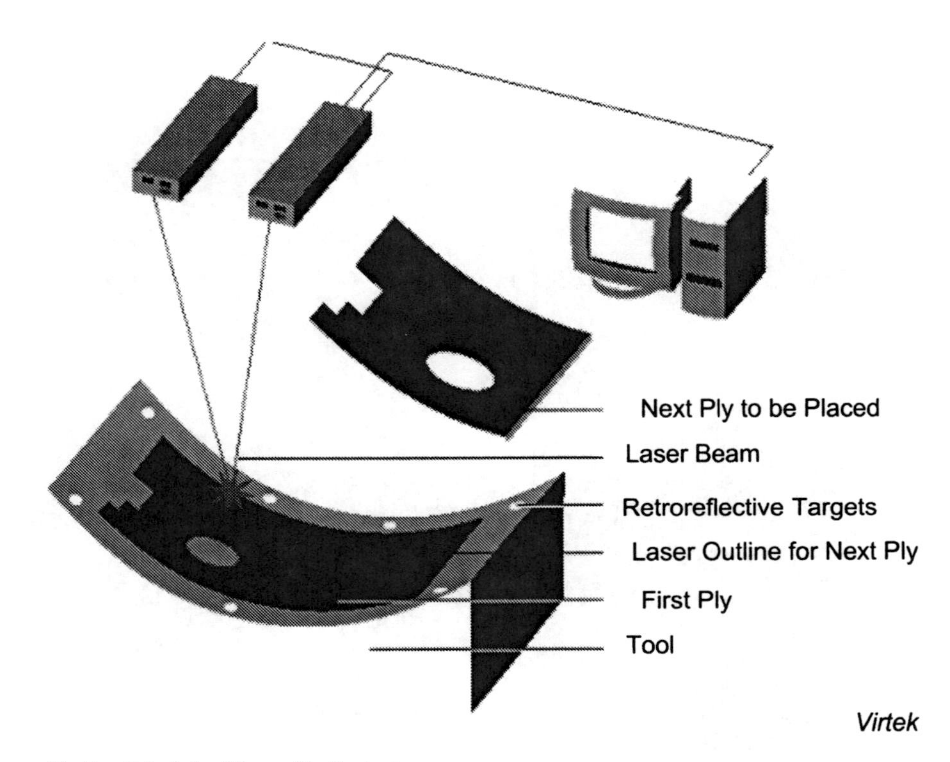

Fig. 7. Principle of Laser Ply Projection

ply periphery on the lay-up. They are programmed offline using the CAD data for each ply and with advanced software are capable of projecting ply locations on both flat and highly contoured lay-ups. The accuracy is generally in the ±0.015-0.040 in range, depending on the projection distance required for the part.[2] An actual projection system is shown in Fig. 8.

Ply location accuracy requirements are normally specified on the engineering drawing or applicable process specification. For unidirectional material, gaps between plies (Fig. 9) are normally restricted to 0.030 inch and overlaps and butt splices are not permitted. For woven cloth, butt splices are usually permitted but require an overlap of 0.5–1.0 inch. The engineering drawing should control the number of ply drop-offs at any one location in the lay-up.

During collation, the operator should make sure that the plies are tacked down uniformly and not distorted, wrinkled or bridged. Teflon squeegees or rollers are often used during lay-up to facilitate removal of wrinkles and air bubbles. Hot air guns or irons that produce material temperatures less than 150 °F can be used to help tack the plies together during collation. For parts with concave corners, it is important that the plies fit down into the radius and do not bridge. If they bridge during

The Boeing Company

Fig. 8. Laser Ply Projection Being Used to Locate Plies

lay-up, it will usually result in resin-rich corners that can contain porosity and voids after cure.

The lay-up should be vacuum debulked every three to five plies or more often if the shape is complex. Vacuum debulking consists of covering the lay-up with a layer of porous release material, applying several layers of breather material, applying a temporary vacuum bag and pulling a vacuum for a few minutes. This helps to compact the laminate and remove entrapped air from between the plies. For some complex parts, hot debulking or prebleeding in an oven under a vacuum or autoclave pressure at approximately 150–200 °F can be useful for reducing the bulk factor. Prebleeding is similar to hot debulking except that in prebleeding some of the resin is intentionally removed with the addition of bleeder cloth, while in hot debulking no resin is intentionally removed.

Dermatitis can be a problem when handling epoxy and other polymeric resin systems. Either protective gloves or specialized hand creams can be used to protect the worker's hands during lay-up. However, any material

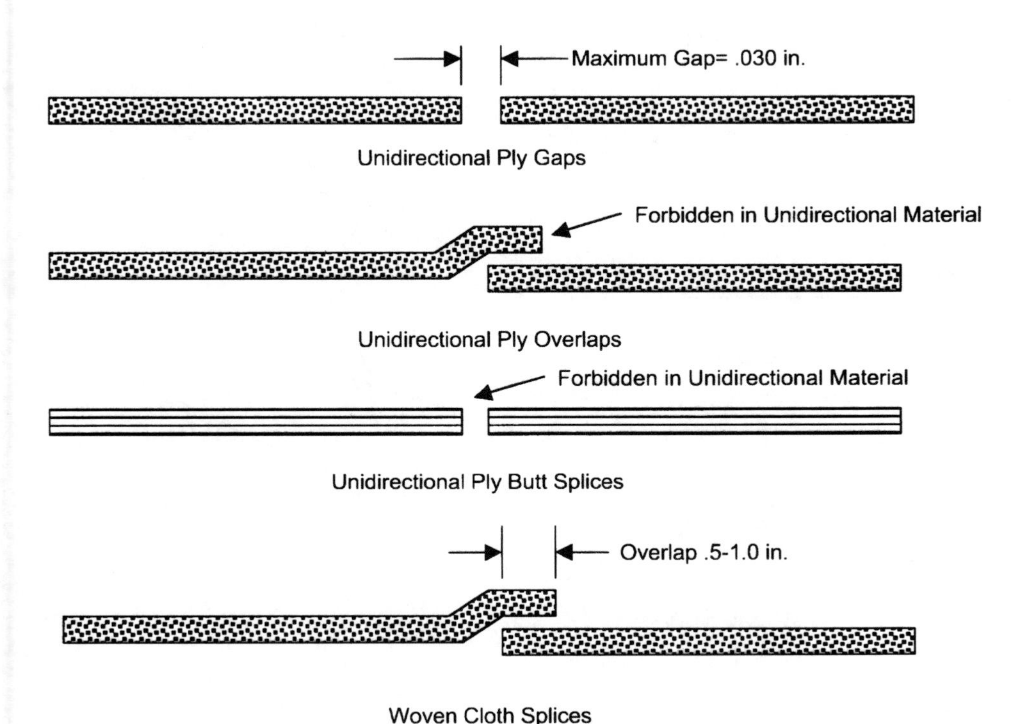

Fig. 9. Composite Lay-Up Allowances

that can potentially come in contact with the lay-up should be chemically tested to make sure that they will not cause a contamination problem.

5.6 Flat Ply Collation and Vacuum Forming

To lower the cost of ply-by-ply hand collation directly to the contour of the tool, a method called flat ply collation was developed in the early 1980s. This method, shown schematically in Fig. 10, consists of manually collating the laminate in the flat condition and then using a vacuum bag to form it to the contour of the tool. If the laminate is thick, this process may have to be done in several steps (Fig. 11) to prevent wrinkling and buckling of the ply packs. Heat (<150 °F) can be used to soften the resin and aid in forming if the contour is severe. An example of flat ply collation for a rather large part is the AV-8B upper and lower wing skins. This part, shown in Fig. 12, is 28 ft tip-to-tip and has a maximum chord dimension of 14 ft. Thickness ranges from 0.104 inch at the tips to 0.478 inch at the inboard buildups. To successfully flat ply collate this part, the skin laminate was divided into 48 precollated ply packs, ranging in thickness from 3 to 17 plies.[3] To accommodate the severe transition at the centerline, an overlapping splice design was incorporated as shown in Fig. 13. Each

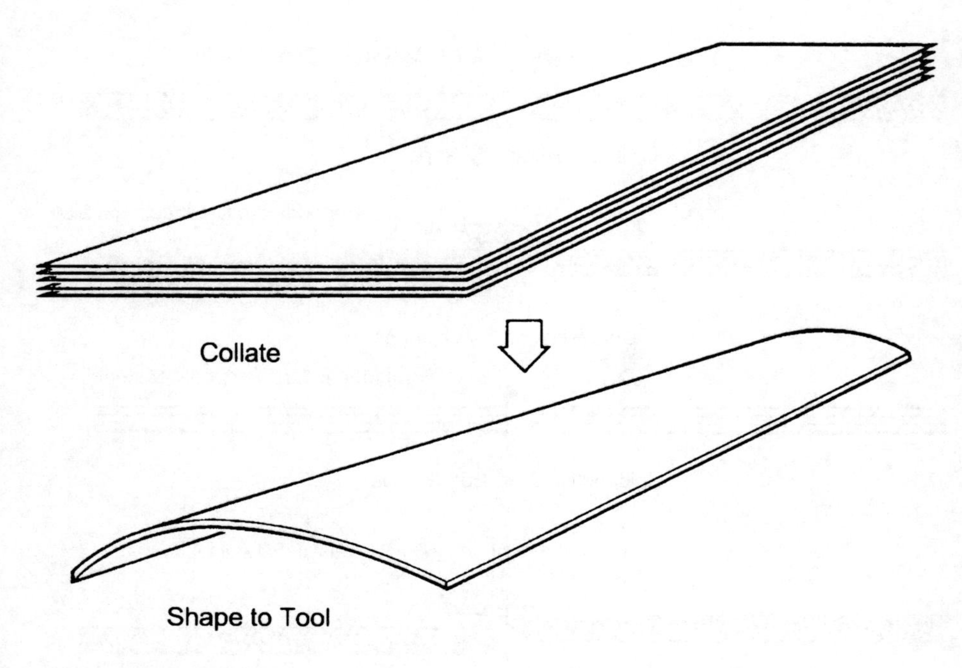

Fig. 10. Flat Ply Collation

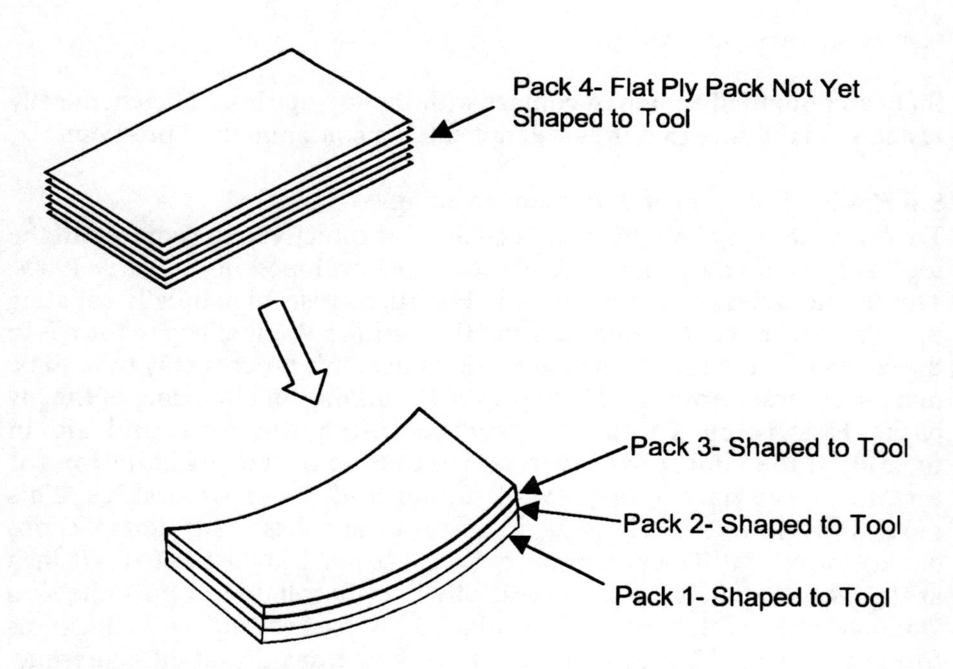

Fig. 11. Step Wise Flat Ply Collation

The Boeing Company

Fig. 12. AV-8B Wing Skin Undergoing Ultrasonic Inspection

pack was flat ply collated on a bench and then moved to the tool where it was vacuum formed and debulked. Low-temperature hot-air guns were used to facilitate forming in some of the areas containing higher contour.

This process has also been used to successfully make substructure parts, such as C-channels as shown in Fig. 14. Normally woven cloth is used and the parts are again flat ply collated, placed on a form block tool, covered with a release film and then vacuum formed to shape using a silicone rubber vacuum bag. Note that it is important to keep the fibers in tension during the forming process. If compression occurs the fibers will wrinkle and buckle. To maintain uniform tension during the forming operation, a double diaphragm forming technique[4] can be used in which the plies are sandwiched between two thin flexible diaphragms pulled together with a vacuum. Again, the application of low heat to soften the resin and aid in forming is quite prevalent. After cure, these long parts may be trimmed into shorter lengths, thus saving the cost of laying up each individual part on its individual tool.

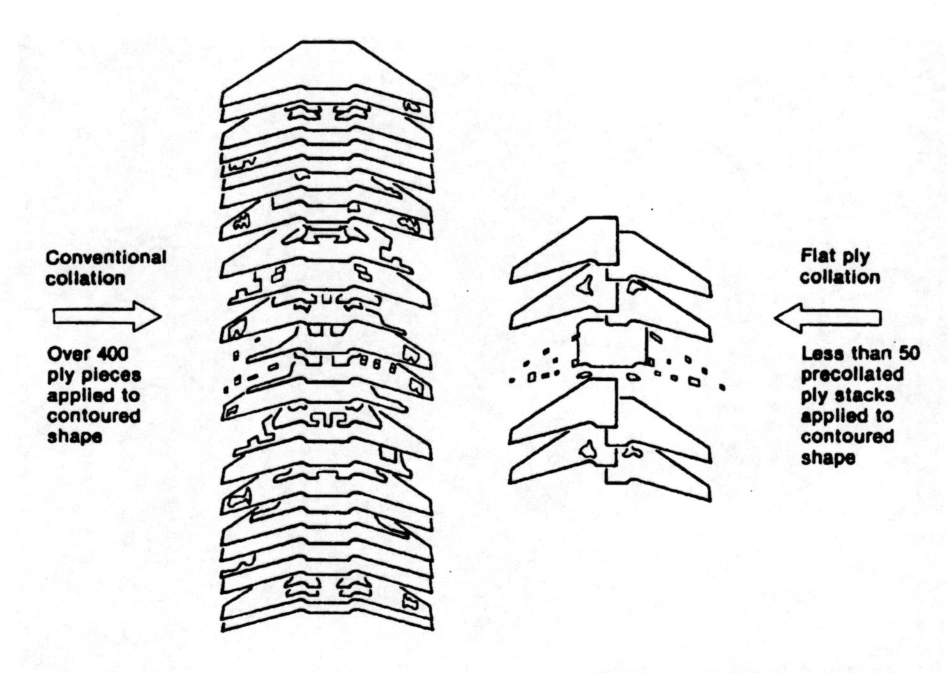

Fig. 13. Ply Pack Splice Configuration for AV-8B Wing Skin

5.7 Automated Tape Laying

Automated tape laying (ATL) is a process that is very amenable to large flat parts, such as wing skins. Tape layers usually lay down either 3, 6 or 12 inch wide unidirectional tape, depending on whether the application is for flat structure or mildly contoured structure. Automated tape layers are normally gantry style machines (Fig. 15) that can contain up to 10 axes of movement.[5] Normally five axes of movement are associated with the gantry itself and the other five axes with the delivery head movement. Commercial tape layers can be configured to lay either flat or mildly contoured parts. A typical tape layer consists of a large floor mounted gantry with parallel rails, a cross-feed bar that moves on precision ground ways, a ram bar that raises and lowers the delivery head, and the delivery head that is attached to the lower end of the ram bar. Flat tape laying machines (FTLMs) can be configured as either a fixed bed machine or open bay gantries, while contour tape laying machines (CTLMs) are normally configured as open bay gantries. The tool is rolled into the working envelope of the gantry, secured to the floor and the delivery head is initialized onto the working surface.

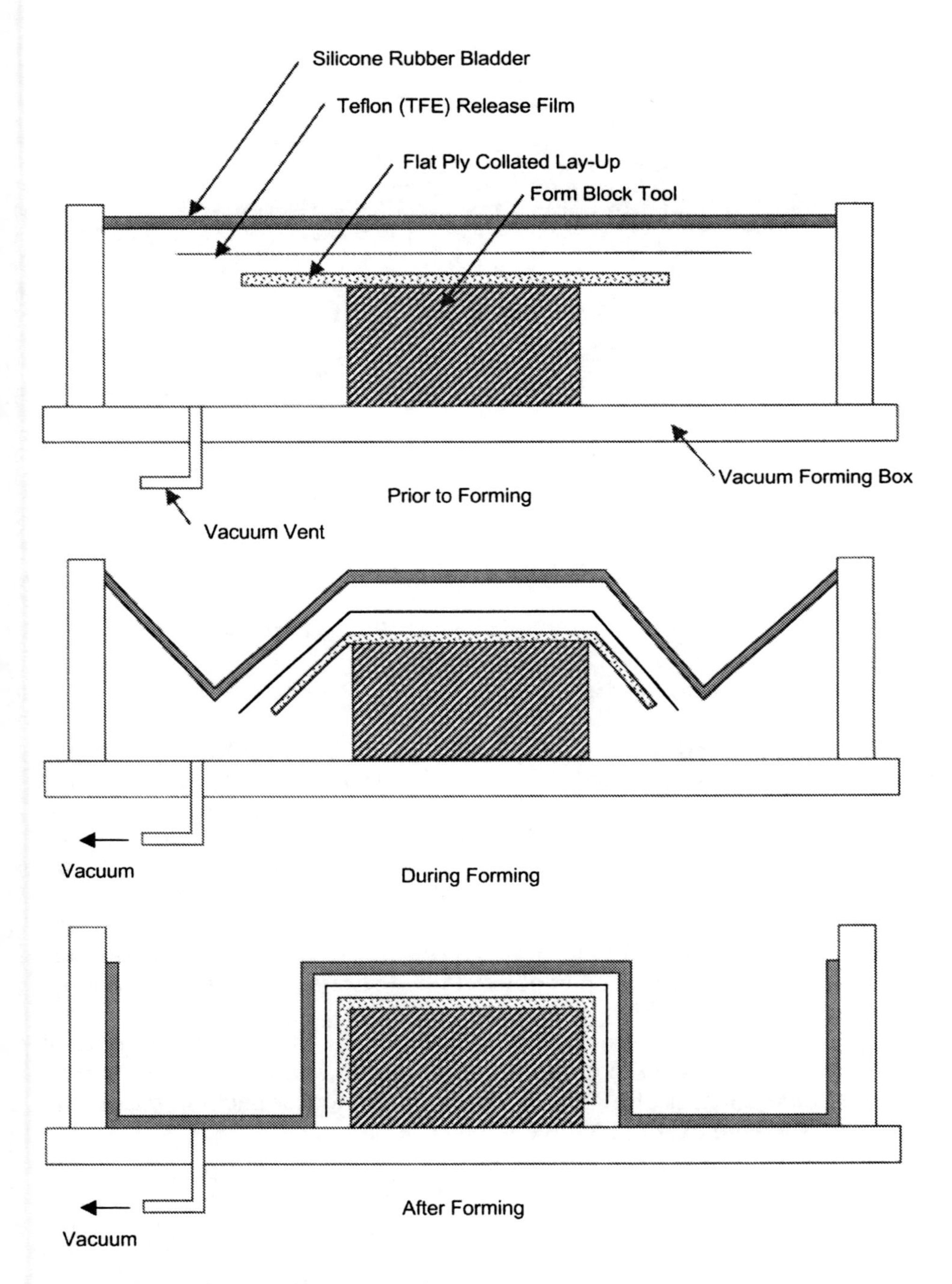

Fig. 14. Vacuum Forming of Substructure Components

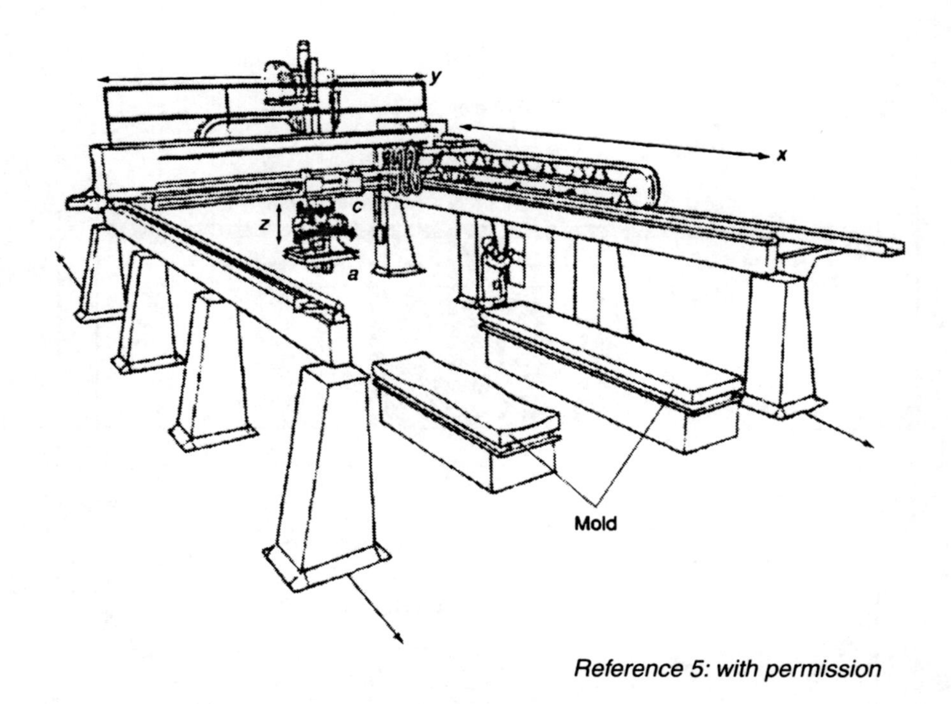

Fig. 15. Typical Gantry Style Tape Laying Machine

The delivery heads (Fig. 16) for both FTLMs and CTLMs are basically the same configuration and will normally accept 3, 6 or 12 inch wide unidirectional tape. The unidirectional tape purchased for ATL applications is closely controlled for width and tack to facilitate the tape laying process. FLTMs use either 6 or 12 inch wide tape to maximize material deposition rates for flat parts, while most CTLMs are restricted to 3 or 6 inch wide tape to minimize tracking errors (gaps and overlaps) when laying contoured parts. The term CTLM currently applies to mild contours that rise and fall at angles up to about 15%. Processes such as hand lay-up, filament winding or fiber placement, depending on the geometry and complexity of the part, are normally used to make more highly contoured parts. Material for ATL comes in large diameter spools, some containing almost 3,000 lineal ft of material. The tape contains a backing paper that must be removed during the tape laying operation.

The spool of material is loaded onto the delivery head supply reel (reels as large as 25 inches in diameter are used) and threaded through the upper tape guide shoot and past the cutters. The material then passes through the lower tape guides under the compaction shoe and onto a backing paper take-up reel. The backing paper is separated from the prepreg and wound onto a take-up roller. The compaction shoe makes contact with the tool surface and the material is laid onto the tool with compaction pressure. To

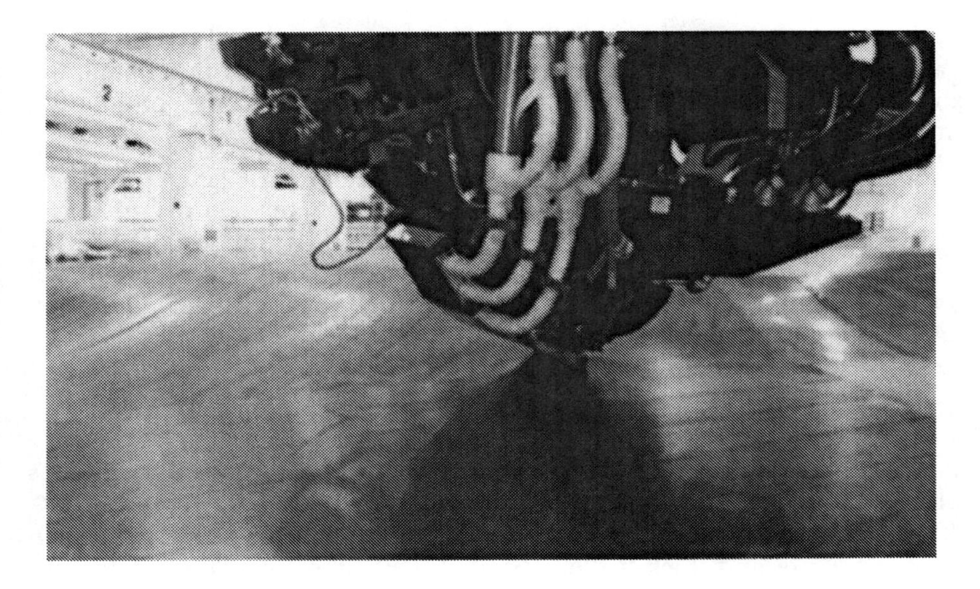

The Boeing Company

Fig. 16. Composite Tape Layer Delivery Head

insure uniform compaction pressure, the compaction shoe is segmented so that it follows the contour of the lay-up. The segmented compaction shoe is a series of plates that are air pressurized and conform to lay-up surface deviations, maintaining a uniform compaction pressure. The machine lays the tape according to the previously generated NC program, cuts the material at the correct length and angle, and when a length (course) is completed, lays out tail, lifts off the tool, retracts to the course start position and begins laying the next course.[6]

Modern tape laying heads have optical sensors that will detect flaws during the tape laying process and send a signal to the operator. In addition, machine suppliers now offer a laser boundary trace in which the boundary of a ply can be traced by the operator to verify the correct position. Modern tape laying heads also contain a hot air heating system that will preheat the tape (80–110 °F) to improve the tack and tape-to-tape adhesion. Computer-controlled valves maintain the temperature in proportion to the machine speed, i.e., if the head stops, the system diverts hot air flow to prevent overheating the material.

Software to drive modern tape layers has improved dramatically in the last 10 years. All modern machines are programmed offline, with systems that automatically compute the "natural path" for tape laying over a contoured surface. As each ply is generated, the software updates the surface geometry, eliminating the need for the designer to redefine the surface for each new ply. The software can also display detailed information about the fiber orientation of each course and the predicted gaps between adjacent courses. Once the part has been programmed, the software will generate NC programs that optimize the maximum quantity of composite tape laid per hour.

Part size and design are key drivers for composite tape layer efficiency. As a general rule of thumb, bigger parts and simpler lay-ups are more efficient. This is illustrated in Fig. 17 for FTLMs[6]. If the design is highly sculptured (lots of ply drop-offs) or the part size is small, the machine will spend a significant amount of time slowing down, cutting and then accelerating back to full speed.

5.8 Filament Winding[7]

Filament winding is a high rate process in which a continuous fiber band is placed on a rotating male mandrel. Lay-down rates as high as 100–400 lbs h^{-1} are not uncommon. It is a highly repeatable process that can fabricate large and thick-walled structure. Filament winding is a mature process, having been in continuous use since the mid-1940s. It can be used to fabricate almost any body of revolution, such as cylinders, shafts, spheres and cones. Filament winding can also fabricate a large range of part sizes–

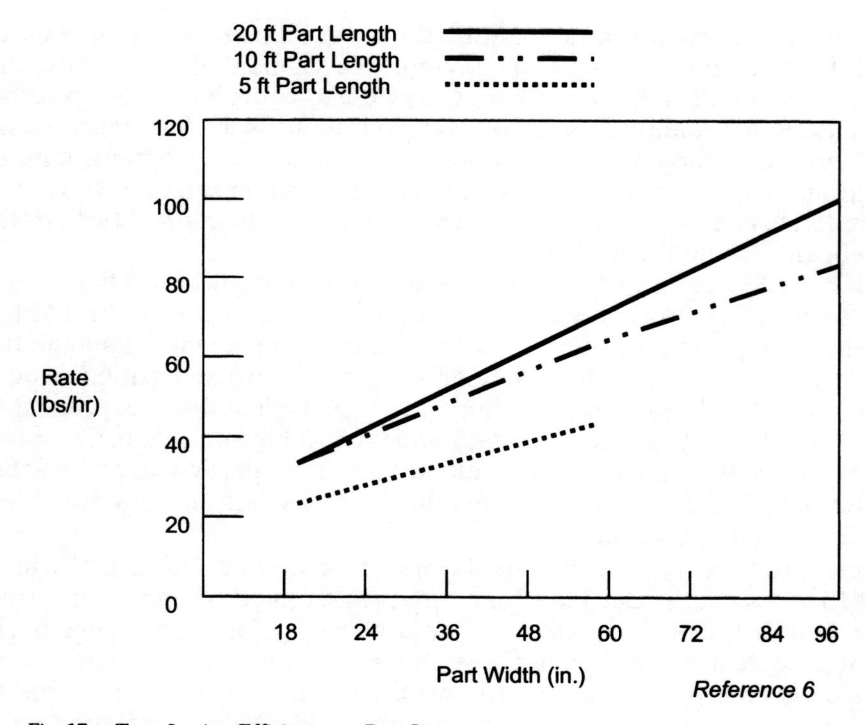

Fig. 17. Tape Laying Efficiency vs. Part Size

parts smaller than 1 inch in diameter (e.g., golf club shafts) and as large as 20 ft in diameter have been wound. The major restriction on geometry is that concave contours cannot be wound, because the fibers are under tension and will bridge across the cavity. Typical applications for filament winding are cylinders, pressure vessels, rocket motor cases and engine cowlings. End fittings are often wound into the structure producing strong and efficient joints.

A typical filament winding process is shown in Fig. 18 in which dry tows are drawn through a bath containing liquid resin, collimated into a band and then wound on a rotating mandrel[8]. Actually, there are three main variants of the filament winding process: (1) wet winding, in which the dry reinforcement is impregnated with a liquid resin just prior to winding; (2) wet rolled prepreg winding, in which the dry reinforcement is impregnated with the liquid resin and then rewound prior to filament winding; and (3) towpreg winding, in which a commercially impregnated tow is purchased from a material supplier. Although many types of filaments are amenable to the filament winding process, the most prevalent are glass, aramid and carbon. There are also a number of different winding methods or patterns, primarily helical winding, polar winding and hoop or circumferential winding.

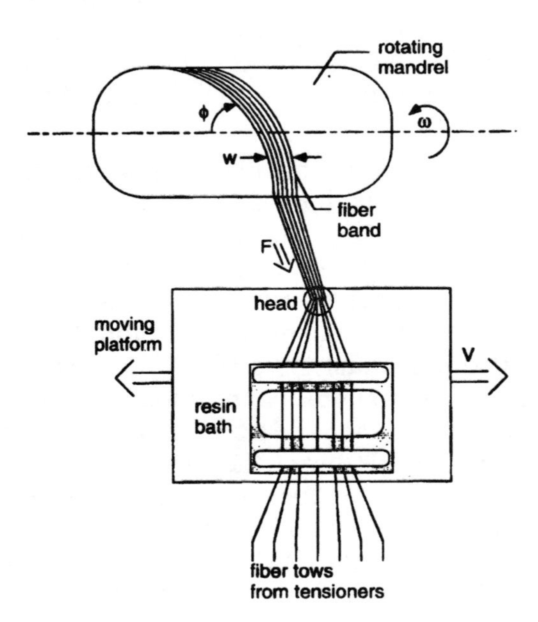

Fig. 18. Filament Winding Process

Filament winding equipment costs can be low, moderate or high, depending on part size, the type of winder and the sophistication of the control system (mechanical or NC control). For high rate applications, some winders have been designed with multiple pay-out systems and multiple mandrels so that several parts can be fabricated simultaneously. Filament wound parts are usually cured in ovens rather than autoclaves, the compaction pressure being provided by mandrel expansion and fiber tension in the lay-up. Circular windings, separated from the part by caul plates or separator sheets, can also be used to provide compaction during cure. Forced air convection ovens are the most common method. Others, such as microwave curing, are faster but result in higher equipment costs. Mandrel costs can be moderate to high, depending on part size and complexity. The mandrel must be able to be removed from the part. This is often accomplished by shrinkage of the mandrel during cool-down, incorporation of a slight draft or taper, wash-out mandrels, plaster break-out mandrels, inflatable mandrels, or, for complex parts, segmented mandrels that can be taken out from the inside of the part in sections. While the inner surface (mandrel side) of the part is usually smooth, the outer surface can be quite rough. If this presents a problem, it is possible to wind sacrificial plies on the outer surface and then grind or machine the outer surface smooth after cure.

Fiber orientation can be a problem for some filament wound designs–the minimum fiber angle that can usually be wound is 10–15° due to slippage of the fiber bands at the mandrel ends; however, schemes such as temporary pins inserted in the mandrel ends during winding can sometimes be used to overcome this limitation.

Glass fibers are the most prevalent fibers used for filament winding. Glass is low cost, fairly easy to handle, has good strength and impact properties, a moderate modulus, and is compatible with epoxy resin systems. E-glass is widely used in commercial and industrial products. For aerospace and high-performance pressure vessels, S-2 glass is often used because of its higher strength and modulus. Typical tow sizes contain between 400 and 1,000 individual glass fibers that have been sized for handling and surface treated to improve the bond to the matrix. It should be noted that the use of the largest band size available will improve winding productivity as the band spreads during the winding process; a larger band size covers a wider area, thus reducing winding time. Other inorganic fibers that have successfully been used include alumina, silicon carbide and quartz. For commercial applications, twisted yarns are often used because they are easier to handle; however, for aerospace and other strength critical applications, untwisted fibers are more common because they result in higher mechanical properties. Aramid fibers are also attractive choices for filament wound parts. Aramid, being an organic fiber, has a low density (about one-half of that of glass fiber) coupled with a high tensile strength and a higher modulus than glass, resulting in parts with high specific strengths and moduli. In addition, aramid is also a very tough fiber and an excellent choice for parts that may be subject to impact abuse. The main disadvantages of aramid are low compressive and shear strength and can often be circumvented in filament wound parts that are loaded primarily in tension (e.g., pressure vessels). For applications requiring the ultimate in performance, carbon fibers are usually specified due to their combination of low density, high strength and modulus. However, carbon fibers are considerably more expensive than either glass or aramid fibers. Both polyacrylonitrile (PAN)- and pitch-based carbon fibers have been used in filament wound parts. PAN-based fibers are usually specified where high strength is required, while pitch-based fibers are often employed where high modulus (i.e., stiffness) is required. However, one of the problems frequently encountered with very high modulus graphite fiber laminates is matrix microcracking, due to the large difference in the coefficients of thermal expansion between the fibers and the matrix. One of the advantages of carbon and graphite fibers is that they are available in a large range of strength and stiffness combinations. Also, there are quite a variety of different tow sizes available. While large tow sizes result in wider bandwidths and increased winding productivity, the mechanical properties usually decrease as the tow size increases. For the best mechanical properties, the tows should not be twisted and remain well collimated during the winding process.

The majority of filament-winding fabricators formulate their own resin systems for both wet winding and wet-rolled prepreg. If a prepreg product form is specified, they will normally purchase the pre-impregnated tow from one of the major prepreg suppliers. Epoxies are the most prevalent matrix resins used for high-performance filament wound parts, while polyesters and vinyl esters are often used for less demanding commercial applications. However, many different types of resins have been successfully used for filament winding, including cyanate esters, phenolics, bismaleimides, polyimides and others. Viscosity and pot life are two of the main factors in selecting a resin for wet winding. Low viscosity, generally around 2,000 cps, is desirable to help wet the fibers spread the band and lower the friction over the guides during the winding process. Pot life is primarily a function of the time it will take to wind the part–large and thicker parts will require a longer pot life than smaller thinner parts. Other factors that need to be considered when selecting a resin include: toxicity, cure temperature and time and the resultant physical, mechanical and environmental properties. In addition, samples should be fabricated and tested prior to committing full-size parts to production. This testing should replicate the actual environment that the part will eventually encounter. A number of premixed wet winding resin systems are also available from material suppliers. While pre-impregnated tow (towpreg) is more expensive than wet winding resin systems, it does offer several important advantages: (1) a qualified fiber and resin system that the end user is familiar with and can often be prepregged onto a tow; (2) it allows the best control of resin content; (3) it allows the highest winding speeds, because there is no wet resin that will be thrown off during winding; and (4) the tack can be adjusted to allow less slippage when winding shallow angles.

Wet winding is accomplished by either pulling the dry tows through a resin bath or directly over a roller that contains a metered volume of resin controlled by a doctor blade. The resin content of wet wound parts is difficult to control, being affected by the resin reactivity, the resin viscosity, the winding tension, the pressure at the mandrel interface and the mandrel diameter. For example, too low a viscosity resin will impregnate the strands thoroughly but will tend to squeeze out during the pressure of the winding operation resulting in an excessively high fiber content. At the other extreme, too high a viscosity will not sufficiently impregnate the strands and there will be a tendency for the cured part to contain excessive porosity. Due to the generally low viscosity of wet winding resins, it is not uncommon to have parts with higher fiber volume percentages (70 vol.% and sometimes higher) than are normally found in composite parts fabricated with lower viscosity prepreg (60 vol.%). Formulating resins for wet winding is usually based on years of experience, where many fabricators have done extensive testing with various combinations of resins, curing agents, diluents and other additives. The goal is to formulate a resin system with the desired viscosity that is non-toxic, has sufficient pot life to wind the required part size and will cure to yield the desired physical, mechanical and environmental performance.

To circumvent some of the problems with controlling a direct wet winding process, wet rolled prepreg is sometimes manufactured by wet impregnating the strands in the normal manner and then respooling them prior to winding. There are two main advantages to this process: the fabricator can conduct offline quality assurance on the wet wound prepreg prior to use, and the viscosity and tack can be controlled somewhat by room-temperature staging. Staging at room or slightly elevated temperature is commonly called B-staging. The objective is to advance the resin to reduce the viscosity and increase the tack. On the negative side, the wet wound prepreg has to be packaged and refrigerated for storage unless it is immediately used for winding.

Commercially supplied prepregs offer the best control of resin content, uniformity and bandwidth control but are also the most expensive of the product forms, usually 1.5–2 times the cost of wet winding materials. While prepreg tows are the predominant prepreg form using in filament winding, some aerospace manufacturers specify slit prepreg tape for fiber placement to insure extremely tight control of the bandwidth and the resultant gaps, during the fiber placement of flight critical hardware. Prepreg tows for filament winding generally (1) have the longest pot lives; (2) allow higher winding speeds because there is less chance of "resin throw" during the winding process; and (3) allow winding angles closer to longitudinal (0°) because they contain higher tack than most wet winding systems and will not tend to slip as much at the ends.

There are several design criteria[7] that need to be considered when designing a mandrel for filament winding:

- *Strength and stability* – The mandrel must be able to support its weight and the weight of the part during both the winding operation and subsequent cure. It must also be thermally stable at the highest temperature encountered during the curing operation.
- *Weight* – The mandrel should be as light as possible. The heavier the mandrel, the harder it will be to handle and it will exhibit slower heat-up and cool-down rates during cure.
- *Thermal expansion* – The mandrel should be designed to expand and exert pressure on the part during cure and then shrink during cool-down to make part removal easier.
- *Removable* – The mandrel must be removable after cure unless the design dictates a fly-a-way mandrel.
- *Cost* – Mandrel cost should be dictated by part complexity, part value and number of parts to be fabricated.

The choice of a mandrel material and design is, to a great extent, a function of the design and size of the part to be built. A large number of materials have been used for filament winding mandrels. Dissolvable mandrels are often used for parts with only small openings. This type of mandrel includes water-soluble sand, water soluble or breakout plaster, low-temperature eutectic salts and occasionally low-melting-point metals. After cure, the disposable mandrel is dissolved out with hot water, melted or broken into small pieces for removal. An alternate to these approaches would be to use an inflatable mandrel that can be either left inside the part as a liner or extracted through an opening.

Reusable mandrels may either be segmented or non-segmented. Segmented mandrels are required when the part geometry does not allow the part to be removed by simply sliding the part off the mandrel after cure. Segmented mandrels are generally more expensive to fabricate and use than non-segmented mandrels. Non-segmented mandrels usually have a slight draft or taper to ease part removal after cure. Reusable mandrels can be fabricated from steel, aluminum or other durable metals that will give the desired expansion and contraction during cure. Large mandrels are usually constructed with thin walls supported by substructure as shown schematically in Fig. 19. This reduces the cost, weight and improves heat-up and cool-down rates during cure. In designing mandrels for filament

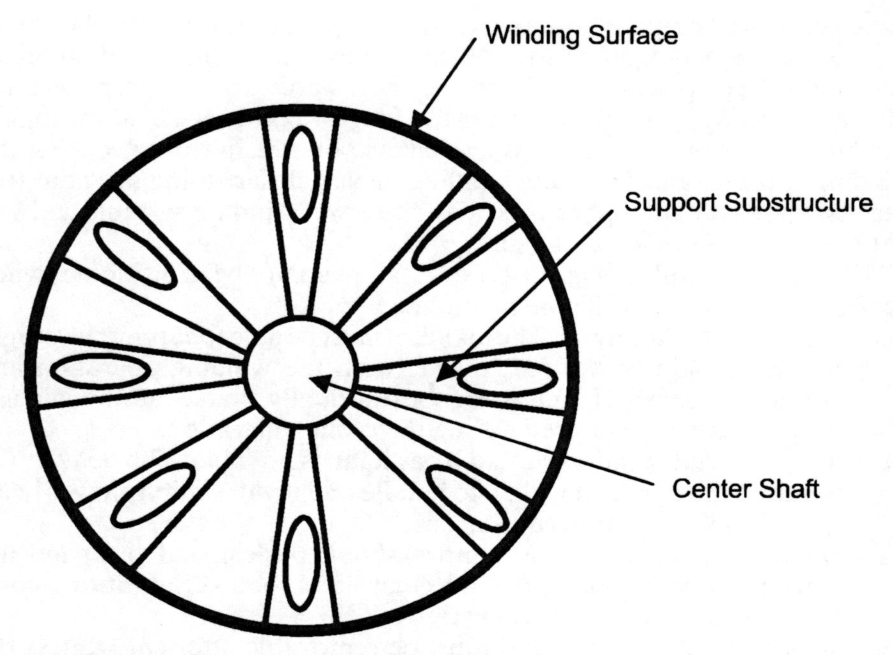

Fig. 19. Filament Winding Mandrel With Substructure

winding, the coefficient of thermal expansion of the mandrel is an important consideration. As the part is heated up to the cure temperature, the mandrel expands and effectively applies pressure to the part, helping to compact the laminate and eliminate voids and porosity. Likewise, during cool-down the mandrel contracts away from the part, making part removal easier. Prior to winding, mandrels are usually prepared by first coating the mandrel with mold release and then applying a layer of gel coat resin. The gel coat provides a tacky base so that the initial bands will adhere to the surface and provide a smoother surface on the cured part.

Most winding machines (Fig. 20) operate similar to a lathe, the mandrel is mounted horizontally and rotates at a constant speed while the carriage delivering the fiber band traverses the length. The carriage speed must be synchronized with the mandrel rotational speed to deliver the band at the correct winding angle. Helical, polar and hoop are the three dominant winding patterns used in filament winding.

Helical winding (Fig. 21) is a very versatile process that can produce almost any combination of length and diameter. In helical winding the mandrel rotates while the fiber carriage traverses back and forth at the speed necessary to generate the desired helical angle (α).[9] As the band is wound, the circuits are not adjacent and additional circuits must be applied before the surface begins to be covered with the first layer. This winding pattern produces band cross-overs at periodic locations along the part,

Fig. 20. Large Filament Winder

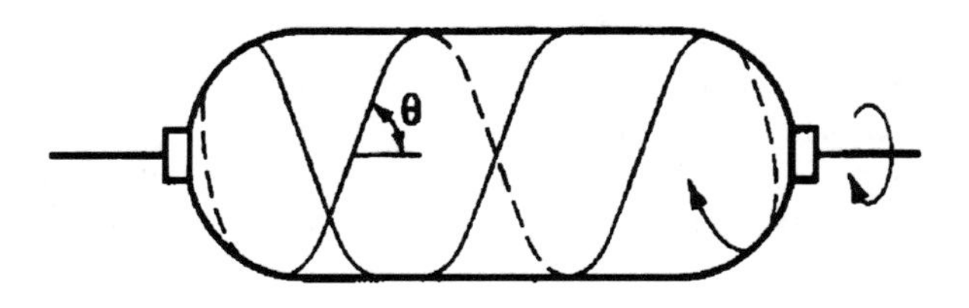

Fig. 21. Principle of Helical Winding

which can be somewhat controlled by the newer NC-controlled winding machines. Due to this cross-over winding pattern, a layer is made up of a two-ply balanced laminate. In helical winding, the relationship between the helical angle (α) and the machine parameters can be expressed as $\tan \alpha = \frac{v}{\pi D N}$ where α is the helical angle, v is the speed at which the carriage traverses the axial direction in inch s^{-1}, D is the diameter of the mandrel in inches, and N is the mandrel rotational speed in rev s^{-1}.

As indicated in this equation, the ratio $v/\pi DN$ must remain fixed to maintain a constant helical angle α; therefore, either the carriage traverse velocity v and/or the mandrel rotational speed N, must be adjusted during winding to compensate for changes in the mandrel diameter D. The carriage normally speeds up and goes into a cross-feed motion perpendicular to the mandrel axis as it goes around the ends, as the length of travel is quite large at the ends.

If the end dome openings are the same size, a geodesic wind pattern may be used. This pattern produces the shortest band path possible and results in uniform tension in the filaments throughout their length. An additional advantage of the geodesic pattern is that it produces a no-slip condition, i.e., there is no tendency for the bands to slip or shift on the mandrel surface.

Helical winders are the most common types of equipment in use today. In helical winding the mandrel can be located either horizontally or vertically. The biggest advantage of a vertically mounted mandrel is less mandrel deflection, since the load imposed by the weight of the mandrel passes through the center axis of the mandrel; however, vertical mounting makes wet winding more difficult. It is not unusual to find modern helical winding machines with as many as six axes of rotation:[7] (1) mandrel rotation, (2) carriage traverse, (3) horizontal cross-feed, (4) vertical cross-feed, (5) eye rotation and (6) yaw.

- *Mandrel rotation.* Mandrel rotational speed is generally constant throughout the winding process.
- *Carriage traverse.* Carriage traverse is also generally constant in the cylindrical portion of the winding process and then changes and reverses at the end domes.
- *Horizontal cross-feed.* This axis is used to position the winding pay-out band closer to the part at the dome ends.
- *Vertical cross-feed.* Vertical cross-feed works in conjunction with the horizontal cross-feed to better position the pay-out band vertical to the dome ends.
- *Pay-out rotation.* This axis allows the winding pay-out to keep the band normal to the winding surface.
- *Yaw.* Yaw allows the pay-out band to be rotated in a 90° plane to give additional control over the band placement.

To adequately control filament winders with this degree of sophistication, it is necessary to use NC programming techniques where each axis can be individually controlled by their own microprocessors. Many modern filament winding machines come with offline programming systems in which the winding patterns and parameters are programmed prior to the actual part fabrication. In addition, there are simulation systems that allow the programmer to simulate the winding process on the computer prior to part fabrication.

To deliver the fiber tows from the spool to the part requires that the tow pass through a series of guides, redirects and spreader bars. For wet filament winding, steel or ceramic eyelets are used to direct the fiber tows as they come off the spool. For prepregged tow, rollers are used instead of eyelets to prevent resin from being scrapped off the towpreg. During the entire delivery process, tension on the tows is minimized to preferably 1 lb or less. Low tension helps to reduce abrasion in the fibers, minimizes the possibility of tow breakage and helps to spread the band as it passes over the spreader bar. Many modern filament winding machines are equipped with automatic tensioning devices to help control the amount of tension during the winding process.

Polar winding (Fig. 22) is somewhat simpler than helical winding in that: (1) a constant winding speed can be used; (2) it is not necessary to reverse the carriage during winding; and (3) the bandwidths are laid adjacent to each other as the part is wound. This is an excellent method for fabricating spherical shapes. In this process, the bands pass tangentially to the opening at one end of the part, reverse direction and then pass tangentially to the opening at the opposite end of the part. The lay-down is planar with the bandwidths adjacent to each other due to the winding arm generating a great circle during each pass. Simple polar winders have only two axes of motion, the mandrel and the winding arm.

Polar winding machines are generally much simpler than helical winding machines but are also limited in their capabilities. The length-to-diameter ratio must be less than 2.0. They are frequently used to wind spherical shapes by utilizing a continuous step-out pattern. A variation of the polar winder is the tumble winder, in which the mandrel is mounted at an inclined axis and tumbles in a polar path while the roving strands remain stationary. While tumble winders are very efficient for spherical shapes, they are usually limited to diameters of 20 inches or less.

Hoop winding, also known as circumferential or circ winding, is the simplest of all of the processes. The winding action, shown in Fig. 23, is similar to a lathe where the mandrel speed is much greater than the carriage travel[10]. Each full rotation of the mandrel advances the carriage one full bandwidth so that the bands are wound adjacent to each other. During part fabrication, hoop winding is often combined with longitudinal (helical or polar) winding to provide adequate part strength and stiffness. The hoop windings can be applied to the cylindrical portion of the part, while the longitudinal windings are applied to both the cylindrical and domed portions of the part. It should again be pointed out that the

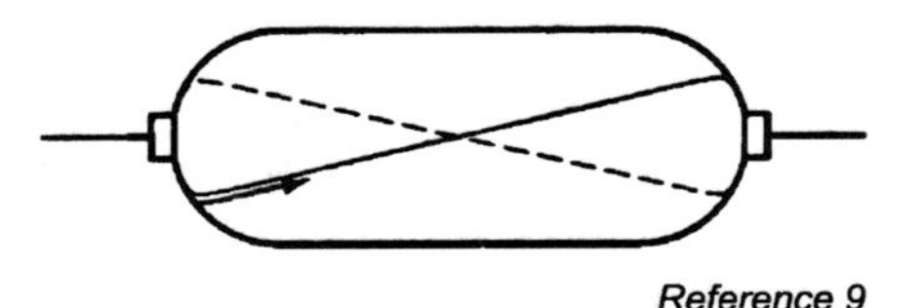

Reference 9

Fig. 22. Polar Winding

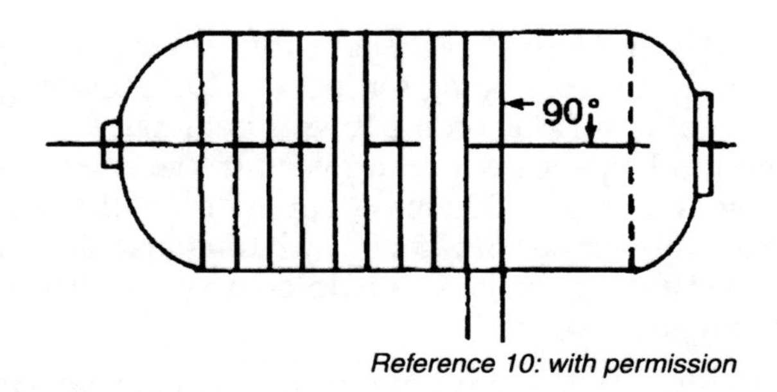

Fig. 23. Hoop or Circumferential Winding

minimum wind angle for longitudinal winding is generally about 10–15° to preclude slippage of the bands at the ends of the mandrel.

After the winding operation is complete, wet wound parts are often B-staged prior to final cure to remove excess resin. This process involves heating to a slightly elevated temperature but below the resin gel temperature. Frequently, the part is heated with heat lamps and the excess resin is removed as the part rotates. The great majority of filament wound parts are cured in an oven (electric, gas fired or microwave) without a vacuum bag or any other supplemental method of applying pressure. As the part heats up to the curing temperature, the mandrel expands but is constrained by the fibers in the wound part. This creates pressure that helps to compact the laminate and reduce the amount of voids and porosity. Since the majority of filament wound parts are cured in ovens rather than autoclaves, filament winding is capable of making very large structures, limited only by the size of the winder and the curing oven available.

Autoclave curing may also be used to further reduce the amount of porosity; however, the compaction pressure applied by an autoclave can also induce fiber buckling and even wrinkles in the part. The use of thin caul plates that are allowed to slip over the surface may help to alleviate some wrinkling on cylindrical surfaces but are prone to leaving marks on the part surface where they terminate. Caul plates with circumferential wound plies have also been used in oven cured parts to improve compaction and provide a smoother surface finish. Occasionally, the part will be wrapped with shrink tape to provide compaction pressure; this is a common method employed in manufacturing carbon fiber golf club shafts.

5.9 Fiber Placement

In the late 1970s Hercules Aerospace Co. (now Alliant Techsystems) developed the fiber placement process. Shown conceptually in Fig. 24, it is a hybrid between filament winding and tape laying. A fiber placement or tow placement machine allows individual tows of prepreg to be placed by the head. The tension on the individual tows normally ranges from 0 up to

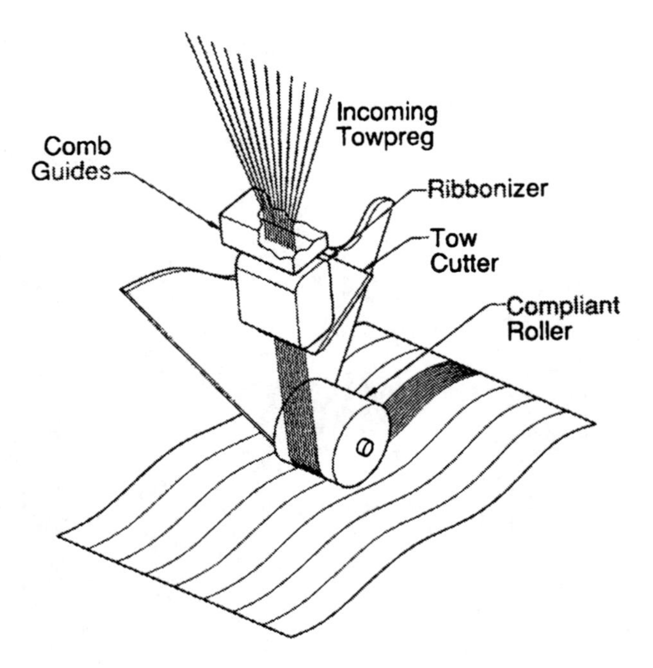

Fig. 24. Fiber Placement Process

about 2 lb. Therefore, true 0° (longitudinal) plies pose no problems. In addition, a typical fiber placement machine (Fig. 25) contains either 12, 24 or 32 individual tows that may be individually cut and then added back in during the placement process. Since the tow width normally ranges from 0.125 inch to 0.182 inch, bands as wide as 1.50–5.824 inches can be applied, depending on whether a 12 or 32 tow head is used. The adjustable tension employed during this process also allows the machine to lay tows into concave contours, limited only by the diameter of the roller mechanism. This allows complicated ply shapes, similar to those that can be obtained by hand lay-up. In addition, the head (Fig. 26) contains a compliant compaction roller that applies pressure in the range of 10–400 lb during the process, effectively debulking the laminate during lay-up. Advanced fiber placement heads also contain heating and cooling capability: cooling is used to decrease the towpreg tack during cutting, clamping and restarting processes, whereas heating can be used to increase the tack and compaction during lay-down. For the current generation of fiber placement heads, a minimum convex radius of approximately 0.124 inch and a minimum concave radius of 2 inches are obtainable. One limitation of the fiber placement process is that there is a minimum course (or ply) length, normally about 4 inches. This is a result of the cut-and-add process.

The Boeing Company

Fig. 25. Fiber Placement Machine

A ply that is cut or added must then pass under the compliant roller, resulting in a minimum length that is dependent on the roller diameter.

Typical applications for fiber placement are engine cowls, inlet ducts, fuselage sections, pressure tanks, nozzle cones, tapered casings, fan blades and C-channel spars. The aft section of a V-22 is shown in Fig. 27. Other applications include rocket fairings and aircraft fairings (Fig. 28). The V-22 aft fuselage contains cocured stiffeners on the inside moldline, whereas the rocket and aircraft fairings shown contain honeycomb core with fiber placed inner and outer skins.

The Boeing Company

Fig. 26. Fiber Placement Head

The Boeing Company

Fig. 27. V-22 Aft Fuselage

The Boeing Company

Fig. 28. Fiber Placed Aircraft Fairing

Extensive testing has shown that the mechanical properties of fiber placed parts can be essentially equivalent to hand laid-up parts.[11] Gaps and overlaps are typically controlled to 0.030 inch or less. One difference between fiber placed and hand layed-up plies are the "stair-step" ply terminations obtained with fiber placement since each tow is cut perpendicular to the fiber direction. Again, this stair-step ply termination (Fig. 29) has been shown to be equivalent in properties to the smooth transition one obtains with manual lay-up. In fact, some parts have been designed so that either fiber placement or manual hand lay-up may be used for fabrication. Since the tows are added in and taken out as they are needed, there is very little wasted material–scrap rates of 2–5% are common in lay-up. In addition, since the head can "steer" the fiber tows, offering the potential for the design of highly efficient load-bearing structure. The software required to program and control a fiber placement machine is even more complex than that required for an automated tape layer or modern filament winder. The software translates CAD part and tooling data into seven-axis commands, developing the paths and tool rotations for applying the composite tows to the part's curved and geometric features, while keeping the compaction roller normal to the surface. A simulator module confirms the part program with three-dimensional (3D) animation, while integrated collision avoidance post-processing of the NC program automatically detects interferences.

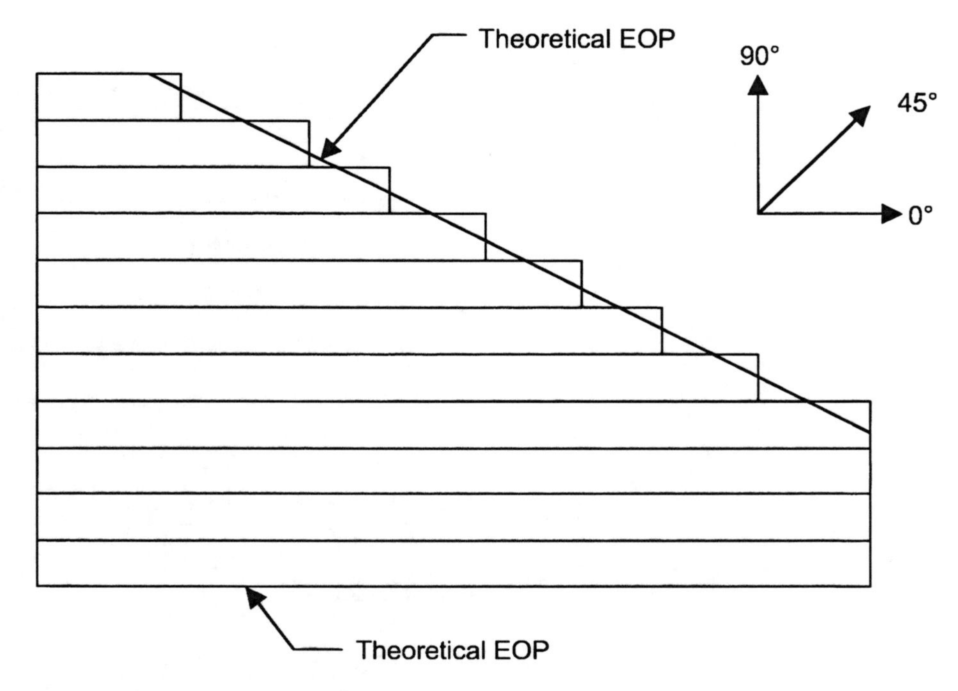

Fig. 29. Stair Step Ply Terminations

Modern fiber placement machines are extremely complex and can be very large installations. Most machines contain seven axes of motion (cross-feed, carriage traverse, arm tilt, mandrel rotation, and wrist yaw, pitch and roll). The larger machines are capable of handling parts up to 20 ft in diameter and 70 ft long with mandrel weights up to 80,000 lb. They typically contain refrigerated creels for the towpreg spools, towpreg delivery systems, redirect mechanisms to minimize twist, and tow sensors to sense the presence or absence of a tow during placement.

Although complex part geometries and lay-ups can be fabricated using fiber placement, the biggest disadvantages are that the current machines are very expensive and complex and the lay-down rates are slow compared to most conventional filament winding operations. Fiber placed parts are usually autoclave cured on carbon/epoxy, steel, or low-expansion invar tools to provide dimensionally accurate parts.

5.10 Vacuum Bagging

After ply collation the laminate is sealed in a plastic bag for curing. A typical bagging scheme is shown in Fig. 30. To prevent resin from escaping from the edges of the laminate, dams are placed around the periphery of the lay-up. Typically, cork, silicone, rubber or metal dams are used. The

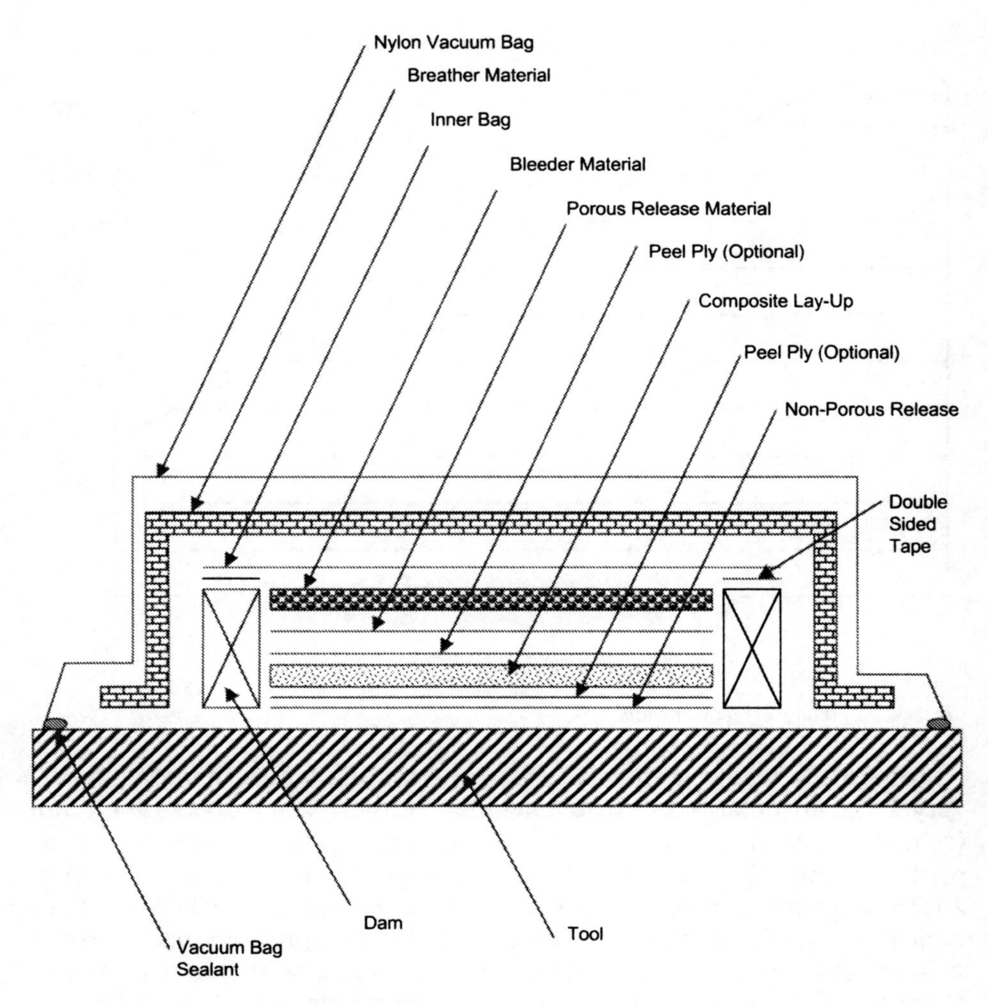

Fig. 30. Typical Vacuum Bagging Schematic

dams should be butted up against the edge of the lay-up to prevent resin pools from forming between the laminate and dams. The dams are held in place with either double-sided tape or Teflon pins.

A peel ply may be applied directly to the laminate surface if the surface is going to be subsequently bonded or painted. Then, a layer of porous release material, usually a layer of porous glass cloth coated with Teflon, is placed over the lay-up. This layer allows resin and air to pass through the layer without letting the bleeder material bond to the laminate surface. The bleeder material can be a synthetic material (e.g., polyester mat) or dry fiberglass cloth, such as 120 or 7781 style glass. The amount of bleeder material depends on the laminate thickness and the desired amount of resin to be removed. The use of a finer weave, such as style-120 glass next to the laminate surface, will produce a smoother surface finish as compared to the coarser weave 7781 glass. The amount of bleeder needed can be calculated as follows.

For a normal 42 ± 3% resin content prepreg, use a bleeder ratio of 0.3 for style-120 glass cloth. If 7781 glass cloth is used, one layer of 7781 glass is equivalent to two layers of 120 glass. Finally, if the heavier style-1000 glass is used, it is equivalent to three layers of 120 glass. The following example illustrates calculating the amount of bleeder required: for a 24-ply thick tape laminate with a resin content of 42 ± 3%, possible bleeder packs would consist of: $24 \times 0.3 = 7.2$ or 7 layers of style-120 glass or one layer of 120-glass followed by three layers of 7781-glass or one layer of 120-glass followed by two layers of 1000-glass. Other bleeder ratios for different resin content prepregs and different bleeder materials can be calculated or determined empirically.

After the bleeder is placed on the lay-up, an inner bag made of Mylar (polyester), Tedlar (PVF) or Teflon (TFE) is placed on the lay-up. The purpose of the inner bag is to let air escape while containing the resin within the bleeder pack. The inner bag is sealed to the edge dams with double-sided tape and then perforated with a few small holes to allow air to escape into the breather system. The breather material is similar to the bleeder material, either a synthetic mat material or a dry glass cloth can be used. If dry glass cloth is used, the last layer next to the vacuum bag should not be coarser than 7781 glass. Heavy glass fabrics, such as style 1000, have been known to cause vacuum bag ruptures during cure due to the nylon bagging material being pushed down into the coarse weave of the fabric and rupturing. The purpose of the breather is to allow air and volatiles to evacuate out of the lay-up during cure. It is important to place the breather over the entire lay-up and extend it to pass the vacuum ports.

The vacuum bag, which provides the membrane pressure to the laminate during autoclave cure, is normally a 3–5 mil thick layer of nylon-6 or nylon-66. It is sealed to the periphery of the tool with a butyl rubber or chromate rubber sealing compound. Nylon vacuum bags can be used at temperatures up to 375 °F. If the cure temperature is higher than 375 °F, a polyimide material called Kapton can be used to approximately 650 °F along with a silicone bag sealant. Higher temperatures usually require the use of a metallic bag (e.g., aluminum foils) and a mechanical sealing system. It should be noted that Kapton bagging films are stiffer and harder to work with than nylon. During the bagging process, it is important that the bagging materials, including the breather, do not bridge in corners (Fig. 31) and that pleats or "dog ears" (Fig. 32) are included to provide extra material to help prevent bag ruptures during the high-temperature–high-pressure autoclave cure. In addition, high elongation

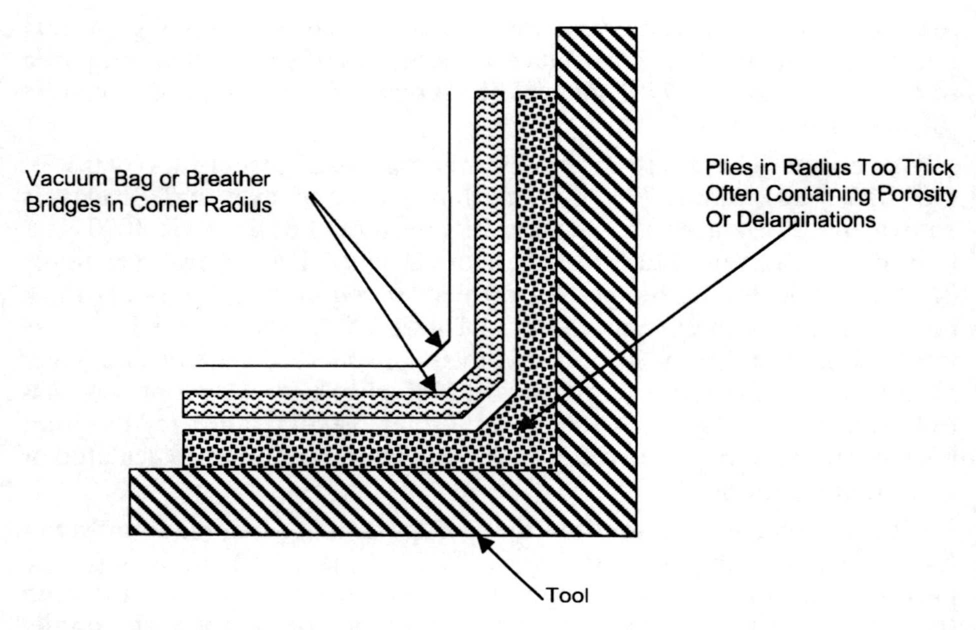

Fig. 31. Corner Bridging of Piles

"stretch" films[12] having elongations close to 500% rather than the standard 300% can be used to help the vacuum bag conform to the contours of the part and tool. All vacuum ports and protrusions on the tool should also be covered with extra breather material to help evacuation and prevent bag rupture.

Some manufacturers have invested in reusable silicone rubber vacuum bags to reduce cost and reduce the chance of a leak or bag rupture during cure. These normally require some type of mechanical seal to the tool. Several examples are shown in Fig. 33. Also, if the part is large, reusable rubber vacuum bags become heavy and difficult to handle so they may require a handling system to facilitate their installation and removal, as shown for the extremely large bag in Fig. 34. There are suppliers who sell both the materials to make silicone rubber vacuum or will provide a complete bag and sealing system ready for use.

After the vacuum bag is sealed to the tool surface, a vacuum is slowly applied and any bridges or wrinkles are worked out to the edges. Once a full vacuum is attained (i.e., 26-29 inches of mercury vacuum), the vacuum source is isolated for several minutes and a static gauge is used to determine if the bag contains any leaks. If the bag is leaking, an ultrasonic sniffer that amplifies the sound of the air rushing through the bag leak can be used to isolate and repair any leaks.

Thermocouples, used to measure the part temperature during cure, can be applied inside the vacuum bag, although a common practice is to weld or

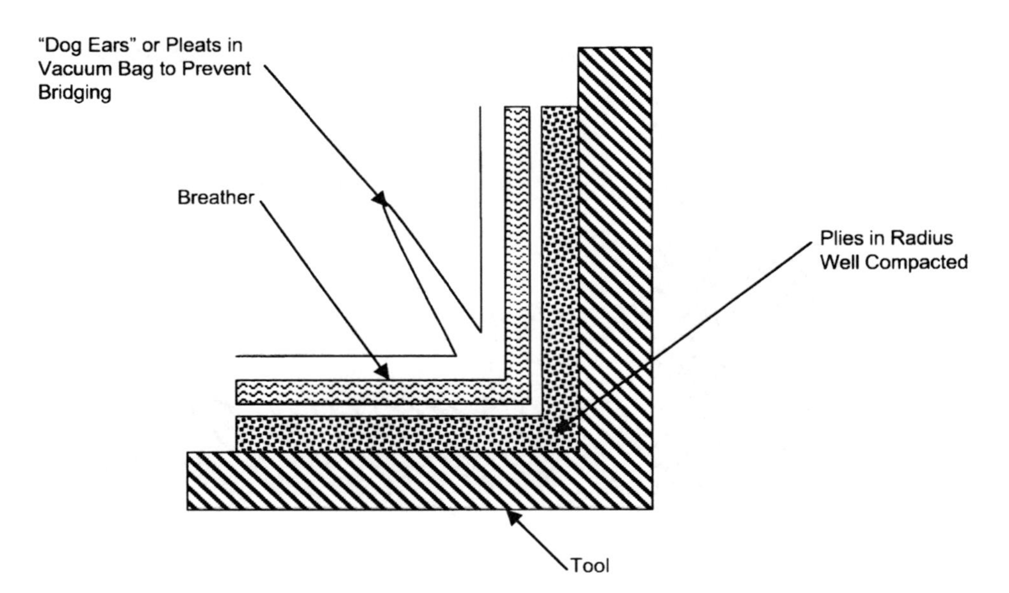

Fig. 32. "Dog Ears" or Vacuum Bag Pleats

attach the thermocouples to the underside of the tool. Thermocouples placed inside the vacuum bag create another potential leak path during cure.

Caul plates or pressure plates can be used to provide a smooth surface on the bag side. Caul plates are frequently made of mold release coated aluminum, steel, fiberglass or glass-reinforced silicone rubber. They range in thickness from as thin as 0.060 inch up to about 0.125 inch. The design of a caul plate and its location in the lay-up are important considerations in achieving the desired surface finish. As shown in Fig. 35, the caul plate may be placed above the bleeder pack or within the bleeder pack, but close to the laminate surface to provide a smooth surface. However, it will then require a series of small holes (e.g., 0.060–0.090 inch diameter) to allow resin to pass through the caul plate into the top portion of the bleeder pack. It should be noted that the caul plate is usually not placed next to the laminate surface because the holes will mark-off on the laminate surface. In general, the further the caul plate is from the laminate surface, the less effective it is in producing a smooth surface due to the cushioning effect of more and more bleeder material. Caul plate material and thickness are also important. Caution should be exercised when using extremely stiff and thick caul plates, because they can result in localized bridging and low-pressure areas during cure. In addition, the caul plate should fit within the dam periphery; if it rides or bridges on top of the dams, a low-pressure area can result at the edge of the part.

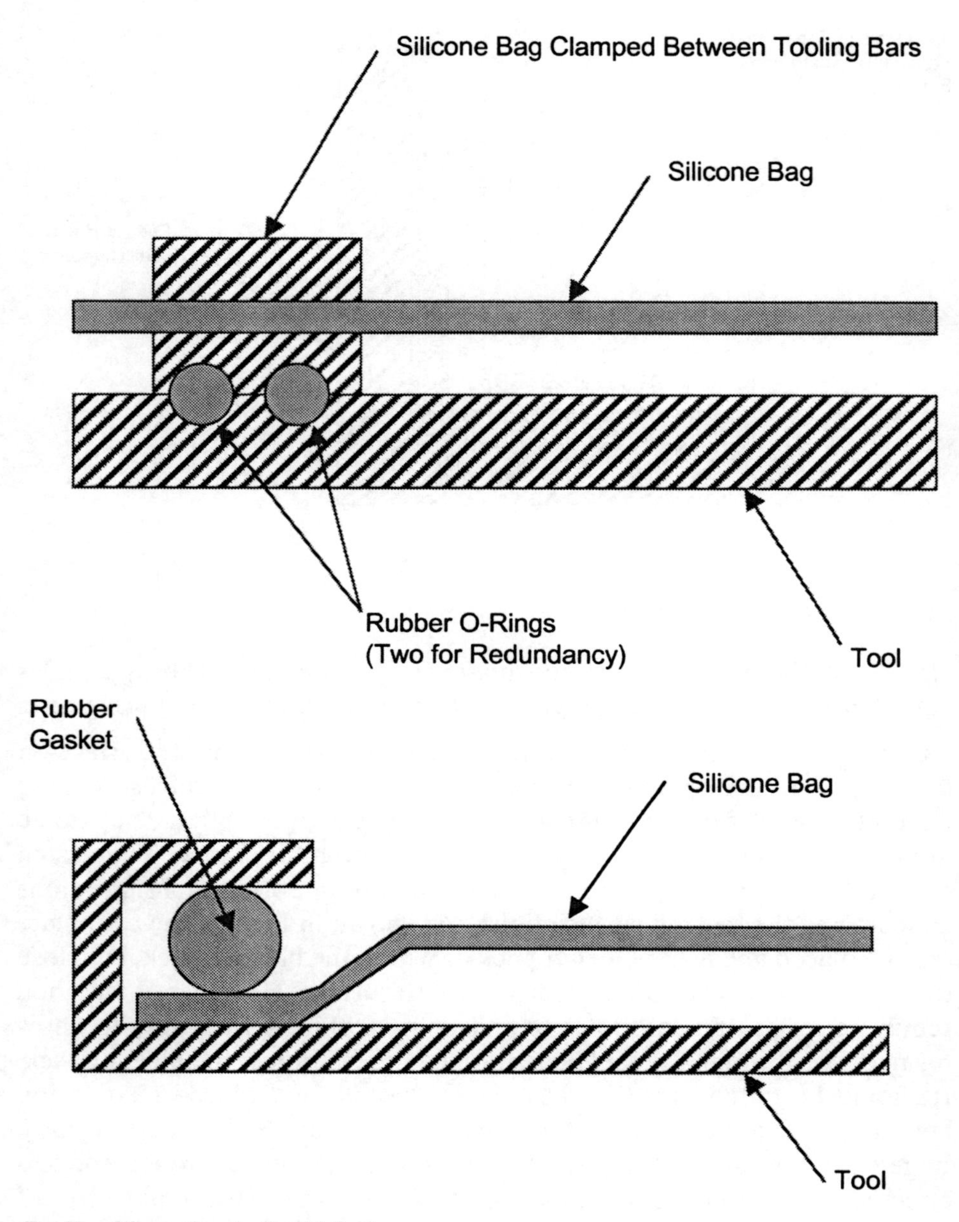

Fig. 33. Silicone Bag Sealing Methods

The current trend in the composites industry is towards net or near-net resin systems (32–35% resin by weight) that require little or no bleeding, in contrast to the more traditional 40–42% resin systems. As shown in Fig. 36, this simplifies the bagging system since the labor and cost of the

The Boeing Company

Fig. 34. Large Reusable Silicon Rubber Vacuum Bag

bleeder material is eliminated. However, when using this type of material, it is even more important to properly seal the inner bag system to prevent resin loss during cure, because it may result in resin starved laminates. The edges are a particularly critical area where excessive gaps or leaks in the dams can result in excessive resin loss and thinner than desired edges. In addition to the elimination of the need for a bleeder pack, net resin content prepregs produce laminates with a more uniform thickness and resin content. The problem with the traditional 40–42% resin content prepregs is that as the laminate gets thicker (i.e., larger number of plies), the ability to bleed resin through the thickness decreases. As more and more bleeder is added, the plies closest to the surface become overbled while those in the middle and on the tool side of the laminate are underbled as shown in Fig. 37.

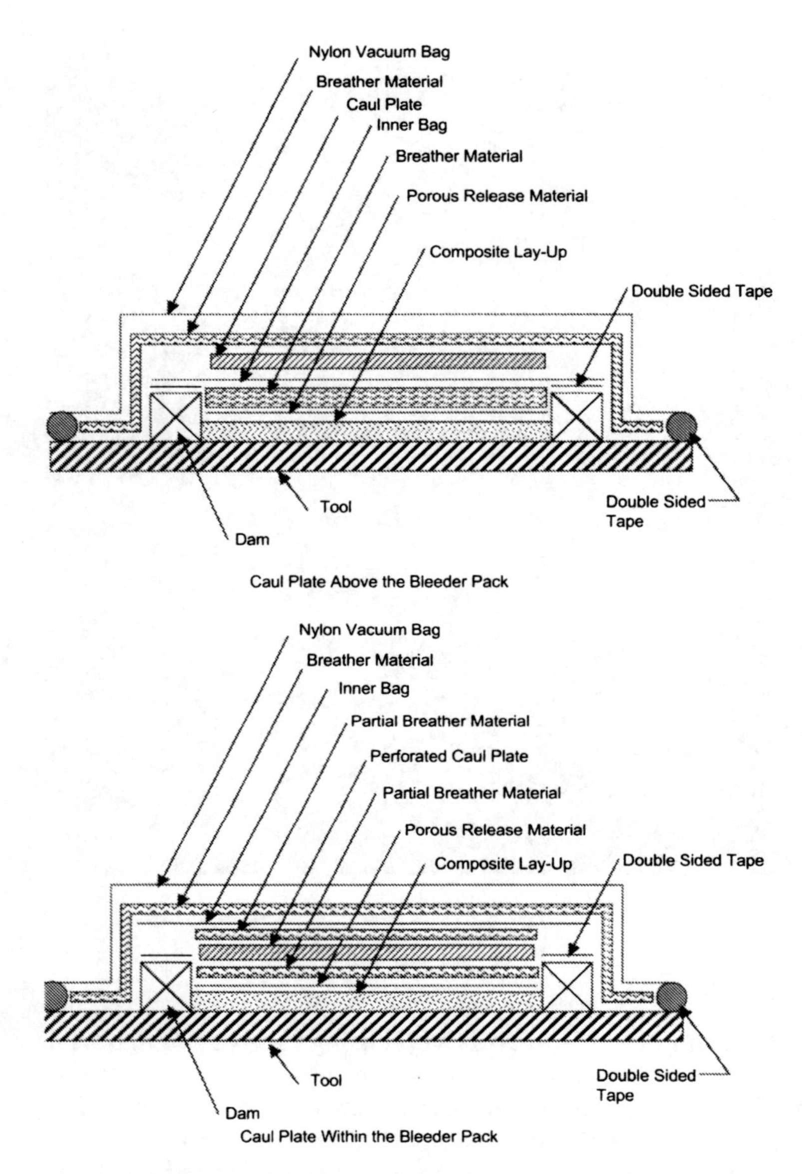

Fig. 35. Possible Caul Plate Arrangements

Once the vacuum bag has been successfully leak checked and the thermocouples applied, it is ready for autoclave cure. A slight vacuum should be maintained on the bag while it is waiting for autoclave cure to make sure that nothing shifts or that wrinkles will not form when the full vacuum is applied when placed in the autoclave. If the lay-up contains honeycomb core, the maximum vacuum that should be applied during leak

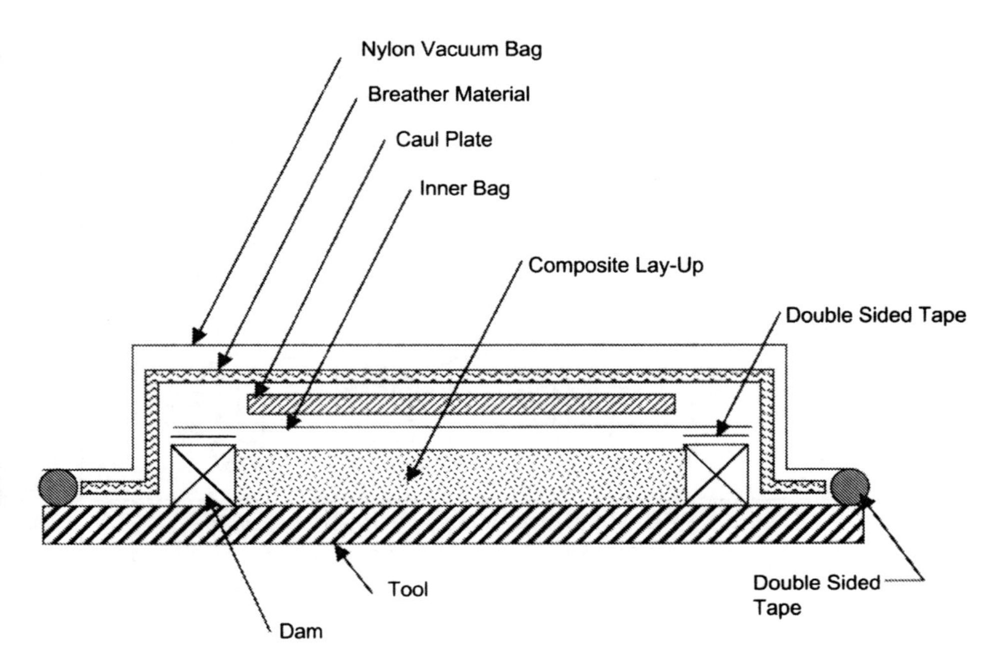

Fig. 36. Bagging Schematic for No Bleed Resin System

checking or cure is 8–10 inches of mercury vacuum. Higher vacuums have been known to cause core migration and even crushing due to the differential pressure that can develop in the core cells.

5.11 Summary

Manual lay-up is a labor intensive process where each ply is laid up on the tool one ply at a time. To make the job less costly, wide material product forms (broadgoods) were developed that can be cut on NC programmed ply cutters, such as reciprocating knife or ultrasonically driven knife cutters. In addition, laser ply projection systems are used to show the exact location of each ply. Flat ply collation followed by vacuum forming to final shape can further reduce costs if the stacks can be formed to the contour of the tool.

ATL is a process that is amenable to large, flat or mildly contoured parts. Two types of machines are currently in use: (1) FTLMs and (2) CTLMs capable of handling contours approaching 15°. Tape laying machines normally use either 3, 6, or 12 inch wide unidirectional tape.

Filament winding is a process in which a fiber band is placed on a rotating mandrel. It is capable of placing as much as 100–400 lbs h^{-1}. It is an excellent lay-up process for bodies of revolution such as cylinders and can produce parts smaller than 1 inch in diameter up to about 20 ft in diameter.

Number Of Plies	Autoclave Cure Pressure (psig) 25	50	100
10	Bag A: 29.0 / B: 28.5 Tool	Bag A: 27.4 / B: 26.3 Tool	Bag A: 26.6 / B: 26.7 Tool
25 (1)	Bag A: 29.4 / 30.7 B: 28.6 / 30.8 Tool	Bag A: 29.5 / 30.1 B: 28.6 / 30.8 Tool	Bag A: 27.5 / 29.9 B: 26.5 / 30.6 Tool
60 (2)	Bag A: 33.7 / 38.7 / 39.6 B: 33.2 / 38.4 / 39.3 Tool	Bag A: 30.4 / 38.7 / 39.7 B: 30.8 / 39.2 / 40.1 Tool	Bag A: 28.5 / 37.1 / 39.6 B: 29.6 / 38.2 / 38.9 Tool

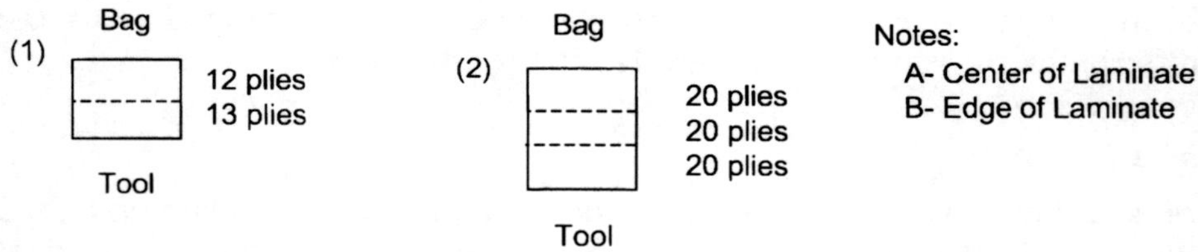

Fig. 37. Through-The-Thickness Resin Content Variations For 40–42 % Resin Content Systems

It cannot lay material into concave cavities because of fiber tension. The three methods of filament winding are helical, polar and hoop winding.

Fiber placement is a hybrid process combining some of the features of ATL and filament winding. Since there is only minimal tension on the fibers during fiber placement, material can be placed into concave cavities and true 0° plies can be placed.

After lay-up, the part is sealed in a vacuum bag for oven or autoclave curing. If the prepreg contains excess resin, then an absorbent bleeder material must be used to remove the excess resin during cure. Breather materials are used to help evacuate all air and volatiles from the part and bag. Caul plates can be placed on the bag side surface to provide smoother surface finishes.

Ply collation remains one of the highest cost areas of composite part fabrication, normally representing 40–60% of the part fabrication cost. Although several automated processes, such as tape laying, filament winding and fiber placement, have been developed to reduce the manual labour costs, manual or hand collation of plies still remains the primary method for high-performance composite parts. Even though hand collation may at first seem the most expensive of the available methods, wide broadgoods combined with automated ply cutting, flat ply collation methods and laser projection systems have helped to reduce the costs. Prior to investing or committing a part to an automated manufacturing method, a careful analysis should be conducted on a part-by-part basis to determine the most cost efficient process.

References

[1] Carpenter J.F., Juergens R.J., "Viscosity/Flow Behavior of Composite Resins", SAMPE Symposium, April 1986.

[2] Virtek Laser Edge product literature

[3] Huttrop M.L., "Cost Reduction Through Design for Automation", 5th Conference on Fibrous Composites in Structural Design, January 1981.

[4] Young M., Paton R., "Diaphragm Forming of Resin Pre-Impregnated Woven Carbon Fibre Materials", 33rd International SAMPE Technical Conference, November 2001.

[5] Grimshaw M.N., "Automated Tape Laying", In *ASM Handbook Vol. 21 Composites*, ASM International, 2001.

[6] Grimshaw M.N., Grant C.G., Diaz J.M.L., "Advanced Technology Tape Laying for Affordable Manufacturing of Large Composite Structures", 46th International SAMPE Symposium, May 2001, pp. 2484–2494.

[7] Peters S.T., Humphrey W.D., Foral R.F., *Filament Winding Composite Structure Fabrication*, SAMPE, 2nd Edition, 1999.

[8] Mantel S.C., Cohen D., "Filament Winding", In *Processing of Composites*, Hanser, 2000.

[9] Grover M.K., *Fundamentals of Modern Manufacturing: Materials, Processes, and Systems*, Prentice-Hall, 1996.

[10] Grimshaw M.N., Grant C.G., Diaz J.M.L., "Advanced Technology Tape Laying for Affordable Manufacturing of Large Composite Structures", 46th International SAMPE Symposium, May 2001, pp. 2484–2494.

[11] Adrolino J.B., Fegelman T.M., "Fiber Placement Implementation for the F/A-18 E/F Aircraft", 39th International SAMPE Symposium, April 1994, pp. 1602–1616.

[12] Dobrowolski A., White N., "Re-usable Customized Vacuum Bags", 33rd International SAMPE Technical Conference, November 2001.

Chapter 6

Curing: It's a Matter of Time (t), Temperature (T) and Pressure (P)

Autoclave curing is the most widely used method of producing high-quality laminates in the aerospace industry. Autoclaves are extremely versatile pieces of equipment. Since the gas pressure is applied isostatically to the part, almost any shape can be cured in an autoclave. The only limitation is the size of the autoclave and the large initial capital investment to purchase and install an autoclave. A typical autoclave system, shown in Fig. 1, consists of a pressure vessel, a control system, an electrical system, a gas generation system and a vacuum system. Autoclaves lend considerable versatility to the manufacturing process. They can accommodate a single large composite part, as illustrated in Fig. 2, for a large wing skin or numerous smaller parts loaded onto racks and cured as a batch. While autoclave processing is not the most significant cost driver in total part cost, it does represent a culmination of all the previously performed manufacturing operations, because final part quality (per ply thickness, degree of cross-linking, and void and porosity content) is determined during this operation.

Autoclaves are normally pressurized with inert gas, usually nitrogen or carbon dioxide. Air can be used, but it increases the danger of a fire within the autoclave during the heated cure cycle. A schematic of the gas flow in a typical autoclave is shown in Fig. 3. The gas is circulated by a large fan at

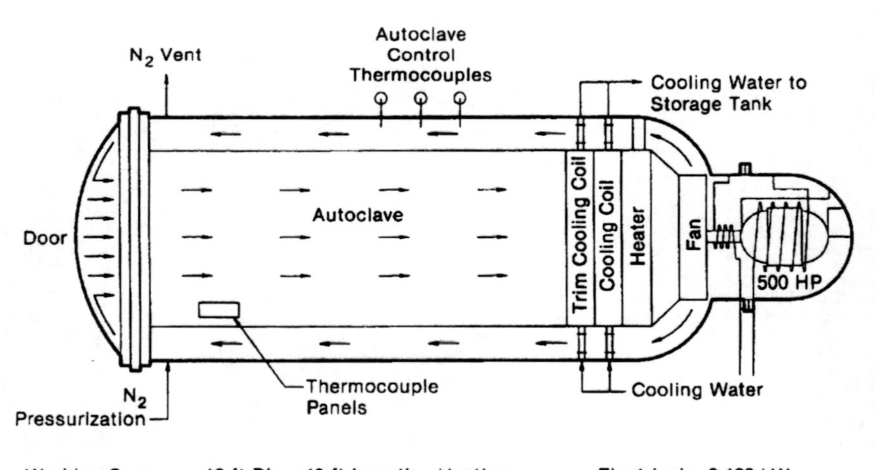

Fig. 1. Typical Production Autoclave Schematic

The Boeing Company

Fig. 2. Loading a Large Wing Skin for Autoclave Cure

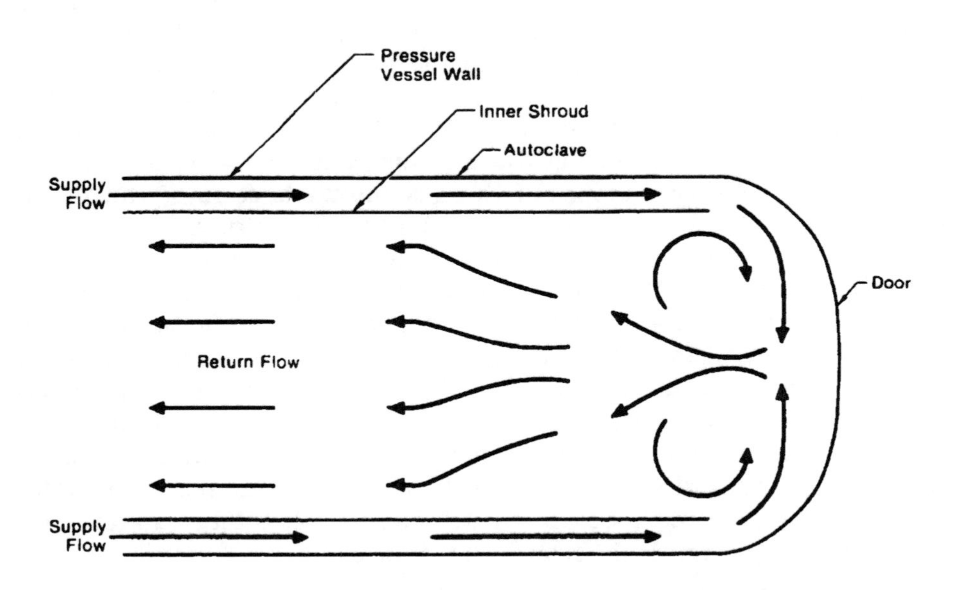

Reference 1: with permission

Fig. 3. Gas Flow Velocity Turbulence Near Door

the rear of the vessel and passes down the walls next to a shroud containing the heater banks, usually electrical heaters, but steam heating is sometimes used for older autoclaves. The heated gas strikes the front door and then flows back down the center of the vessel to heat the part. There is considerable turbulence in the gas flow near the door,[1] which produces higher velocities that stabilize as the gas flows towards the rear. The practical effect of this flow field is that you can often encounter higher heating rates for parts placed close to the door; however, the flow fields are dependent on the actual design of the autoclave and its gas flow characteristics.

Another problem that can be encountered is blockage, in which large parts can block the flow of gas to smaller parts located behind them. Manufacturers typically use large racks (Fig. 4) to insure even heat flow and maximize the number of parts that can be loaded for cure.

At least three non-tool-related variables are known to affect convective heat transfer in an autoclave: (1) increasing the gas pressure improves the convective heat transfer; (2) increasing the gas velocity improves the convective heat transfer; and (3) increasing the gas flow turbulence improves the convective heat transfer. In one study[1], a standard production size autoclave was used that was 40 ft in length and 12 ft in diameter with a heating capability to 650 °F and a maximum pressure of 150 psig. The autoclave was pressurized with nitrogen gas circulated by a 600 rpm fan producing gas flow of 60,000 cfm. The results showed that: (1) the highest velocities occurred near the front and along the centerline of the autoclave. As a result, a low or zero velocity region occurred at some outer radial positions. Analysis showed that recirculation occurs in this region. The high-velocity region is caused by the supply flow striking the autoclave door at the edges and moving towards the center to form a backward flowing high-velocity plume. As the gas moves backward, the plume dissipates and the flow becomes slower and more uniform. This velocity drop over the tools in the middle of the autoclave causes their convection coefficient to be lower than for parts located near the door. (2) The velocity was noticeably higher at the top of the autoclave than at the sides. (3) The average axial velocity was between 10 and 16 fps; however, velocities as high as 44 fps occurred near the door. (4) The axial variation in velocity started to decrease approximately 13 ft from the door (entrance) and became more and more uniform downstream. (5) The turbulence intensity was very high at the door region, typically 13–15%. It dampened out at the same time that the radial velocity distribution became more uniform and kept decreasing towards the rear of the autoclave.

Tool design can also dramatically affect part heat-up rates. Recommendations for the design of an individual tool are fairly obvious and well understood in the industry (e.g., thin tools heat faster than thick

GKN Aerospace Services

Fig. 4. Production Batching Using Racks

tools); materials with high thermal conductivity heat faster than those with lower thermal conductivity; and tools with well-designed gas flow paths heat up faster than those with restricted flow paths (e.g., tools with open egg-crate support structures heat up faster than those without open support structures). Match die tooling, often used for complex parts and large unitized structures, presents its own special set of problems. Tool dimensions and fit become critical. If the tool is not dimensionally correct, it will not be possible to obtain quality parts. Mismatches and incorrect dimensions will result in high- and low-pressure areas during cure, resulting in excessive thin-out and voids and porosity.

Composite parts can also be cured in presses or ovens. The main advantage of a heated platen press is that much higher pressures (e.g., 500–1,000 psig) can be used to consolidate the plies and minimize void formation and growth. Presses are often used with polyimides that give off water, alcohol or high-boiling-point solvents, such as N-methylpyrrolidone (NMP). On the other hand, presses usually require matched metal tools for each part configuration and are limited by platen size to the number of parts that can be processed at one time. Ovens, usually heated by convective forced air, can also be used to cure composite structures; however, since pressure is provided by only a vacuum bag (–14.7 psia), the void contents of the cured parts are normally much higher (e.g., 5–10%) than those of autoclave cured parts (1–2%).

6.1 Curing of Epoxy Composites

The thermal portion of a classic cure cycle for a 350 °F curing thermoset epoxy part is shown schematically in Fig. 5. It contains two ramps and two isothermal holds. The first ramp and isothermal hold, usually in the range of 240–280 °F, is used to allow the resin to flow (bleed) and volatiles to escape. The imposed viscosity curve in the figure shows that the semi-solid resin matrix melts on heating and experiences a dramatic drop in viscosity. The second ramp and hold is the polymerization portion of the cure cycle. During this portion, the resin viscosity initially drops slightly due to the application of additional heat and then rises dramatically as the kinetics of the resin start the cross-linking process. The resin gels into a solid and the cross-linking process continues during the second isothermal hold, usually at 340–370 °F for epoxy resin systems. The resin is held at this cure temperature for normally 4–6 h, allowing time for the cross-linking process to be completed. It should be noted that as the industry has moved towards net resin content systems, the use of the first isothermal hold which allows time for resin bleeding has been eliminated by many manufacturers resulting in a straight ramp-up to the curing temperature as illustrated in Fig. 6.

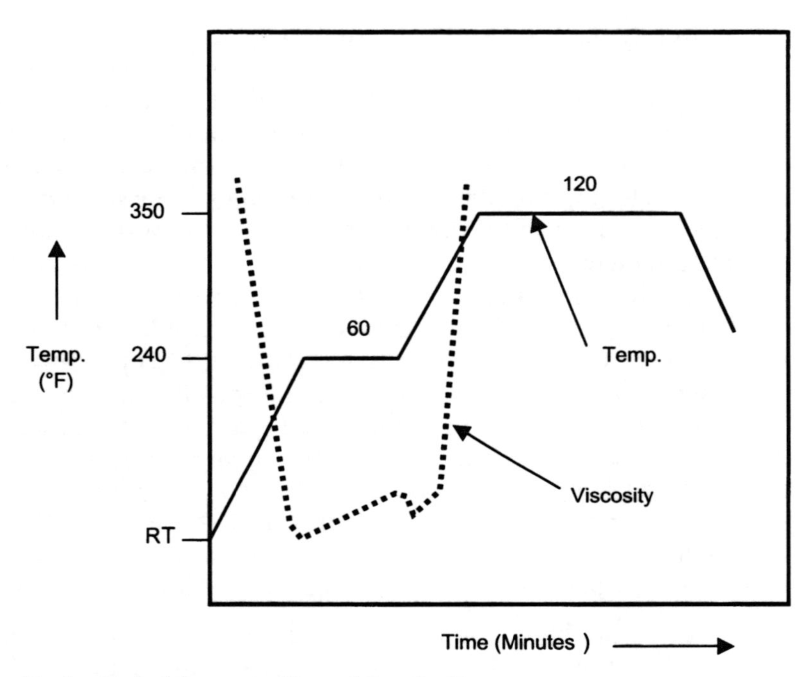

Fig. 5. Typical Composite Thermal Cure Profile

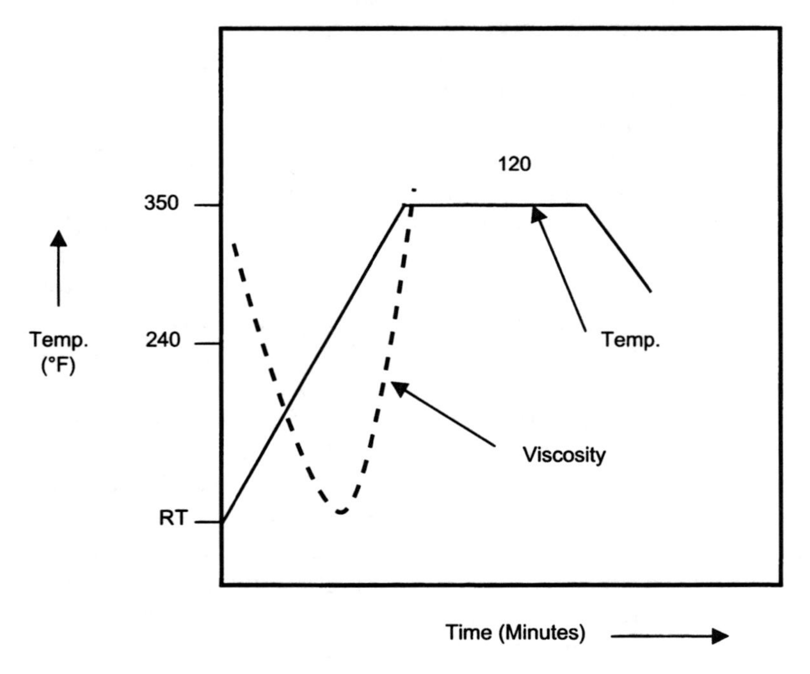

Fig. 6. Straight Ramp-Up Cure Cycle

High pressures (e.g., 100 psig) are commonly used during autoclave processing to provide ply compaction and suppress void formation. Autoclave gas pressure is transferred to the laminate due to the pressure difference between the autoclave environment and the vacuum bag interior. Translation of the autoclave pressure to the resin depends on several factors, including the fiber content, laminate configuration and the amount of bleeder used. The classical approach to applying autoclave pressure during the cure cycle is shown in Fig. 7. In this approach, during the ramp-up to the first hold, only vacuum pressure is applied and maintained until the end of the first isothermal hold. At that point, autoclave pressure is applied, normally 85–100 psig for epoxies and the vacuum pressure is removed by venting to the atmosphere. The rationale behind this approach is that the vacuum will help to remove volatiles from the melting resin while the application of the higher autoclave pressure would tend to trap them in the laminate. At the end of the first isothermal hold, full autoclave pressure is applied to insure that the laminate is well compacted before the resin viscosity rises to gel, otherwise the laminate would be poorly compacted and contain numerous voids and porosity.

This approach to applying autoclave pressure can cause problems in a production environment. If the autoclave contains a large number of parts with varying heat-up rates, the actual point in time to vent the bag to atmosphere and apply autoclave pressure can be questionable. A typical example of the dilemma facing the autoclave operator is shown in Fig. 8. It is not clear when the hold period should start or when is the proper point to vent the bag and apply full autoclave pressure. Again, if the resin gels during this first isothermal hold with only vacuum pressure applied to the laminate, then the probability of gross porosity is very high. It should also be pointed out that unpressurized autoclaves are very inefficient at the heating parts, i.e., the higher the autoclave pressure, the more efficient the heat transfer. A second problem with applying only vacuum pressure during the initial portion of the cure cycle deals with hydrostatic resin pressure,[2] as illustrated in Fig. 9. Even though a relatively high autoclave pressure (e.g., 100 psig) may be used during the cure cycle, the actual pressure on the resin, the hydrostatic resin pressure, may be significantly less. Because of the load-carrying capability of the fiber bed in a composite lay-up, the hydrostatic resin pressure is typically less than the applied autoclave pressure. With only vacuum pressure applied during the initial part of the cure cycle, the hydrostatic resin pressure on the resin can be extremely low, even negative. This is an ideal condition for void formation and growth if allowed to persist to high enough temperatures. The hydrostatic pressure is critical because it is the pressure that helps to keep volatiles dissolved in solution. If the resin pressure drops below the volatile vapor pressure, then the volatiles will come out of solution and form voids.

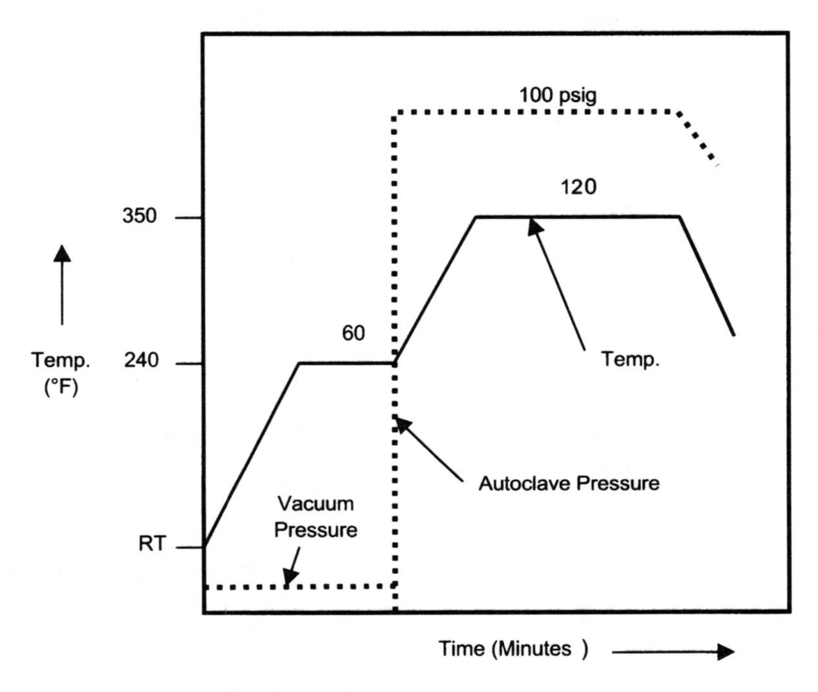

Fig. 7. Classical Carbon/Epoxy Cure Cycle

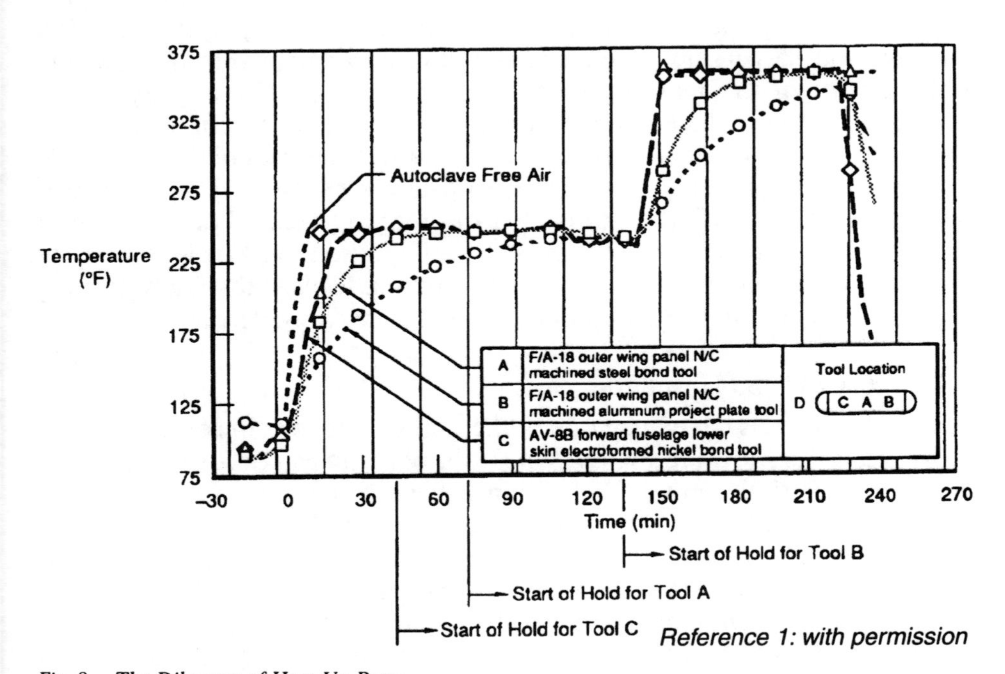

Fig. 8. The Dilemma of Heat-Up Rates

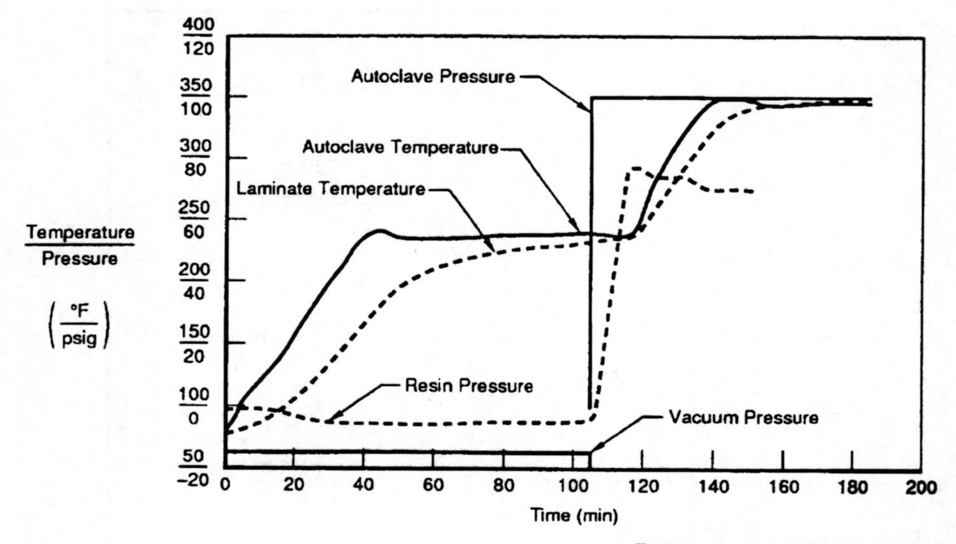

Fig. 9. Vacuum Only Can Cause Negative Pressure

To circumvent both of these problems in a production environment, a significant portion of the autoclave pressure can be applied immediately before initiating the heat-up cycle. For standard epoxy systems, a full vacuum and 85 psig autoclave pressure can be applied through the first hold, and then the bag vented to atmosphere and 100 psig autoclave pressure applied before ramping up to the final cure temperature. This approach, shown in Fig. 10, applies full vacuum at the start of the cure cycle and also applies autoclave pressure, 85 psig for example. The vacuum is again maintained until the end of first isothermal hold and then vented to atmosphere while the autoclave pressure is increased to 100 psig. This cycle was developed when a large number of parts on tools with widely varying heat-up rates had to be loaded in an autoclave for a single cure.

6.2 Theory of Void Formation

Porosity and voids have been one of the major problems in composite part fabrication. As shown in Fig. 11, voids and porosity can occur at either the ply interfaces (interlaminar) or within the individual plies (intralaminar). The terms "voids" and "porosity" are used fairly interchangeably in industry; however, the term "void" usually implies a large pore, whereas "porosity" implies a small pore. Void formation and growth in addition curing composite laminates is primarily due to entrapped volatiles.[3] Higher temperatures result in higher volatile pressures. Void growth will potentially occur if the void pressure (i.e., the volatile vapor pressure)

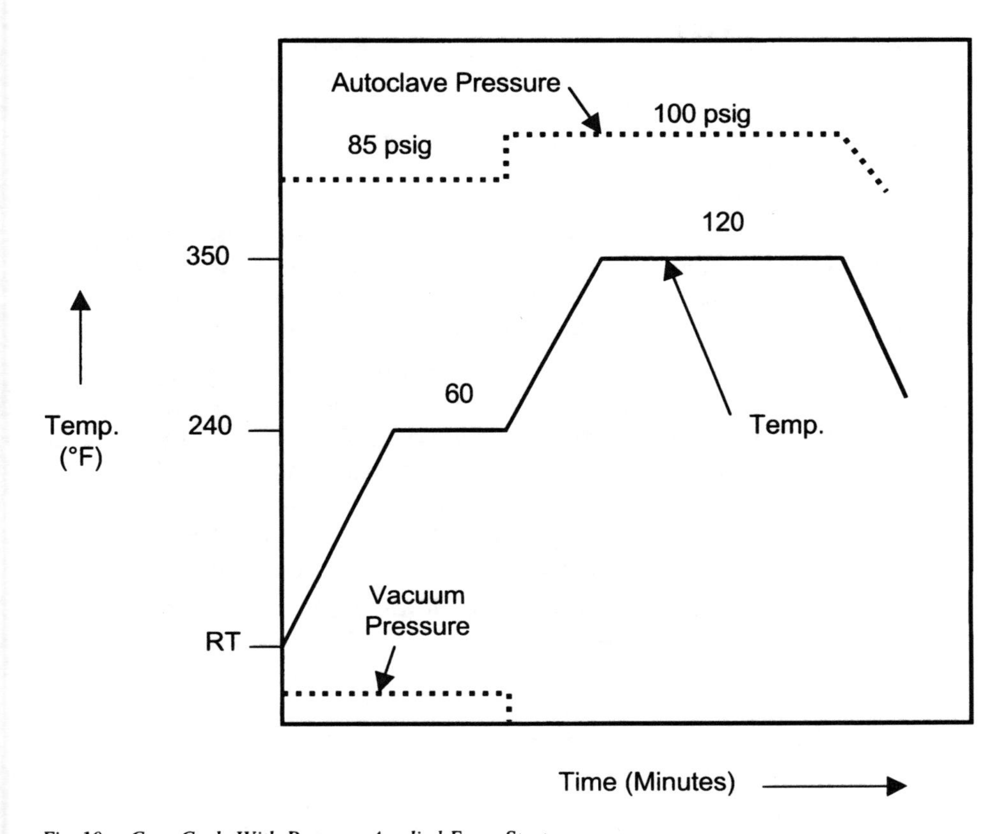

Fig. 10. Cure Cycle With Pressure Applied From Start

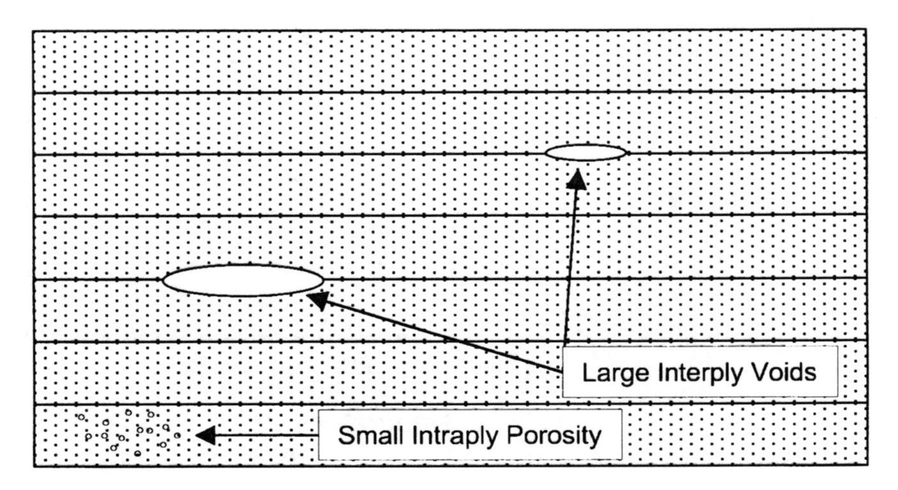

Fig. 11. Interply and Intraply Voids and Porosity

exceeds the actual pressure on the resin (i.e., the hydrostatic resin pressure) while the resin is a liquid (Fig. 12). The prevailing relationship, therefore, is:

$$P_{\text{Void}} > P_{\text{Hydrostatic}} \rightarrow \text{void formation and growth}$$

When the liquid resin viscosity dramatically increases, or gellation occurs, the voids are locked into the resin matrix. Note that the applied pressure on the laminate is not necessarily a factor. As will be shown later, the hydrostatic resin pressure can be low even though the applied autoclave pressure is high, leading to void formation and growth.

Composite prepregs, like most organic materials, absorb moisture from the atmosphere. The amount of moisture absorbed is dependent on the relative humidity of the surrounding environment, while the rate of moisture absorption is dependent on the ambient temperature. While the carbon fibers themselves absorb minimal moisture, epoxy resins readily absorb moisture. Thus, the final prepreg moisture content is a function of the relative humidity, ambient temperature and prepreg resin content.

Since moisture is typically the most predominate volatile present in hot melt addition-curing prepregs, the amount of absorbed moisture in the

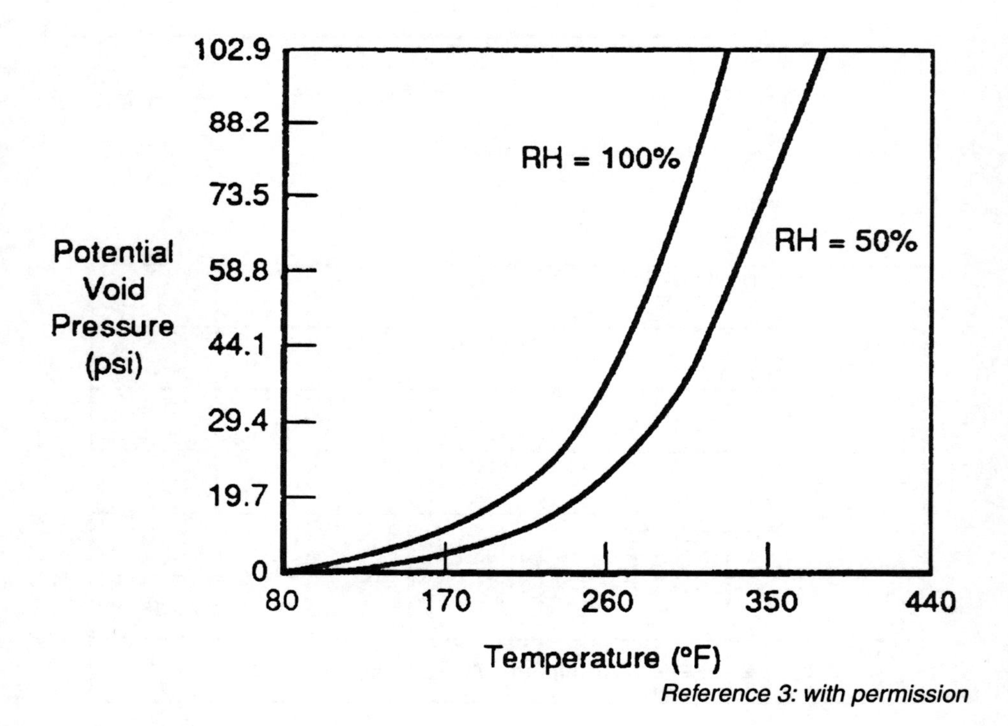

Fig. 12. Potential Void Formation

prepreg determines the resultant vapor pressure of volatiles generated during the cure cycle. An examination of Fig. 12 explains why composite fabricators control the lay-up room environment, i.e., higher moisture contents result in higher vapor pressures, increasing the propensity for void formation and growth. Even though we are dealing with relatively small amounts of moisture (e.g., 1%) as opposed to the larger amounts of moisture, other polymers such as nylons can absorb; even these small percentages can generate large gas volumes and pressures when heated. It should also be pointed out that other types of volatiles, such as solvents used in a solvent impregnation process or volatiles resulting from condensation curing reactions, can greatly complicate this problem, leading to much higher vapor pressures and a much greater propensity for void formation.

To fully understand void growth, an appreciation of the importance of hydrostatic resin pressure must be developed. Because of the load-carrying capability of the fiber bed in a composite lay-up, the hydrostatic resin pressure needed to suppress void formation and growth is typically only a fraction of the applied autoclave pressure.[3] The hydrostatic resin pressure is critical, because it is this pressure that helps to keep volatiles dissolved in solution. If the resin pressure drops below the volatile vapor pressure, the volatiles will come out of solution and form voids. To develop a better understanding of the interactions of the resin flow process, hydrostatic resin pressure and the load-carrying capability of the fiber bed, a mechanical analogy,[2] is presented in Fig. 13. In this analogy, a laminate undergoing cure is simulated as a piston-spring-valve setup. The spring represents the fiber bed and is assumed to have a load-carrying capability. Just like a spring, the fiber bed will support larger and larger loads as it undergoes compression. The liquid contained in the piston represents the ungelled resin. Finally, the valve is the means by which the liquid resin can leave the system, i.e., it could be representative of bleeder, a poorly dammed part or any other system leak.

A description of each step in this simplified model is given below:

Step 1. Initially, there is no load on the system. The liquid hydrostatic pressure and the load carried by the fiber bed is zero.

Step 2. A 100 lb load is applied to the system but no liquid has escaped (closed valve). The liquid carries the entire load and the load on the fiber bed is zero. Note that the downward force (in this case 100 lb) is equal to the upward force (again 100 lb). This upward force is the sum of the load carried by the liquid (100 lb) and the spring-like fiber bed (0 lb).

Step 3. The valve is now opened allowing resin to escape (e.g., resin bleeding). However, at this point the resin still carries the entire 100 lb load.

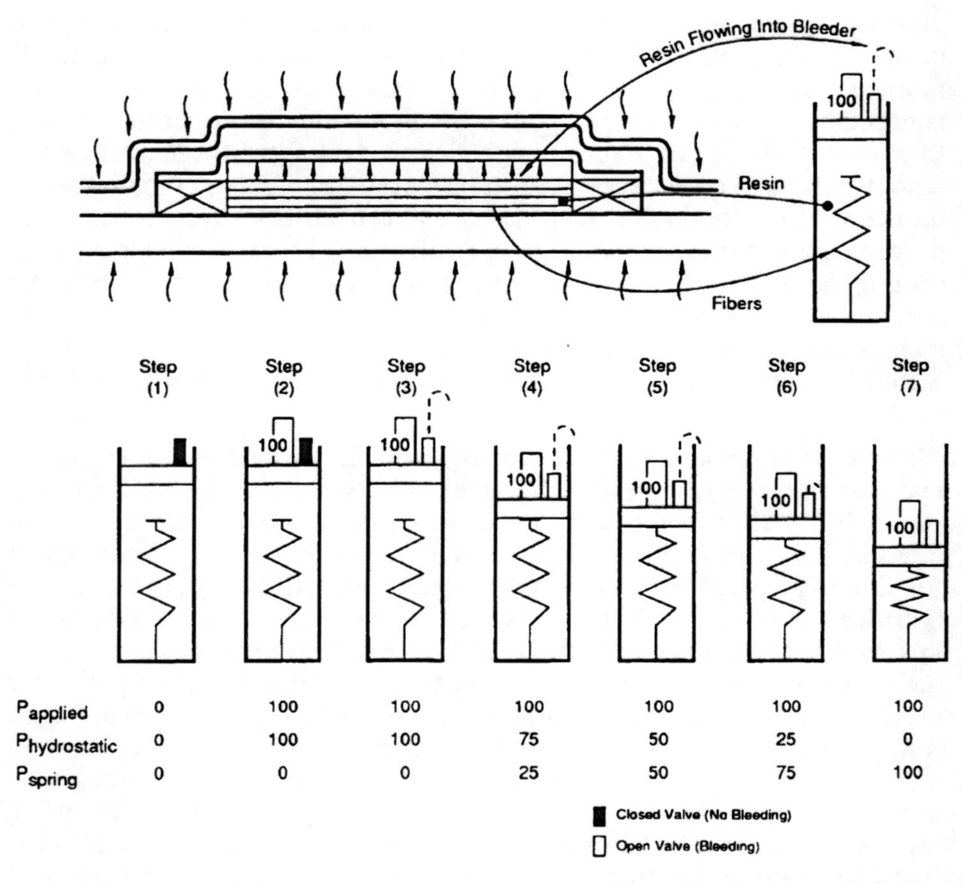

Fig. 13. Resin Flow Analogy

Step 4 . Liquid continues to escape but at a decreasing rate due to a portion of the load now being carried by the spring (25 lb). This is analogous to bleeding in a laminate occurring rapidly until the fiber bed starts supporting a portion of the applied autoclave pressure.

Steps 5 and 6 . Liquid continues to escape, but at an ever decreasing rate since a greater portion of the load is borne by the spring. In an actual laminate, the rate of bleeding would be retarded both by the increasing load-carrying capability and the reduced permeability of the fiber bed as it is compacted.

Step 7 . No further bleeding occurs because the pressure on the resin has now dropped to zero and the entire load (100 lb) is being carried by the spring. If this condition occurs during actual autoclave processing before the resin gels (solidifies), it would be quite easy for dissolved volatiles to vaporize out of solution and form voids.

Although this analogy greatly simplifies the composite flow process, it does illustrate several key points about the resin flow process. In the early stages of the cure cycle, the hydrostatic resin pressure should be equal to the applied autoclave pressure. As resin flow occurs, the resin pressure drops. If a laminate is severely overbled, the resin pressure could drop to a low enough value to allow void formation. Thus, the hydrostatic resin pressure is directly dependent on the amount of resin bleeding that occurs. As the amount of bleeding increases, the fiber volume increases, resulting in an increase in the load-carrying capability of the fiber bed. It should be noted that resin flow and bleeding can be intentional or unintentional. Intentional bleeding is of course the bleeder cloth used to remove the excess resin from the prepreg during cure. Examples of unintentional bleeding are excessive gaps between the dams and laminate, tears in the inner bag that allow resin to flow into the breather material, and mismatched tooling details that allow escape paths for the liquid resin. Therefore, the hydrostatic resin pressure is directly dependent on the amount of resin bleeding that occurs. As the amount of resin bleeding increases, the fiber volume increases, resulting in increase in the load-carrying capability of the fiber bed and a decrease in the hydrostatic resin pressure.

The rate of bleeding is dependent on several factors, including the permeability of the fiber bed, both vertically and horizontally, and the viscosity of the liquid resin. The permeability of the fiber bed will depend on the weave of the fabric, the fiber diameter and fiber volume fraction. The resin viscosity is determined by the chemistry of the resin and the thermal profile of the cure cycle. The cure cycle greatly affects the resin viscosity and the flow process, directly through the pressure application and indirectly through the effect of the thermal profile on the resin viscosity. Referring back to the cure cycle shown in Fig. 5, the second ramp portion of this cure cycle is critical from a void nucleation and growth standpoint. During this ramp, the temperature is high, the resin pressure can be near its minimum and the volatile vapor pressure is high and rises with the temperature. These are the ideal conditions for void formation and growth.

Unfortunately, the void problem cannot be resolved simply by maintaining the hydrostatic resin pressure above the potential void pressure of the volatiles (although this is a good start). During collation or ply lay-up, air can become entrapped between the prepreg plies. The amount of air entrapped depends on many variables: the prepreg tack, the resin viscosity at room temperature, the degree of impregnation of the prepreg and its surface smoothness, the number of intermediate debulk cycles used during collation and geometrical factors such as ply drop-offs, radii, and so forth. The obvious places where entrapped air pockets form are the terminations of the internal ply drop-offs. In addition, air can be entrained in the resin itself during the mixing and prepregging operations. This entrained air can also lead to voids or at least serve as nucleation sites. To summarize, the void formation and growth process is complex and yet to be fully understood. However, a number of basic principles are fairly well understood and have been investigated through several research studies. Several of these studies are summarized in the next section.

6.3 Hydrostatic Resin Pressure Studies

Considerable research[2] has been conducted to develop a better understanding of resin pressure and the many variables that influence resin pressure. The majority of this work has been done with unidirectional carbon/epoxy; however, woven and unidirectional carbon/bismaleimide have also been studied. The experimental setup used for the initial studies is shown in Fig. 14. To measure the hydrostatic resin pressure, a transducer was recessed into the tool surface and filled with an uncatalyzed liquid resin. To assure that the laminate did not deflect under pressure and contact the transducer, a stiff wire screen was placed over the transducer recess.

Studies were initially conducted on unidirectional carbon/epoxy composites to evaluate the effects of (1) the type of epoxy resin (Hexcel's 3501-6 and 3502), (2) laminate bleeding (normal and overbleed), (3) pressure application (normal applied autoclave and internally pressurized bag (IPB)) and (4) laminate thickness (10 and 40 plies). In all cases, the lay-ups were cross-plied containing 0°, 90° and ±45° plies. For

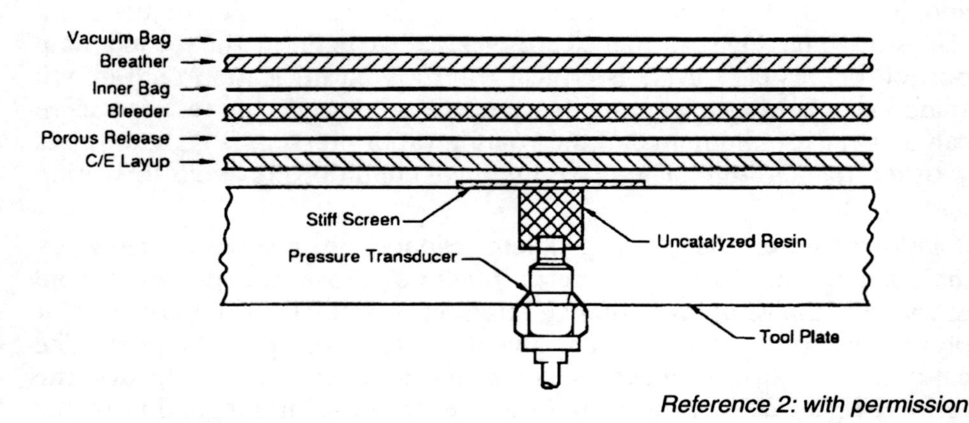

Fig. 14. Resin Pressure Setup

uniformity, the thermal cure cycle previously described in Fig. 5 was used for all of the tests.

These initial tests showed the following.

Ahigh-flowresinsystemexperienceslargerpressuredropsthanalow-flow system. A comparison of the resin pressures for 3501-6 and 3502 (Fig. 15) shows that the higher-flow 3502 resin system experiences a larger pressure drop than the lower-flow 3501-6 system. The 3501-6 system is a lower-flow system because it contains a boron trifluoride catalyst that significantly alters the cure behavior, resulting in a lower-flow system that gels at a lower temperature. Since high-flow resin systems are more prone to bleeding, additional care must be taken when they are tooled or bagged. High-flow laminates should be tightly bagged and sealed to eliminate leak paths. Since high-flow resin systems also typically have high gel temperatures, additional care must be taken to insure that the potential void pressure does not exceed the hydrostatic resin pressure prior to resin gellation.

Overbleeding a laminate causes a large drop in resin pressure. A comparison of normal bleeding with overbleeding is shown in Fig. 16. Overbleeding was accomplished by using 3 times the normal amount of glass bleeder and by removing the inner bag to create a free-bleeding condition. While this would normally not be done in composite part fabrication, overbleeding can and does occur because of leaky damming systems or leaky matched-die molds. The analysis of these cured laminates included nondestructive testing (NDT), thickness measurements, resin

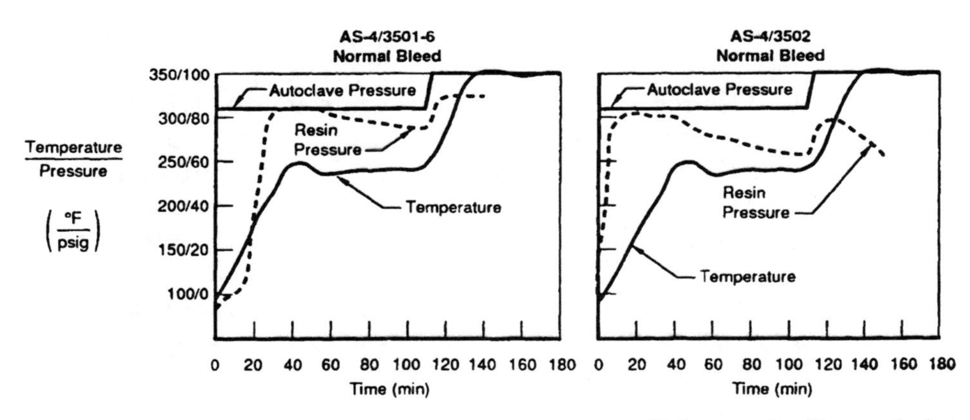

Fig. 15. Resin Flow Comparison

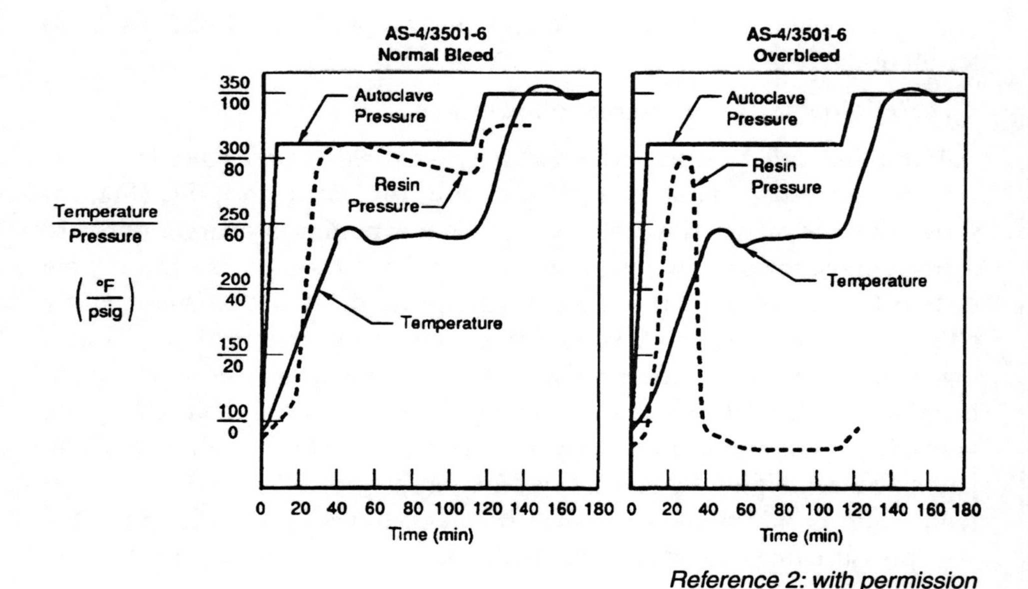

Fig. 16. Overbleeding Causes Pressure Loss

content determinations and materiallographic cross-sections. The resin contents and thickness measurements showed the dramatic effect of overbleeding. The resin contents and thickness values of the overbled laminate were significantly lower than those for standard bled laminate. Since the resin pressure actually dropped to below 0 psig, due to the vacuum pulled underneath the bag, little resistance to void growth existed. As expected, the ultrasonic NDT results and materiallographic cross-sections revealed gross voids and porosity in the overbled laminate.

Internal bag pressure can be used to maintain hydrostatic resin pressure and reduce resin flow. IPB curing was originally developed during the Ref. 4 program. In this process, two separate pressure sources are used: (1) a normal external applied autoclave pressure that provides the ply compaction and (2) a somewhat lesser internal bag pressure that applies pressure directly to the liquid resin to keep volatiles in solution and thereby prevent void nucleation and growth. An autoclave setup for IPB curing is shown in Fig. 17. In the cycle used for this experiment (Fig. 18), an external applied autoclave pressure of 100 psig was used along with an internal bag pressure of 70 psig. This results in a compaction, or membrane, pressure of 30 psig (applied autoclave pressure – internal bag pressure = 100 psig – 70 psig) on the plies and a hydrostatic pressure of 70 psig minimum on the resin. There is nothing magical about these pressure selections. If desired, the compaction pressure could be raised back up to

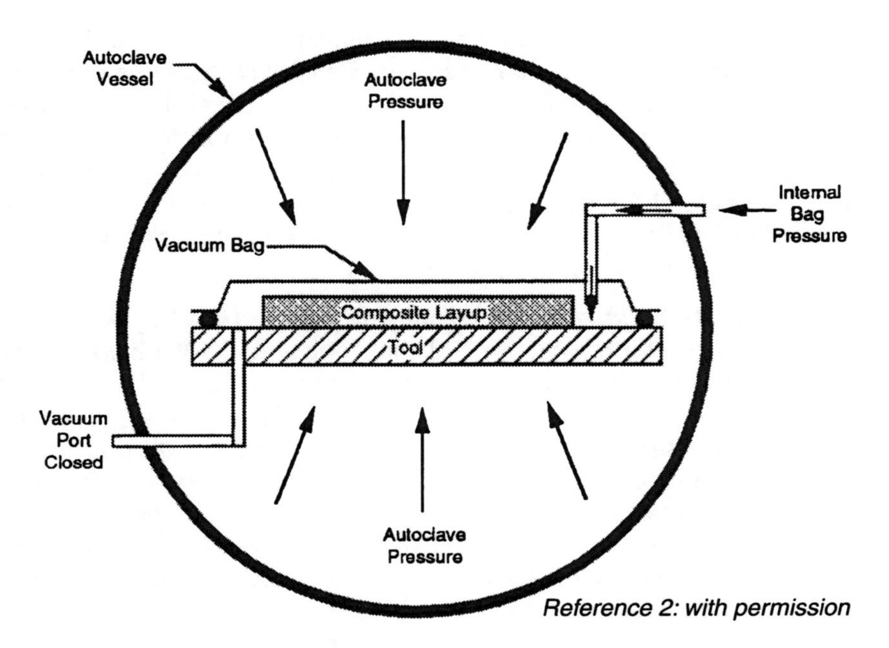

Fig. 17. Schelatic of Internally Pressurized Bag Curing

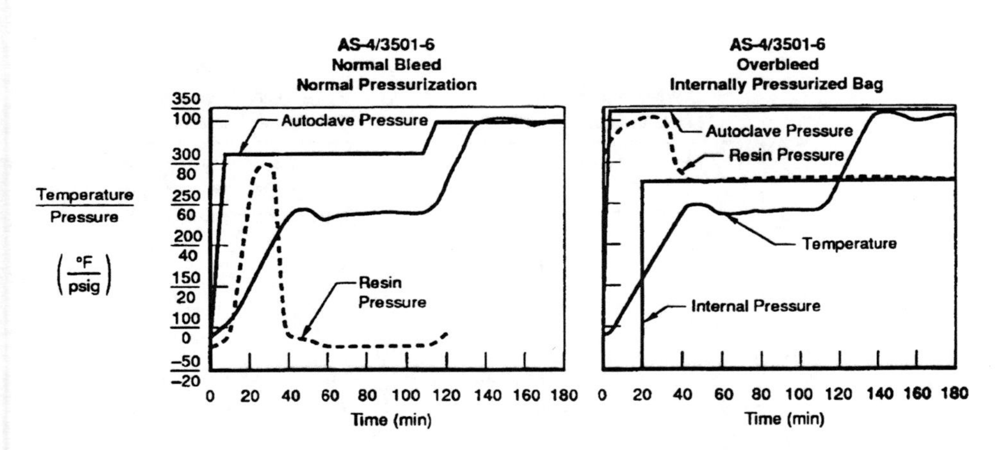

Fig. 18. IPB Maintains Resin Pressure

100 psig by simply increasing the applied autoclave pressure to 170 psig. The only restriction is that the applied autoclave pressure must be greater than the internal bag pressure to prevent blowing the bag off of the tool. Of course, the internal bag pressure needs to be high enough to keep the

volatiles in solution to prevent void nucleation and growth. To test a worst-case condition, the IPB laminate was bagged in the same manner as the previous over-bleed laminate in which the resin pressure had dropped to zero. Even though this bagging procedure resulted in a severely over-bled and porous laminate in the previous test, the addition of internal bag pressure prevented both over-bleeding and porosity. The absence of over-bleeding was a result of the lower membrane pressure (30 psig for the IPB cure versus 100 psig for the normal cure).

Thicker laminates maintain a higher resin pressure than thin laminates. The resin pressure results for a thin (10 plies) and a thick (40 plies) laminate are shown in Fig. 19. Note that the resin pressure for the thicker laminate essentially follows the applied autoclave pressure. However, since the resin pressure was measured at the tool surface, it was not known if a pressure gradient existed through the thickness of the thicker laminate. The higher resin pressure exhibited for the thicker laminate could be due to the inability to bleed thicker laminates. This has been qualitatively observed with materiallographic sections taken from 0.5 inch thick laminates, in which the surface plies appear to be overbled (i.e., thin per ply thicknesses), and the center and tool side plies appear underbled (i.e., thick per ply thicknesses).

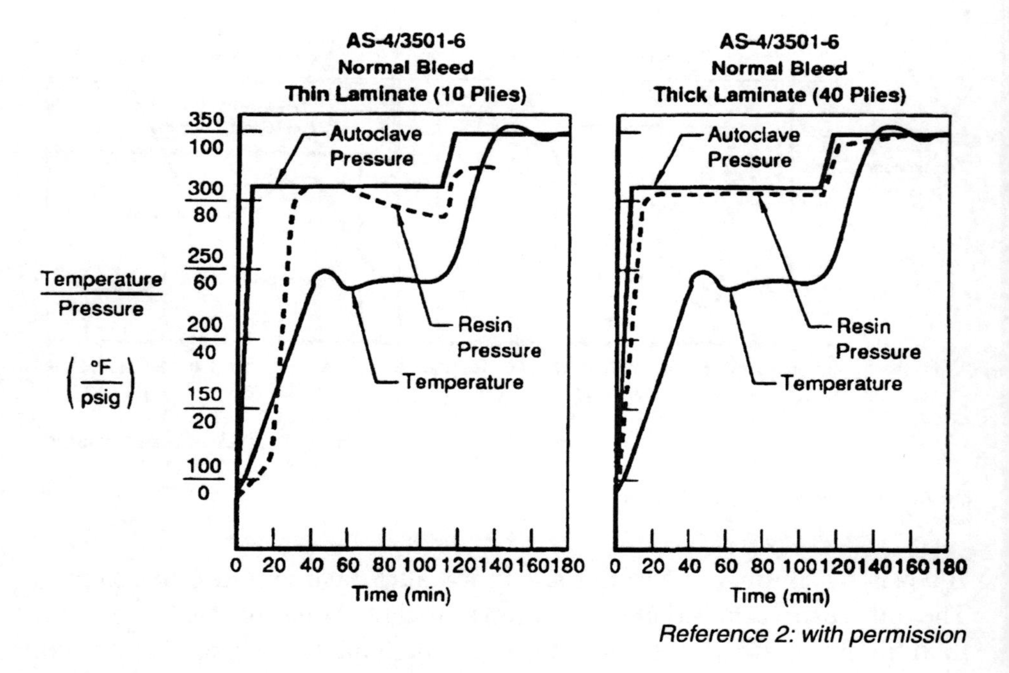

Fig. 19. Laminate Thickness Comparaison

To investigate the potential pressure gradients that can exist within a laminate during curing, miniature pressure transducers capable of measuring the hydrostatic resin pressure have been embedded at multiple locations within laminates to study the effects of vertical and horizontal pressure gradients.[2] Because the previous thick laminate test (40 plies) showed almost no pressure drop at the tool surface, a 60 ply thick laminate was collated with miniature pressure transducers embedded at multiple locations within the laminate and bleeder. Three tool mounted transducers were also used to monitor the resin pressure at several locations along the surface of the tool. The pressure curves from this vertical flow test (Fig. 20) confirmed that a significant pressure gradient can exist within a laminate. Again, the resin pressure at the tool surface was essentially equal to the applied autoclave pressure. These results also illustrate the vertical compaction process. Initially, the resin pressure within the entire laminate is near the applied autoclave pressure. As resin bleeding occurs, the hydrostatic resin pressure at the top of the laminate drops (transducer 3). At this point, resin begins to bleed from the middle of the laminate towards the top and the pressure drops here too (transducer 2) but remains above the pressure at the top. The opposite process occurs in the bleeder. As resin fills the bleeder, the pressure in the bleeder rises. Resin contents and photomicrographs were taken to confirm the resin pressure results. One resin content specimen was split into three pieces, one each from the top, middle and bottom 20 plies. The resin content at the top was 24.6%, the middle was 31.0% and the bottom was 33.0%, confirming the resin pressure results. The photomicrographs showed the same results, i.e., the

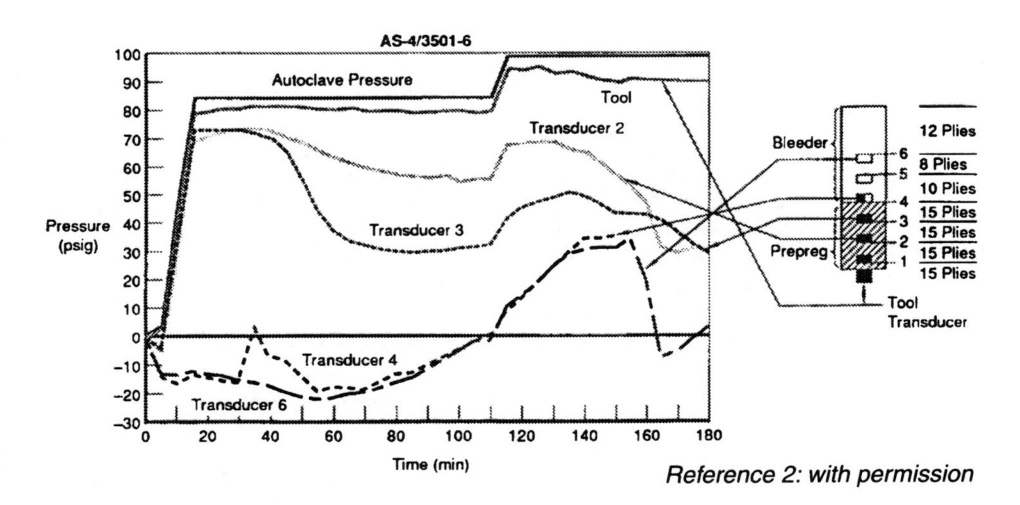

Fig. 20. Vertical Flow Laminate

top plies were compacted considerably more than those at the laminate bottom near the tool surface.

Just as the thick laminate experienced a pressure gradient, it was suspected that a horizontal pressure gradient also existed. Previously, all measurements had been conducted at the center of the laminate. Two laminates were processed to investigate horizontal pressure gradients. The first had cork dams located as close as possible to the laminate edges, while the second had a large (0.5 inch) gap between the laminate edges and the dams. Both laminates (Fig. 21) contained non-porous release film against the top surface to prevent vertical flow. The pressure curves for both laminates (Fig. 22) showed the existence of a horizontal pressure gradient and that the magnitude of the gradient depends on the amount of horizontal flow, i.e., the larger gap distance between the laminate edges and the dams resulted in more horizontal flow. The pressure curves also

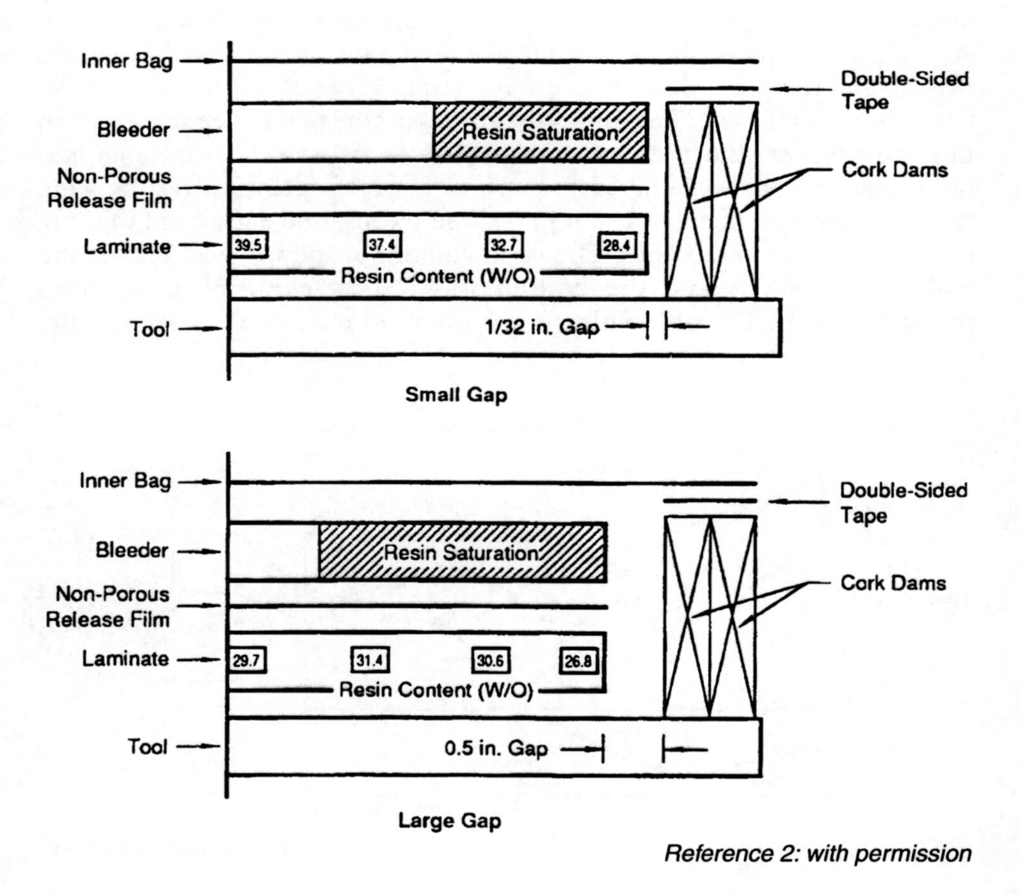

Fig. 21. Horizontal Flow Bagging Arrangement Showing Resultant Resin Contents

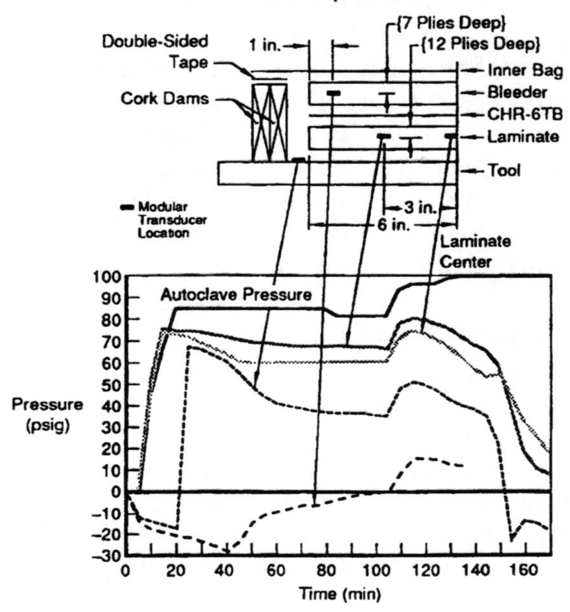

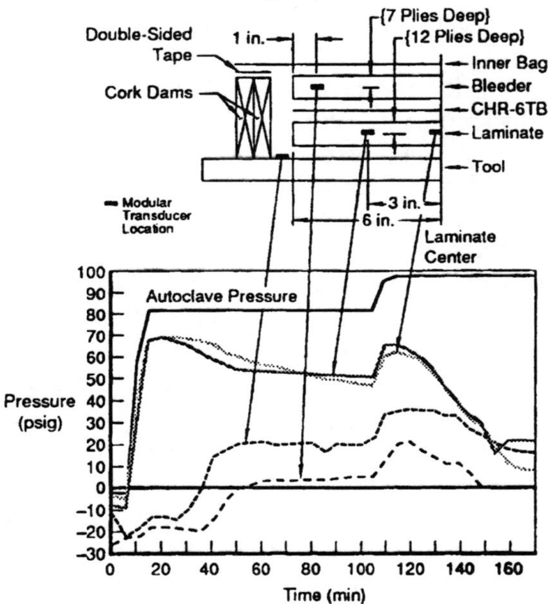

Reference 2

Fig. 22. Horizontal Flow Laminates

illustrate the horizontal flow process. Initially, the resin pressure approaches the applied autoclave pressure and then decreases as bleeding occurs. The opposite occurs in the bleeder. Initially, the applied vacuum is measured and the pressure increases as resin begins to fill the bleeder. Note that the horizontal pressure gradient is very small for a majority of the laminate but becomes large near the edges. Resin content results confirmed the resin pressure results, showing that a large resin content gradient (Fig. 21) existed at the edge of the laminate. The figure also illustrates that resin bled further into the bleeder when a large gap was used between the laminate edges and the dams.

6.4 Chemical Composition Variables

The chemical composition of a thermoset resin system can dramatically affect volatile evolution, resin flow and reaction kinetics. Addition curing polymers, in which no reaction by-products are given off during cross-linking, are in general much easier to process than the condensation systems, which can give off large amounts of water or alcohol as by-products during curing. Further, many condensation systems are impregnated using solvents, and the resultant prepreg typically contains large amounts of high-boiling-point solvents, which are extremely difficult or impossible to remove during cure. Because the reaction by-products and solvents evolve during processing, condensation systems such as phenolics and polyimides are extremely difficult to process without voids and porosity.

Although not as sensitive to processing changes, addition curing systems can also present difficulties. Changes in composition can drastically affect their processability. While Hexcel's 3501-6 and 3502 epoxy resins are typical of resins used for current production prepregs, slight differences in chemical composition do affect their processability. The major difference between the two is the boron trifluoride (BF_3) catalyst in 3501-6. Although present in very small amounts (nominally 1.1% by weight), BF_3 significantly reduces the flow, gel time and gel temperature of 3501-6. It is generally recognized that resin flow prior to gellation is a critical variable in processing thermoset laminates. Too much flow can result in resin-starved laminates, while too little flow can produce resin-rich laminates that can exceed thickness specifications and cause assembly fit-up problems.

A real world example of the effect of chemical composition on flow was experienced during the early development work for the co-cured rib shown in Fig. 23. Due to its geometrical complexity, it was manufactured in match die tooling. The resin system was a new toughened epoxy that had a quite high viscosity (i.e., very little resin flow). The result was that the resin did not flow enough to adequately fill the tool. This yielded unacceptable parts

Fig. 23. Complex Cocured Rib

in which the surfaces contained numerous "dry" areas that were resin starved and the net molded edges were rough and poorly impregnated. Ultrasonic inspection revealed numerous areas of porosity and voids. The solution selected was to have the material supplier reformulate the resin so that it had a lower viscosity. Of course, if the viscosity is too low, this can also create problems with excessive resin leaking out of the tool. Since matched-die tools normally consist of a number of tooling details that must be fitted together during lay-up, they contain multiple leak paths that can allow a very low viscosity resin to escape from during the cure cycle. Thus, complex part processing requires not only resin systems amenable to matched-die tooling but also well-designed tooling.

6.5 Net Resin and Low-flow Resin Systems

Prepregs introduced in the 1970s were considered moderate-to-high flow systems that facilitated the intentional bleeding of resin during autoclave processing. As part applications became larger and more complex through the 1980s, the use of net resin and low (controlled) flow resin prepregs became more prevalent. Carbon/epoxy prepregs for resin bleed processing are usually specified at resin contents ranging from 37% to 42% by weight resin, while the desired or nominal resin contents for these materials are

31% resin content for unidirectional prepregs and 35% resin content for woven cloth prepregs. A 57–60% final fiber volume is typically desired to balance both mechanical performance and part quality. For net resin systems, it is critical to insure that little or no resin bleeding occurs in order to maintain resin pressure during autoclave processing. Many net resin systems utilize controlled flow resins that have higher minimum viscosities and lower flow numbers than the moderate-to-high flow systems. Net resin, controlled-flow prepregs offer improved dimensional control, reduced debulking requirements, reduced consumables (bleeder materials) and reduced ply movement during processing over resin bleed systems. In addition, resin flow into honeycomb core, tooling details and bagging regions are reduced.

6.6 Resin and Prepreg Variables

The resin mixing and prepregging operations can also influence the processability of the final prepreg. During normal mixing operations, air can be easily mixed into the resin. This entrained air can later serve as nucleation sites for voids and porosity. However, many mixing vessels are equipped with seals that allow vacuum degassing during the mixing operation, a practice that has been found to be effective in removing entrained air and may be beneficial in producing superior quality laminates.

Prepreg physical properties can also influence final laminate quality. Prepreg tack is one such property. Prepreg tack is a measure of the stickiness or self-adhesive nature of the prepreg plies. Many times, prepregs with a high tack level have resulted in laminates with severe voids and porosity. This could be due to the potential difficulty of removing entrapped air pockets during collation with tacky prepreg. Again, moisture can be a factor. Prepregs with a high moisture content have been found to be inherently tackier than low moisture content material. Previous work[4] has indicated a possible correlation between prepreg tack and resin viscosity, i.e., prepregs that are extremely tacky also have high initial resin viscosities. Resins with such high viscosities will be less likely to cold flow and eliminate voids at ply terminations.

Prepreg physical quality can greatly influence final laminate quality. Ironically, prepreg that appears "good" (i.e., smooth and well impregnated) may not necessarily produce the best laminates. Several material suppliers have determined that only partially impregnating the fibers during prepregging results in a prepreg that consistently yields high-quality parts, whereas a "good" (i.e., smooth and well-impregnated) prepreg can result in laminates with voids and porosity. Partially impregnated prepregs have the same resin content and fiber areal weight as the fully impregnated material. The only difference is the placement of the resin with respect to the fibers. The partial impregnation process provides an evacuation path for air and low-temperature volatiles entrapped in the lay-up. As the resin melts and flows, full impregnation occurs during cure.

One series of studies[5,6] showed that partially impregnated prepregs can yield high-quality laminates even with moisturized prepreg. They successfully applied partial impregnation to five different resin systems and eight different fibers. A closely related phenomenon is the surface condition of the prepreg. A fully impregnated prepreg will not cause a problem if a surface has the impressions of the fibers (sometimes called a corduroy texture), again providing an evacuation path. These three prepreg conditions are summarized in the highly idealized schematic shown in Fig. 24.

Prepreg aging can also affect final laminate quality. Even at room temperature, polymerization occurs. Eventually, most thermoset prepregs reach a point where little or no resin flow will occur. Thus, lay-up rooms are environmentally controlled not only to reduce moisture absorption, but also to reduce prepreg aging. For production composite epoxy systems such as 3501-6 and 3502, aging only mildly reduces the amount that the resin will flow. The current out-time limit for epoxy prepregs is usually 20–30 d. Good-quality laminates can be produced even after 30 days. However, for V378A, a bismaleimide resin, out-time significantly affects processability. The specification limit is 10 d and even this is pushing the upper limit. Laminates fabricated beyond this limit have exhibited unacceptable quality and an increased tendency towards microcracking, primarily due to the presence of an extremely reactive ingredient.

6.7 Lay-up Variables

A number of lay-up variables, such as prepreg moisture and prepreg age, have already been discussed. In this section, the effects of two other lay-up

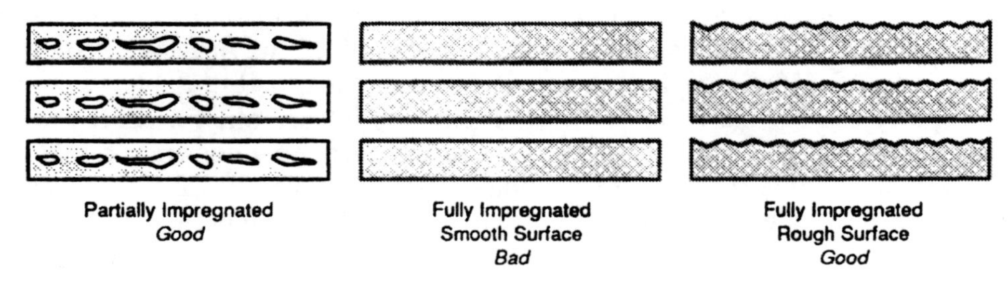

Fig. 24. Effects of Prepreg Physical Quality

variables will be addressed: laminate configuration (i.e., thickness, stacking sequence and ply drop-offs) and final bagging for autoclave cure.

Laminate configuration can influence final quality. In the section on hydrostatic resin pressure, it was shown that thin laminates are more prone to bleeding than thick laminates. In other words, it is quite easy to remove too much resin from a thin laminate that can result in voids and porosity. On the other hand, thick laminates tend to be too thick if they are bled from only one surface. Porosity in thick laminates has been observed in the surface plies, next to the bleeder pack, which can undergo severe overbleeding, while the center and tool plies of the laminate undergo no or minimal bleeding. Due to their thickness, it is easier to trap air during ply collation of thick laminates that cannot escape during "debulking" or "processing". Furthermore, it is much more difficult to evacuate volatiles from the center of large thick laminates where air and volatiles cannot escape horizontally to the edges.

Unidirectional laminates can be difficult to process. Unidirectional plies tend to nest and seal off at the edges, limiting the effectiveness of debulking. This leaves trapped air pockets in the laminate. In addition, it has been observed that volatiles will attempt to travel down the fiber lengths to evacuate out the edges. This has led to the term "linear porosity" for microscopic tubes of porosity that run parallel to the fiber orientation. In cross-plied laminates, several plies oriented in the same direction have the same effect. Internal ply drop-off areas also tend to be porosity prone. Many times voids are present at ply terminations. This is especially true in low-flow systems where the resin does not flow and properly fill the gaps at ply terminations.

Once a laminate is collated, it must be bagged for autoclave curing. The bagging operation contains several variables that can affect part quality, a number of which are illustrated in Fig. 25. If too much bleeder is used, overbleeding can occur which could result in a large drop in the hydrostatic resin pressure, a condition conducive to porosity and void formation. If the resin has a low viscosity (i.e., high flow), the danger of overbleeding is even greater. Even if the correct amount of bleeder is used, an improperly sealed inner bagging system can allow resin to escape into the breather system. For example, if the caul plate severs the inner bag or the dams are not properly sealed, the resin will escape and the hydrostatic resin pressure could fall below the volatile vapor pressure leading to voids and porosity. Another variable that will occasionally cause a problem is that if the caul plate is misallocated or slips and bridges over the top of a dam, resulting in a localized low-pressure area along the edge of the laminate that will experience voids or possibly even large delaminations. Finally, if the outer vacuum bag (usually nylon film) bridges and ruptures during the autoclave cycle, a partial or total loss of the compaction force can result. If the resin

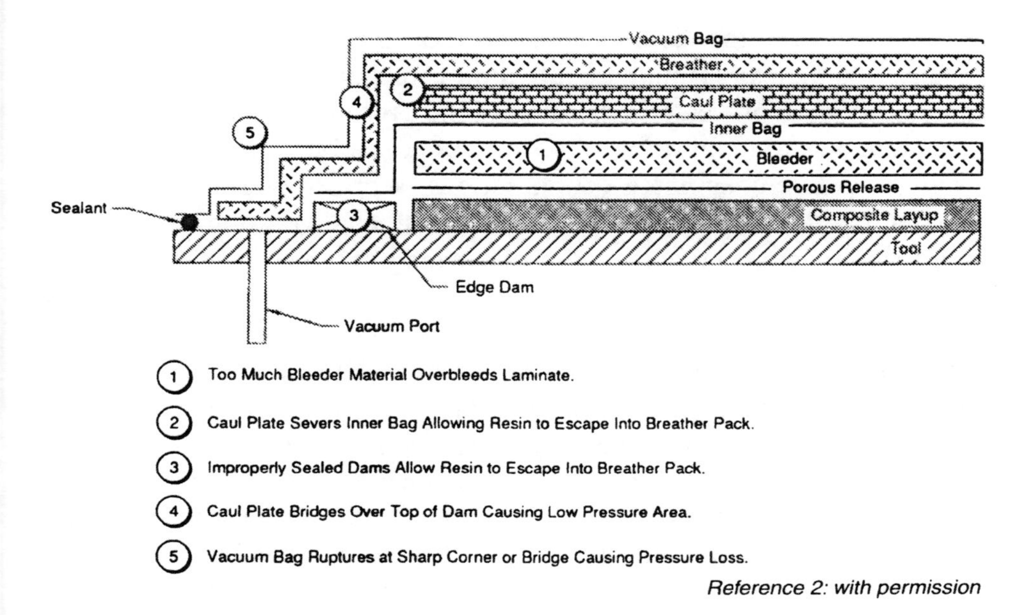

Fig. 25. Potential Bagging Problems

has not already reached the point of gellation in the cure cycle, massive amounts of voids and porosity can form.

6.8 Debulking Operations

Debulking is the process of removing air and compacting the plies during the lay-up operation. A typical debulking operation might consist of pulling a vacuum on the lay-up for 5–15 min. Debulking operations usually depend on the geometry of the part and the thickness; however, a frequency of about every three to five plies is fairly typical. Although considerable controversy exists over the merits of debulking, a general guideline is that a more compacted laminate will have fewer void nucleation sites. This must be weighed against the additional cost of interrupting the collation process to conduct the debulk operation.

Although debulking is routinely conducted at room temperature using vacuum pressure, hot debulking in an autoclave is sometimes required to sufficiently reduce the bulk factor when close tolerance matched-die tools are used (i.e., the tool will not fit together unless the bulk factor is reduced to near final dimensions). It is unfortunately a nuisance to debulk every part every several plies and it can dramatically increase the collation costs. A trade-off between part quality and production cost must once again be made in a logical manner.

To investigate the combined effects of initial debulking, prepreg moisture content and laminate configuration on final laminate quality, four laminates[7] were processed according to the procedures shown in Fig. 26. The laminates were fabricated from a standard carbon/epoxy prepreg. To evaluate different laminate configurations, a cross-plied laminate with internal ply drop-offs was used. This laminate configuration contained a thin area (36 plies), a thick area (48 plies) and a tapered area with internal ply terminations. All four laminates were ultrasonically inspected prior to final autoclave processing. The ultrasonic results shown in Fig. 27 illustrate that essentially no sound was transmitted through the laminates debulked only once but the hot precompacted laminates transmitted a much larger percentage of the original signal. It is also worth noting that the ply drop-off and thick areas of the laminate were of much poorer quality. Of the two hot precompacted laminates (laminates 3 and 4), the wet exposed laminate was of significantly poorer quality. The ultrasonic inspection attenuation for laminate 3 was from 30 to 62 dB, whereas that for laminate 4 was from 10 to 42 dB. Because the hot precompaction temperature was only 150 °F, the additional porosity cannot be due to voids caused by the vaporization

Laminate 1	***Laminate 2***
Moisturized prepreg	• Dry prepreg
• Poor compaction	• Poor compaction
Laminate 3	***Laminate 4***
• Moisturized prepreg	• Dry prepreg
• Good compaction	• Good compaction

Where:

Poor compaction	=	No vacuum debulks
Good compaction	=	Vacuum debulk every 5 plies during collation, Followed by 66°C (150°F) @ 6.8 kg/cm^2 (100 psi) Autoclave debulk for 2 hours (Bagged for No Resin Bleed)
Moisturized prepeg	=	Plies exposed to 32°C (90°F), 85% RH for 72 hours Prior to collation
Dry prepeg	=	Plies exposed to 24°C (75°F), 35% RH for 72 hours Prior to collation

Reference 2: with permission

Fig. 26. Laminate Collation Procedures

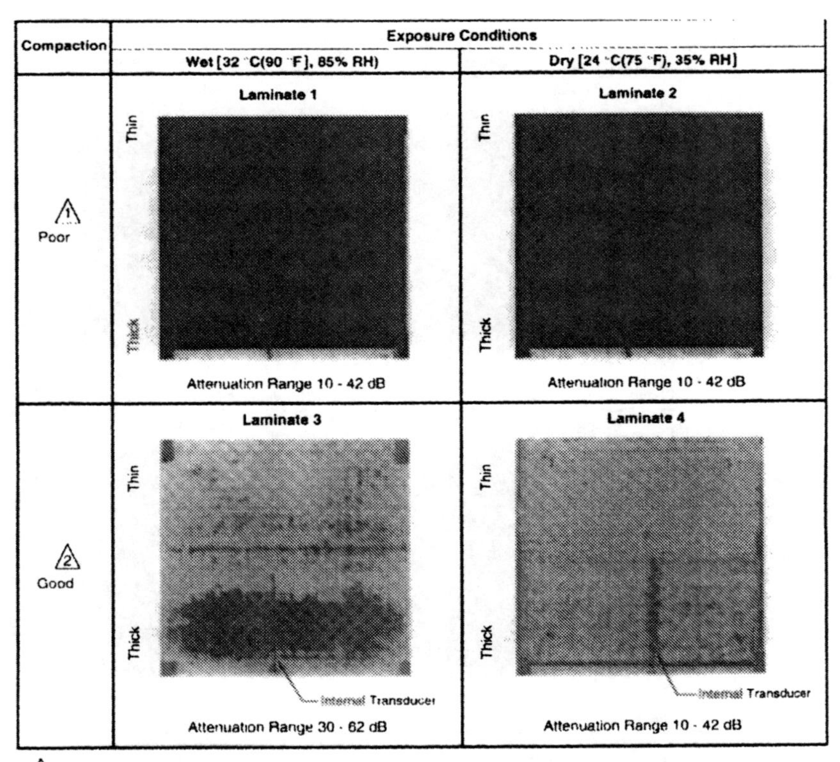

Fig. 27. Precure NDT Results

of absorbed moisture, but is more likely due to voids caused by air trapped during collation. The wet exposed prepreg exhibited a noticeable increase in tack that could contribute to increased air entrapment during collation. All four laminates were autoclave processed using the standard cure cycle discussed previously. After processing, all four laminates were ultrasonically inspected. The results are presented in Fig. 28. As expected, the dry-hot precompacted laminate (laminate 4) was of the highest quality. It was also of interest that the extent of compaction did not significantly affect laminate quality for dry prepreg. Good initial compaction, however, did significantly reduce the amount of porosity in the wet exposed laminates. The results of this limited study showed the significant impact of absorbed prepreg moisture on final laminate quality. Although initial laminate precompaction helped to reduce the amount of porosity when moisture was present, considerable porosity was still present after curing, particularly in the thick section and at internal ply terminations. In addition

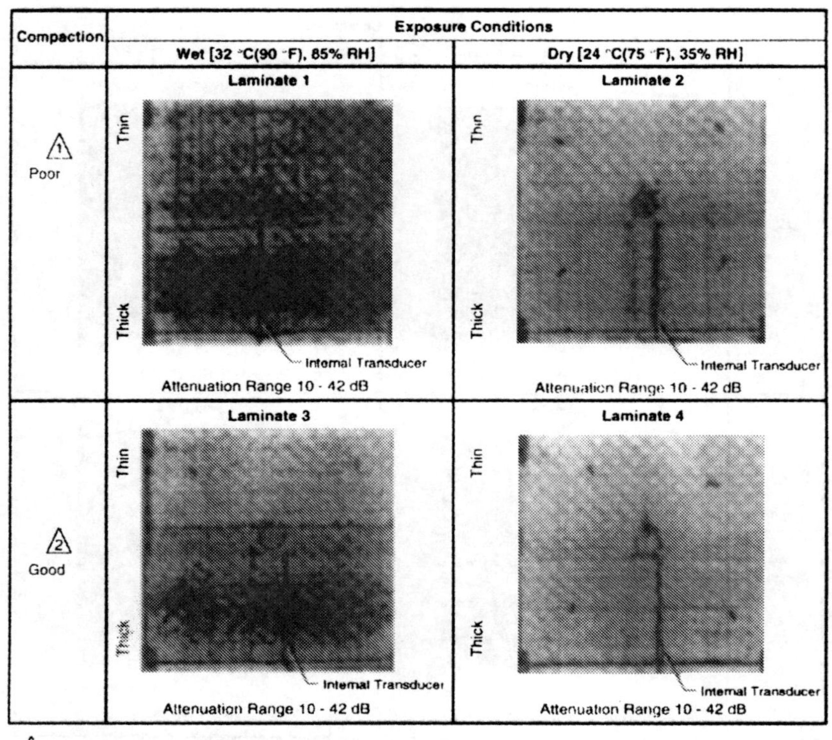

Fig. 28. After Cure NDT Results

to being a volatile that can cause void formation, absorbed moisture can also dramatically increase the prepreg tack level, which in turn increases the amount of entrapped air during collation. This further enhances the opportunity for additional void formation during autoclave processing.

6.9 Caul Plates and Pressure Intensifiers

Many composite applications utilize a caul plate or pressure intensifier. Caul plates result in a greatly improved part surface finish compared with a bag surface, improved dimensional control and improved radius quality. Caul plates are also used to reduce ply movement during processing of honeycomb reinforced parts. Caul plates may be semi-rigid or rigid in nature. Semi-rigid caul plates, which are the most common type, are typically constructed of thin metal, composite or rubber materials so that they are flexible in nature. Pressure intensifiers are used at part corners to reduce ply thinning at male radii and ply bridging at female radii.

Rigid caul plates are typically constructed of thick metal or composite materials. Thick caul plates are used on very complex part applications or cocured parts where dimensional control is critical. Many rigid caul plates result in a matched-die configuration similar to compression or resin transfer molding. Parts processed in this manner are extremely challenging, because resin pressure is much more dependent on tool accuracy and the difference in thermal expansion between the tool and the part. Tool accuracy is critical to insure that no pinch points are encountered that would inhibit a tool from forming to the net shape of the part. Both semi-rigid and rigid caul plates dramatically influence resin flow, resin pressure distribution and final part quality. Semi-rigid caul sheets and pressure intensifiers reduce vacuum bag bridging around tight comers, resulting in a more uniform pressure distribution; however, rigid caul plates completely alter resin pressure distributions by minimizing laminate thinning. A rigid caul plate applies pressure only at laminate high points.

6.10 Condensation Curing Systems

Condensation curing systems, such as polyimides and phenolics, give off water and alcohol as part of their chemical cross-linking reactions. In addition, the reactants are often dissolved in high-temperature boiling-point solvents, such as dimethylformamide (DMF), dimethylactamide (DMAC), *N*-methylpyrrolidone (NMP) or dimethylsufoxide (DMSO) to allow prepregging. Even the addition curing polyimide PMR-15 uses methanol as a solvent for prepregging. The eventual evolution of these volatiles during cure creates a major volatiles management problem that can result in high void and porosity percentages in the cured part. Unless a heated platen press with extremely high pressures (e.g., 1,000 psig) is used to keep the volatiles in solution until gellation, they must be removed either before the cure cycle or during cure heat-up. In addition, since these materials boil or condensate at different temperatures during heat-up, it is important to know the point(s) during the cycle when the different species will evolve. An example of the complex volatile evolution for the thermoplastic polyimide K-IIIB is shown in Fig. 29.[8]

There are three strategies for volatile management: (1) use a press with an applied hydrostatic resin pressure greater than the volatile vapor pressure to keep the volatiles in solution until the resin gels; (2) remove the volatiles by laying up only a few plies at a time and hot debulking under vacuum bag pressure at a temperature higher than the volatile boiling point; or (3) use slow heat-up rates and vacuum pressure during cure with intermediate holds to remove the volatiles again, before resin gellation. It should be noted that more than one of these strategies can be used at the same time. The advantage of a heated platen press is that high pressures can be used to suppress volatile evolution. However, the tooling must be

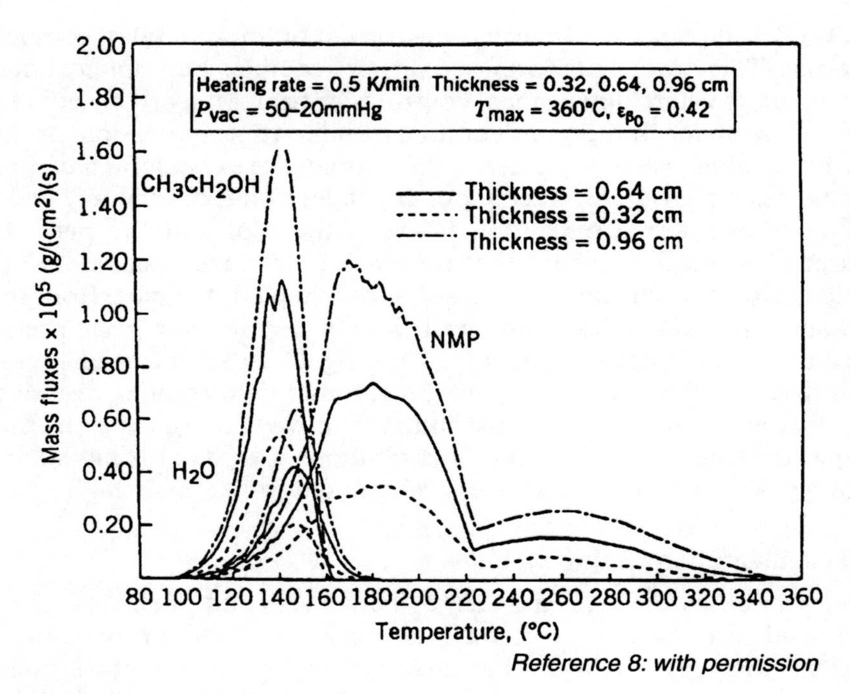

Reference 8: with permission

Fig. 29. Volatile Evolution of K-IIIB Prepreg

designed to withstand the higher pressures, and special damming systems must be incorporated to prevent excessive resin squeeze-out. The second method, intermediate hot debulks under vacuum pressure, is effective but is very costly and labor intensive since the ply collation operation has to be interrupted every several plies; then bagged and moved to an oven; hot debulked; and then cooled before further collation. The last method, as shown in Fig. 30[9] for a typical autoclave cure cycle for PMR-15, incorporates multiple holds under vacuum during heat-up to evacuate the volatiles during various points in the cure cycle. It should be noted that some manufacturers use a 600 °F cure and post-cure rather than the 575 °F as shown in the figure. In addition, some use only a partial vacuum during the early stages of cure and apply a full vacuum in the latter stages. Although the final cure of PMR-15 is an addition reaction, it undergoes condensation reactions early in the cure cycle during the imidization stage that creates a volatile management problem. The tricky part to this approach is determining the optimum time and temperatures for the hold periods and the heat-up rates to use. Some of the physiochemical methods described in Chapter 3 can be used in helping to design these cure cycles. To obtain full cross-linking, polyamides often require extended post-cure cycles. In a typical post-cure cycle for PMR-15 shown in Fig. 31,[9] note that

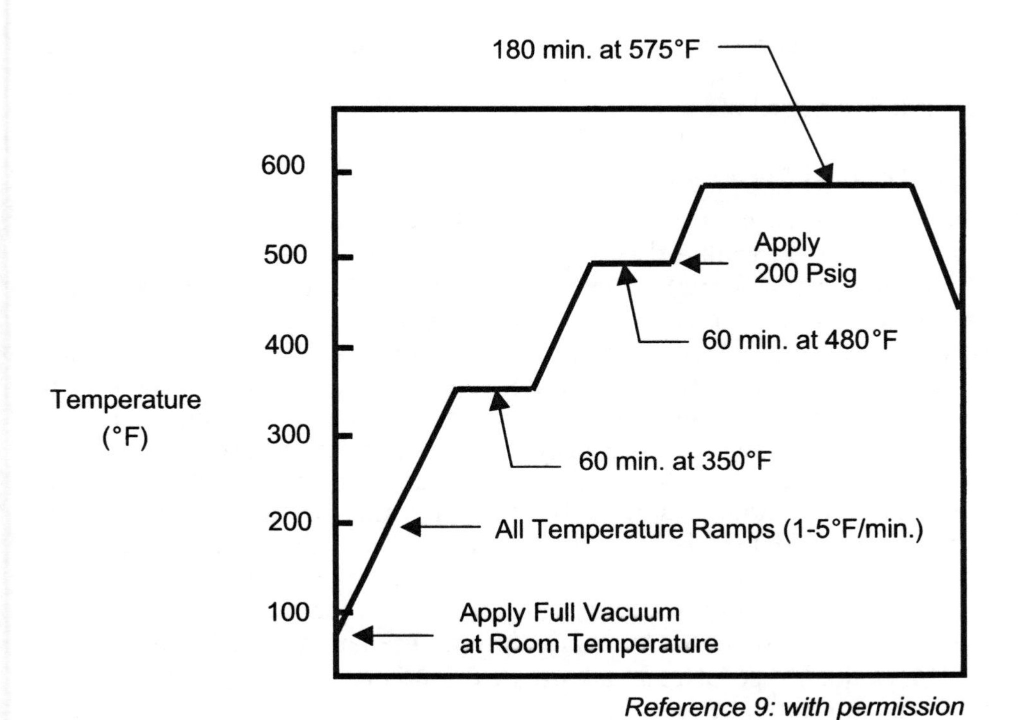

Fig. 30. Typical PMR-15 Cure Cycle

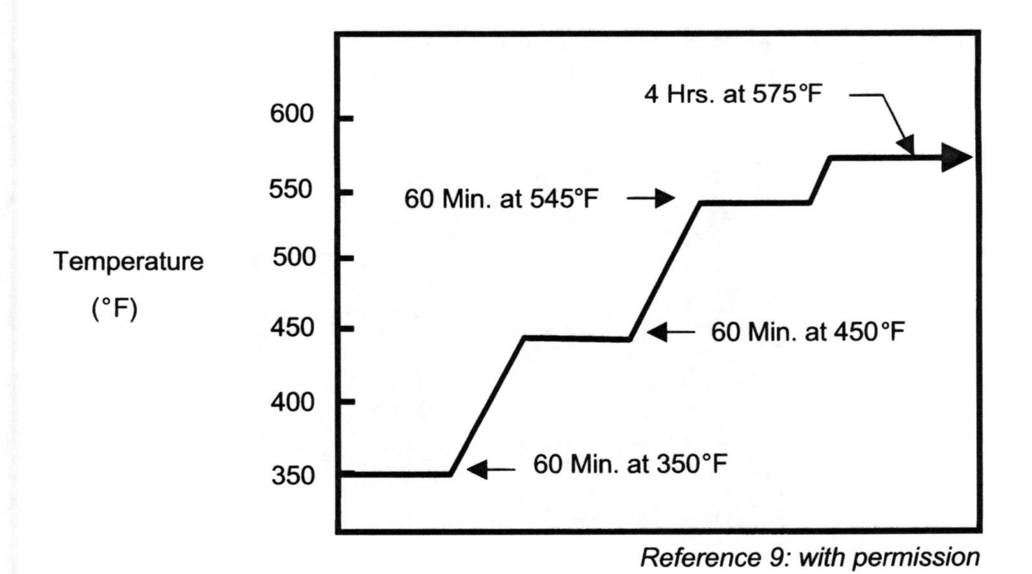

Fig. 31. Typical PMR-15 Post Cure Cycle

even the post-cure cycle incorporates multiple hold periods during heat-up to help minimize residual stress buildup and thus reduce the likelihood of matrix microcracking.

6.11 Residual Curing Stresses

Residual stresses develop during elevated temperature curing of composite parts. They can result either in physical warpage or distortion of the part (particularly thin parts) or in matrix microcracking either immediately after cure or during service. Distortion and warpage causes problems during assembly and is more troublesome for composite parts than metallic ones. While the distortion in thin sheet metal parts can often be pulled out during assembly, composite parts run the danger of cracking and even delamination if they are stressed during assembly. Microcracking is known to result in degradation of the mechanical properties of the laminate including the moduli, Poisson's ratio and the coefficient of thermal expansion (CTE).[10] Microcracking (Fig. 32[11]) can also induce secondary forms of damage such as delaminations, fiber breakage and the creation of pathways for the ingression of moisture and other fluids. Such damage modes have been known to result in premature laminate failure.[12,13]

The major cause of residual stresses in composite parts is due to the thermal mismatch between the fibers and the resin matrix. Recalling that the residual stress on a simple constrained bar is

$$\sigma = \alpha E \Delta T$$

where σ = residual stress, α = CTE and ΔT = temperature change, a rather simplified analogy for a composite part is that the CTE difference between

Reference 11: with permission

Fig. 32. Matrix Microcracking

the fibers ($\approx$0 for carbon fiber) and the resin is large $\approx 20 - 35 \times 10^{-6}$ per °F for thermoset resins). The modulus difference between the fibers (30–140 msi) and the resin (0.5 msi) is also large. The temperature difference (ΔT) is the difference from when the resin becomes a solid gel during cure and the usage temperature. The so-called stress-free temperature is somewhere between the gel temperature and the final cure temperature as the cross-linking structure develops strength and rigidity. The usage temperature for epoxy composites usually ranges anywhere from –67 to 250 °F. There are several observations we can make from this simplified analogy: High-modulus carbon, graphite and aramid fibers have negative CTEs. Normally, the higher the fiber modulus, the more negative the CTE becomes which leads to increases in residual stresses and helps to explain why more matrix microcracking is observed with high-modulus graphite fibers than high-strength carbon fibers. Carbon/epoxy resin systems are usually cured at either 250 or 350 °F. Since there will be a smaller ΔT for the systems cured at 250 °F, they should experience less microcracking than the systems cured at 350 °F. Very high temperature polyimides that are often cured at temperatures in the range of 600–700 °F develop very high residual stresses and are very susceptible to microcracking. Since the ΔT difference becomes larger when the use temperature is lowered, e.g., when the temperature is –40 to 67 °F for a cruising airliner at 30,000–40,000 ft, more microcracking is normally observed after cold exposures than elevated temperature exposures. The analogy presented above greatly oversimplifies the residual stress problem in composite structures. In fact, analysis of residual stresses in composites is probably one of the most complex problems analysts have tried to address. There is quite a bit of conflicting data in the literature over the various causes of residual stresses and the effects of material, lay-up, tooling and processing variables on residual stresses.

Composites by their very nature are anistropic materials and residual stresses result due to differences in ply orientations. For example, as shown in Fig. 33, a 0° ply expands very little during cure because it has a very low CTE, whereas a 90° ply expands significantly because it is dominated by the thermal expansion of the matrix.[14] Similar types of residual stresses are created at all ply interfaces having a different orientation, e.g., at +45° and –45° ply interfaces. If the laminate is not balanced and symmetric, macro-warpage will certainly occur during cool-down. A balanced laminate is one that for every $+\theta$ ply in the lay-up there is an equivalent $-\theta$ ply in the lay-up. An example of a balanced laminate is 0°, +45°, –45°, 90°, –45°, +45°, 0°, whereas an unbalanced laminate would be 0°, +45°, –45°, 90°, –45°, +45°, 90°. A symmetric laminate is one that is balanced at its centerline and forms a mirror image on both sides of the centerline. For example, a symmetric laminate would be 0°, +45°, –45°, –45°, +45°, 0°, whereas a non-symmetric

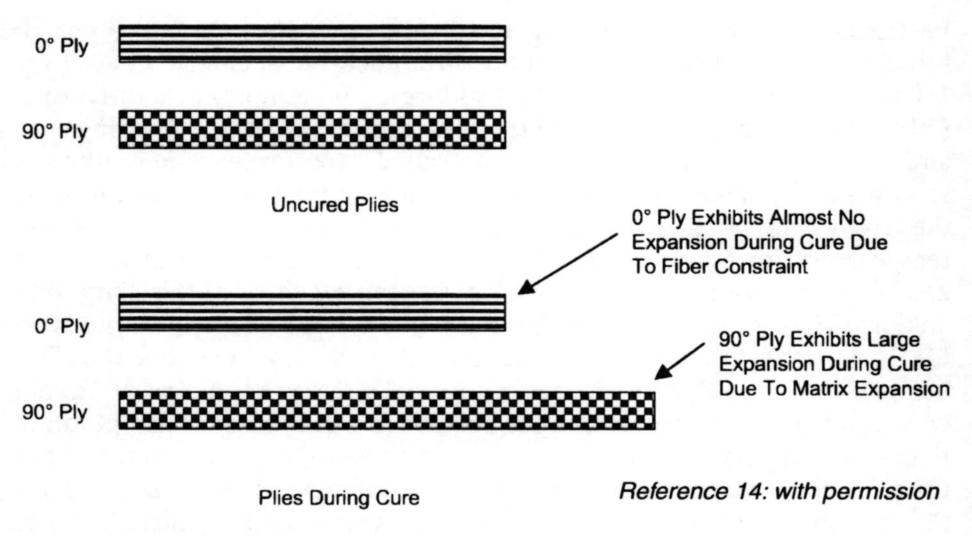

Fig. 33. Differential Ply Expansion During Cure

laminate would be 0°, +45°, –45°, +45°, –45°, 0°. To complicate the situation even further, small deviations in ply alignment, even a couple of degrees, can produce warpage in thin laminates. While the warpage may not show up in thicker laminates, it is still there as residual stress but is just constrained by the thickness of the laminate.

In the simplified analogy at the beginning of this section, the concept of ΔT and stress-free temperature was introduced. There is considerable controversy in the literature whether the gel temperature or the cure temperature should be taken as the stress-free temperature. Actually, a lot depends on the cure cycle employed. It is possible to heat many carbon/epoxy systems to their cure temperatures using fairly aggressive heat-up rates and the resin does not gel until the 350 °F cure temperature is reached. On the other hand, one can use a slow heat-up rate (e.g., 1–2 °F min^{-1}) and the resin gels at temperatures as much as 75–100 °F below the final 350 °F cure temperature.

A resin fraction gradient through the thickness of the laminate has been shown to influence the distortion of both flat and complex-shaped symmetric parts.[15,16] Since it is common practice to bleed laminates from the bag side, the side of the laminate closest to the bag will generally have a lower resin content in a thick laminate than the side closest to the tool. The bleeder materials cause the bag side of the laminate to be resin poor, while the center and tool side may have too high a resin content. Fiber volumes of 52% and 59% have been observed for the bag and tool side respectively, whereas the middle had a fiber volume of 57%.[15] This produces an asymmetric laminate condition with more shrinkage in the areas with more resin because of the high CTE of the resin and the chemical shrinkage of the resin due to cross-linking.[17] The high resin CTE, together with the chemical shrinkage of the resin that occurs during cure, causes the bottom of the laminate to shrink more than the top portion because of the higher resin fraction in the bottom portion, resulting in distortion in a convex-up curvature shape.

In general, the more complex the shape, the more complex the residual stress state becomes. The spring-in that occurs during curing in a simple 90° angled part is shown in Fig. 34. This is somewhat analogous to the spring-back phenomena observed in sheet metal forming although the causes are totally different. Volumetric shrinkage usually occurs in epoxies (≈1–6%) as they cure. While the fiber reinforcement tends to limit this effect in the in-plane direction, shrinkage through the thickness is largely unrestrained. This has little effect on flat symmetric laminates but contributes to the spring-in of curved parts. This effect can be illustrated by considering a laminate with a bend that is compressed through its thickness. If the bend angle is constrained during this compression, the inside ply will be stretched while the outside ply is compressed. When the constraint is removed after cure, spring-in results. A further complicating factor is that it is quite common for male corners to thin out at the radius due to pressure intensification while female corners are often too thick due to lack of pressure (Fig. 35). Taking a ply of 0°, 90° woven cloth and draping it over any shape makes it obvious to even the casual observer that the fibers move

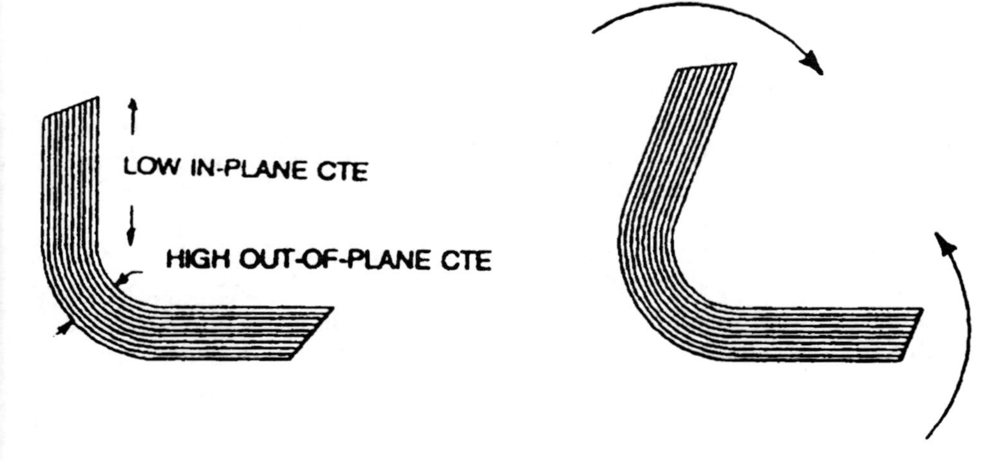

Fig. 34. Composite Part Spring-In

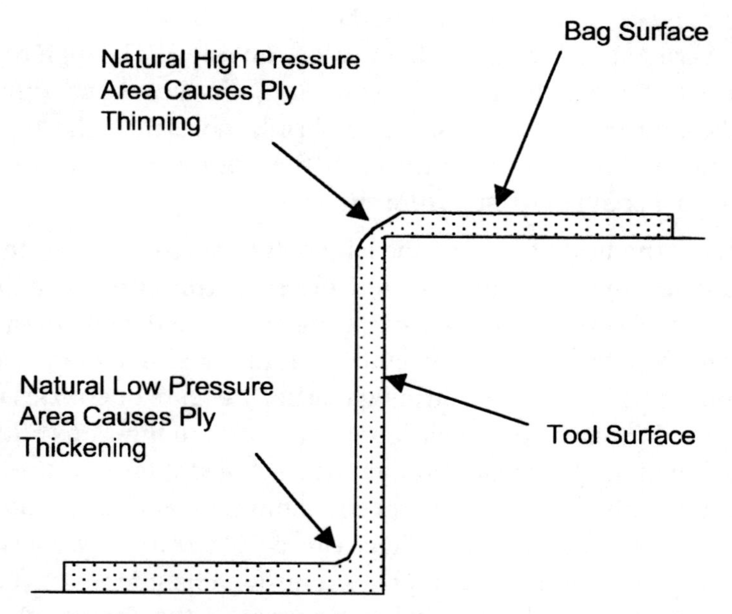

Fig. 35. Low and High Pressure Areas During Cure

and adjust to accommodate the shape; the more complex the shape, the more the fibers must move to accommodate the shape. Tool dimensions must be sized to account for these dimensional changes that occur in both the tool and the part during curing. As it was discussed in Chapter 4 on Cure Tooling, it is a common practice to accommodate spring-in by adjusting the tooling angle outwards by 1–3°.

Differences in thermal expansion between the tool material and the composite part must also be considered. For example, an aluminum tool that has a very high thermal expansion will require more compensation than a composite tool that has almost the same CTE as the composite part. In addition, there is the potential for part–tool interaction or mold stretching[17] (Fig. 36) due to CTE differences between the tooling and composite part. It should be noted that while some researchers have found that part–tool interactions are dependent on the type of tooling material,[18] others have failed to find any significant effects.[19] According to part–tool interaction theory, large differences in the CTE of metal tools (e.g., aluminum with a CTE of 23.6 $\mu\varepsilon$ $°C^{-1}$) and carbon/epoxy composite parts with a CTE of ≈ 0 for a 0° ply results in shear stresses in the part surface as the tool heats up. Based on friction between the part and the tool surface, the tool pulls the fibers on the tool surface layer of the composite as it expands. A state of residual tensile strain develops in the tool surface plies as the surrounding resin cures at elevated temperature. Upon removal of

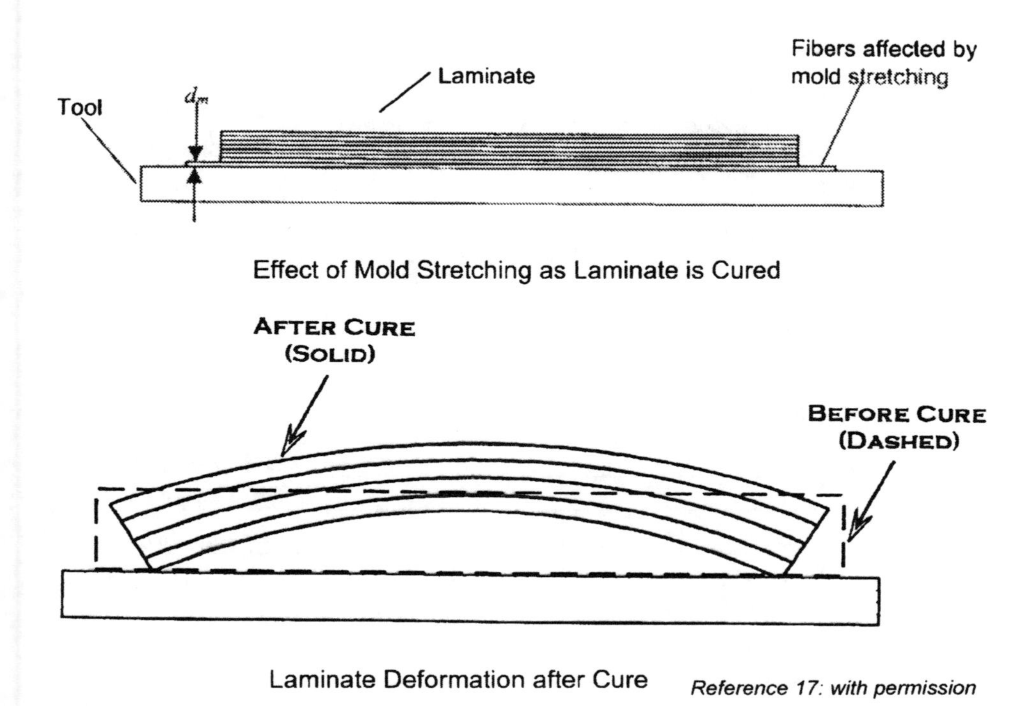

Fig. 36. Mold Stretching During Cure

the autoclave pressure and separation of the part from the tool after cool-down, part curvature results as the tensile strains cause the plies that were adjacent to the tool surface contract.[17] However, test results have shown that the surface roughness of the tool does not have a strong effect on distortion, while the tooling material does as shown in Fig. 37 for thin and thick laminates cured on aluminum, steel and glass tooling plates.[20] Note that the thicker laminates displayed less distortion than the thin laminates. However, this does not necessarily mean that the residual stress state was lower for the thick laminates. The fact that they were thicker would help to minimize the visible distortion. In fact, there is some experimental data[11] that suggests that thicker laminates, while not displaying as much distortion, have a greater tendency to microcrack during thermal cycling.

One study[19] that was conducted on the effects of processing temperature and lay-up on T300/976 carbon/epoxy laminates found that spring-in was lower when lower cure temperatures were used and when the degree of cure $\propto$ was lower. They proposed a cure cycle in which the part is first cured at a low temperature to develop a $\propto$ of 0.5–0.7 followed by a final elevated temperature cure. For such a curing cycle, the spring-in was less than when cured at a constant elevated temperature. They also found that the spring-

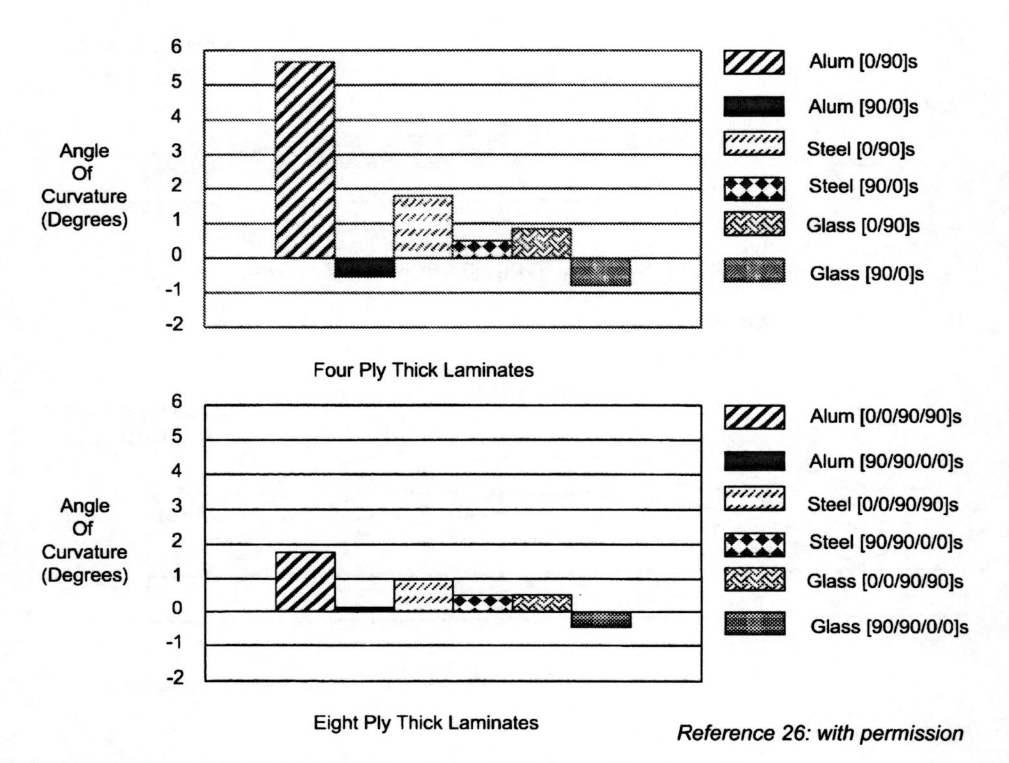

Fig. 37. Effects of Tool Material, Orientation and Thickness of Laminate Distortion

back was somewhat less with a slow cooling rate, although others have found no such effect. Other researchers[21] also found that higher cure temperatures caused the formation of microcracks that were more tortuous and wider than those in laminates cured at lower temperatures. They concluded that larger thermal stresses were generated in laminates when cured at higher temperatures and that fracture processes relieved these stresses in the laminates when cycled to cryogenic temperatures (liquid nitrogen) resulting in delaminations, wider microcracks and higher microcrack densities. A concurrent study[22] found that adding rubber toughening agents to the resin increased the cryogenic microcracking resistance.

While residual stresses in composites are extremely complicated and there is considerable conflicting data on the effects of different variables, the following guidelines are offered for minimizing their effects:

- Use only balanced and symmetric laminates. Minimize ply lay-up misorientation or distortion whenever possible.

- Design tools with compensation factors to account for thermal growth and angular spring-in. The use of low CTE tools will probably help to minimize residual stresses when curing carbon fiber composites.
- The use of lower-modulus fibers and tougher resin systems helps to accommodate residual stresses and microcracking.
- Slow heat-up rates during cure with intermediate holds and lower curing temperatures probably help in minimizing residual stresses by balancing the rate of chemical resin shrinkage with the rate of thermal expansion.[23,24] Likewise, there is some evidence that slow cool-down rates help.

6.12 Exotherm

When a thermoset polymer cures, the chemical reactions give off heat and the reaction is termed exothermic. The concern when curing a composite laminate in an autoclave is that the heat-up rate can be too fast, resulting in a significant temperature rise in a thick laminate as shown in Fig. 38. This

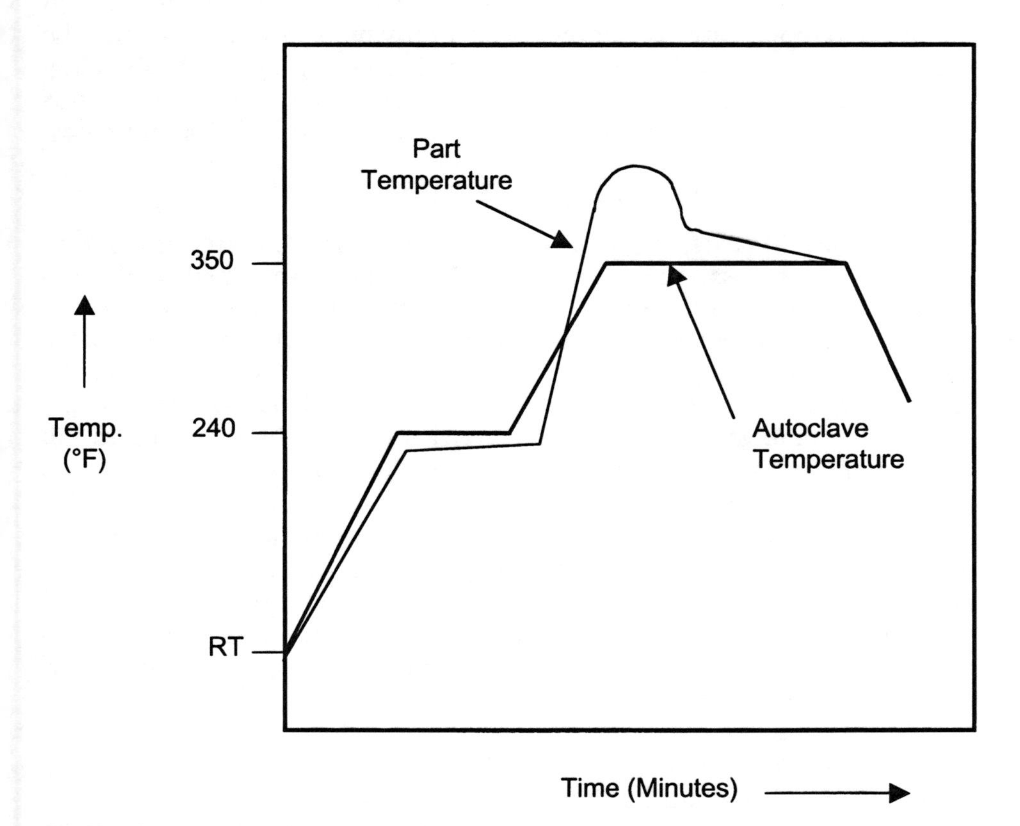

Fig. 38. Conceptual Exotherm During Composite Curing

could result in degradation of the laminate by overheating or cause uneven curing through the thickness. In a worst of the scenarios, the exotherm could be so great that the laminate and bagging materials could catch fire or at least become charred.

This condition rarely, if ever, occurs in production for epoxy resins, because (1) thick laminates usually drive thick tools that exhibit very slow heat-up rates, and (2) thick laminate parts are usually made in matched metal tools. In order to get the lay-up to fit within these tools, the plies or ply packs either have to be hot-debulked if the material is a net resin content prepreg or prebled to remove resin and compact the lay-up if the material is an excess resin content prepreg. These two factors, namely large massive matched metal dies with slow heat-up rates and compacted net resin content plies that have been advanced during the hot debulking or prebleeding operation, are the main reasons runaway exothermic reactions are not observed in industry. If a part is experiencing an excessive temperature rise due to an exothermic reaction, then it may be necessary to use net resin content prepregs and slow the heat-up rate down and even put in intermediate holds that allow the autoclave and part to stabilize at the same temperature. If exotherm still remains a problem, it may be necessary to use a resin with a less reactive curing agent. Embedding thermocouples within a test laminate and monitoring their temperatures during cure can measure the potential for an exotherm in thick laminates.

6.13 In-process Cure Monitoring

Over the past 30 years, there has been a considerable amount of research done on in- process cure monitoring. The most researched method is dielectrics, in which a sensor is placed either on the laminate, or sensors are placed on both surfaces of the laminate. As the resin melts and flows during the initial part of the cure cycle, the molecules will fluctuate within the alternating dielectric field and give an indication of the viscosity of the resin. The readings taken by the dielectrometer can then be used to tell when the resin viscosity starts to rise and full autoclave pressure should be applied. There is also some body of evidence that the end of cure can also be detected so one would know when the cure is complete and to start cool-down; however, as curing proceeds, molecular activity decreases drastically and the dielectric response becomes weaker. Other methods researched include: acoustic emission, ultrasonics, acousto-ultrasonics, fluorescence techniques and mechanical impedance analysis. In spite of all of this research, thermocouples, usually attached to the bottom of the tool, remain the industry standard.

In-process cure monitoring is a good tool for resin characterization and cure cycle development, but has been rarely used in production for two reasons: (1) the incorporation of sensors during lay-up adds to both the collation and bagging costs; and (2) when multiple parts are cured in an autoclave, they may experience different heat-up rates and it becomes difficult to know which part should be the controlling one for pressure application. As stated previously, for addition curing thermosets, it is usually better to apply the pressure at the start of the cure cycle, thus removing any decision making from the autoclave operator. However, if condensation curing systems are being used where one needs to delay full pressure application until all the volatiles are removed from the resin, then the use of in-process cure monitoring may be warranted, at least during the early stages of part production.

6.14 Cure Modeling

There have been a number of mathematical models of the curing process developed which can quite accurately predict resin kinetics, resin viscosity, resin flow, heat transfer, void formation and residual stresses. Like in-process cure monitoring, these models can be invaluable tools for characterizing new materials and for cure cycle development. They can also be used for modeling the heat-up and cool-down rates for candidate tool designs. Models have been developed for many processes, including thermoset curing, thermoplastic consolidation, filament winding, pultrusion and liquid molding. In addition, for many years models have been successfully used for injection and compression molding.

When using any model, it is important to understand: (1) the mathematics of the equations making up the model; (2) all assumptions that were made in the model formulation; (3) the boundary conditions for the model; and (4) the solution method used for the model. Excellent overviews of the different models developed for curing can be found in Refs. 25 and 26.

6.15 Summary

Autoclave processing remains the mainstay for processing continuous fiber-reinforced thermoset composite parts. Considerable research has been conducted to establish a scientific understanding of the many complex and interrelated phenomena that occur during autoclave processing. This research has yielded tremendous insight into the fundamental principles that guide heat transfer, resin viscosity, resin flow, chemical cross-linking, void formation, growth and transport during processing.

Much of this chapter focused on final part quality through control of resin pressure and void management. Maintaining the resin hydrostatic pressure above the potential void pressure is key to minimizing void formation. The resin pressure, however, is typically lower than the autoclave pressure due to resin flow, bagging, tooling concept and support materials such as honeycomb core. Other variables influence void formation, such as material surface texture, entrained air in the resin, prepreg moisture content and prepreg tack. Whereas these variables are much less understood, their impact is significant. In many instances, the resin pressure can be high throughout the entire cure cycle, but voids will remain that are present before, or created during the collation operation.

(i) Void management is critical to the reproducible fabrication of high-quality composite parts.

(ii) Much information is available to allow for a more scientific approach to devising suitable materials, lay-ups, tooling, and cure cycles:

(a) Void formation in addition curing composite laminates is primarily due to entrapped volatiles. High temperatures result in high volatile pressures. Void growth will occur if the potential void pressure (volatile vapor pressure) exceeds the hydrostatic resin pressure while the resin is still a liquid.

(b) Overbleeding can cause a dramatic drop in the hydrostatic resin pressure, a condition conducive to void formation and growth.

(c) The bagging arrangement for each part strongly affects its processability and final laminate thickness. Dam placement and bleeder ratios directly affect the hydrostatic resin pressure.

(d) IPB curing can be used to help maintain the hydrostatic resin pressure and control resin flow.

(e) High prepreg tack, which can result in trapped air during lay-up, can be a contributor to void formation and growth.

(f) Chemical composition, resin mixing, prepregging procedures, lay-up and tooling variables can also contribute to void formation.

(iii) Residual stresses resulting from elevated temperature cure processes are extremely complex and difficult to analyze. Use only balanced and symmetric laminates. Design tools with compensation factors to account for thermal growth and spring-in. Use low-modulus fibers and toughened resin systems where possible. Slow heat-up and cool-down rates and low maximum cure temperatures will probably help minimize residual stresses.

(iv) Condensation curing systems are much more difficult to process than addition curing systems since they can give off water, alcohol and solvents during cure. If condensation systems must be used, then volatile management during cure becomes the main challenge.

References

[1] Griffith J.M., Campbell F.C., Mallow A.R., "Effect of Tool Design on Autoclave Heat-up Rates," *Society of Manufacturing Engineers, Composites in Manufacturing 7 Conference and Exposition*, 1987.

[2] Campbell F.C., Mallow A.R., Browning C.E., "Porosity in Carbon Fiber Composites: An Overview of Causes," *Journal of Advanced Materials* **26**(4), July 1995, pp. 18–33.

[3] Kardos J.L., "Void Growth and Dissolution," In *Processing of Composites*, Hanser, 1999, pp. 182–207.
[4] Brand R.A., Brown G.G., McKague E.L., "Processing Science of Epoxy Resin Composites," Air Force Contract No. F33615-80-C-5021, Final Report for August 1980–December 1983.
[5] Thorfinnson B., Bierrinann T.F., "Production of Void Free Composite Parts without Debulking," *31st International SAMPE Symposium and Exposition*, April 1986.
[6] Thorfinnson B., Bierrnann T.F., "Measurement and Control of Prepreg Impregnation for Elimination of Porosity in Composite Parts," Society of Manufacturing Engineers, Fabricating Composites 88, September 1988.
[7] Browning C.E., Campbell F.C., Mallow A.R., "Effect of Precompaction on Carbon/ Epoxy Laminate Quality," *AIChE Conference on Emerging Materials*, August 1987.
[8] Kardos J.L., "The Processing Science of Reactive Polymer Composites," In *Advanced Composites Manufacturing*, Wiley, 1997, pp. 68–77.
[9] Mace W.C., "Curing Polyimide Composites," In *ASM Vol. 1 Engineered Materials Handbook Composites*, ASM International, 1987, pp. 662—663.
[10] Thompkins S.S., Shen J.Y., Lavoie, *Proceedings of the 4th International Conference on Engineering, Construction, and Operations in Space*, 1994, p. 326.
[11] Dharia A.K., Hays B.S., Seferis J.C., "Evaluation of Microcracking in Aerospace Composites Exposed to Thermal Cycling: Effect of Composite Lay-up, Laminate Thickness and Thermal Ramp Rate," *33rd International SAMPE Technical Conference*, November 2001.
[12] Swanson S.R., *Introduction to Design and Analysis with Advanced Composite Materials*, Prentice-Hall, 1997.
[13] Mallick P.K., *Fiber Reinforced Composites: Materials, Manufacturing and Design*, Marcel Dekker, 1993.
[14] Tsai S.W., *Composites Design--1986*, Think Composites, 1986, pp.15-1–15-21.
[15] Radford D.W., *Composites Engineering* **5**(7), 1995, p. 923.
[16] Yang S.Y., Huang C.K., *Journal of Advanced Materials* **28**(2), 1997, p. 47.
[17] Darrow D.A., Smith L.V., "Evaluating the Spring-in Phenomena of Polymer Matrix Composites," *33rd International SAMPE Technical Conference*, November 2001.
[18] Melo J.D.D., Radford D.W., *31st SAMPE Technical Conference*, 1999.
[19] Sarrazin H., Kim B., Ahn S.H., Springer G.E., "Effects of Processing Temperature and Layup on Springback," *Journal of Composite Materials* **29**(10), 1995, pp. 12378–12394
[20] Cann M.T., Adams D.O., "Effect of Part–Tool Interaction on Cure Distortion of Flat Composite Laminates," *46th International SAMPE Symposium*, May 2001, pp. 2264–2277.
[21] Timmerman J.F., Hayes B.S., Seferis J.C., "Cryogenic Cycling of Polymeric Composite Materials: Effects of Cure Conditions on Microcracking," *33rd International SAMPE Technical Conference*, November 2001.
[22] Nobelen M., Hayes B.S., Seferis J.C., "Low-temperature Microcracking of Composites: Effects of Toughness Modifier Concentration," *33rd International SAMPE Technical Conference*, November 2001.
[23] Karkkainen R., Madhukar M., Russell J., Nelson K., "Empirical Modeling of In-cure Volume Changes of 3501-6 Epoxy," *45th International SAMPE Symposium*, May 2000, pp. 123–135.
[24] Kim R., Rice B., Crasto A., Russell J., "Influence of Process Cycle on Residual Stress Development in BMI Composites, *45th International SAMPE Symposium*, May 2000, pp. 148–155.
[25] *Processing of Composites*, Hanser, 1999.
[26] *Advanced Composites Manufacturing*, Wiley, 1997.

Chapter 7

The Effect of Chemical Composition and Processing on Carbon/Epoxy Laminate Quality: A Combination of Effects

The chemical composition and test methods for epoxy resin systems were discussed in Chapter 3 on Thermoset Resins. Now that curing has been covered in Chapter 6, the combined influence of both chemical composition and processing conditions on final composite properties will be covered in this chapter. Chemical composition and processing of epoxy resin systems are known to affect the flow behavior and reaction kinetics of carbon/epoxy prepregs. A study[1] was conducted to assess the influence of catalyst content, resin mixing and advancement, and prepreg resin content on the properties of both neat resin and prepreg. Testing included physical, chemical, thermal and viscosity characterizations. Next, carbon/epoxy laminates were fabricated and evaluated to determine the impact of both prepreg variations and lay-up variability on final laminate quality and properties.

It is generally recognized that resin flow prior to gellation is a critical variable in processing carbon/epoxy laminates. Too much flow can result in resin-starved laminates which often contain excessive porosity, while too little flow can produce resin-rich laminates that can exceed thickness tolerances and cause assembly fit-up problems.

Formulators of epoxy resin systems often use catalysts to control or alter the flow behavior of their base resin systems. Since a catalyst increases the reaction rate, it is normal for a catalyzed resin to exhibit less total flow during processing than an uncatalyzed system.

Hexcel's 3501-6 epoxy resin was used as the baseline system. Since 3501-6 nominally contains a 1.1% BF_3 catalyst, extremes of high and low flow were achieved by varying the catalyst content from 0% to 2.2%. Varying the degree of advancement during the resin mixing operations further altered the flow behavior. As shown in Table 1, Hexcel prepared subbatch

Table 7.1 Starting Material Variations

Batch Number	*Variation*	*Prepreg Resin Content*	*BF_3 Content (wt. %)*	*Advancement (1)*
1	Low Flow	32	2.2	High
2 (2)	Normal Flow	32	1.1	Standard
3	High Resin Content	42	1.1	Standard
4	High Flow	32	0	Low
5 (3)	Normal Flow	32	1.1	Standard
6	Normal Flow (Vacuum Degassed)	32	1.1	Standard

(1) Low- Minimum Processing Time
Standard- Normal Processing Time
High- Maximum Processing Time
(2) Control Batch for Batches 1, 3 and 4
(3) Control Batch for Batch 6

quantities of high flow (0% catalyst, low advancement), normal flow (1.1% catalyst, normal advancement) and low flow (2.2% catalyst, high advancement) resins. In addition, a vacuum degassed batch was prepared to study the possible effects of entrained air on final laminate quality (i.e., porosity). The subbatches of resin were prepregged onto AS-4 carbon fiber. All batches, except batch 3, were prepregged to yield a 32% net resin content prepreg. In batch 3, the 42% high resin content prepreg was included to assess the effects of resin content on laminate processing. Note that batch 2 was the control batch for batch 1 (low flow), batch 3 (high resin content) and batch 4 (high flow). These batches were all prepared from the same lots of raw chemical ingredients and the same lot of AS-4 fiber. Batch 6, the vacuum degassed batch, was a replacement batch. Therefore, a new control batch (batch 5) was prepared for comparison purposes. Different raw ingredients and fiber lots were used for these last twobatches.

The overall test plan for this study is shown in Fig. 1. Initially, prepreg physical properties and neat resin physiochemical properties were

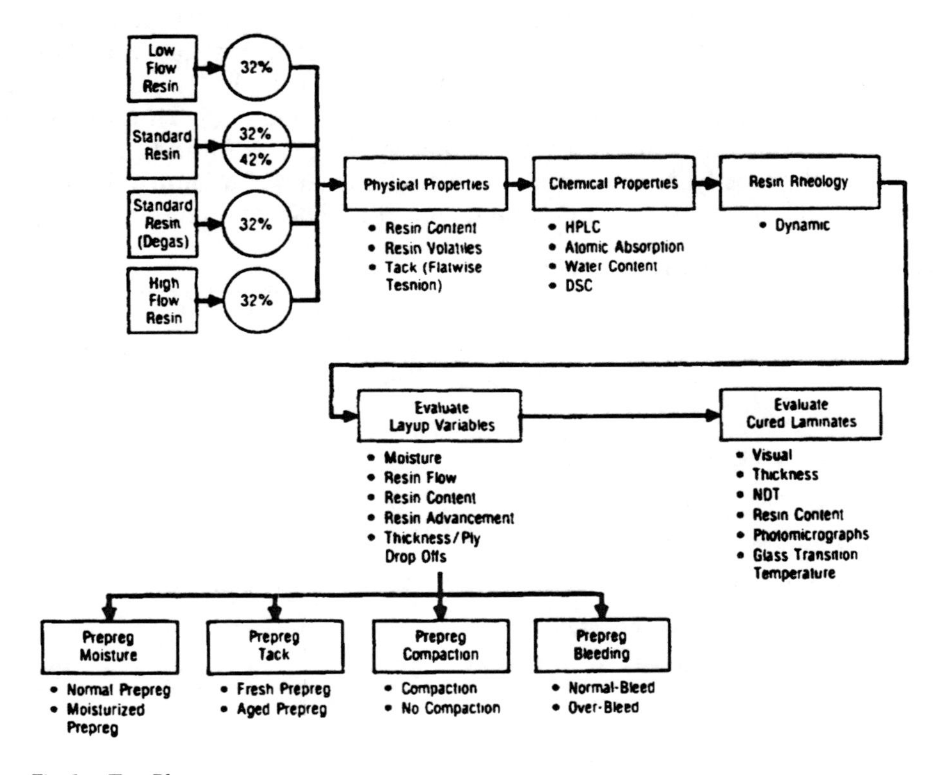

Fig. 1. Test Plan

determined. Next, laminates were fabricated to study the combined effects of material variations and lay-up variability on final cured laminate quality.

7.1 Prepreg Physical Properties

The results of the physical properties testing are summarized in Table 2. Prepreg resin content and volatiles, fiber areal weight and per ply thickness measurements were made using standard laboratory procedures. Prepreg tack was measured using a flatwise tension test previously developed for prepreg.[2] Prior to testing, ten ply thick 0°, 90° prepreg lay-ups were collated and then vacuum debulked between 2 inch × 2 inch square flat-wise tension blocks for 5 min under 20–29 inches of mercury vacuum. After debulking, they were immediately loaded to failure in a mechanical test frame. Both freshly received, fresh prepreg and prepreg aged for 30 days in the lay-up room were tested. However, the 30-day-aged prepreg contained essentially no tack and would not adhere to the loading blocks.

Since previous work[2] indicated a possible correlation between prepreg tack and room-temperature viscosity, the initial neat resin viscosity at 30 °C (86 °F) was determined on a rheometrics dynamic spectrometer (RDS 7700). As shown in Fig. 2, the data also indicates a possible correlation between room-temperature viscosity and prepreg tack. These data are significant because of the potential difficulty of removing entrapped air pockets during collation with tacky prepreg materials. In addition, resins with a high viscosity will be less likely to cold flow and eliminate voids at ply terminations. It should also be noted that the high resin content prepreg

Table 7.2 Prepreg Physical Property Test Results

Batch Number	1	2 (1)	3	4	5 (2)	6
Variation	Low Flow	Normal Flow	High RC	High Flow	Normal Flow	Vacuum Degas
BF_3 Content (wt. %)	2.2	1.1	1.1	0	1.1	1.1
Prepreg Resin Content (wt. %)	32.5	31.0	41.2	33.3	33.9	33.7
Prepreg Resin Volatiles (wt. %)	0.66	1.02	1.12	0.16	0.18	0.22
Prepreg Fiber Areal Wt. (g/m^2)	154	155	156	155	148	149
Prepreg Per Ply Thickness (in.)	.0062	.0062	.0073	.0062	.0059	.0058
Prepreg Tack by Flatwise Tension Strength (psi)	6.3	(3)	8.3	3.5	12.3	21.5
Resin Viscosity at 86 °F (10^5 pose)	4.5	3.6	3.1	4.3	4.6	6.4

(1) Control for Batches 1, 3 and 4
(2) Control for Batch 6
3 Separated due to weight of loading blocks
4 All values are averages of hree specimens

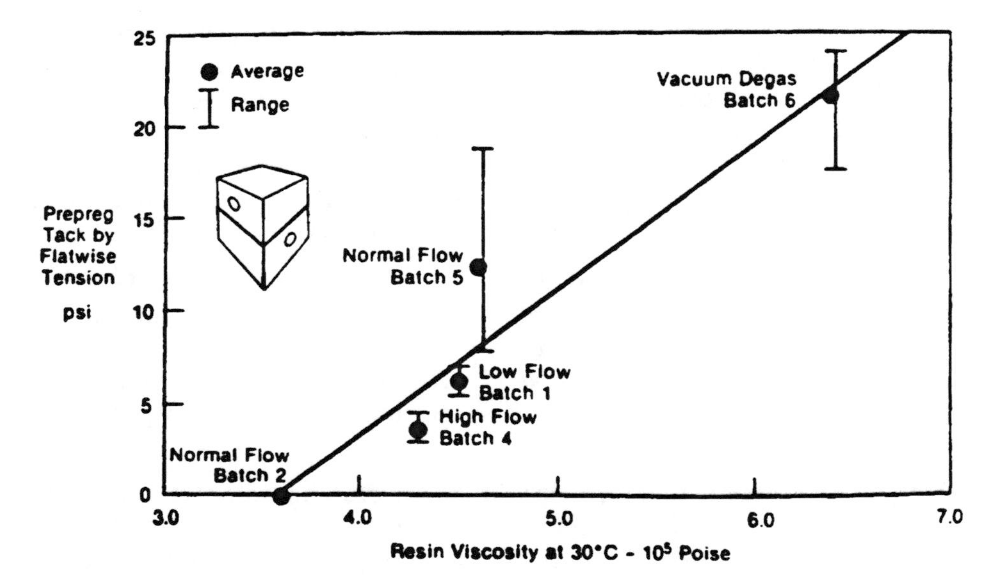

Fig. 2. Prepreg Tack Versus Viscosity

Table 7.3 Prepreg Chemical Property Test Results

Batch Number	1	2 (1)	3	4	5 (2)	6
Variation	Low Flow	Normal Flow	High RC	High Flow	Normal Flow	Vacuum Degas
BF_3 Content (wt. %)	2.2	1.1	1.1	0	1.1	1.1
Water Content by GC (wt. %)	0.44	0.32	0.31	0.21	0.44	0.31
Neat Resin Density (g/cc)	1.07	1.06	–	1.09	1.13	1.24
BF_3 Content by AA (wt. %)	2.32	1.41	1.37	0	1.26	1.25

(1) Control for Batches 1, 3 and 4
(2) Control for Batch 6

(batch 3) exhibited significantly higher flat-wise tension strengths than the net resin content prepregs from batches 1, 2 and 4.

7.2 Chemical Properties

The prepreg water content, neat resin density and BF_3 catalyst content measurements are summarized in Table 3. Water content measurements were made by gas chromatography (GC), while atomic absorption (AA) was used for the BF_3 measurements. Neat resin density was measured by weighing the sample in both air and water. The results showed two interesting points: (1) the higher the catalyst content (BF_3) within each batch fabricated from the same raw materials, the higher the water content, and (2) the density of the vacuum degassed batch was significantly greater than that of the batch fabricated using normal processing (i.e., no degassing). The correlation between catalyst content and water content could be a result of the hygroscopic nature of the BF_3 salt. The batch with no catalyst (batch 4) had the lowest water content, while the batch with twice the normal catalyst content (batch 1) had the highest water content

The density results indicated that: vacuum-degassing the resin during mixing might indeed be effective in removing entrapped air and/or moisture from the resin. Note that although the catalyst content of the vacuum degassed batch (batch 6) and its control (batch 5) were essentially the same, the water content of the vacuum degassed material was approximately 25% less.

Hexcel's results for high-pressure liquid chromatography (HPLC) tests conducted on the finished prepreg are shown in Table 4, along with the initial mix compositions for the major epoxy (TGMDA–tetraglycidyl methylenedianiline), the curing agent (DDS–4,4′-diaminodiphenyl sulfone) and minor epoxy no. 1 (novolac). Since HPLC gives the "free" or unreacted quantities for TGMDA, DDS and the novolac, the remaining amounts of these quantities are combined or "reacted," during neat resin cooking and subsequent prepregging operations. Therefore, the percent change from the initial composition to that after prepregging should indicate the amount of reaction occurring during the mixing and

Table 7.4 High Pressure Liquid Chromatography Results

Batch Number	1	2 (1)	3	4	5 (2)	6
Variation	Low Flow	Normal Flow	High RC	High Flow	Normal Flow	Vacuum Degas
BF_3 Content (wt. %)	2.2	1.1	1.1	0	1.1	1.1
Major Epoxy						
• Initial %	56.6	56.5	56.5	56.5	56.5	56.5
• Prepreg %	50.3	55.1	53.3	53.4	53.0	53.5
• % Decrease	6.2	1.4	3.2	3.1	3.5	3.0
Curing Agent						
• Initial %	25.0	25.0	25.0	25.0	25.0	25.0
• Prepreg %	21.1	24.2	23.5	23.2	23.6	23.3
• % Decrease	3.9	0.8	1.5	1.8	1.4	1.7
Minor Epoxy No. 1						
• Initial %	8.50	8.5	8.5	8.50	8.5	8.50
• Prepreg %	6.85	8.87	8.63	8.16	7.36	6.67
• % Decrease	1.7	+0.37	+0.13	0.34	1.14	1.83

(1) Control for Batches 1, 3 and 4
(2) Control for Batch 6

prepregging operations. All batches showed decreasing amounts of unreacted TGMDA and DDS for the prepreg samples. The data for the novolac epoxy did not show as clear a trend. However, HPLC is not as sensitive a test for the novolac as it is for the TGMDA and DDS.

Doubling the BF_3 catalyst content (batch 1) had the greatest effect. The greater catalyst content, along with the longer processing time for this batch, resulted in larger reductions in both the TGMDA major epoxy and DDS curing agent. These data suggest that the BF_3 catalyst significantly accelerates the curing reaction, even at the relatively moderate temperatures used during the mixing and prepregging operations.

7.3 Thermal Properties

Both dynamic and isothermal DSC scans were performed on neat resin samples. Dynamic scans were made at three different heating rates (2, 5 and 10 °C min^{-1}). In addition, scans were made at 5 °C min^{-1} on resin samples aged for 30 days at room temperature. The dynamic DSC results are summarized in Table 5 and representative scans (at 5 °C min^{-1}) are shown in Fig. 3 for the high, low and normal flow resins.

The normal flow 3501-6 dynamic DSC curve exhibits two distinct exotherm peaks. The first or minor peak is a direct result of the BF_3 catalyst.

Table 7.5 Dynamic DSC Test Results

Batch Number	1	2 (1)	4	
Variation	Low Flow	Normal Flow	High Flow	
BF_3 Content (wt. %)	2.2	1.1	0	
		Scan Rate of 2 °C/min		
Second exotherm (°C)	185	191	205	
Enthalphy (cal/gm)	112	120	129	
		Scan Rate of 5 °C/min		
Second exotherm (°C)	207	213	235	
Enthalphy (cal/gm)	106	111	130	
		Scan Rate of 10 °C/min		
Second exotherm (°C)	–	229	255	
Enthalphy (cal/gm)	103	109	130	
		Scan Rate of 5 °C/min		
Second exotherm (°C)	207	212	241	Prepreg Aged for 30 Days at RT prior To testing
Enthalphy (cal/gm)	97	104	124	

(1) Control for Batches 1 and 4

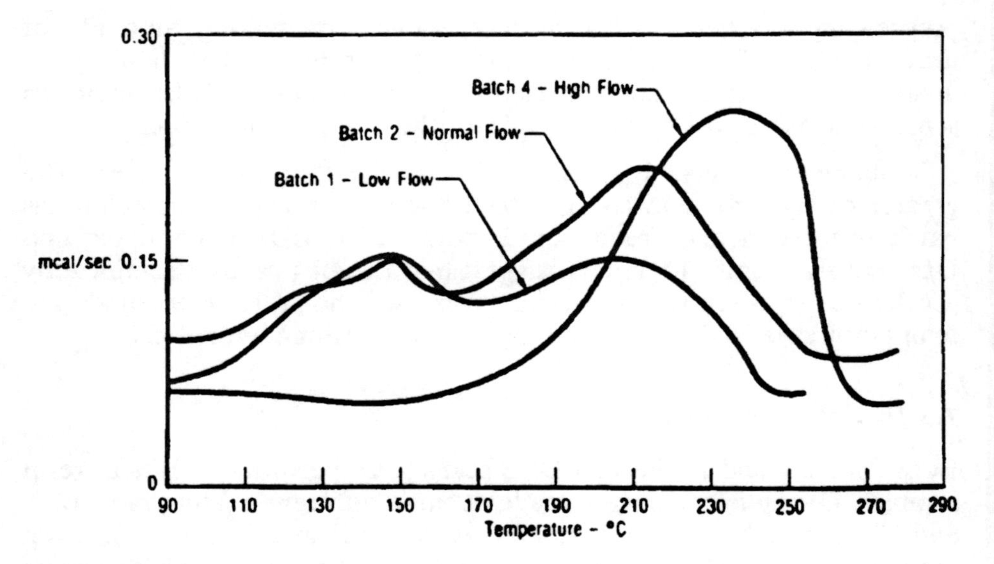

Fig. 3. Dynamic DSC Scan Comparison

The high-flow batch with no catalyst exhibited only the single large exothermic peak characteristic of resins lacking a BF_3 catalyst, such as Hexcel's 3502 and Cytec's 5208. This batch also had much higher total heat of reaction, ΔH. Again, this was a function of the absence of the catalyst, since the catalyst reacts rapidly at low temperatures, thereby reducing the total heat of reaction. On the other hand, for the low-flow batch with twice the normal catalyst content, the first exothermic peak (usually the minor) was larger than the second peak. When DSC scans were run on the prepreg material, the second peak again became the predominate peak. Since the BF_3 catalyst reacts vigorously at relatively low temperatures, the heat from the prepregging operation probably resulted in enough reaction to suppress the minor exothermic peak. Further evidence of the importance of the BF_3 catalyst can be seen by examining the heat of reaction data. The total heat of reaction for the low-flow batch (2.2% BF_3) was lower than the normal-flow batch (1.1% BF_3), while that of the high-flow batch (0% BF_3) was much higher. In addition, the batch with no catalyst was scan-rate independent, while all batches with BF_3 catalyst exhibited a decrease in the total heat of reaction with increasing scan rates.

7.4 Rheological Properties

Dynamic viscosity measurements were performed at 1, 2 and 5 °C min^{-1} on each batch of neat resin using RDS 7700. In addition to viscosity vs. time curves, flow numbers:

Table 7.6 Neat Resin Dynamic Viscosity Test Results (1 °C/min)

Batch Number	1	2 (1)	3
Variation	Low Flow	Normal Flow	High Flow
BF_3 Content (wt. %)	2.2	1.1	1.1
Minimum Viscosity (poise)	10.6	6.4	1.3
Temperature at Minimum Viscosity (°C)	100	104	148
Gel Temperature at 1,000 poise (°C)	151	165	180
Flow Number (min/poise)	3.45	8.30	36.75

(1) Control for Batches 1and 4

$$\text{flow number} = \int_{t_0}^{t_{gel}} dt/\eta$$

were calculated from each curve. Flow is the reciprocal of viscosity when integrated as a function of time between the starting time (t_0) and the time to gellation (t_{gel}). The test results for the viscosity tests at a heating rate of 1 °C min^{-1} are summarized in Table 6.

The dramatic effect of the BF_3 catalyst was again evident. The low flow system (batch 1) exhibited a flow number of approximately one-half that of the normal-flow material (batch 2) and an order of magnitude less than that of the high-flow system (batch 4).

The rest of the viscosity data further emphasize the effect of the BF_3 catalyst. For example, the minimum viscosity for the high-flow material (batch 4) was much lower than that of the low-flow material (batch 1). The gel temperature also showed the effect of varying the catalyst content. The low-flow system (batch 1) gelled at a lower temperature than the control (batch 2), while the high-flow system (batch 4) gelled at a much higher temperature. The temperature at the minimum viscosity was also significantly higher. Examining the viscosity curves shown in Fig. 4 show this difference in flow behavior. These results can be explained by the fact that the BF_3 catalyst greatly accelerates the reaction at relatively low temperatures, resulting in higher reaction rates, lower total flow, higher minimum viscosities and lower gellation temperatures. When the catalyst is absent, higher temperatures are necessary to initiate the chemical reactions, thereby providing more time for total flow and higher gellation temperatures.

7.5 Laminate Evaluations

The objectives of the laminate evaluations were to determine the combined effects of prepreg and lay-up variations on final cured laminate quality. Significant variables are summarized below:

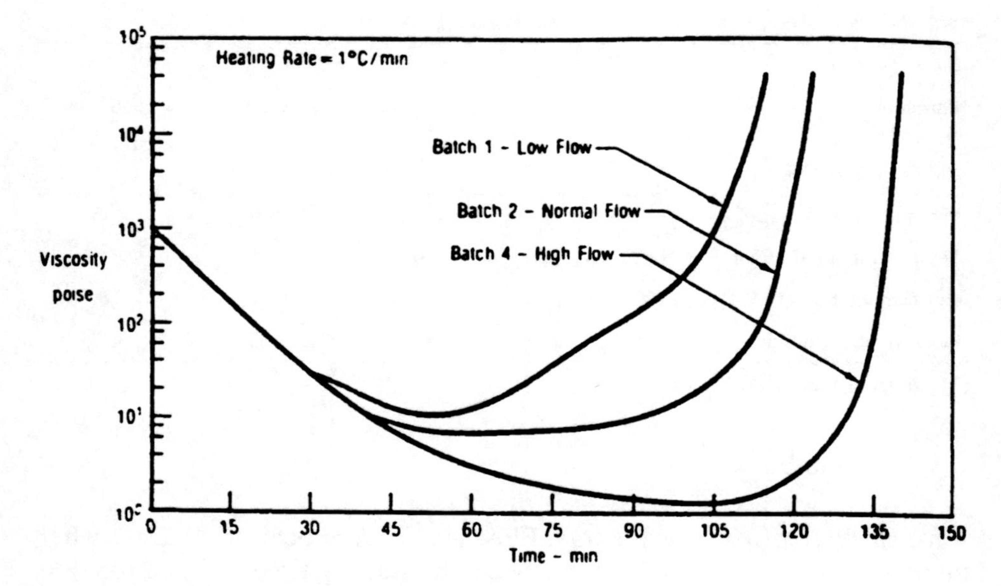

Fig. 4. RDS Dynamic Viscosity Comparison

- *Base line laminate* – use fresh prepreg, debulk every five plies and standard amount of bleeder material;
- *"Wet" prepreg* – start with fresh prepreg, condition prepreg until prepreg moisture content is 1.0%, debulk every five plies, standard amount of bleeder material;
- *"Aged" prepreg* – start with fresh prepreg, expose prepreg to lay-up room environment for 30 days, debulk every five plies, standard amount of bleeder material;
- *No debulking* – use fresh prepreg, do not debulk plies, standard amount of bleeder material;
- *Overbleed* – use fresh prepreg, debulk every five plies, three times the standard amount of bleeder material.

A total of 26 laminates were cured using these lay-up variables and the six altered batches of prepreg. The laminate configuration (Fig. 5) contained a thin area (24 plies), a thick area (54 plies) and a tapered area with internal ply drop-offs ranging from one to five plies. The tapered area allowed an assessment of laminate quality (i.e., voids and porosity) for varying numbers of internal ply drop-offs. The prepreg for the "wet" and "aged" laminates was conditioned in a temperature-and humidity-controlled lay-up room. Moisturization for the "wet" prepreg laminates consisted of precutting the plies, laying them out on lab benches and then elevating the temperature and humidity of the lay-up room until the desired 1.0% by weight of moisture was absorbed. The prepreg for the

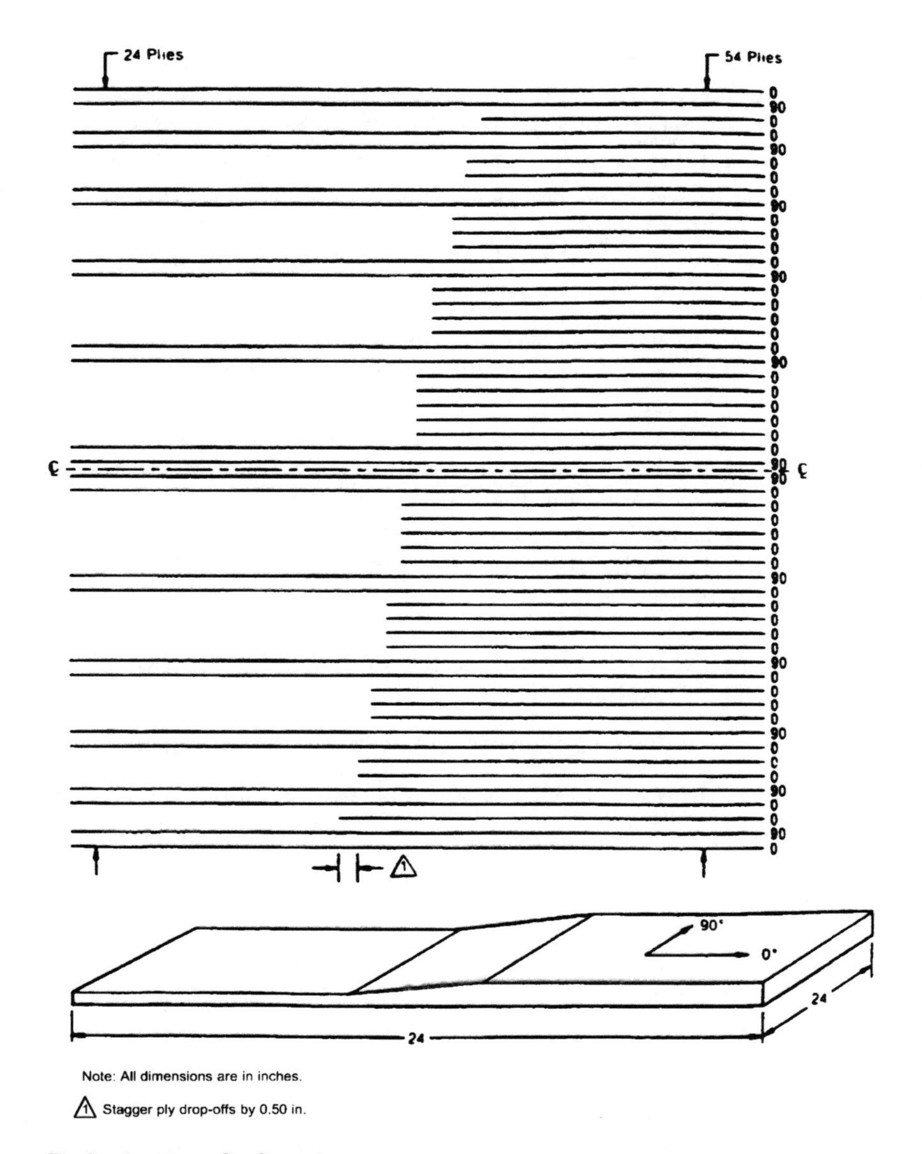

Fig. 5. Laminate Configuration

"aged" laminates was also precut and exposed on lab benches for 30 days; however, the temperature and humidity of the lay-up room was maintained within the normal operating range for the "aged" ply sets.

Unless otherwise specified, standard fabrication procedures were used. Hand collation was used for all laminates. If debulking was specified, the lay-up was debulked every five plies for a maximum of 10 min. at 20–

29 inches of mercury vacuum. After collation, the laminates were bagged and autoclave cured. During bagging, different amounts of bleeder cloth were used to control resin bleeding. For normally bled laminates fabricated with net resin content prepreg, no glass bleeder was used. The amount of bleeder cloth was increased as the amount of resin to be removed during cure increased. For example, in the case of the high resin content (42%) prepreg laminates were to be overbled, and three times the normal amount of glass bleeder was used during bagging.

All laminates were autoclave cured under 85–100 psig autoclave pressure. The laminates were initially heated to 240 ± 10 °F at a rate of approximately 4 °F min^{-1}. They were then held at 240 ± 10 °F for 70 min. After the 240 °F hold, they were heated to 350 ± 10 °F at a rate of approximately 4 °F min^{-1}. All laminates were then cured at 350 ± 10 °F for 130 min. After cure, the laminates were evaluated for quality by through-transmission ultrasonic C-scanning. Materiallographic sections were then taken to confirm the ultrasonic results. Finally, additional sections were taken for resin content analyses and glass transition temperature (T_g) measurements. The through-transmission ultrasonic results are summarized in Table 7. Several conclusions can be drawn from the nondestructive test results:

- In all cases, standard baseline processing produced acceptable laminates.
- In general, the vacuum-degassed batch produced the highest-quality laminates.

Table 7.7 Through–Transmission Ultrasonic C–Scan Indications

Batch Number	1	2 (1)	3	4	5 (2)	6
Variation	Low Flow	Normal Flow	High RC	High Flow	Normal Flow	Vacuum Degas
BF_3 Content (wt. %)	2.2	1.1	1.1	0	1.1	1.1
Baseline Laminate	Clean	Clean	Clean	Clean	Clean	Clean
"Wet" Prepreg	Moderate Porosity	Moderate Porosity	Extreme Porosity	Gross Porosity	Gross Porosity	Minor Porosity
"Aged" Prepeg	Clean	Clean	Clean	Clean	Clean	Clean
No Debulking	–	Clean	Clean	–	Clean	Minore Porosity
Over Bleed	Moderate Porosity	Clean	Clean	Minor Porosity	–	–

(1) Control for Batches 1, 3 and 4
(2) Control for Batch 6
Extreme–Gross–Moderate–Minor–Clean
(Bad) (Good)

- The highest-quality laminate for the "wet" prepreg lay-up variation was produced from the vacuum-degassed material. This could be due to the removal of trapped air from the resin that may serve as nucleation sites for voids. In all other cases, moisture resulted in moderate to extreme amounts of porosity.
- There was also a noticeable color difference between the vacuum-degassed resin and the other resins. The vacuum-degassed batch was a translucent dark brown color, while the other resin batches were an opaque tan color due to the entrained air. The poorest-quality laminate for the no-debulking lay-up variation was the vacuum-degassed batch. This batch of material also had the highest tack as measured by flat-wise tension strength. As mentioned earlier, previous studies have shown that high tack may be detrimental to removing entrapped air that can result in voids during cure. Collating the laminates with no debulking would only magnify the difficulty of removing entrapped air.
- The "aged" prepreg lay-up variation was the least detrimental lay-up variation addressed.
- The thick sections and ply drop-off areas were of much poorer quality than the thin sections. This could be due to the increased difficulty of removing volatiles (i.e., moisture) from thicker sections during curing.

In general, the materiallographic examination confirmed the results of the ultrasonic nondestructive testing. The baseline laminate photomicrographs were relatively free of porosity, confirming the NDT results. The most obvious result was that the high-resin-content laminates were significantly thicker than the net-resin-content laminates.

Photomicrographs from the "wet" prepreg laminates exhibited quite a large amount of porosity, particularly in the case of the high-resin-content "wet" prepreg laminate. The nondestructive test results had identified this lay-up variable as the most severe. The photomicrographs also showed the differences between the thick and thin sections. The thick sections contained significantly more porosity than the thin sections. The ply drop-off areas also contained a considerable amount of porosity at the end of the ply terminations. However, looking at the low magnification photomicrographs, most of the porosity associated with the ply drop-offs did not occur at the ply drop-offs themselves, but at areas just below the ply drop-offs. The vacuum-degassed prepreg, which gave the highest-quality laminate for the "wet" prepreg, gave the poorest no-debulking laminate. Many of the voids in the photomicrographs from the vacuum-degas–no-debulking laminate were large and on the surfaces of the plies. Voids of this type are believed to be typical of voids created by air trapped during lay-up, probably due to the extreme tackiness of this material.

Table 7.8 Cured Resin Content Results

Batch Number	1		2 (1)		3		4		5 (2)		6	
Variation	Low Flow		Normal Flow		High RC		High Flow		Normal Flow		Vacuum Degas	
BF_3 Content (wt. %)	2.2		1.1		1.1		0		1.1		1.1	
Baseline Laminate	Thin	Thick	Thin	Thick	Thin	Thick	Thin	Thick	Thin	Thick	Thin	Thick
	31.7	31.3	30.3	30.9	35.6	38.5	29.8	28.7	32.3	31.4	31.4	31.5
"Wet" Prepreg	Thin	Thick	Thin	Thick	Thin	Thick	Thin	Thick	Thin	Thick	Thin	Thick
	31.3	31.8	30.39	31.6	42.2	41.9	30.2	30.7	32.5	32.5	31.7	32.7
"Aged" Prepeg	Thin	Thick	Thin	Thick	Thin	Thick	Thin	Thick	Thin	Thick	Thin	Thick
	32.6	31.7	31.5	31.6	37.7	39.5	30.8	30.5	32.4	32.0	32.6	33.0
No Debulking			Thin	Thick	Thin	Thick			Thin	Thick	Thin	Thick
			32.1	31.6	36.9	38.6			32.4	32.3	33.5	33.0
Over Bleed	Thin	Thick	Thin	Thick	Thin	Thick	Thin	Thick				
	27.9	29.7	24.9	26.9	28.7	34.4	24.8	22.6				

(1) Control for Batches 1, 3 and 4
(2) Control for Batch 6
3 All values reported in wt. % resin

The cured laminate resin content results are summarized in Table 8. Two specimens were taken from both the thick and thin section of each laminate. The high-resin-content prepreg resulted in the highest-resin-content laminates. The laminate thickness was too great and the bleeder amount insufficient to obtain the desired cured resin content (28–32%). These panels illustrate one benefit that can be gained by using a net-resin-content prepreg. The amount of BF_3 catalyst had only a slight effect on the final resin content. Much of this was due to the prepregs being net-resin-content materials. However, for the overbled laminates, the amount of BF_3 catalyst did have a significant effect on the cured laminate resin content. The high-flow system exhibited a much lower resin content than either the normal or the low-flow systems.

Except for the high-flow resin system, the panel thickness (i.e., the number of compacted plies) was also affected by resin flow. The thick sections of the low-flow, normal-flow and high-resin-content laminates had noticeably high resin contents. These high resin contents in thick sections were due to the increase in flow resistance as each ply compacted. Eventually, the flow resistance became so large that it restricted flow from the bottom section of the panel resulting in higher resin contents. A similar result would be obtained for the high-flow system if a thicker laminate had been used.

"Wet" prepreg did not seem to affect resin flow except for the high-resin-content panel. This panel had the highest resin content of all panels. "Aged" prepreg had only a slight effect on the final resin content. As expected, the resin contents for the aged laminates were slightly higher than for the baseline laminates. This was due to the reduction in flow associated with resin advancement. No debulking during lay-up did not have any noticeable affect on the final resin content. Over-bleeding yielded the expected result. All laminates were lower in resin content than the baseline laminates.

The T_g for each laminate was determined by DSC. The results are presented in Table 9. Also reported are the residual heats of reaction, if measurable. The baseline laminate values show the effect of chemistry changes (catalyst content) on the glass transition temperature. The lowest T_g value was for the laminate with no BF_3 catalyst, while the highest value was from the panel with double the catalyst content. The four panels with the normal catalyst content fell between the two extremes. The residual heats of reaction showed the same trend. The zero-catalyst-content panel had the largest residual ΔH, while the double-catalyst-content panel had the lowest. The T_g values from the "wet" prepreg panels show how moisture can change the reaction kinetics. Moisture deactivates the BF_3 catalyst, thus slowing the reaction. The T_g for all the panels that contained catalyst dropped 27–40 °F. For the high-flow system that did not contain a BF_3 catalyst, the water acted as a mild catalyst and increased the T_g by 14 °F. Moisture also affected the residual heat of reaction. For the

Table 7.9 Glass Transition Temperature (T_g) Results

Batch Number	1	2 (1)	3	4	5 (2)	6
Variation	Low Flow	Normal Flow	High RC	High Flow	Normal Flow	Vacuum Degas
BF_3 Content (wt. %)	2.2	1.1	1.1	0	1.1	1.1
Baseline Laminate	401 (0.32)	382 (0.53)	379 (0.82)	361 (4.00)	392 (.74)	376 (0.65)
"Wet" Prepreg	356	354	346	365 (1.29)	364	349
"Aged" Prepeg	395 (0.51)	385 (.62)	386 (0.90)	363 (3.24)	387 (0.57)	390 (0.81)
No Debulking		386 (0.74)	390 (1.05)		387 (0.76)	386 (0.76)
Over Bleed	399 (0.44)	373 (0.47)	392 (0.73)	362 (3.20)		

(1) Control for Batches 1, 3 and 4
(2) Control for Batch 6
3 All values reported in °F
4 Values in parentheses are residual ΔH in cal/gm

laminates that contained BF_3, no residual ΔH was measurable, although the T_g was lower. The residual ΔH for the zero-catalyst-content laminate was also reduced significantly. For the "aged" prepreg panels, the T_g's appeared to increase slightly. This could be due to the reaction that occurs during the aging period. As was the case for the baseline laminates, the T_g for the double-catalyst-content laminate was the highest.

7.6 Summary

Several conclusions can be drawn from this study:

- *Catalyst content* – The amount of catalyst in an epoxy resin profoundly affects the thermal and rheological properties. The BF_3 catalyst greatly accelerates the reaction at low temperatures; resulting in high reaction rates, lower total flow, higher minimum viscosities and lower gellation temperatures. The high reaction rates associated with a high catalyst content resulted in higher glass transition temperatures during laminate processing. For the two-hour cure used during this study, the laminates made from prepreg with no catalyst exhibited residual heats of reaction, indicating that additional cure time or higher temperatures would be required to complete the reaction.
- *Resin mixing* – Vacuum degassing during mixing is an effective means of removing entrained air from the resin. The vacuum-degassed resin exhibited a higher density and a translucent dark brown color, indicating that air had been effectively removed. When moisture was present in the prepreg, the vacuum-degassed prepreg produced the highest-quality laminate containing only minor porosity, while the other moisture-containing prepregs yielded laminates with greater amounts of porosity. This could be due to the presence of entrained air serving as nucleation sites for porosity. However, the vacuum-degassed prepreg supplied for this study was extremely tacky and viscous, which contributed to some trapped air pockets during ply collation and resulted in some minor porosity when the laminate was not debulked.
- *Prepreg resin content* – Net (32%) resin content prepreg proved easier to process than high (42%) resin content prepreg. During laminate processing, it was extremely difficult to bleed the high-resin-content prepreg down to the desired thickness and resin content. This was particularly true as the laminate thickness increased. Although not evaluated during this study, the reader should be cautioned that thin-section laminates fabricated from a net-resin-content prepreg could quite easily be inadvertently overbled, particularly if the part is not properly dammed during cure.
- *Lay-up variability* – Prepreg moisture was the most severe lay-up variable evaluated during this study. In every case, prepreg moisture

resulted in some porosity in the cured laminates. In general, the porosity was more severe in the thicker sections and ply drop-off areas than in the thinner sections. In all cases, baseline processing (i.e., dry prepreg, debulking, standard bleeding) produced acceptable laminates, even when the prepreg was aged for 30 days in an environmentally controlled lay-up room.

References

[1] Campbell F.C., Mallow A.R., Carpenter J. F., "Chemical Composition and Processing of Carbon/Epoxy Composites," *American Society of Composites, Second Technical Conference*, September 1987.

[2] Brand R.A., Brown G.G., McKague E. L., "Processing Science of Epoxy Resin Composites," Air Force Contract No. F33615-80-C-5021, Final Report for August 1980–December 1983.

Chapter 8

Adhesive Bonding and Integrally Cocured Structure: A Way to Reduce Assembly Costs through Parts Integration

Adhesive bonding is a method of joining composite structures together that eliminates some or all of the cost and weight of mechanical fasteners. There are two main methods of bonding composite structures: secondary bonding and cocuring. In secondary bonding, cured composites are adhesively bonded to other cured composites, honeycomb core, foam core or metallic pieces. Cocuring is a process in which uncured composite plies are cured and bonded simultaneously during the same cure cycle to either core materials or to other composite parts. The ability to make large bonded and cocured unitized structure can eliminate a significant portion of the assembly costs.

8.1 Adhesive Bonding

Adhesive bonding is a widely used industrial joining process in which a polymeric material (the adhesive) is used to join two separate pieces (the adherends). There are many types of adhesives; some cure while others do not. Some are strong and rigid, while others are weak and flexible. Adhesives used for structural bonding of composite structures are always cured at either room temperature or elevated temperatures and must posses adequate strength to transfer the loads through the joint. There are many types of structural adhesives; however, epoxies, nitrile phenolics and bismaleimides are the most prevalent. In addition to fabricating large bonded composite components, adhesive bonding is frequently used for repairing damaged composite parts.

Bonded joints may be preferred if thin composite sections are to be joined when bearing stresses in bolted joints would be unacceptably high, or when the weight penalty for mechanical fasteners is too high. In general, thin structures with well-defined load paths are good candidates for adhesive bonding, while thicker structures with complex load paths are better candidates for mechanical fastening.[1]

8.1.1 Advantages of Adhesive Bonding

The advantages of adhesive bonding include:[2]

(1) Bonding provides a more uniform stress distribution than mechanical fasteners by eliminating the individual stress concentration peaks caused by mechanical fasteners. As shown in Fig. 1, the stress distribution across the joint is much more uniform for an adhesive-bonded joint than for a mechanical joint, leading to better fatigue life than for the mechanically fastened joint. Bonded joints also provide superior vibration and damping capability.

(2) Due to the elimination of mechanical fasteners, bonded joints are usually lighter than mechanically fastened joints and are cheaper in some applications.

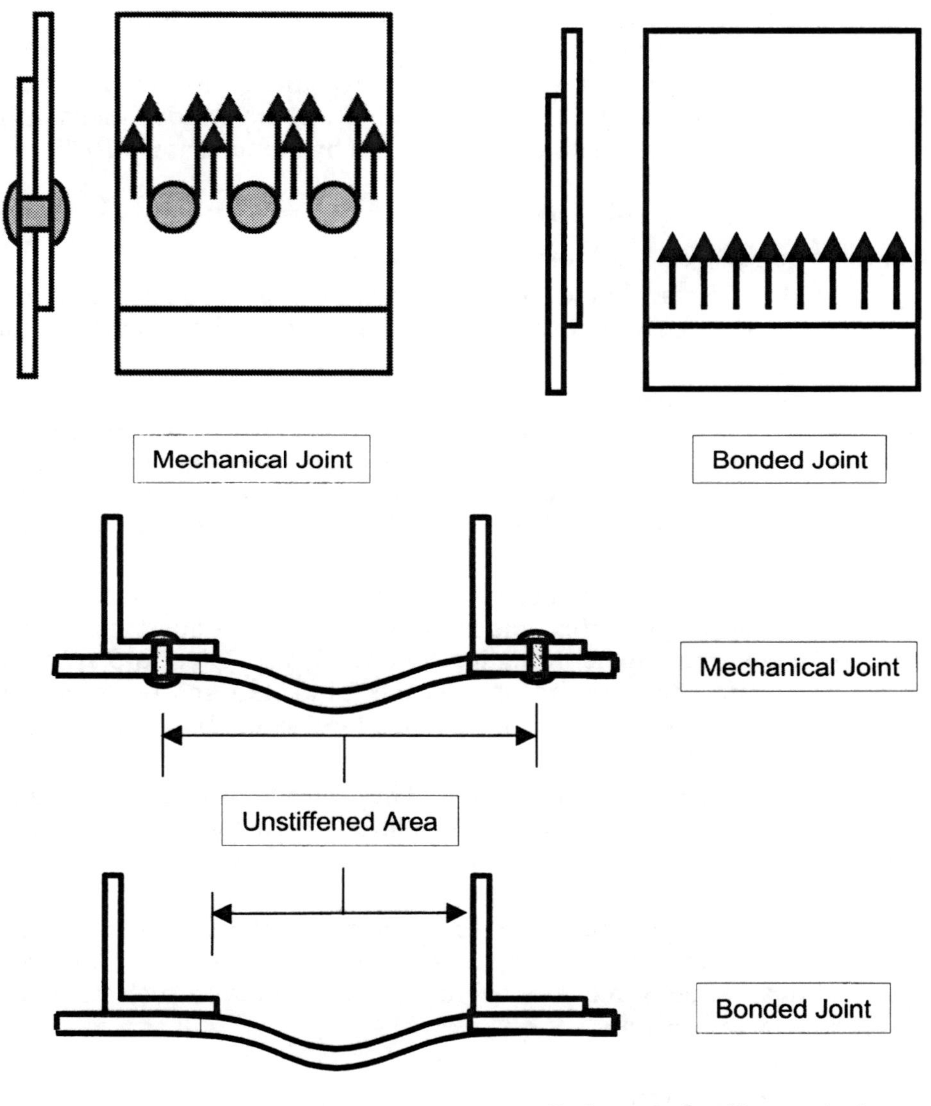

Fig. 1. Load Distribution Comparison for Mechanically Fastened and Bonded Joints

(3) Bonded joints enable the design of smooth external surfaces and integrally sealed joints with minimum sensitivity to fatigue crack propagation. Dissimilar materials can be assembled with adhesive bonding and the joints are electrically insulating, which prevents galvanic corrosion of metal adherends.

(4) Bonded joints provide a stiffening effect compared to riveted or spot-welded constructions. While rivets or spot welds provide local point stiffening, bonded joints provide stiffening over the entire bonded area. The significance of this effect is shown in Fig. 1, where bonded joints may increase the buckling strength of the structure by as much as 30–100%.

8.1.2 Disadvantages of Adhesive Bonding

Adhesive bonding also has some disadvantages, including:

(1) Bonded joints should be considered to be permanent joints. Disassembly is not easy and often results in damage to the adherends and surrounding structure.

(2) Adhesive bonding is much more sensitive to surface preparation than mechanical fastening. Proper surface preparation is absolutely essential to producing a strong, durable bond. For field repair applications, it can be extremely difficult to execute proper surface preparation. For original manufacturing, adhesive bonding requires clean rooms with temperature and humidity control.

(3) Adhesively bonded joints can be nondestructively tested for voids and unbonds; however, at this time there is no reliable nondestructive test method for determining the strength of a bonded joint. Therefore, traveler or process control test specimens must be fabricated and destructively tested using the same surface preparation, adhesive and bond cycle as the actual structure.

(4) Adhesive materials are perishable. They must be stored according to the manufacturer's recommended procedures (often refrigerated). Once mixed or removed from the freezer, they must be assembled and cured within a specified time.

(5) Adhesives are susceptible to environmental degradation. Most absorb moisture and exhibit reduced strength and durability at elevated temperatures, while some are degraded by chemicals, such as paint strippers or other solvents.

8.2 Theory of Adhesion

There are a number of theories on the nature of adhesion during adhesive bonding but there is some general agreement about what leads to a good adhesive bond. Surface roughness plays a key role. The rougher the surface the more surface area available for the liquid adhesive to penetrate and lock onto. However, for this to be effective, the adhesive must wet the surface, which is a function of adherend cleanliness and adhesive viscosity and surface tension. The importance of surface cleanliness cannot be overemphasized; surface cleanliness is one of the cornerstones of successful adhesive bonding.

In metals, coupling effects as a result of chemical etchants/anodizers or other treatments can also play a role in adhesion by providing chemical end-groups that attach to the metal adherend surface and provide other chemically compatible end-groups with the adhesive.

Therefore, for the best possible adhesive joint, the following areas must be addressed: the surface must be clean; the surface should have maximum surface area through mechanical roughness; the adhesive must flow and thoroughly wet the surface; and the surface chemistry must be such that there are attractive forces on the adherend surface to bond to the adhesive.

8.3 Joint Design[1]

In a structural adhesive joint, the load in one component is transferred through the adhesive layer to another component. The load transfer efficiency depends on the joint design, the adhesive characteristics, and the adhesive/substrate interface. To effectively transfer loads through the adhesive, the substrates (or adherends) are overlapped so that the adhesive is loaded in shear. Typical joint designs are shown in Fig. 2.

As shown in the shear stress distribution for a typical joint (Fig. 3), the loads peak at the joint ends, while the center portion of the joint carries a much lower portion of the load. Therefore, adhesives designed tocarry high

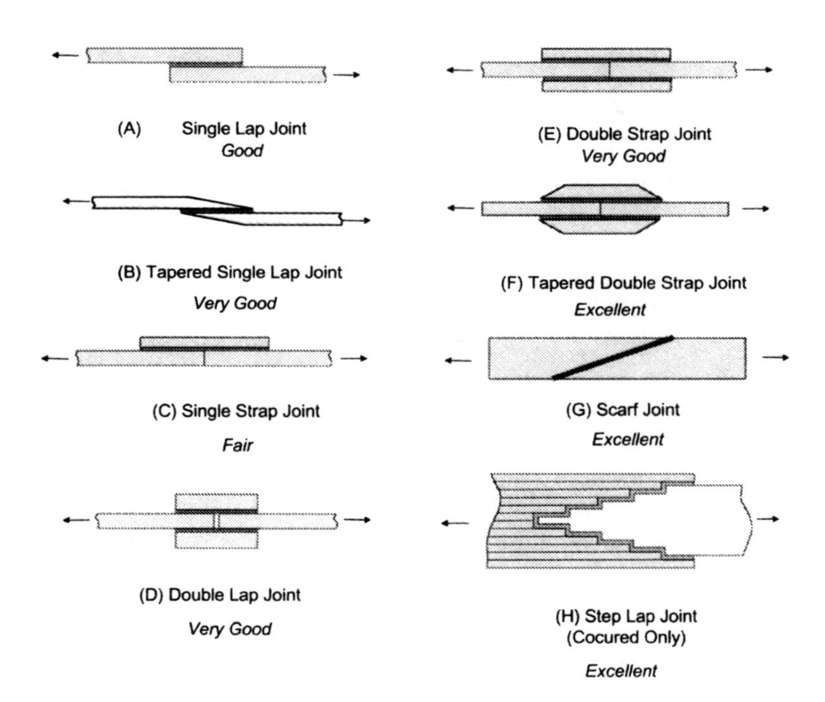

Fig. 2. Typical Adhesively Bonded Joint Configurations

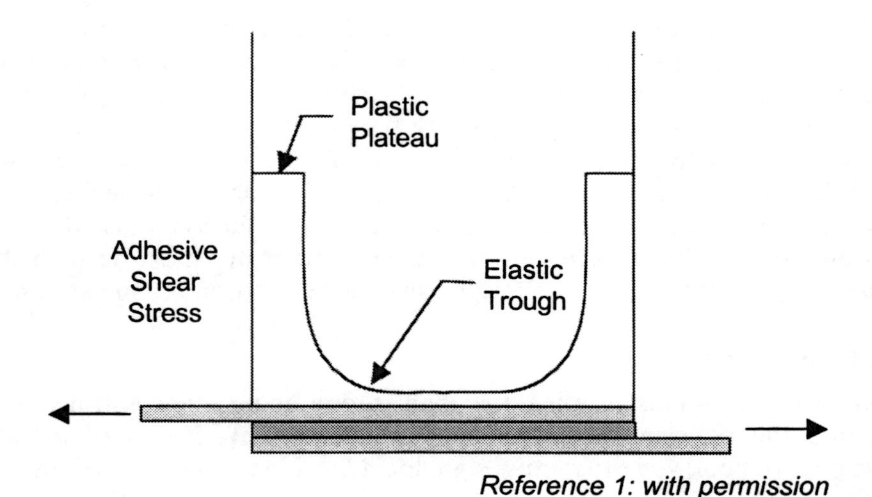

Fig. 3. Typical Bondline Shear Stress Distribution

loads need to be strong and tough, especially if there is any bending in the joint that would induce peel loads. Adhesives are frequently modified with rubber or other elastomers, which reduces the adhesive modulus in order to improve fracture toughness and fatigue life. A comparison of a "brittle" high-strength, high-modulus adhesive with a "ductile" lower-strength, lower-modulus adhesive is shown in Fig. 4. The ductile adhesive, providing it has adequate shear strength, is a much more forgiving adhesive than the higher-strength brittle adhesive, particularly in real joints that often experience peel and bending loads. However, the joint design must ensure that the adhesive is loaded in shear as much as possible. Tension, cleavage and peel loading (Fig. 5) should be avoided when using adhesives. Actually, tension loading is acceptable as long as there is appreciable surface area but certainly not in the butt joint shown. A variety of joint designs are shown in Fig. 6. Fig. 6A-D joints are all loaded in shear, while Fig. 6E is loaded in tension and Fig. 6F is loaded in peel. Note that Figs. 6E and 6F exhibit large tensile stress peaks in the adhesive. These peaks can be much higher than those in shear loading and are far less forgiving. Also, these peaks are very sensitive to small eccentricities. These problems make tension and peel, at best, very questionable load paths. Fig. 6D shows a scarf joint that is often used for damage repair. It is impressive in that there are no shear peaks in the adhesive stress distribution (i.e., a constant-shear stress). Further considerations for joint design are summarized in Table 1.

Bonding to composites rather than metals introduces significant differences in criteria for adhesive selection for two reasons: (1) composites have a lower interlaminar shear stiffness compared to metals, and (2)

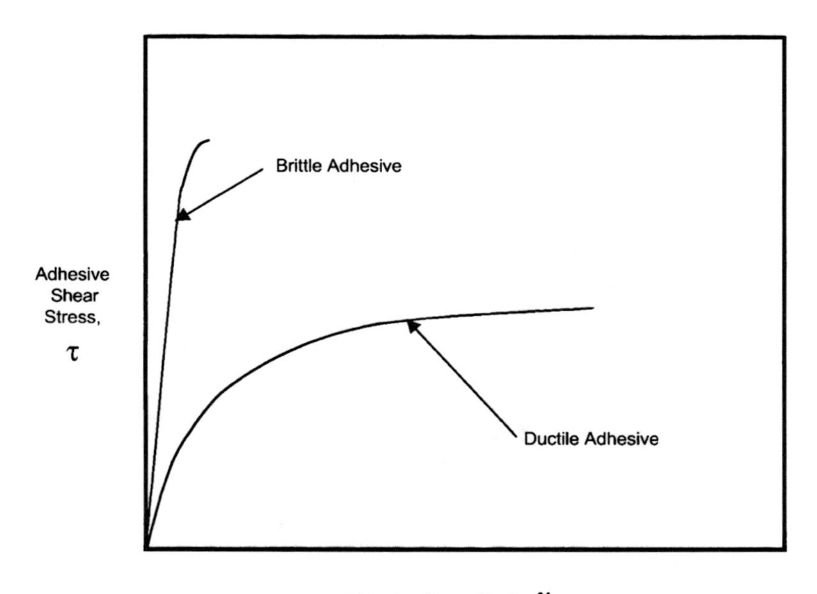

Fig. 4. Typical Stress-Strain Behavior for Brittle and Ductile Adhesives

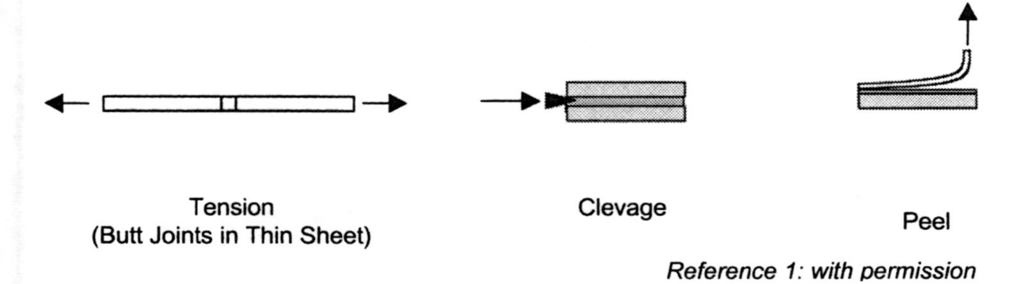

Fig. 5. Load Paths to Avoid in Bonded Structure

composites have much lower shear strength than metals. This occurs because the interlaminar shear stiffness and strength depends on the matrix properties and not the higher properties of the fibers. The exaggerated deformations in a composite laminate bonded to a metal sheet and placed under tension are shown in Fig. 7. The adhesive passes the load from the metal into the composite until, at some distance *L*; the strain in each material is equal. In the composite, the matrix resin acts as an adhesive to pass the load from one fiber ply to the next. Because the matrix shear stiffness is low, the composite plies deform unequally in tension, as shown

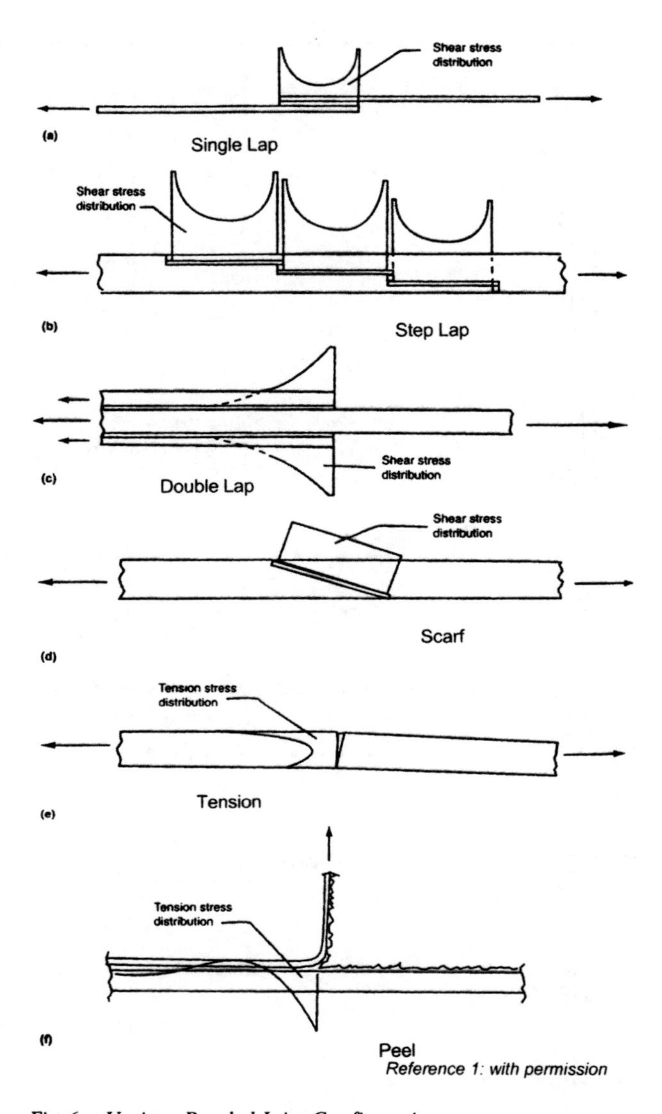

Fig. 6. Various Bonded Joint Configurations

in the figure. Failure tends to initiate in the composite ply next to the adhesive near the beginning of the joint, or in the adhesive in the same neighborhood. The highest failure loads are achieved by an adhesive with a low shear modulus and high strain to failure, as previously shown in Fig. 4. This results because L will be the largest and the maximum shear stress will be lowest. It should also be noted that there is a limit to the thickness of the composite that can be loaded by a single bondline; however, multiple steps

Table 8.1 Considerations for Designing Adhesively Bonded Joints

• The adhesive must be compatible with the adherends and able to retain its required strength when exposed to in-service stresses and environmental factors.
• The joint should be designed to ensure a failure in one of the adherends rather than a failure within the adhesive bondline.
• Thermal expansion of dissimilar materials must be considered. Due to the large thermal expansion difference between carbon composite and aluminum, adhesively bonded joints between these two materials have been known to fail during cool down from elevated temperature cures as a result of the thermal stresses induced by their differential expansion coefficients.
• Proper joint design should be used, avoiding tension, peel or cleavage loading whenever possible. If peel forces cannot be avoided, a lower modulus (nonbrittle) adhesive having a high peel strength should be used.
• Tapered ends should be used on lap joints to feather out the edge-of-joint stresses. The fillet at the end of the exposed joint should not be removed.
• Selection tests for structural adhesives should include durability testing for heat, humidity (and/or fluids), and stress, simultaneously.

Reference 1

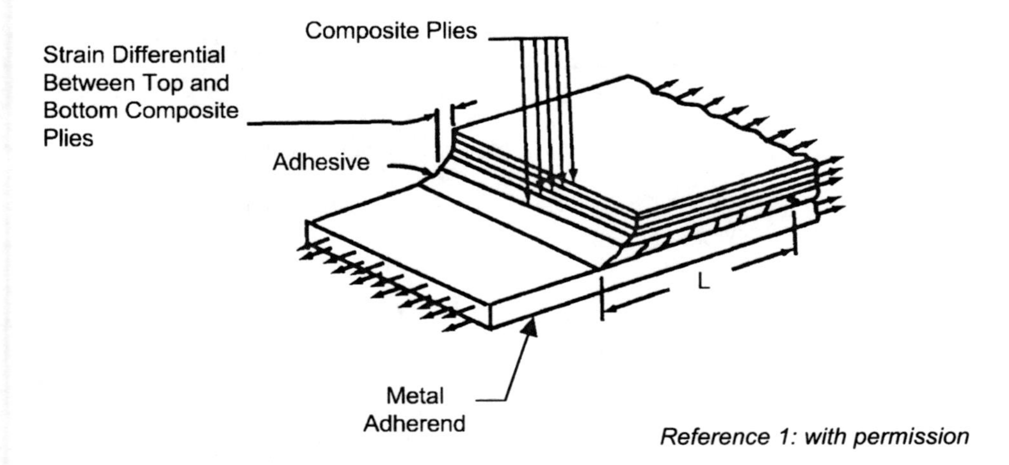

Fig. 7. Uneven Strain Distribution in Composite Plies

in the composite thickness giving multiple bondlines can be used for thick material, as shown in the cocured step lap joint in Fig. 2. The effects of adherend thickness and joint configuration on failure mode are shown in Fig. 8. For thick adherends, it is necessary to go to either a cocured scarf or step-lap joint to carry the load. The other option for thick joints is to use mechanical fasteners. Note that the double scarf joint shown in Fig. 8 is rarely used because it is extremely difficult to fabricate. The step-lap joint configuration, while not easy to fabricate, contains discrete steps that can be used for accurate ply location during fabrication.

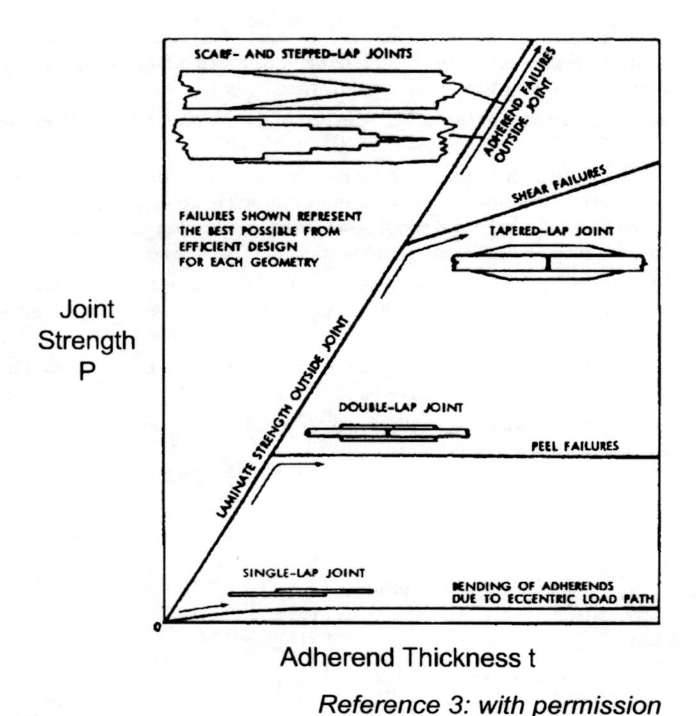

Fig. 8. Effect of Adherend Thickness on Failure Modes of Adhesively Bonded Joints

Basic design practice for adhesive bonded composite joints should include making certain that the surface fibers in a joint are parallel to the direction of load to minimize interlaminar shear, or failure, of the bonded substrate layer. In designs in which joint areas have been machined to a step-lap configuration, for example, it is possible to have a joint interface composed of fibers at an orientation other than the optimal 0° orientation to the load direction. This tends to induce substrate failure more readily than would otherwise occur.

8.4 Adhesive Testing

Adhesive bond strength is usually measured by the simple single lap shear test as shown in Fig. 9. The lap shear strength is reported as the failure stress in the adhesive, which is calculated by dividing the failing load by the bond area. Since the stress distribution in the adhesive is not uniform over the bond area (it peaks at the edges of the joint as previously shown in Fig. 3), the reported shear stress is lower than the true ultimate strength of the adhesive. While this test specimen is relatively easy to fabricate and

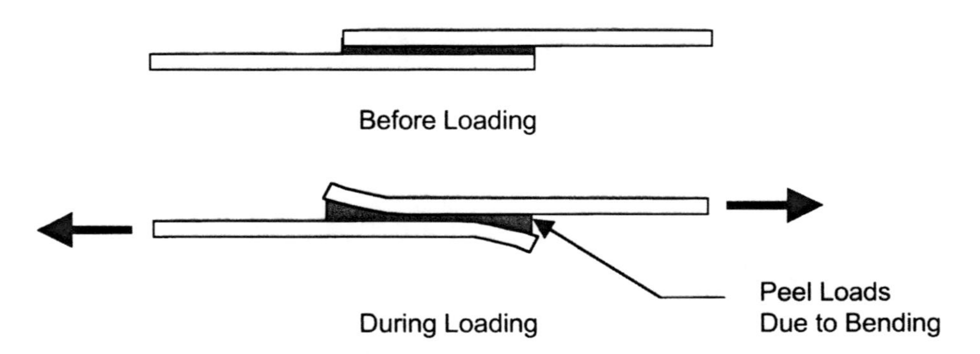

Fig. 9. Typical Single Lap Shear Test Specimen

test, it does not give a true measure of the shear strength due to adherend bending and induced peel loads. In addition, there is no method of measuring the shear strain and, thus, of calculating the adhesive shear modulus required for structural analysis. To measure the shear stress versus shear strain properties of an adhesive as previously shown in Fig. 4, an instrumented thick adherend test can be run where the adherends are so thick that the bending forces are negligible. However, the single lap shear test is an effective screening and process control test for evaluating adhesives, surface preparations, and for in-process control. There are many other tests for characterizing adhesive systems, many of which are summarized in Ref. 4.

When testing or characterizing adhesive materials, there are several important points that should be considered: (1) all test conditions must be carefully controlled including the surface preparation, the adhesive and the bonding cycle; (2) tests should be run on the actual joint(s) that will be used in production; and (3) a thorough evaluation of the in- service conditions must be carried out, including temperature, moisture and any solvents or fluids that the adhesive will be exposed to during its service life. The failure modes for all test specimens should be examined. Some acceptable and unacceptable failure modes are shown in Fig. 10. For example, if the specimen exhibits an adhesive failure at the adherend-adhesive interface rather than a cohesive failure within the adhesive, it may be an indication of a surface preparation problem that will result in decreased joint durability.

8.5 Surface Preparation[1]

Surface preparation of a material prior to bonding is the keystone upon which the adhesive bond is formed. Extensive field service experience with structural adhesive bonds has repeatedly demonstrated that adhesive durability and longevity depends on the stability and bondability of the adherend surface.

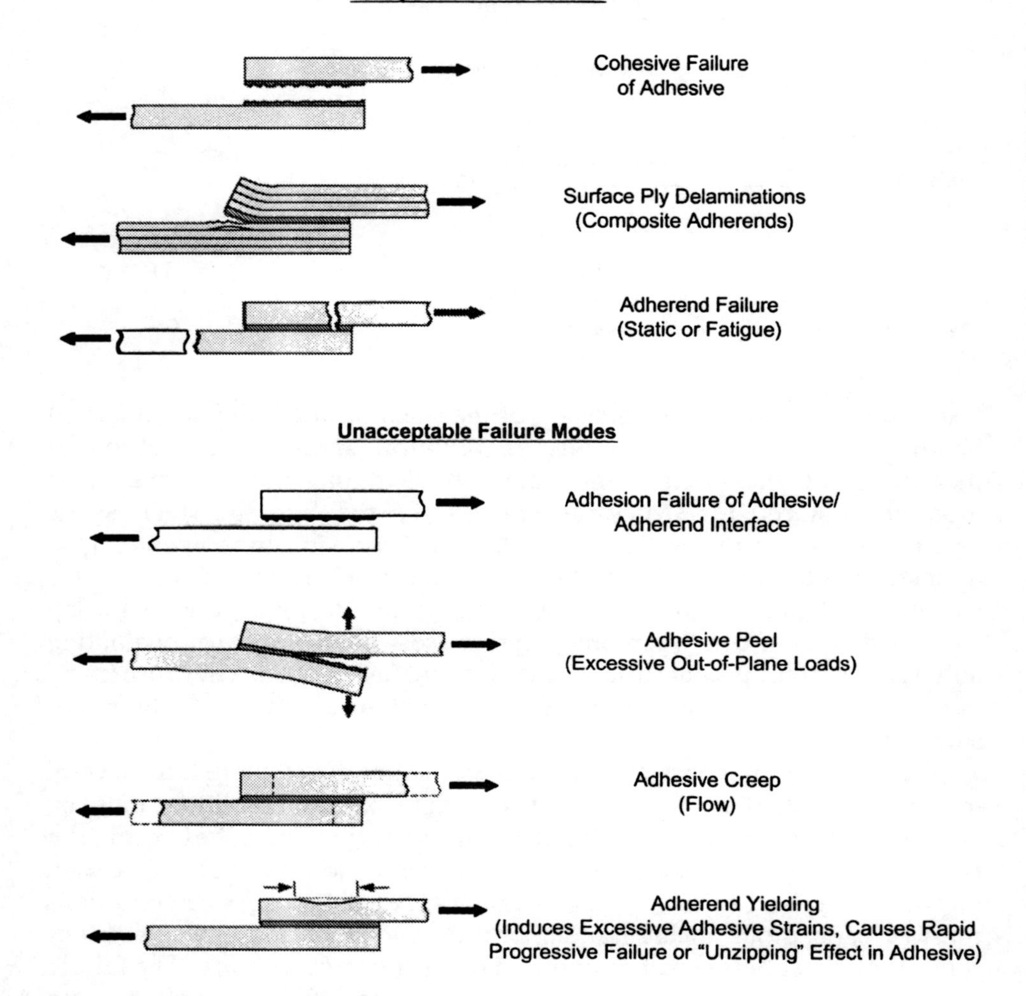

Fig. 10. Typical Failure Modes of Bonded Joints

In general, high-performance structural adhesive bonding requires that great care be exercised throughout the bonding process to ensure the quality of the bonded product. Chemical composition control of the adhesive; strict control of surface preparation materials and process parameters; and control of the adhesive lay-up, part fit-up, tooling, and the curing process are all required to produce durable structural assemblies.

The first consideration for preparing a composite part for secondary adhesive bonding is moisture absorption of the laminate itself. Absorbed laminate moisture can diffuse to the surface of the laminate during elevated-temperature cure cycles resulting in weak bonds or porosity or voids in the adhesive bondline, and in extreme cases where fast heat-up rates are used, actual delaminations within the composite laminate plies. If honeycomb is used in the structure, moisture can turn to steam, resulting in node bond failures or blown core. Relatively thin composite laminates (0.125 inch or less in thickness) may by effectively dried in an air circulating oven at 250 °F for 4 h minimum. Drying cycles for thicker laminates should be developed empirically using the actual adherend thicknesses. After drying, the surface should be prepared for bonding and then the actual bonding operation conducted as soon as possible. It should be noted that prebond thermal cycles, such as those using encapsulated film adhesive to check for part fit-up prior to actual bonding, can also serve as effective drying cycles. In addition, storage of dried details in a temperature- and humidity-controlled lay-up room can extend the time between drying and curing.

Numerous surface preparation techniques are currently used prior to the adhesive bonding of composites. The success of any technique depends on establishing comprehensive material, process and quality control specifications and adhering to them strictly. One method that has gained wide acceptance is the use of a peel ply. In this technique, a closely woven nylon or polyester cloth is used as the outer layer of the composite during lay-up; this ply is torn or peeled away just before bonding or painting. The theory is that the tearing or peeling process fractures the resin matrix coating and exposes a clean, virgin, roughened surface for the bonding process. The surface roughness attained can, to some extent, be determined by the weave characteristics of the peel ply. Some manufacturers advocate that this is sufficient, while others maintain that an additional hand sanding or light grit blasting is required to adequately prepare the surface. The abrasion increases the surface area of the surfaces to be bonded and may remove residual contamination, as well as removing fractured resin left behind from the peel ply. The abrading operation should be conducted with care, however, to avoid exposing or rupturing the reinforcing fibers near the surface.

The use of peel plies on composite surfaces to be structurally bonded certainly deserves careful consideration. Factors that need to be considered include: the chemical makeup of the peel ply (e.g., nylon versus polyester) as well as its compatibility with the composite matrix resin; the surface treatment used on the peel ply (e.g., silicone coatings that make the peel ply easier to remove also leave residues that inhibit structural bonding); and the final surface preparation (e.g., hand sanding versus light grit blasting) employed. The reader is referred to Refs. 5 and 6 for a more in-depth analysis of the potential pitfalls of using peel plies on surfaces to be bonded. The authors of Refs. 5 and 6 maintain that the only truly effective method of surface preparation is a light grit blast after peel ply removal. Nevertheless, peel plies are very effective in preventing gross surface contamination that could occur between laminate fabrication and secondary bonding.

A typical cleaning sequence would be to remove the peel ply and then lightly abrade the surface with a dry grit blast at approximately 20 psig. After grit blasting, any remaining residue on the surface can be removed by dry vacuuming or wiping with a clean dry chessecloth. Although hand-sanding with 120-240 grit silicon carbide paper can be substituted for grit blasting, hand sanding is not as effective as grit blasting in reaching all of the impressions left by the weave of the peel ply on the composite surface. In addition, the potential for removing too much resin and exposing the carbon fibers is actually higher for hand sanding than it is for grit blasting.

If it is not possible to use a peel ply on a surface requiring adhesive bonding, the surface can be precleaned (prior to surface abrasion) with a solvent, such as methyl ethyl ketone, to remove any gross organic contaminants. In cases where a peel ply is not used, some type of light abrasion followed by a dry wipe (or vacuum) is then required to break the glazed finish on the matrix resin surface. The use of solvents to remove residue after hand sanding or grit blasting is discouraged due to the potential of recontaminating the surface.

Another method can be used to avoid abrasion damage to fibers. When the carbon composite is first laid up, a ply of adhesive is placed on the surface where the secondary bond is to take place. This adhesive is then cured together with the laminate. To prepare for the secondary bond, the surface of this adhesive ply is abraded with minimal chance of fiber damage; however, this sacrificial adhesive ply adds weight to the structure.

Surface conditioning techniques can be automated for use in high-production situations. All surface treatments should have the following principles in common: (1) the surface should be thoroughly cleaned prior to abrasion to avoid smearing contamination into the surface: (2) the glaze on the matrix surface should be roughened without damaging the reinforcing fibers or forming subsurface cracks in the resin matrix; (3) all residue should be removed from the abraded surface in a dry process; and (4) the prepared surface should be bonded as soon as possible after preparation.

Aluminum and titanium are often bonded in composite assemblies, although aluminum should not be bonded directly to carbon/epoxy because the large differences in the coefficients of thermal expansion result in significant residual stresses, and because carbon fiber in contact with aluminum forms a galvanic cell that corrodes the aluminum. Although seemingly adequate bond strength can often be obtained with rather simple surface treatments (e.g., surface abrasion or sanding of aluminum adherends), long-term durable bonds under actual service environments can suffer significantly if the metal adherend has not been processed using the proper chemical surface preparation.

Several different methods are used to prepare aluminum alloys for adhesive bonding and all have advantages and disadvantages that should be considered, including cost, cycle time, bond durability, performance and environmental compliance. Aluminum alloys can be precleaned by vapor degreasing followed by alkaline cleaning. The main objective of aluminum etching or anodizing procedures is to create a clean surface that contains a porous oxide layer that the adhesive can flow into and become mechanically interlocked. The surface morphologies of the three most prevalent commercial processes are shown in Fig. 11. Forest Products Laboratory (FPL) etching is a chromic-sulfuric acid etch and is one of the earliest methods developed for preparing aluminum for bonding.Chromic

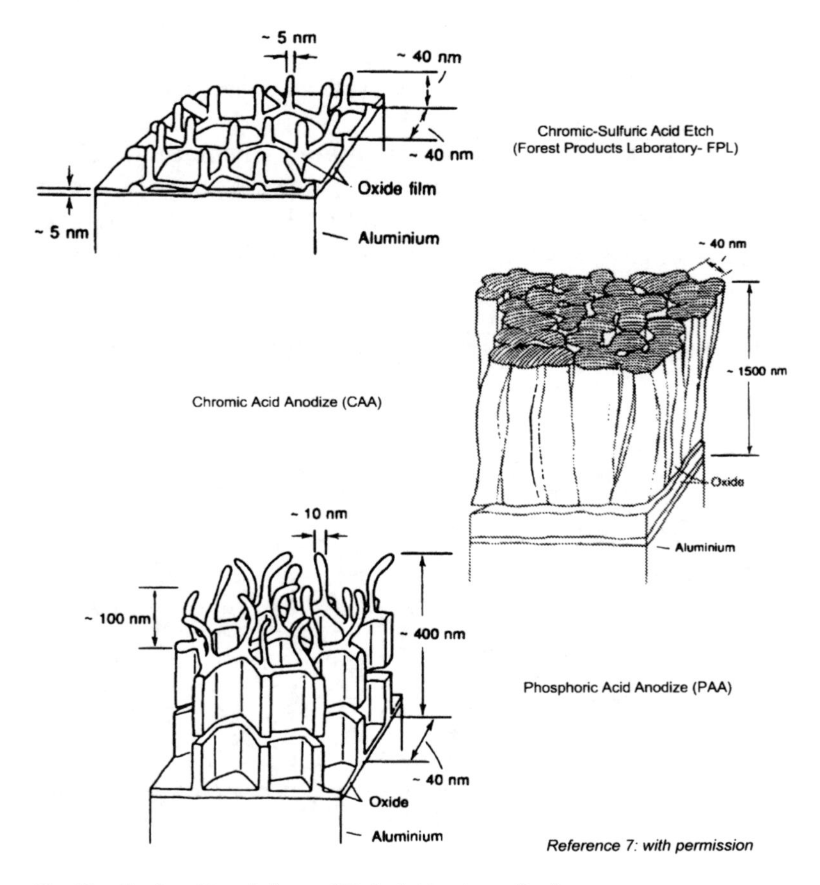

Fig. 11. Surface Morphology of Etched Aluminum Surfaces

acid anodizing is a later method and is perhaps more widely used than the FPL etch. Chromic acid etching produces a thicker, more robust oxide film than the FPL process. Different manufacturers use minor variations of this method, usually in the sealing steps after anodizing. Phosphoric acid anodize (PAA) is the most recent of the well-established procedures and has an excellent service record for environmental durability. It also has the advantage of being very forgiving of minor variations in procedure. The PAA process produces a more open oxide film and thinner oxide film than that produced by the CAA process. It also results in a bound phosphate that improves the durability of the bond.

A more recent environment-friendly process being developed for both aluminum and titanium is the sol-gel process that does not use acids that have to be disposed of as hazardous materials. The process works by producing a tailored gradient interphase coating (Fig. 12) in which one side is molecularly bonded to the oxide structure on the metallic surface and the other side is molecularly cross-linked to the adhesive or primer.[8] The aqueous based sol-gel solution can be brushed or sprayed on the surface and does not require rinsing. This process is particularly attractive for field repair applications[9] and can also be used as a surface preparation for painting.[10]

Several methods are also used with titanium. Any method developed for titanium should undergo a thorough test program prior to production implementation and then must be monitored closely during production usage. A typical process used in the aerospace industry involves:

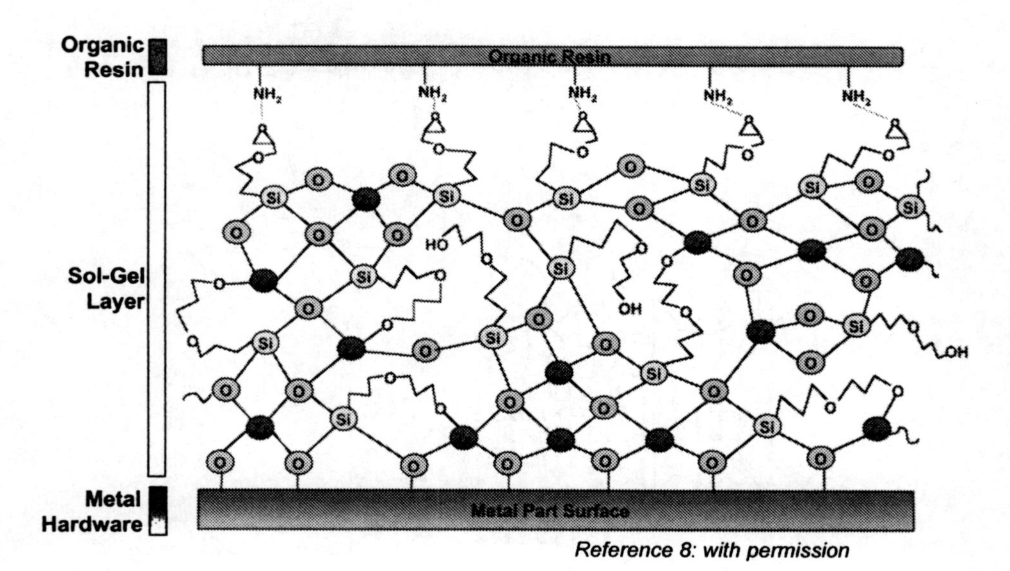

Fig. 12. Notional Sol-Gel Interface

- solvent wiping to remove all grease and oils;
- liquid honing at 40–50 psig pressure;
- alkaline cleaning in an air agitated solution maintained at 200–212 °F for 20–30 min;
- thoroughly rinsing in tap water for 3–4 min;
- etching for 15–20 min in a nitric-hydrofluoric acid solution maintained at a temperature below 100 °F;
- thoroughly rinsing in tap water for 3–4 min followed by rinsing in deionized water for 2–4 min;
- inspecting for a water break free surface;
- oven drying at 100–170 °F for 30 min minimum;
- adhesive bonding or applying primer within 8 h of cleaning.

The combination of liquid honing, alkaline cleaning and acid etching results in a complex chemically activated surface topography containing a large amount of surface area that the adhesive can penetrate and adhere to. The adhesive bond strength is a result of both mechanical interlocking and chemical bonding. Other methods, such as dry chromic acid anodizing are also used.

Because metallic cleaning is such a critical step, dedicated processing lines (Fig. 13) are normally constructed and chemical controls, as well as periodic lap shear cleaning control specimens, are employed to ensure in-process control. Automated overhead conveyances are used to transport the parts from tank-to-tank under computer-controlled cycles to ensure the proper processing time in each tank.

Due to the rapid formation of surface oxides on both titanium and aluminum, the surfaces should be bonded within 8 h of cleaning or primed with a thin protective coat (0.0001–0.0005 inches) of epoxy primer. Primer thickness is important. Actually, thinner coatings give better long-term durability. Color chips are often used in production to determine primer thickness. For parts that will undergo a severe service environment, priming is always recommended because today's primers contain corrosion-inhibiting compounds (strontium chromates) that enhance long-term durability. The two critical variables in corrosion of metal bonds are the metal surface preparation treatment and the chemistry of the primer. Some primers contain phenolics, which have been found to produce outstanding bond durability.[11] Once the primer has been cured (e.g., 250 °F), the parts may be stored in an environmentally controlled clean room for quite long periods of time (e.g., up to 50 days or longer would not be unusual).

All cleaned and primed parts should be carefully protected during handling or storage to prevent surface contamination. Normally, clean white cotton gloves are used during handling and wax-free Kraft paper may be used for wrapping and longer storage. Gloves, which are used to handle

GKN Aerospace Services

Fig. 13. Automated Chemical Cleaning Line

cleaned and/or primed adherends, should be tested to ensure that they are not contaminated with silicones or hydrocarbons which can contaminate the bondline, or sulfur which can inhibit the cure of the adhesive.

8.6 Epoxy Adhesives[1]

Epoxy-based adhesives are by far the most commonly used materials for bonding or repair of composite structures. The existence of a large variety of materials, to fit nearly any handling, curing or performance requirement, results in an extensive list from which to choose. Epoxy adhesives impart high-strength bonds and long-term durability over a wide range of temperatures and environments. The ease with which formulations can be modified makes it fairly easy for the epoxy adhesive fabricator to employ various materials to control specific performance properties, such as density, toughness, flow, mix ratio, pot life/shelf life, shop handling characteristics, cure time/temperature and service temperature.

Advantages of epoxy adhesives include excellent adhesion, high strength and modulus, low or no volatiles during cure, low shrinkage and good chemical resistance. Disadvantages include cost, brittleness unless modified, moisture absorption that adversely affects properties and relatively long cure times. A wide range of one-part and two-part epoxy systems are available. Some systems cure at room temperature, while others require elevated temperatures.

Epoxy resins used as adhesives are generally supplied as liquids or low-melting-temperature solids. They commonly contain bifunctional epoxy groups, although higher functionalities are available. They can be cured by a variety of curing procedures, including an admixture with a stoichiometric proportion of polyfunctional primary amines or acid anhydrides. The amine or anhydride groups react with the epoxy groups by an addition reaction to give a densely cross-linked structure. Some epoxy compositions can be cured through a homopolymerization reaction initiated by strong organic bases (or rarely, acids). These compositions are less sensitive to mix ratio, but are seldom encountered as two-part systems. The rate of the reaction may be adjusted by adding accelerators in the initial formulation or by increasing the temperature. To improve structural properties, particularly at elevated temperatures, it is common to cure temperatures close to (or preferably above) the maximum use temperature for the structure.

Epoxy resin systems are usually modified by a wide range of additives that control particular properties, including accelerators, viscosity modifiers and other flow control additives, fillers and pigments, flexibilizers and toughening agents. Epoxy-based adhesives are available in two basic cure chemistries: room temperature and elevated temperature. Within each cure type, there is a wide range of formulated resins to cover specific application and performance requirements.

One-partelevated-temperaturecuringepoxyliquidandpasteadhesives . These materials typically require an elevated temperature cure of 250–350 °F. The primary chemistry of one-part systems usually consists of a mixture of bifunctional and multifunctional resins with non-catalyzed or imidizole catalyzed dicyanimide. As such, the normal room temperature shelf-life ranges from 15–30 days for catalyzed materials, and up to six months for non-catalyzed systems. Service temperatures for these materials are generally close to their respective curing temperatures; however, the actual service temperature should always be determined by testing at the expected service conditions.

Typical packaging for the one-part adhesives includes pint, quart, gallon, and five-gallon containers. In addition, for ease of application, most are supplied in cylindrical polyethylene sealant gun cartridges that, depending on their size, can contain >500 g of material. Since the chemistry and performance of some one-part paste adhesives are similar to that of film adhesives, these materials are often referred to as "film adhesive in a tube."

Two-part room-temperature curing epoxy liquid and paste adhesives . These systems are most commonly used when a room temperature cure is desired. They are available as clear liquids or as filled pastes with a consistency ranging from low-viscosity liquids to heavy-duty putties. Typical cure times are 5–7 days; however, in most cases 70–75% of the ultimate cure can be achieved within 24 h and, if needed, the pressure can usually be released at that point. Under normal bondline thickness conditions (0.005-0.010 inches), cure can be accelerated with heat without fear of exotherm. A typical cure would be one hour at 180 °F.

Two-part systems require mixing a part A (the resin and filler portion) with a part B (the curing agent portion) in a predetermined stoichiometric ratio. Two-part epoxy adhesives usually require mixing in precise proportions, to avoid a significant loss of cured properties and environmental stability. The amount of material to be mixed should be limited to the amount needed to accomplish the task. In general, the larger the mass, the shorter the pot, or the working life of the material. The pot life is defined as the period between the time of mixing the resin and curing agent and the time at which the viscosity has increased to the point when the adhesive can no longer be successfully applied as an adhesive.

Many of the bifunctional and multifunctional resin types formulated into one-part systems are also employed in the part A's of the two-part systems. However, the ability to cure, or cross-link, at room temperature is due to different curing agent (part B) chemistries. These are generally mixtures, to various degrees, of modified and unmodified aliphatic amines, polyamides, and modified cycloaliphatic amines. Curing of higher-temperature service two-part systems, which requires an elevated-temperature cure, is usually accomplished singularly or with mixtures of aromatic and unmodified cycloaliphatic amines. A number of these materials are primary skin sensitizers, and some caution is necessary to avoid direct contact.

Some of the more reactive aliphatic amines will react with ambient water and carbon dioxide, and, if left exposed too long in the mixed condition, prior to part mating, a carbonate layer may form on the adhesive surface. If this occurs, it will inhibit good substrate to adhesive contact and will significantly decrease mechanical properties. The opportunity for carbonate formation can be limited by avoiding high-humidity bonding conditions and mating the parts as soon as possible after adhesive application. In cases where carbonate formation cannot be avoided, covering the exposed area with polyethylene film until mating the parts will help minimize carbonate formation. Troweling the surface, prior to mating, can also be an effective method of disrupting any carbonate formation.

In addition to the packaging methods described for the one-part materials, meter mix equipment is available for continuous application where applicable, usually involving large areas. Mixing of the two parts can be accomplished by pumping material through either a hydraulic or static mixer. For smaller applications, dual-cartridge kits, in which both parts A and B are manually pushed through a static mixer, are also available for most two-part systems. Two-part adhesives can usually be stored at room temperature; however, some A's contain resins that can self-polymerize and require cold storage.

Two-part resin systems are frequently used to repair damaged composite assemblies. Low-viscosity versions can be used to impregnate dry carbon cloth for repair patches or inject into cracked bondlines or delaminations. Thicker pastes are used to bond repairs where more flow control is required. For example, if the material has too low a viscosity and is cured under high pressure, the potential for bondline starvation exists due to excessive flow and squeeze out. Viscosity control of two-part adhesives is usually done with metallic and/or non-metallic fillers. Fumed silica is frequently added to provide slump and flow control.

Many adhesives are of the same resin and curing chemistry family; however, different versions are manufactured (non-filled, metallic or non-metallic filled, thixotroped, low density and toughened) for specific performance requirements. For example, a non-metallic filled adhesive may be preferred over a metallic filled adhesive if there is concern for possible galvanic corrosion in the joint. In thin composite structures where bending or flexing is a concern, a toughened adhesive is usually warranted.

In addition to composite bonding and repair applications, two-part epoxy paste adhesives are also used for liquid shim applications during mechanical assembly operations. The ability to tailor flow, cure time and compressive strength has made these materials ideal for use in areas of poor fit-up.

Epoxy film adhesives. Structural adhesives for aerospace applications are generally supplied as thin films supported on a release paper and stored under refrigerated conditions (0 °F). Film adhesives are preferred to liquids and pastes because of their uniformity and reduced void content. Film adhesives are available using high-temperature aromatic amine or catalytic curing agents with a wide range of flexibilizing and toughening agents. Rubber-toughened epoxy film adhesives are widely used in the aircraft industry. The upper temperature limit of 250–350 °F is usually dictated by the degree of toughening required and by the overall choice of resins and curing agents. In general, toughening a resin results in a lower usable service temperature. Film materials are frequently supported by fibers (scrim cloth) that serve to improve handling of the films prior to cure, control adhesive flow during bonding and assist in bondline thickness control and provide galvanic insulation. Fibers can be incorporated as short-fiber mats with random orientation or as woven cloth. Commonly encountered fibers are polyesters, polyamides (nylon), and glass. Adhesives containing woven cloth may have slightly degraded environmental properties because of wicking of water by the fiber. Random mat scrim cloth is not as efficient for controlling film thickness as woven cloth, because the unrestricted fibers move during bonding, although spun-bonded non-woven scrims do not move and are, therefore, widely used.

8.7 Bonding Procedures[1]

Some general guidelines for adhesive bonding are summarized in Table 2. The basic steps in the adhesive bonding process are:

- collection of all the parts in the bonded assembly, which are then stored as kits;
- verification of the fit to bondline tolerances;
- cleaning of the parts to promote good adhesion;
- application of the adhesive;
- mating of the parts and adhesive to form the assembly;
- application of force concurrent with application of heat to the adhesive to promote cure, if required;
- inspection of the bonded assembly.

Prekitting of adherends . Many adhesives have a limited working life at room temperature, and adherends, especially metals, can become contaminated by exposure to the environment. Thus, it is normal practice to kit the adherends so that application of the adhesive and buildup of the bonded assemblies can proceed without interruption. The kitting sequence is determined by the product and production rate. Prefitting of the details is also useful in determining locations of potential mismatch such as high and low spots. A prefit-check fixture is often used for complex assemblies containing multiple parts. This fixture simulates the bond by locating the various parts in the exact relationship to one another, as they will appear in the actual bonded assembly. Prefitting is usually conducted prior to cleaning so that the details can be reworked if necessary.

Prefit evaluation. For complex assemblies, a prefit evaluation ("verifilm") is frequently conducted as depicted in Fig. 14. The bondline thickness is simulated by placing a vinyl plastic film or the actual adhesive encased in plastic film in the bond lines. The assembly is then subjected to the heat and pressure normally used for curing. The parts are disassembled, and the vinyl film or cured adhesive is then visually or dimensionally evaluated to see what corrections are required. These corrections can include sanding the parts to provide more clearance, reforming metal parts to close the gaps or applying

Table 8.2 General Considerations for Adhesive Bonding

• When received, the adhesive should be tested for compliance with the material specification. This may include both physical and chemical tests.
• The adhesive should be stored at the recommended temperature.
• Cold adhesive should always be warmed to room temperature in a sealed container.
• Liquid mixes should be degassed, if possible, to remove entrained air.
• Adhesives which evolve volatiles during cure should be avoided.
• The humidity in the lay-up area should be below 40% relative humidity for most formulations. Lay-up room humidity can be absorbed by the adhesive and is released later during heat cure as steam, yielding porous bondlines and possibly intervering whith the cure chemistry.
• Surface preparation is absolutely critical and should be conducted carefully.
• The recommended pressure and the proper alignment fixtures should be used. The bonding pressure should be great enough to ensure that the adherends are in intimate contact with each other during cure.
• The use a vacuum as the method of applying pressure should be avoided whenever possible, since an active vacuum on the adhesive during cure can lead to porosity or voids in the cured bondline.
• Heat curing systems are almost always preferred, because they yield bonds that have a better combination of strength, heat and humidity resistance.
• When curing for a second time, such as during repairs, the temperature should be at least 50 °F below the earlier cure temperature. If this is not possible, then a proper and accurate bond form must be used to maintain all parts in proper alignment and under pressure during the second cure cycle.
• Traveler coupons should always be made for testing. These are test coupons that duplicate the adherends to be bonded in material and joint design. The coupon surfaces are prepared by the same method and at the same time as the basic bond. Coupons are also bonded together at the same time with the same adhesive lot used in the basic joint and subjected to the same curing process simultaneously with the basic bond. Ideally, traveler coupons are cut from the basic part, on which extensions have been provided.
• The exposed edges of the bond joint should be protected with an appropriate sealer, such as an elastomeric sealant or paint.

Reference 1

additional adhesive (within permissible limits) to particular locations in the bondline.

Verification of bondline thickness may not be required for all applications. However, the technique can be used to validate the fit of the mating parts prior to the start of production or to determine why large voids are produced in repetitive parts. When using paste adhesives, the adhesive thickness can be simulated by encasing the adhesive in plastic film (as mentioned above) or aluminum foil. Once the fit of mating parts has been evaluated, any necessary corrections can be made. For cases in which the component parts can be dimensionally corrected, it is much more efficient to make the correction than risk having to scrap the parts or having them fail in service.

Adhesive application. The most commonly used adhesives are supplied as liquids, pastes or prefabricated films. The liquid and paste systems may be supplied as one-part or two-part systems. The two-part systems must be

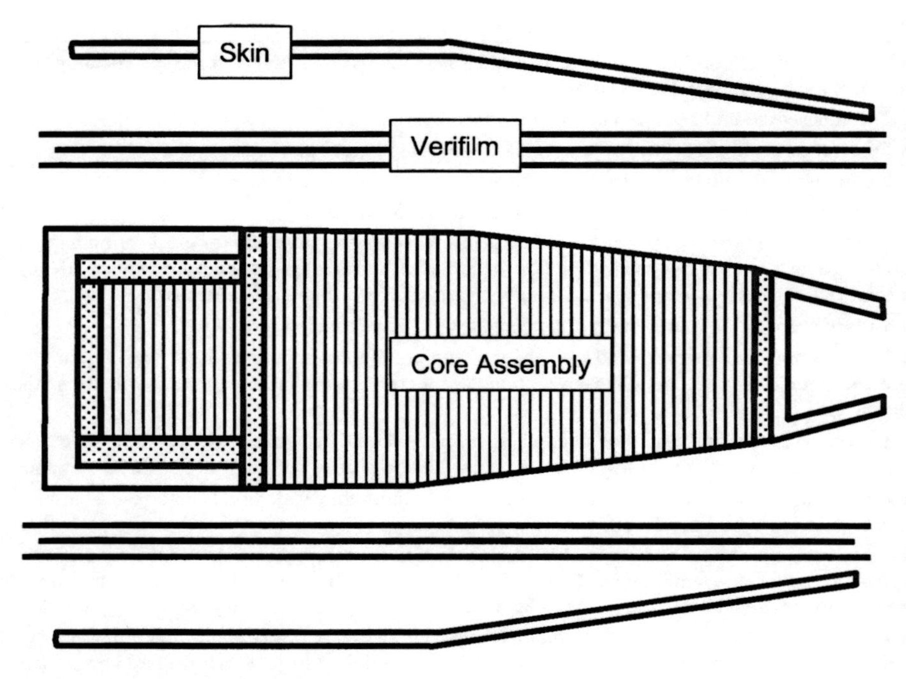

Fig. 14. Prefit of Details Using Verifilm

mixed before use and, thus, require scales and a mixer. The amount of material to be mixed should be limited to the amount needed to accomplish the task. The larger the mass, the shorter the pot or working life of the mixed adhesive. To prevent potential exotherm conditions, excess mixed material should be removed from the container and spread out in a thin film. This will prevent the risk of mass-related heat buildup and the possibility of a fire or the release of toxic fumes.

The storage facility plays an important role in a production bonding operation. Many one-part adhesives and film adhesives must be stored at temperatures below ambient. Adhesives containing organic solvents must be segregated to reduce the risk of fire, and others containing polymers and solvents, must be shaken periodically to prevent settling or gelation.

One factor that must be considered in adhesive application is the time interval between adhesive preparation and final assembly of the adherend. This factor, which is referred to as pot, open, out-time or working life, must be matched to the production rate. Obviously, materials that are ready to bond quickly are needed for high-rate applications, such as those found in the automotive and appliance industries. It should be noted that many two-part systems that cure by chemical reaction often have a limited working life before they become too viscous to apply.

Application of liquid adhesives can be accomplished using brushes, rollers, manual sprays or robotically controlled sprays. A robot can apply tightly controlled quantities of adhesive to specific areas. Solvated two-part systems are sprayed using equipment with two pumps; preset quantities of each component are pumped through the spray head where they are blended into a single stream. Of course, many plants use several different application systems simultaneously for their various job shop requirements.

Application of paste adhesives can be accomplished by brush, by spreading with a grooved tool or by extrusion from cartridges or sealed containers using compressed air. For the latter, the combination of the orifice diameter and the applied pressure controls the size of the bead applied to the work. Robots can move the application head in a constant path at a repetitive surface speed to enhance the accuracy of bead placement and size. The use of robots to apply paste adhesive is analogous to their use to locate spot welds. In the automotive industry, several vans with plastic skins bonded to a steel structural frame are assembled using paste adhesives applied by robots.

Film adhesives are high quality but costly and, thus, are used mainly in aircraft applications. They consist of an epoxy, bismaleimide or polyimide resin film and a fabric carrier. The fabric guarantees a bondline thickness of some minimum because it prevents adherends from contacting each other directly. These adhesives are manually cut to size, usually with knives, and placed in the bondlines. When applying film adhesives, it is important to prevent or eliminate entrapped air pockets between the adherend and adhesive film by pricking bubbles or “porcupine” rolling over the adhesive prior to application.

Bond line thickness control . Controlling the thickness of the adhesive bondline is a critical factor in bond strength. This control can be obtained by matching the quantity of available adhesive to the size of the gap between the mating surfaces under actual bonding conditions (heat and pressure). For liquid and paste adhesives, it is a common practice to embed nylon or polyester fibers in the adhesive to prevent adhesive-starved bondlines. Higher applied loads during bonding tend to reduce bondline thickness. A slight overfill is usually desired to ensure that the gap is totally filled. Conversely, if all the adhesive is squeezed out of a local area due to a high spot in one of the adherends, a disbond can result.

For highly loaded bondlines and large structures, film adhesives are used that contain a calendered film with a thin fabric layer. The fabric maintains the bondline thickness by preventing contact between the adherends. In addition, the carrier acts as a corrosion barrier between carbon skins and aluminum honeycomb core. In the most common case, the bondline thickness can vary from 0.002 to 0.010 inches. Extra adhesive can be used to handle up 0.020 inch gaps. Larger gaps must be accommodated by reworking the parts or producing hard shims to bring the parts within tolerance.

Bonding. Theoretically, only enough contact pressure is required so that the adhesive will flow and wet the surface during cure. In reality, somewhat higher pressures are usually required to squeeze out excess adhesive to provide the desired bondline thickness, and/or to provide sufficient force to ensure all of the interfaces obtain intimate contact during cure.

The position of the adherends must be maintained during cure. Slippage of one of the adherends before the adhesive gels results in the need for costly reworking, or the entire assembly might be scrapped. When a paste or liquid adhesive is used, it is usually helpful to have a load applied to the joint to deform the adhesive to fill the bondline. C-clamps, spring loaded clamps, shot bags and jack screws are frequently used for simple configurations but some care is required that these pressure devices do not become heat sinks if elevated-temperature curing is required. Liquid and paste adhesives that are cured at room temperature will normally develop enough strength after 24 h, so that the pressure can be removed. For these adhesives that require moderate cure temperatures (e.g., 180 °F), heat lamps or ovens are frequently used. When using heat lamps, some degree of caution is necessary to ensure that the part does not get locally overheated. If the contour is complex, it may be necessary to bag the part and employ the isostatic pressure of an autoclave. Instead of using the positive pressure of a vented bag in an autoclave, a vacuum bag (<15 psig) in an oven is quite commonly used. The disadvantage of this process is that the vacuum tends to cause many adhesives to release volatiles and form porous and weak bondlines.[12]

When elevated-temperature (e.g. 250–350 °F) curing film adhesives are used, autoclave pressures of 15–50 psig are normally used to force the adherends together. Autoclave bonded parts are made on bond tools very similar to the ones discussed in Chapter 4 (Cure Tooling). The bagging procedures for autoclave bonding are also very similar to those discussed in Chapter 6 (Curing) except that bleeder is not required since we are not trying to remove any excess resin during cure.

Epoxy film adhesives typically do not have a high flow and they contain a thin scrim fabric to ensure that the bondline thickness is maintained. The majority of these adhesive systems cure at 1–2 h at elevated temperature. Both straight heat-up and ramped (intermediate hold) cure cycles are used. A typical autoclave cure cycle for a 350 °F curing epoxy film adhesive would be:

- Pull a 20–29 inches Hg vacuum on the assembly and check for leaks. If the assembly contains honeycomb core, do not pull more than 8–10 inches of Hg vacuum.

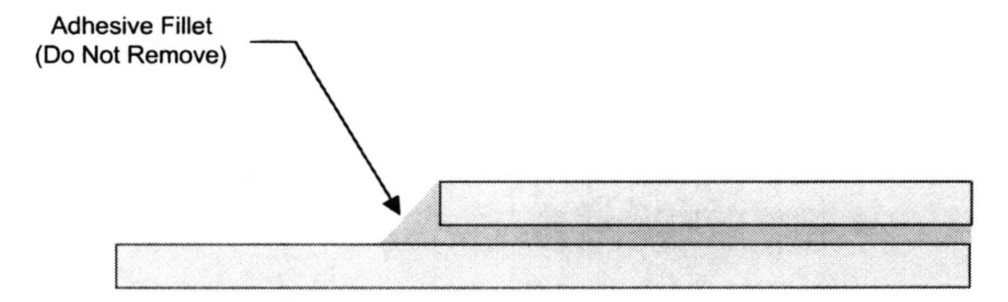

Fig. 15. Typical Adhesive Filleting in Bonded Joint

- Apply autoclave pressure, usually in the range of 15–50 psig. Vent the bag to atmosphere when the pressure reaches 15 psig.
- Heat to 350 °F at a rate of 1-5 °F min^{-1}. (Option: an intermediate hold at 240 °F for 30 min is sometimes used to allow the liquid resin to thoroughly wet the adherend surfaces.[12])
- Cure at 350 ± 10 °F for 1–2 h under 15–50 psig.
- Cool to 150 °F before releasing autoclave pressure.

During cure, the adhesive flows and forms a fillet at the edge of the bond as shown in Fig. 15. It is important not to remove this fillet during clean-up after bonding. Testing has shown that the presence of the fillet significantly improves the joint strength.[13]

8.8 Sandwich Structures

Sandwich construction is used extensively in both the aerospace and commercial industries because it is an extremely lightweight structural approach that exhibits high stiffness and strength-to-weight ratios. The basic concept of a sandwich panel[14] is that the facings carry the bending loads (tension and compression), while the core carries the shear loads, much like the I-beam comparison shown in Fig. 16. As shown in Fig. 17, sandwich construction, especially honeycomb core construction, is extremely structurally efficient,[14] particularly in stiffness critical applications. Doubling the thickness of the core increases the stiffness over 7× with only a 3% weight gain, while quadrupling the core thickness increases stiffness over 37× with only a 6% weight gain. Little wonder that structural designers like to use sandwich construction whenever possible. Sandwich panels are typically used for their structural, electrical, insulation and/or energy absorption characteristics.

Facesheet materials that are normally used are aluminum, glass, carbon or aramid. Typical sandwich structure has relatively thin facing sheets (0.010–0.125 inches) with core densities in the range of 1–30 pcf (pounds per cubic foot). Core materials include metallic and non-metallic honeycomb core,

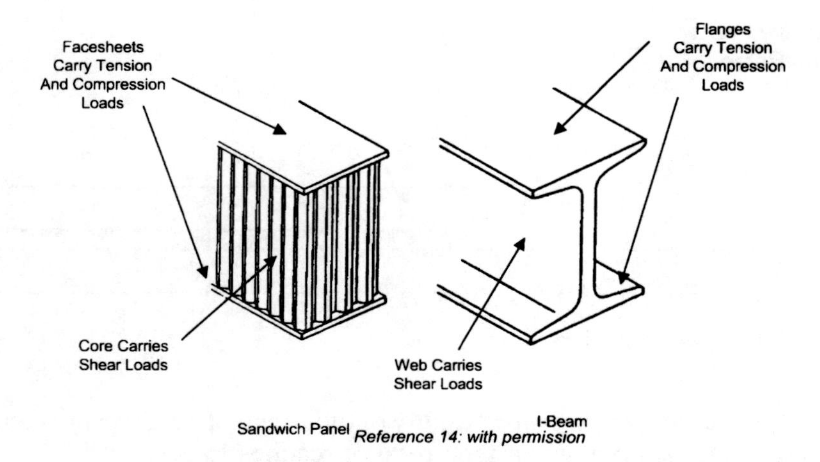

Fig. 16. Why Sandwich Structures Are So Efficient

	Solid Material	Core Thickness t	Core Thickness 3t
	t	2t	4t
Stiffness	1.0	7.0	37.0
Flexural Strength	1.0	3.5	9.2
Weight	1.0	1.03	1.06

Reference 14: with permission

Fig. 17. Efficiency of Sandwich Structure

balsa wood, open and closed cell foams and syntactics. A cost versus performance comparison[15] is given in Fig. 18. Note that, in general, the honeycomb cores are more expensive than the foam cores but offer superior performance. This explains why many commercial applications use foam

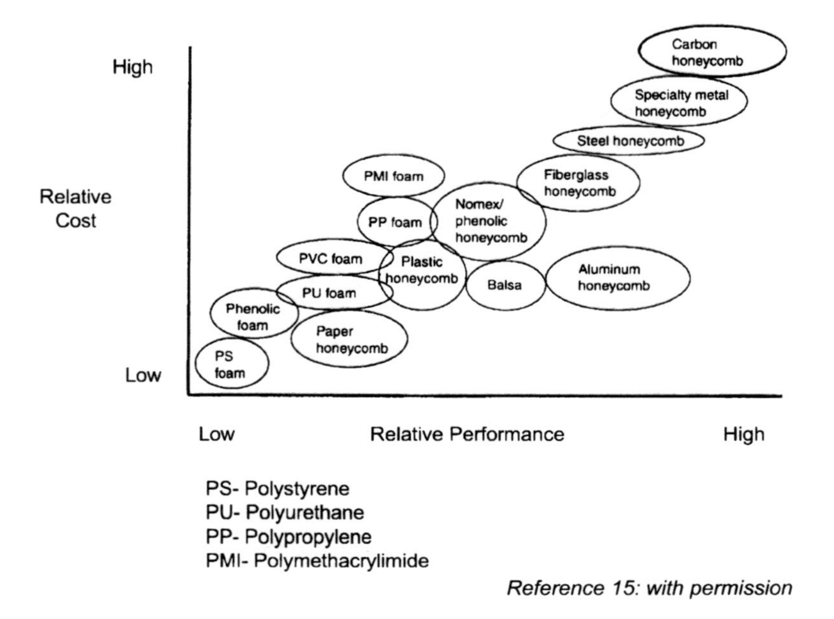

Fig. 18. Cost Versus Performance for Core Materials

cores, while aerospace applications use the higher performance but more expensive honeycombs. It should also be noted that the foam materials are normally much easier to work with than the honeycombs. A relative strength and stiffness comparison of different core materials is given in Fig. 19.

Foam core sandwich assemblies can be bonded together with supported film adhesives but the more common case is to use either liquid/paste adhesives or do wet lay-up of the skin plies directly on the foam surface. More recently, foam cores with dry composite skins are impregnated and bonded with liquid molding techniques, such as resin transfer molding (RTM) or low-pressure vacuum assisted resin transfer molding (VARTM). Supported film adhesives are normally used to bond composite structural honeycomb assemblies.

8.9 Honeycomb Core

The details of a typical honeycomb core panel are shown in Fig. 20. Typical facesheets include aluminum, glass, aramid, and carbon. Structural film adhesives are normally used to bond the facesheets to the core. It is important that the adhesives provide a good fillet at the core-skin interface. Typical honeycomb core terminology is given in Fig. 21. The honeycomb itself can be manufactured from aluminum, glass fabric, aramid paper,

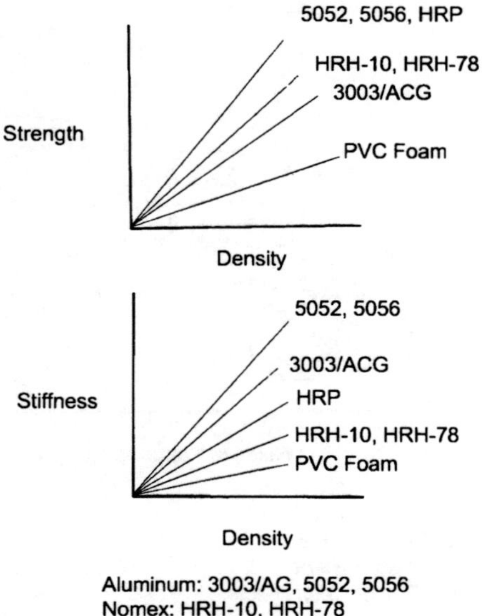

Reference 14: with permission

Fig. 19. Strength and Stiffness of Various Core Materials

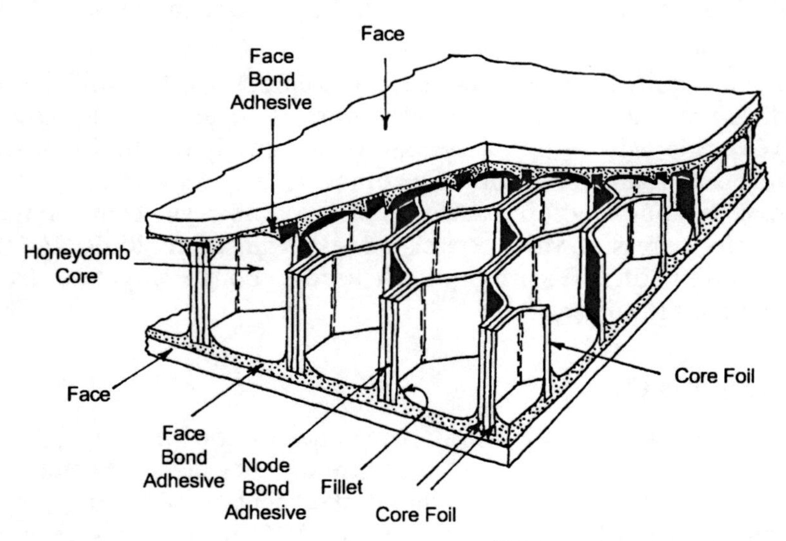

Reference 1: with permission

Fig. 20. Honeycomb Panel Construction

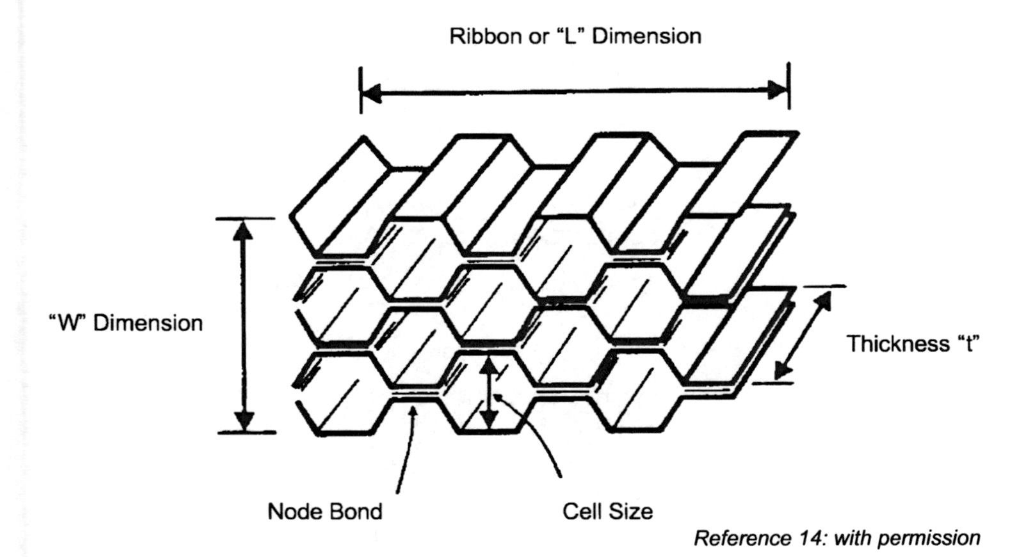

Fig. 21. Honeycomb Core Terminology

aramid fabric or carbon fabric. Honeycomb manufactured for use with organic matrix composites is bonded together with adhesive, called the node bond adhesive. The *L* direction is the core ribbon direction and is stronger than the width (node bond) or "W" direction. The thickness is denoted by *t* and the cell size is the dimension across the cell as shown in the figure.

Although there are a variety of cell configurations available,[16] the three most prevalent (Fig. 22) are hexagonal, flexible-core and overexpanded core. Hexagonal core is by far the most commonly used core configuration. It is available in aluminum and all non-metallic materials. Hexagonal core is structurally very efficient. This configuration can even be made stronger by adding longitudinal reinforcement (reinforced hexagonal) in the *L* direction along the nodes in the ribbon direction. The main disadvantage of the hexagonal configuration is limited formability; aluminum hex-core is typically roll-formed to shape, while non-metallic hex-core must be heat-formed. Flexible core was developed to provide much better formability. This configuration provides for exceptional formability on compound contours without buckling the cell walls. It can be formed around tight radii in both the *L* and *W* directions. Another configuration with improved formability is overexpanded core. This configuration is hexagonal core that has been overexpanded in the *W* direction, providing a rectangular configuration that facilitates forming in the *L* direction. The *W* direction is about twice the *L* direction. This process increases the *W* shear properties

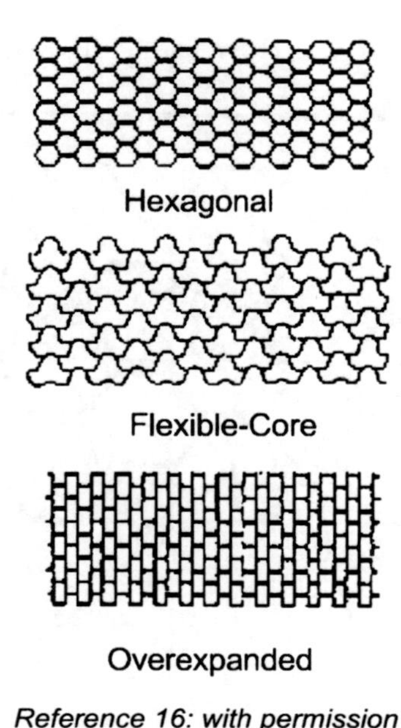

Fig. 22. Types of Honeycomb Core Cell Configurations

and decreases the *L* shear properties compared to hexagonal core. Still a third way to make more formable honeycomb core is with tailored node bond widths and spacings. An excellent source of more detailed information on honeycomb core is Ref. 17.

Honeycomb core is normally made by either the expansion or the corrugation process. The expansion process (Fig. 23) is the one that is the most prevalent for lower-density (≤10 pcf) honeycomb core used for composite assemblies. The foil is cleaned, corrosion protected if it is aluminum, printed with layers of adhesive, cut to length, stacked, and then placed in a press under heat and pressure to cure the node bond adhesive. After curing, the block or honeycomb before expansion (HOBE) is sliced to the correct thickness and expanded by clamping and then pulling on the edges. Expanded aluminum honeycomb retains its shape at this point due to yielding of the aluminum foil during the expansion process. Nonmetallic cores, such as glass or aramid, must be held in the expanded position and dipped in a liquid epoxy, polyester, phenolic or polyimide resin that then must be cured before the expansion force can be released. Several dip-and-cure sequences can be required to produce the desired density.[18] Since phenolics and polyimides are high-temperature condensation curing resins, it is important that they are thoroughly cured to drive off all volatiles. If the volatiles are not totally removed during core

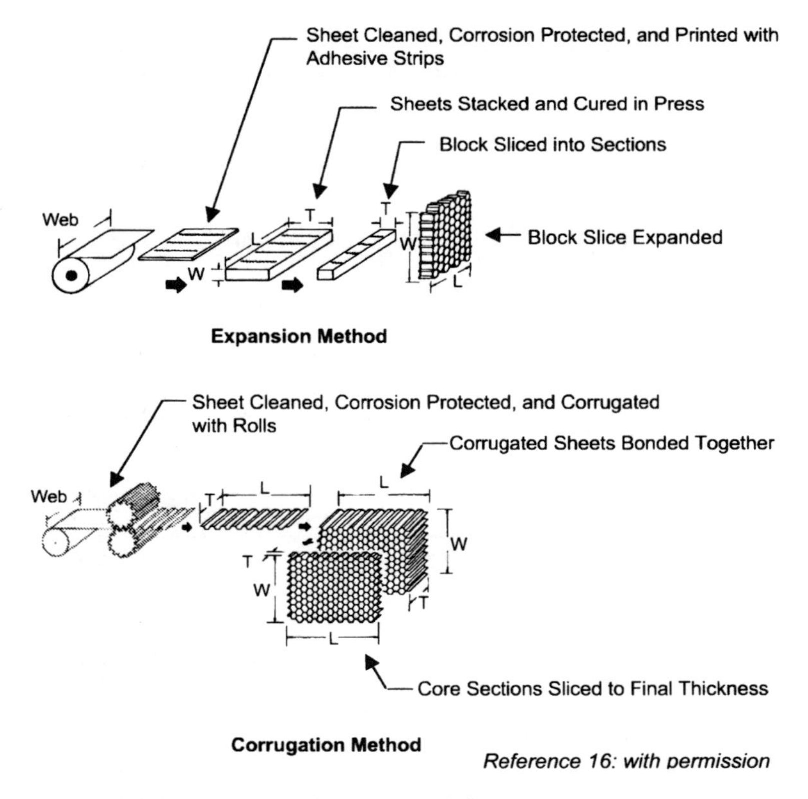

Fig. 23. Fabrication Methods for Honeycomb Core

manufacturing, they can evolve during sandwich curing creating enough pressure to potentially split the node bonds; therefore, after the initial cure, it is common practice to post-cure the phenolic or polyimide core at higher temperatures to ensure that the reactions are complete. Corrugation is a more expensive process reserved for materials that cannot be made by expansion or for higher-density cores such as ≥10 pcf. For example, high-temperature metallic core (e.g., titanium) is made by corrugation and then welded together at the nodes to make the completed sheets.

The comparative properties of some of the commercial honeycomb cores are given in Table 3. Aluminum honeycomb has the best combination of strength and stiffness. The higher performance aerospace grades are 5052–H39 and 5056–H39, while the commercial grade is made from 3003 aluminum. Cell sizes range from 1/16 to 3/8 inches but 1/8 and 3/16 inches are the ones most frequently used for aerospace applications. Glass fabric honeycomb can be made from either a normal bidirectional glass cloth or a bias weave (±45°) cloth. It is usually impregnated with phenolic resin but for high-temperature applications a polyimide resin is used. The advantage

Table 8.3 Characteristics of Typical Honeycomb Core Materials

Name and Type of Core	*Strength/Stiffness*	*Maximum Temperature (°F)*	*Typical Products Forms*	*Density (pcf)*
5052-H39 and 5056-H39 Al Core	High/High	350	Hexagonal Flex-Core	1-12 2-8
3003 Al Commercial Grade Hexagonal Core	High/High	350	Hexagonal	1.8-7
Glass Fabric Reinforced Phenolic	High/High	350	Hexagonal Flex-Core OX	2-12 2.5-5.5 3-7
Bias Weave Glass Fabric Reinforced Phenolic	High/Very High	350	Hexagonal OX	2-8 4.3
Bias Weave Glass Fabric Reinforced Polyimide	High/High	500	Hexagonal	3-8
Aramid Paper Reinforced Phenolic (Nomex)	High/Moderate	350	Hexagonal Flex-Core OX	1.5-9 2.5-5.5 1.8-4
Aramid Paper Reinforced Polyimide (Nomex)	High/Moderate	500	Hexagonal OX	1.5-9 1.8-4
High Performance Aramid Paper Reinforced Phenolic (Korex)	High/High	350	Hexagonal Flex-Core	2-9 4.5
Aramid Fabric Reinforced Epoxy	High/Moderate	350	Hexagonal	2.5
Bias Weave Carbon Fabric Reinforced Phenolic	High/High	350	Hexagonal	4

Reference 19

of the bias weave is that it enhances the shear modulus and improves the damage tolerance of the core. There are three types of aramid core. The original Nomex core is made by impregnating aramid paper with either a phenolic or polyimide resin. However, an issue with Nomex is that the resin cannot fully impregnate the paper; therefore, Dupont developed Korex paper that is thinner and more easily saturated leading to better impregnation, resulting in a core material that has improved mechanical properties and less moisture absorption.[20] Kevlar honeycomb is made by impregnating Kevlar 49 fabric. Finally, bias weave carbon fabric core is a high-performance material but expensive; it was developed for special applications requiring high specific stiffness and thermal stability when bonded with carbon-reinforced facesheets.

The good news about honeycomb core is that it does offer superior performance compared to other sandwich cores. A comparison of strength and stiffness for several core types was previously shown in Fig. 19. Note that aluminum core has the best combination of strength and stiffness, followed by the non-metallic honeycombs and then polyvinyl chloride (PVC) foam. The bad news about honeycomb core is that it is expensive and difficult to fabricate complex assemblies and the in-service experience, particularly with aluminum honeycomb, has not always been good. It is also very difficult to make major repairs to honeycomb assemblies (see Chapter 13, Nondestructive Inspection and Repair).

Aluminum honeycomb assemblies have experienced serious in-service durability problems, the most severe being moisture migrating into the assemblies and causing corrosion of the aluminum core cells. An example[21] of a seriously corroded aluminum honeycomb assembly is shown in Fig. 24. The core was corroded so extensively that pieces of the core were actually "poured" out of the assembly when the skin was removed.

Honeycomb suppliers have responded by producing corrosion inhibiting coatings that have improved durability. The newest corrosion protection system, called PAA core, is shown in Fig. 25. The core foil is first cleaned and then anodized with phosphoric acid. It is then coated with a corrosion inhibiting primer before printing with node bond adhesive. PAA core has demonstrated an approximately threefold increase in corrosion protection when compared to typical (non-PAA) corrosion resistant aluminum honeycomb. However, even the most rigorous corrosion protection methods will not stop core corrosion but only delay its onset.

It should also be noted that the freeze-thaw cycles encountered during a typical aircraft flight can cause node bond failures (Fig. 26) if liquid water is present in the honeycomb cells.[22] At high altitudes the standing waterin

Corroded Core With Skins Removed

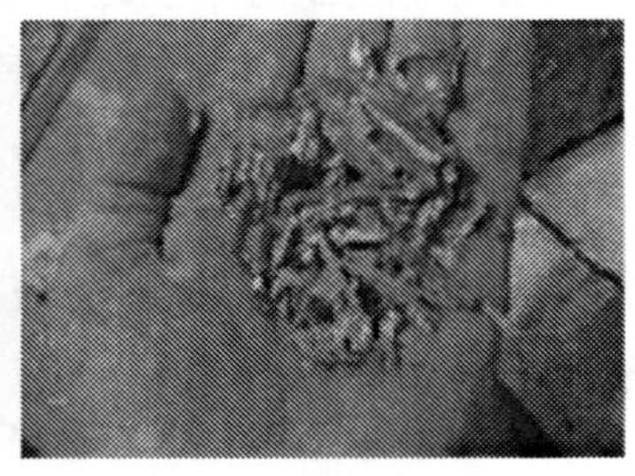

Loose Pieces of Cells

Reference 21: with permission

Fig. 24. Corroded Aluminum Honeycomb Core

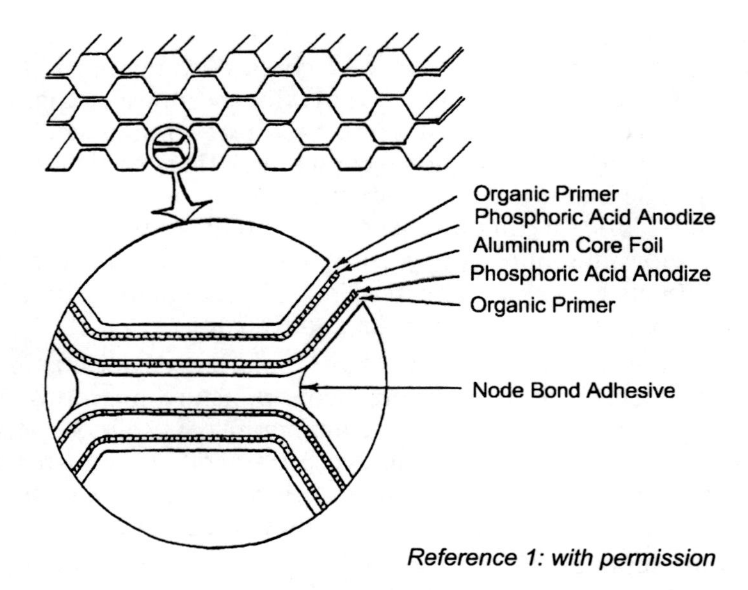

Fig. 25. Construction of Phosphoric Acid Anodize Honeycomb Core

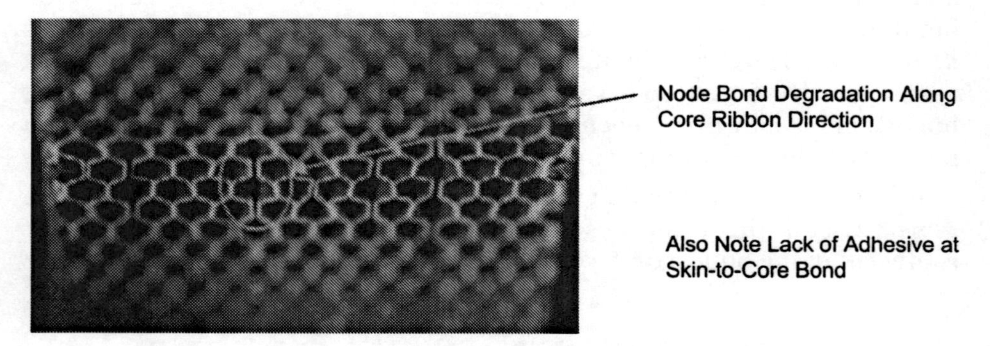

Fig. 26. Degradation of Node and Fillet Bonds in Aluminum Honeycomb Bonded Structure

the core freezes, expands and stresses the cell walls. After landing, the water thaws and the cell walls relax. After a number of these freeze-thaw cycles, the node bonds fail and the damage propagates. This freeze-thaw cyclic damage is not confined to aluminum honeycomb but can also occur in the non-metallic cores. In addition, water in the honeycomb can also cause disbonds and delaminate the facesheets, particularly if the temperatures exceed the boiling point of water (212 °F) as can happen during operation or repair.

Liquid water normally enters the core through exposed edges, such as panel edges, closeouts, door and window sills, attachment fittings, or almost any location that the skin and core bond terminates. The majority of the damage is often found at the edges of panels.[23] Adhesive bond degradation will lower the skin-to-core bond strength, the fillet bond strength and the node bond strength. Node bond degradation can reduce the core shear strength so that the assembly fails prematurely by core failure. In addition, water will enter the assembly through any puncture in the facesheets. Since some honeycomb assemblies contain extremely thin skins, water has been known to pass through the skins and then condense on the cell walls. Interconnected microcracks in thin skin honeycomb panels can also allow water ingression.[24] Although absorbed moisture affects the properties of any composite assembly, it is the presence of liquid water in the cells that does the majority of the damage. Many field reports blame water ingression on "poor" sealing techniques. While there is a great deal of truth to the statement that good sealing practices are important, it is the author's opinion that it is just a matter of time before water will find its way into the core of most honeycomb designs and initiate the damage process. In one study, if corrective actions were not taken to correct water ingression until the assembly delaminated, the cost for repairing the composite honeycomb assemblies for a large commercial aircraft was $280,000 over a 25-year assumed life.[25]

8.10 Honeycomb Processing

Honeycomb processing before adhesive bonding includes: perimeter trimming, mechanical or heat forming, core splicing, core potting, contouring and cleaning.

Trimming. The four primary tools used to cut honeycomb to plan dimensions are serrated knife, razor blade knife, band saw, and a die. The serrated and razor-edge knives and die cutter are used on light-density cores, white heavy-density cores and complex-shaped cores are usually cut with a band saw.

Forming. Metallic, hexagonal honeycomb can be roll- or brake-formed into curved parts. The brake-forming method crushes the cell walls and densify the inner radius. Overexpanded honeycomb can be formed to a cylindrical shape on assembly. Flexible core usually can be shaped to compound curvatures on assembly.

Non-metallic honeycomb can be heat-formed to obtain curved parts. Usually, the core is placed in an oven at high temperature for a short period of time (e.g., 550 °F for 1–2 min). The heat softens the resin and allows the cell walls to deform more easily. Upon removal from the oven, the core is quickly placed on a shaped tool and held there until it cools.

Fig. 27. Complex Structure with Different Core Densities

Splicing . When large pieces of core are required, or when complex shapes dictate, smaller pieces or different densities of core can be spliced together to form the finished part. This is usually accomplished with a foaming adhesive, as shown in Fig. 27. Core splice adhesives normally contain blowing agents that produce gases (e.g., nitrogen) during heat-up to provide the expansion necessary to fill the gaps between the core sections. Different core types, cell sizes or densities can be easily interconnected in this manner. Although foaming adhesives can be supplied as either pastes or films, the film product form is the most prevalent.

Paste foams are one-part epoxy pastes that expand during heating. They are used for core-joining, insert-potting and edge-filling applications. Foaming film adhesives are thick unsupported films (0.04–0.06 inches) that expand 1.5–3 times their original thickness when cured. Although some of these products can damage the core by overexpanding if too much material is used in the joint, most just expand to fill the gap and then stop when they meet sufficient resistance. A normal practice is to allow up to three layers of foam adhesive to fill gaps between core sections. Larger gaps call for rework or replacement of the core sections. It is also important not to

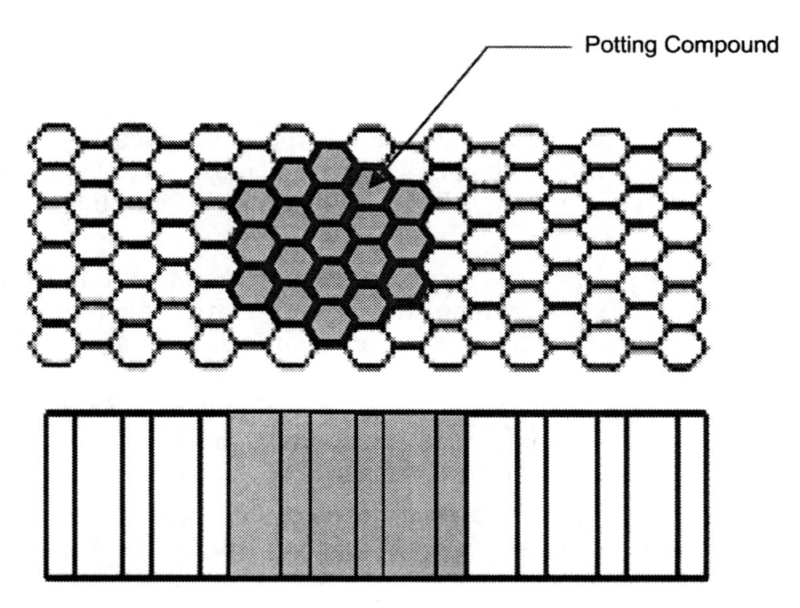

Fig. 28. Core Potting in Honeycomb Core

process foaming adhesives under a vacuum, or excessive frothing of the foam bondline may occur.

Potting. Potting compounds are frequently required for fitting attachment where fasteners must be put through the honeycomb assembly. As shown in Fig. 28, the cells are potted with a high-viscosity paste that is cured either during core splicing operations or during final bonding. These compounds usually contain fillers such as milled glass or aramid fibers, silica, or glass or phenolic microballoons. They can be formulated to cure at room temperature, 250 °F or 350 °F, depending upon the intended use temperature for the structure.

Machining . In many applications, honeycomb must have its thickness machined to some contour. This is normally accomplished using valve-stem-type cutters on expanded core. Occasionally, the solid honeycomb block is machined using milling cutters. Typical machines used for contour machining (carving) are gantry, apex, three-dimensional tracer or numerically controlled (NC) five-axis ones. With five-axis NC machining, the cutting head is controlled by computer programs, and almost any surface that can be described by x, y and z coordinates can be produced. These machines can carve honeycomb at speeds of up to 3,000 inches per minute with extreme accuracy. A standard contour tolerance of an NC machine is ±0.005 inch. Many core suppliers supply cores machined to be contour ready for final bonding.

Cleaning and drying. It is preferable to keep honeycomb core clean during all manufacturing operations prior to adhesive bonding; however, aluminum honeycomb core can be cleaned effectively by solvent vapor degreasing. Some manufacturers require vapor or aqueous degreasing of all aluminum cores prior to bonding; however, most part manufacturers accept "form B" core from the honeycomb suppliers and bond without further cleaning.

Non-metallic core, such as Nomex or Korex (aramid), fiberglass, and graphite core, readily absorbs moisture from the atmosphere. Similar to composite skins, *non-metallic core sections should be thoroughly dried prior to adhesive bonding*. A further complication is that since the cell walls are relatively thin and have a large surface area, they can reabsorb moisture rather rapidly after drying and, therefore, should be bonded into assemblies as soon as possible after drying.

Honeycomb Bonding. Honeycomb bonding procedures are similar to regular adhesive bonding with a few special considerations. Unlike many composite assemblies, honeycomb assemblies require special closeouts, several of which are shown in Fig. 29. During bonding, these require filler blocks in cavities and ramp areas to prevent edge crushing in the cure cycle. Again, closeouts are areas for potential water ingression, so special care is required during both the design and manufacturing process.

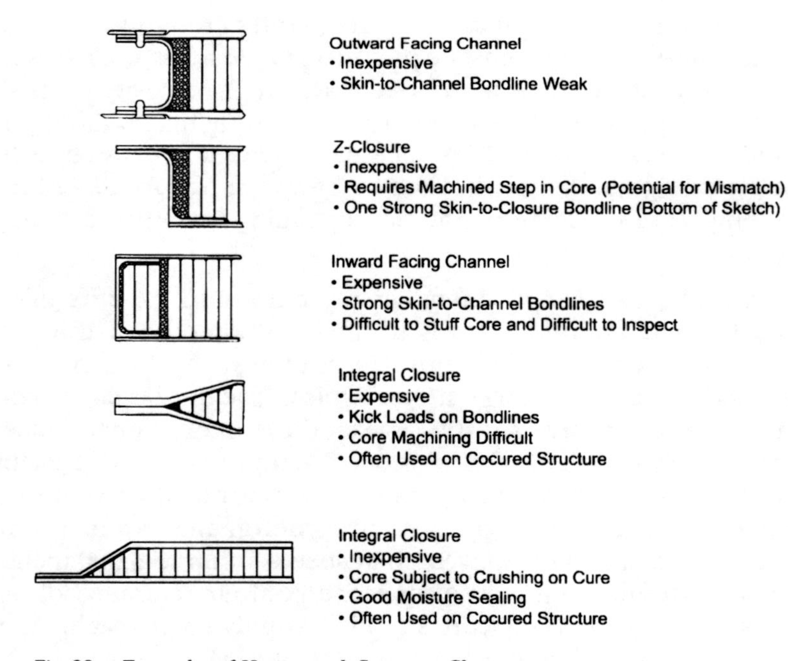

Fig. 29. Examples of Honeycomb Structure Close-outs

Pressure selection is an important consideration during honeycomb bonding. The pressure should be high enough to push the parts together but not be so high that there is danger of crushing or condensing the core. The allowable pressure depends on both the core density and the part geometry. Common bonding pressures can be in the range 15–50 psig for honeycomb assemblies. As previously discussed for adhesive bonding, the positive pressure of an autoclave with a vented bag gives a quality superior to that of a bond produced in an oven under vacuum bag pressure. The amount of pressure, as well as the adhesive selected, are important in forming fillets at the core-skin bondlines. The degree of filleting to a large extent determines the strength of the assembly. Common adhesive filleting conditions are shown in Fig. 30.

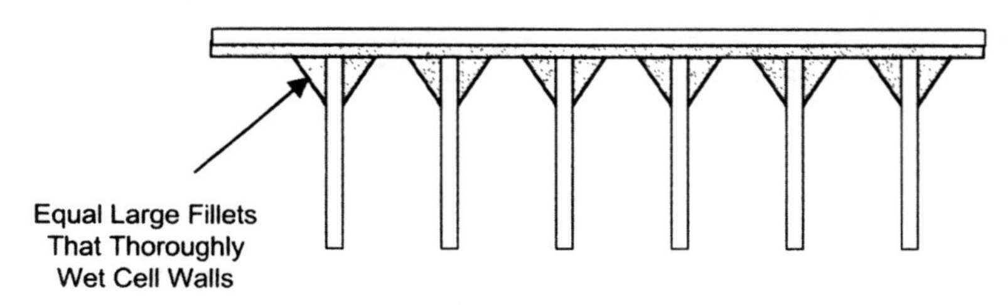

Good-Well Formed Adhesive Fillets

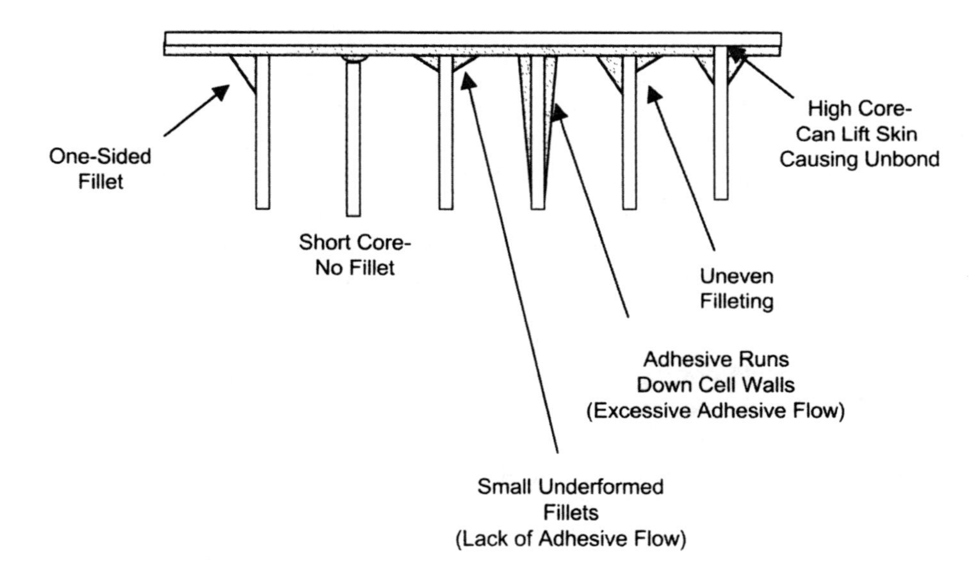

Undesirable Adhesive Filleting Conditions

Fig. 30. Examples of Honeycomb Core Filleting

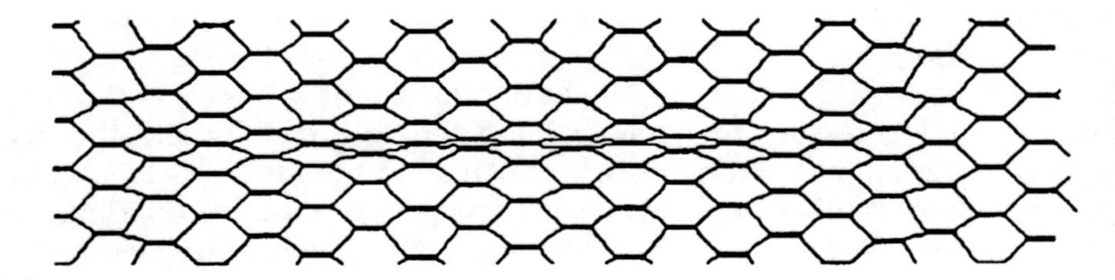

Fig. 31. Crushed or Condensed Core

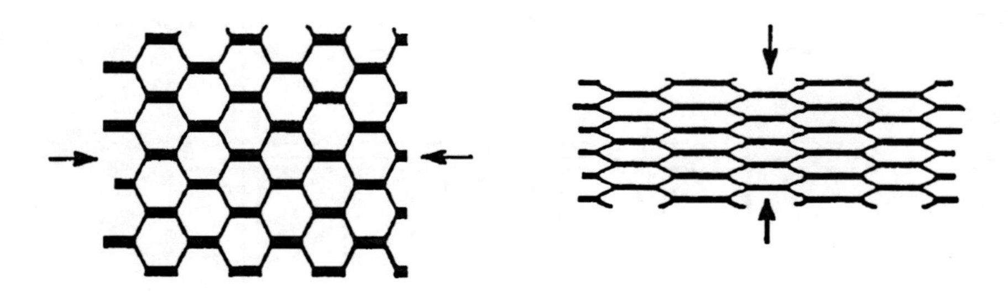

Fig. 32. Honeycomb Core Differential Compression Properties

Pressures that are applied on the sides of the core can easily condense the cells (Fig. 31). Since honeycomb is stronger in the longitudinal (L) direction than the width (W) direction, the core is more prone to crushing in the W direction (Fig. 32). Even when the initial vacuum is pulled, vacuum pressure alone has been known to cause core migration and cell crushing. Some manufacturers limit the vacuum level to 8–10 inches of Hg to help in preventing differential pressures within the cells. Autoclave processing of honeycomb assemblies is more sensitive to bag leaks than regular adhesive bonding. If pressure enters the bag through a leak, it can literally blow the honeycomb apart due to the large differential pressure. An example of some severely "blown" honeycomb core is shown in Fig. 33. This extent of damage would require repairing before the assembly would be useable.

Honeycomb assemblies can also be made by cocuring the composite plies onto the core. In this process, the composite skin plies are consolidated and cured at the same time they are bonded to the skins. Although adhesive is normally used at the skin-core interface, self-adhesive prepreg systems are

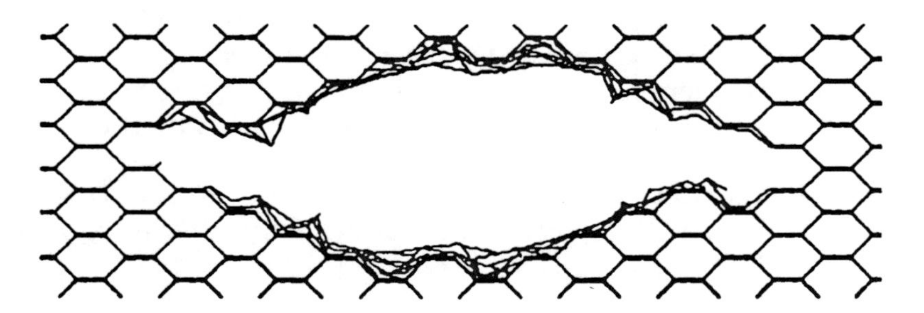

Fig. 33. Example of Severely Blown Core

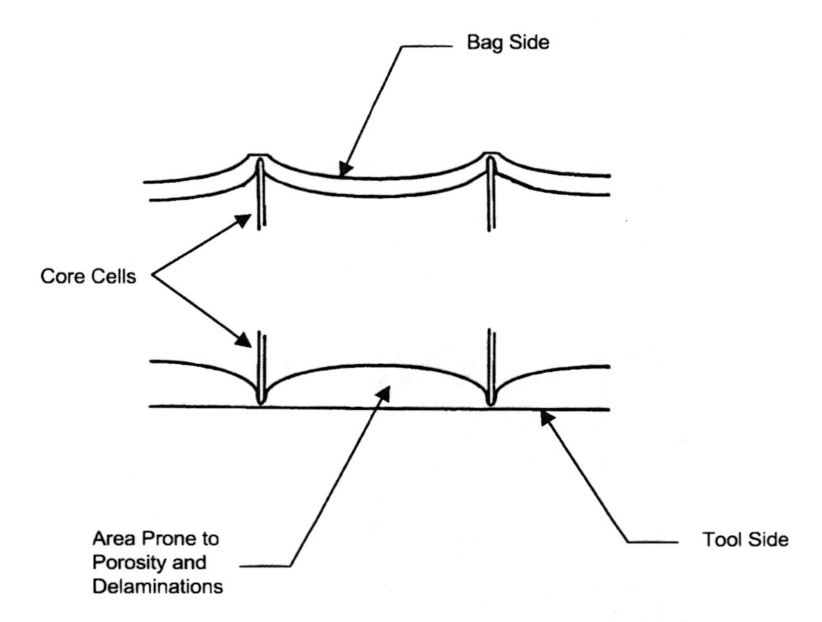

Fig. 34. Pillowing Effect in Composite Cocured Honeycomb Panels

available that do not require a film adhesive. To prevent core crushing and migration, this process is normally conducted at approximately 40–50 psig, as opposed to the normal 100 psig used for regular laminate processing. This can produce skins that are somewhat more porous than those processed under higher pressures, but the biggest drawback is the pillowing or dimpling that occurs in the skins (Fig. 34) due to the skin only being supported at the cell walls. Although the schematic is somewhat exaggerated in the amount of pillowing usually experienced, the pillowing does create a serious knockdown in mechanical properties, as much as 30% in some cases.[26] The amount of pillowing can be reduced by using a smaller cell size (e.g., 1/8 versus 3/16 inches); however, smaller cell sizes usually result in higher core densities and thus increased weight.

Although core migration and crushing can be a problem when bonding pre-cured composite skins, it is an even bigger problem with cocured skins (see Fig. 35). A considerable amount of work has been done to solve this problem.[27-29] Potential solutions include: (1) reducing the ramp angle (20° or less is recommended); (2) increasing the core density; (3) using grip, grit

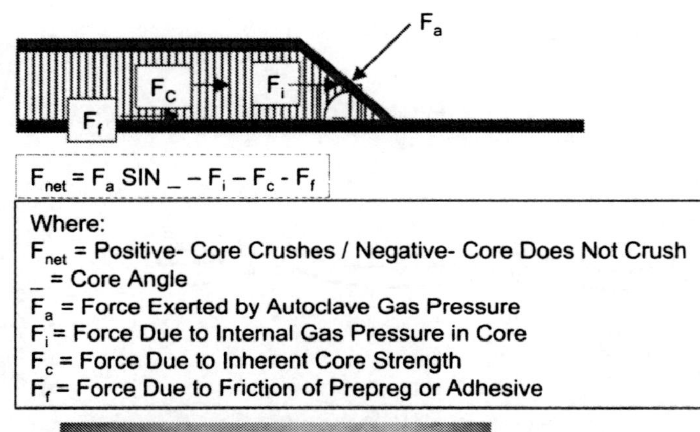

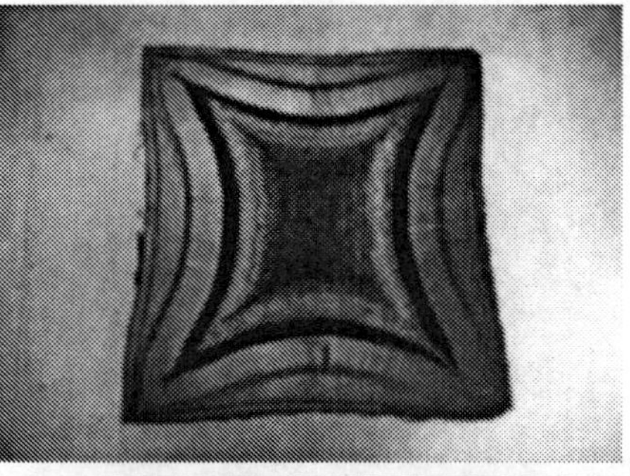

Top View
Of
Panel

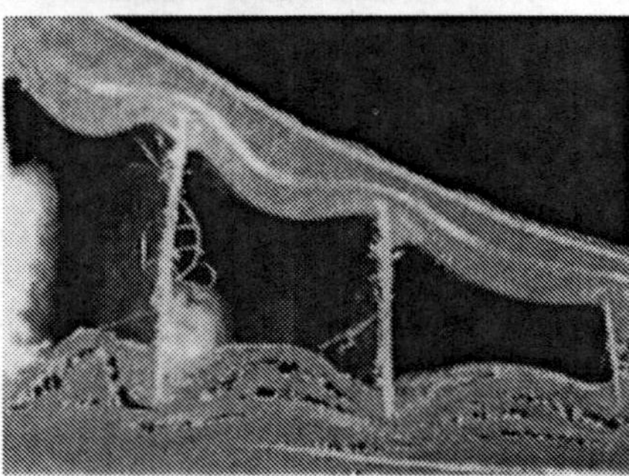

Section Showing
Core and Prepreg
Migration

Reference 27 and 30: with permission

Fig. 35. Core Crush Due to Core Migration During Cure

or hold down strips to restrain the plies; (4) potting of the cells in the ramp area to increase the core rigidity; (5) encapsulating the core with a layer of adhesive prior to cocuring; (6) bonding fiberglass plies into the center of the core (septums) to increase core rigidity; (7) adjusting the temperature and pressure during heat-up; and (8) even using "high friction" prepregs[30,31] that minimize ply movement during curing.

8.11 Balsa Wood

Balsa wood is one of the oldest forms of core materials and was used in early aircraft but is now primarily used in boat industry. The core is manufactured by first cutting sections transverse to the grain direction that are then cut into rectangles and adhesively bonded together. This results in the grains running perpendicular to the facesheets and is known as end-grained balsa. Typical cell sizes are 0.002 inch in diameter with densities in the range of 6–19 pcf. End-grain balsa has good mechanical properties, is fairly inexpensive and is easy to machine and bond to facesheets. The main disadvantages include severe moisture sensitivity, lack of formability and variable mechanical properties, since the block is made by bonding together smaller sections of variable density. Since balsa can absorb large amounts of resin during cure, it is a common practice to seal the surface prior to the lay-up operations.

8.12 Foam Cores

A third type of core material frequently used in adhesively bonded structure is foam core. While the properties of foam cores are not as good as honeycomb core, they are used extensively in commercial applications such as boat building and light aircraft construction. The term polymer foam or cellular polymer refers to a class of materials that are two-phase gas-solid systems in which the polymer is continuous and the gaseous cells are dispersed through the solid. These polymeric foams can be produced by several methods including extrusion, compression molding, injection molding, reaction injection molding and other solid-state methods.[32] Foam cores are made by using a blowing or foaming agent that expands during manufacture to give a porous, cellular structure. The cells may be open and interconnected or closed and discrete. Usually, the higher the density, the greater the percentage of closed cells. Almost all foams used for structural applications are classified as closed-cell, meaning almost all of their cells are discrete. Open-cell foams, while good for sound absorption, are weaker than the higher-density closed-cell foams and also absorb more water, although water absorption in both open and closed cell foams can be problematic. Both uncross-linked thermoplastic and cross-linked thermoset polymers may be foamed with the thermoplastic foams exhibiting better formability and the thermoset foams better mechanical properties and higher temperature resistance. Almost any polymer can be made into a foam material by adding an appropriate blowing or foaming agent.

The blowing agents used to manufacture foams are usually classified as either physical or chemical blowing agents. Physical blowing agents are usually gases mixed into the resin that expand as the temperature is increased, while chemical blowing agents are often powders that decompose on heating to give off gases, usually nitrogen or carbon dioxide.

Although there are foams that can be purchased as two-part liquids that expand after mixing for foam-in-place applications, the majority of structural foams are purchased as pre-expanded blocks that can be bonded together to form larger sections. Sections may be bonded together using either paste or adhesive films. Sections can also be heat-formed to contour using procedures similar to those for non-metallic honeycomb core. Although the uncross-linked thermoplastic foams are easier to thermoform, many of the thermoset foams are only lightly cross-linked and exhibit formability. Core densities normally range from about 2 to 40 pcf. The most widely used structural foams are summarized in Table 4. It is important to thoroughly understand the chemical, physical and mechanical properties of any foam considered for a structural application, particularly with respect to solvent and moisture resistance and long-term durability. Depending on their chemistry, foam core materials can be used in the temperature range 150–400 °F.

Polystyrene cores are lightweight, low-cost and easy to sand but are rarely used in structural applications due to their low mechanical properties. They cannot be used with polyester resins because the styrene in the resin dissolves the core; therefore, epoxies are normally employed.

Polyurethane foams are available as either thermoplastics or thermosets with varying degrees of closed cells. There are polyurethane foams that are available as finished blocks and formulations that can be mixed and foamed in place. Polyurethane foams exhibit only moderate mechanical properties and the resin-core interface bond tends to deteriorate with age, leading to skin delaminations. Polyurethane foams can be readily cut and machined to contours but hot wires should be avoided for cutting since harmful fumes can be released.

Polyvinyl chloride (PVC) foams are one of the most widely used core materials for sandwich structures. PVC foam can be either uncross-linked (thermoplastic) or cross-linked (thermoset). The uncross-linked versions are tougher, more damage resistant and are easier to thermoform, while the cross-linked materials have higher mechanical properties, are more resistant to solvents and have better temperature resistance; however, they are more brittle and more difficult to thermoform than the uncross-linked materials. They can be thermoformed to contours because they are not highly cross-linked like the normal thermoset adhesives and matrix

Table 8.4 Characteristics of Some Foam Sandwich Materials

Name and Type Of core	*Density (pcf)*	*Maximum Temperature (°F)*	*Characteristics*
Polystyrene (Styrofoam)	1.7-3.5	165	Low density, low cost, closed cell foam capable of being thermoformed. Used for wet or low temperature lay-ups. susceptible to attack by solvents.
Polyurethane Foam	3-29	250-350	Low to High density close cell foam capable of thermoforming at 425-450 °F. Both thermoplastic and thermoset foams are available. used for cocured and secondarily bonded sandwich panels with both flat and complex curved geometries.
Polyvinyl Chloride Foam (Klegecell and Dinvinycell)	1.8-26	150-275	Low to high density foam. Low density can contain some open cells. High density is closed cell. Can be either thermoplastic (better formability) or thermoset (better properties and heat resistance). Used for secondarily bonded or cocured sandwich panels with both flat and complex curved geometries.
Polymethacrylimide Foam (Rohacell)	2-18.7	250-400	Expensive high performance closed cell foam that can be thermoformed. high temperature grades (WF) can be autoclaved at 350F/100 psig. Used for secondarily bonded or cocured high performance aerospace structures.

systems. The cross-linked systems can be toughened with plasticizers in which some of the mechanical properties of the normal cross-linked systems are traded for some of the toughness of the uncross-linked materials. PVC foams are often given a heat stabilization treatment to improve their dimensional stability and reduce the amount of off-gassing during elevated-temperature cures. Styrene acrylonitrile (SAN) foams are also available that have mechanical properties similar to cross-linked PVC but have the toughness and elongation of the uncross-linked PVCs. Patterns of grooves can be scribed in the surfaces to act as infusion aids for resin transfer molding processing.

Polymethylmethacrylimides (PMIs) are lightly cross-linked closed-cell foams that have excellent mechanical properties and good solvent and heat resistance. They can be thermoformed to contours and are capable of withstanding autoclave curing with prepregs. These foams are expensive and are usually reserved for high-performance aerospace applications.

Due to the inherently lower mechanical properties for foam cores compared to honeycomb cores, manufactures are looking at methods of reinforcing foam cores to improve the mechanical properties. Two of these methods include stitching through both the skins and foam core[33] and the insertion of high-strength pultruded pins (carbon, glass or ceramic) into the foam core to form a truss configuration.[34]

8.13 Syntactic Core

Syntactic core consists of a matrix (e.g., epoxy) that is filled with hollow spheres (e.g., glass or ceramic microballoons) as shown in Fig. 36. Syntactics can be supplied as pastes for filling honeycomb core or as B-staged formable sheets for core applications. Syntactic cores are generally much higher in density than honeycomb, with densities in the range of 30–80 pcf. The higher the percentage of the microballoon filler, the lighter but weaker the core becomes. Syntactic core sandwiches are used primarily for thin secondary composite structures where it would be impractical or too costly to machine honeycomb to thin gages. When cured against pre-cured

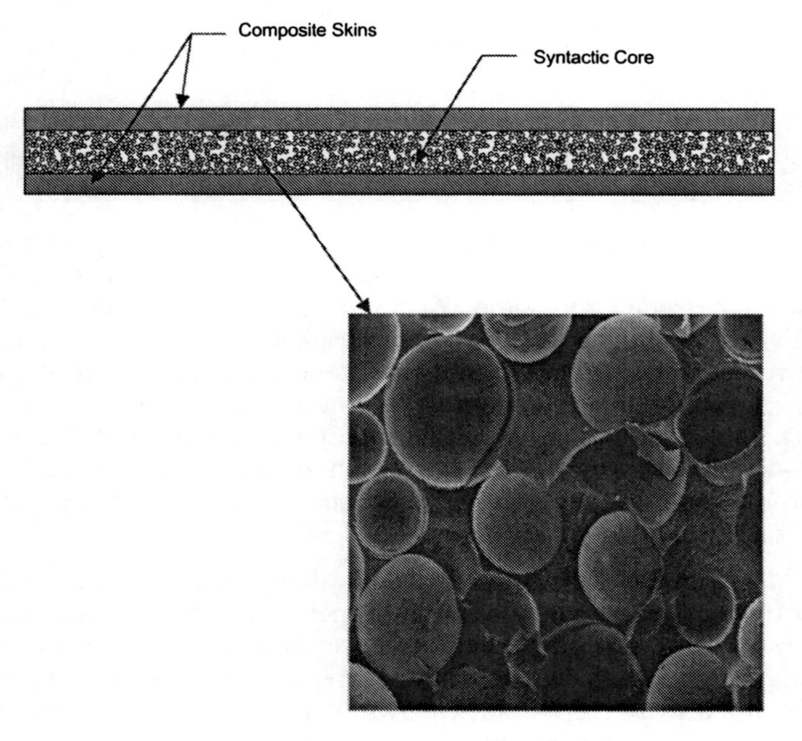

Fig. 36. Syntactic Core Construction

composite details, syntactics do not require an adhesive. However, if the syntactic core is already cured and requires adhesive bonding, it should be scuff-sanded and then cured with a layer of adhesive.

Glass microballoons are the most prevalent fillers used in syntactic core, ranging in diameter from 1 to 350 mm, but typically in the range of 50–100 mm. Glass micoballoons have specific gravities 18 times lower than fillers like $CaCO_3$; however, there have been issues in the past with moisture absorption into the glass microballoons This moisturization problem is currently being addressed by syntactic core manufacturers. Ceramic microballoons have properties similar to glass but better elevated-temperature properties, while polymeric microballoons (e.g., phenolic) are lower density than either glass or ceramic but have lower mechanical properties. The properties of the microballoons can be improved by increasing the wall thickness at the expense of higher densities. In most commercial applications, the microballons have a size distribution to improve the packing density. Packing densities as high as 60–80% have been achieved.

8.14 Inspection

Adhesively bonded joints and assemblies are normally nondestructively inspected after all bonding operations are completed. Radiographic and ultrasonic inspection methods are typically used to look for defects in both the bondlines and the honeycomb core portions of the assemblies. For a description of these test methods, refer to Chapter 13 (Nondestructive Inspection and Repair). In addition to these methods, it is quite common practice to leak-check honeycomb bonded assemblies by immersing the assembly for a short time in a tank of hot water (e.g., 150 °F). The hot-water heats the residual air inside the honeycomb core and any leaks can be detected by air bubbles escaping from the assembly.

A systematic test program on the effects of defects in aluminum honeycomb bonded panels is reported in Refs. 35 and 36. Their results are summarized in Table 5. All tests were conducted with 5052-H39 aluminum core (0.625 inch thick) bonded with a 250 °F curing film adhesive and 2024-T6 aluminum facesheets (0.40 inch thick). Both hexagonal core (density of 3.2–4.4 pcf) and corrugated core (density of 8–10 pcf) were evaluated for static and fatigue properties. In addition, panels were tested after environmental cycling. Of the defects they tested, the specimens with a void or gap at the edge closeout gave the greatest strength reductions. This is a result of the facesheets being unsupported at the closeout and being vulnerable to increased lateral loading when subjected to compression or transverse shear. Another significant finding was that the foaming adhesive used to bond the core to the closeouts is particularly sensitive to moisture and quickly loses its strength.

Table 8.5 Characteristics of Some Foam Sandwich Materials

High Degradation (>30%)	*Moderate Degradation (.30 to 10%)*	*Low Degradation (<10%)*
• Gap between Core and Edge Member	• Unbonded Nodes	• Core Splice Exceeding Separation Limit
• Voids in Foam Adhesive at Edge Member	• Gaps at Machined Core/ Stepped Skin	• Diagonal Line of Collapsed Cells
• Mismatched Nodes (Corrugated Core)	• Crushed Core at Edge Member	• Drilled Vent Hole in Skin
• Incomplete Edge Sea	• Blown Core	• Sideways Condensed Core
	• Over Expanded Core	• Incomplete Core Splice
		• Nest Cell (Corrugated Core)
		• Misaligned Ribbon

Reference 35

8.15 Integrally Cocured Structure

Integrally cocured or unitized structure is another manufacturing approach that can greatly reduce the part count and the final assembly costs for composite structures. The process flow for an integral cocured control surface[37] is shown in Fig. 37. In this particular piece of structure, the spars are cocured to the lower skin. The upper skin is cured at the same time as the spars are cocured to the lower skin but is separated from it and from the spar assembly by a layer of release film. This is necessary to allow mechanical installation of the ribs and center control box components. The plies for the spars are collated and then hot-pressure-debulked on their individual tooling details. Hot-pressure debulking is required to remove excess bulk from the plies so that all of the tooling details fit together. While the spars are being prepared, both the upper and lower skins are collated and hot-pressure-debulked on separate plastic lay-up mandrels (PLMs). The lower skin is placed on the tool first followed by the spar details and the tooling filler blocks that go in the bays between the spars. As mentioned previously, the upper skin is separated from the lower skin and spar assembly by a layer of release material. The completed assembly is bagged, leak-checked and then autoclave-cured. Several of the key lay-up and tool assembly sequences are shown in Fig. 38. Note that since this type of structure does not contain honeycomb core, drain holes can be drilled in strategic locations to allow any water to drain out of the assembly while in-service.

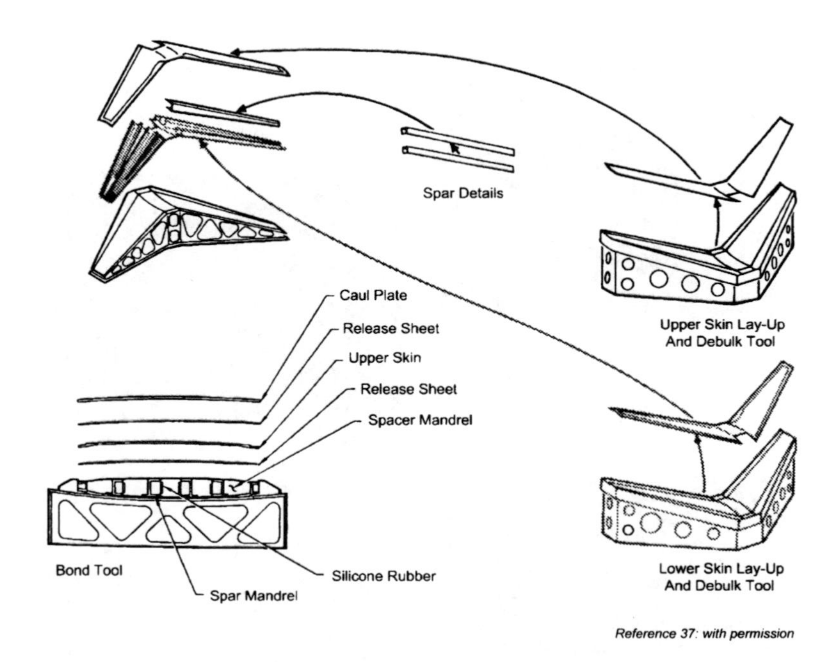

Fig. 37. Process Flow for Integrally Cocured Control Surface

Pressure is provided both by the autoclave and the expansion of the aluminum substructure blocks as shown in Fig. 39. The autoclave applies pressure to the skins and spar caps, while the expansion of the aluminum substructure blocks applies pressure to the spar webs. Note that the expansion of the aluminum substructure blocks can be supplemented by the presence of silicone rubber intensifiers, if required. The unit after cure and removal from the tool is shown in Fig. 40. As shown in Fig. 41, the upper skin is mechanically fastened to the spar caps after the substructure is completed. The advantages of this type of structure are obvious: fewer detail parts, fewer fasteners and less problems with part fit-up on final assembly. The main disadvantages are the cost and accuracy of the tooling required and the complexity of the lay-up that requires a highly skilled workforce. To help control tool accuracy, the substructure tooling is usually machined as a single block (Fig. 42) and then sectioned into the individual tooling details.

One potential problem with this type of structure is spring-in. As shown in Fig. 43, both the spar cap and web spring in during cool down after cure. The web spring-in can largely be eliminated by placing a couple of plies on the backside to support the web. However, the spar cap cannot be compensated by increasing the angle on the tooling block, because it would

Fig. 38. Key Process Steps for Integrally Cocured control Surface

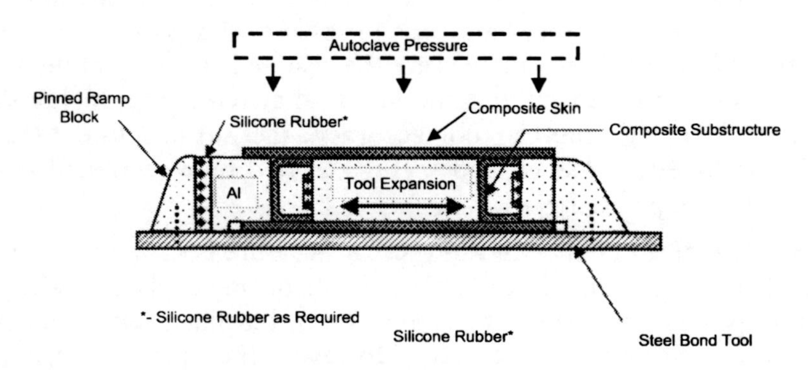

Fig. 39. Pressure Application for Unitized Cocured Structure

Lower Skin and Spars

Matching Upper Skin

The Boeing Company

Fig. 40. Cocured Unitized Control Surface

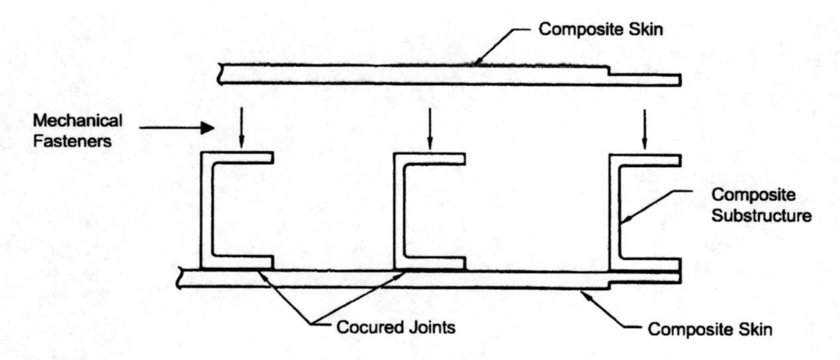

Fig. 41. Final Assembly for Integrally Cocured Control Surface

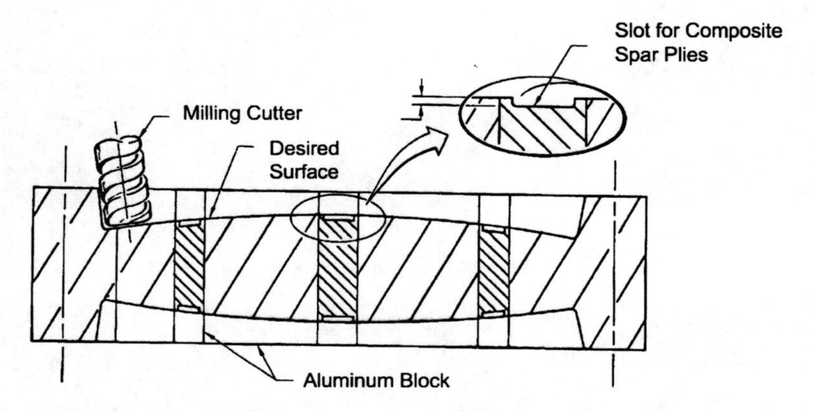

Fig. 42. Machining Substructure Spar Mandrels and Filler Blocks

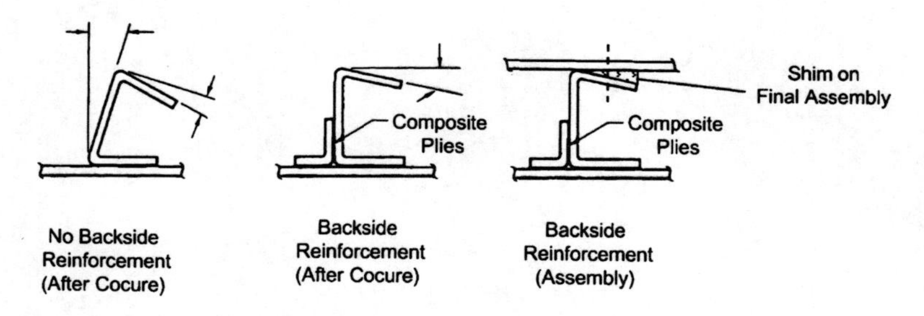

Fig. 43. Spring-In For Cocured Joints

make an indentation in the concurrently cured upper skin. It is, therefore, necessary to shim this joint during final assembly.

Another type of structure that is frequently cocured are skins with cocured hats,[38] such as the one shown in lay-up in Fig. 44. Although matched die

The Boeing Company

Fig. 44. Lay-Up of Fuselage Panel with Cocured Hats

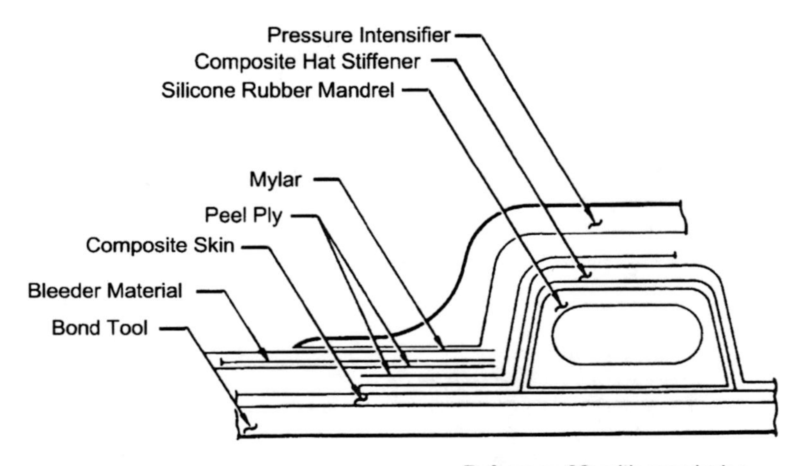

Reference 38: with permission

Fig. 45. Process Flow for Integrally Cocured Control Surface

tooling can be used to make this type of structure, the more common practice is to use localized tooling only at the hat stiffener locations. A typical bagging arrangement, shown in Fig. 45, contains an elastomeric mandrel to support the hat during cure; an elastomeric pressure intensifier to ensure that the radii obtain sufficient pressure; and some plastic shims to minimize mark-off on the skin from the pressure intensifier. In this figure, the mandrel contains a hole in the center to equalize the pressure, although some mandrels are solid elastomer or frequently elastomer reinforced with carbon or glass cloth. If the stiffeners require exact location, it may be necessary to use cavity tooling similar to that shown in Fig. 46.

One of the key design areas for cocured stiffeners concerns terminations. Since the bond holding the stiffener to the skin is essentially a resin or an adhesive bond, any peel loads induced at the stiffener ends could cause the bondline to "unzip" and fail. The most prevalent method for preventing this is by installing mechanical fasteners near the stiffener terminations, as shown in the hat design of Fig. 47. Note that the hat is also thicker at the ends and scarfed to further help reduce the tendency for bondline peeling. Other methods include stitching and pinning in the transverse (Z) direction. However, stitching of pregreg lay-ups is expensive and can be damaging to the fibers. Z-pining is a relatively new technology in which small diameter pre-cured carbon pins are driven through the prepreg lay-up before cure with an ultrasonic gun.

Cobonding is a hybrid process combining cocuring and adhesive bonding. As shown in the joint in Fig. 48, two cured pieces of structure can be joined

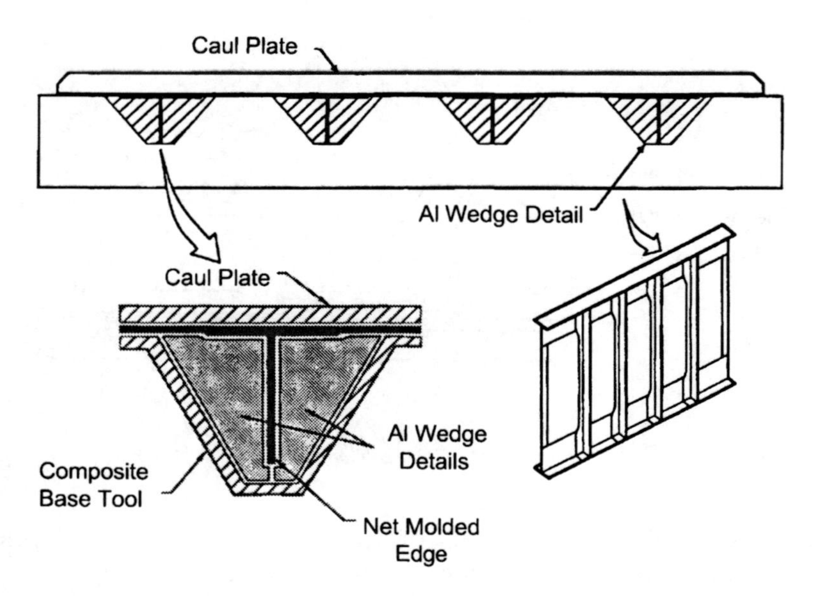

Fig. 46. Cavity Tool for Precise Location of Substructure

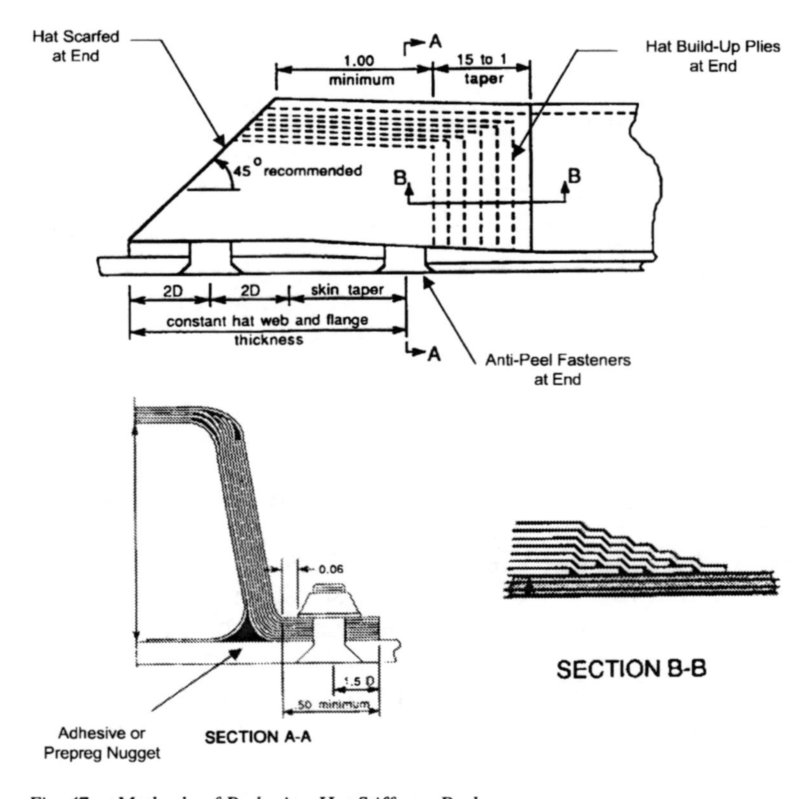

Fig. 47. Methods of Reducing Hat Stiffener Peel

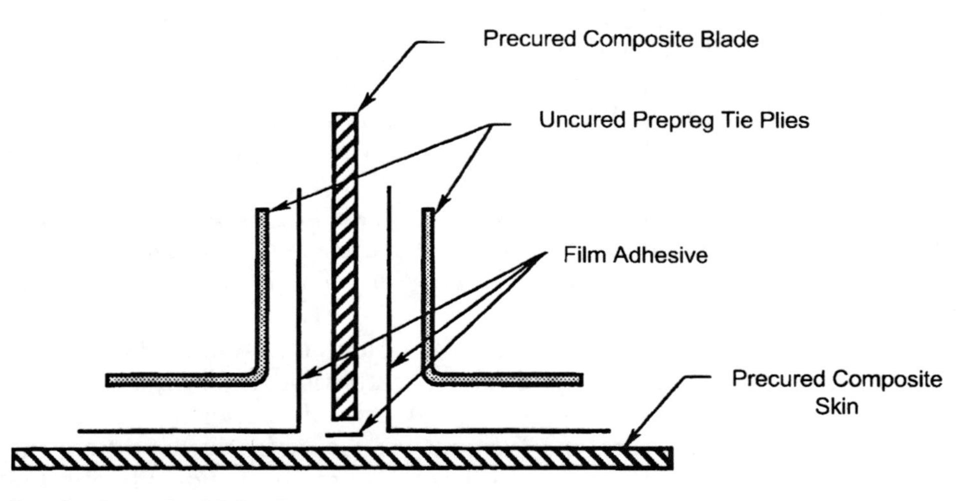

Fig. 48. Principle of Cobonding

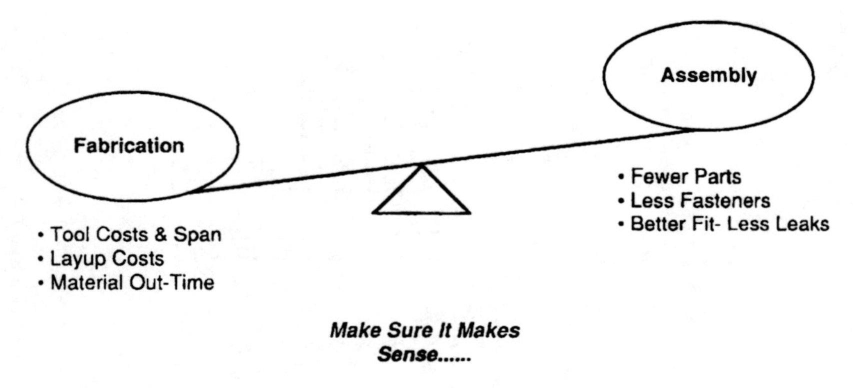

Fig. 49. Advantages and Disadvantages of Unitized Cocured Structure

by using a combination of uncured composite tie plies along with adhesive. The surfaces of the precured composite parts must, of course, be prepared for adhesive bonding just like any other adhesive bonding process. The advantage of this process is that, in certain situations, the amount of tooling required for the cocuring operation can be reduced.

Cocured unitized structure is a trade-off. As shown in Fig. 49, the advantages include fewer parts, less fasteners and better fit-up at assembly. These advantages must be weighed against the additional tooling costs and span time to produce the tooling, the lay-up costs and personnel requirements for large complex assemblies and the material out-time issue if the lay-up requires a long time.

8.16 Summary

Adhesive bonding is often preferred if thin composite sections are to be joined when bearing stresses in bolted joints would be unacceptably high, or when the weight penalty for mechanical fasteners is too high. In general, thin structures with well-defined load paths are good candidates for adhesive bonding, while thicker structures with complex load paths are better candidates for mechanical fastening. It is important to load the adhesive in shear. Tension, cleavage and peel loading should be avoided to prevent premature failures. The joint should always be designed so that failure occurs outside of the joint; i.e., the joint is stronger than the adherends.

Adhesively bonded structure offers the potential to reduce the number of parts and fasteners required for final assembly and, therefore, lower the cost of composite structures. The most critical fabrication aspect of bonding is surface preparation. Without proper surface preparation, the long-term durability of the joint will not be realized. Fortunately, the surface preparation for composites is quite a bit simpler than that for aluminum or titanium. When adhesive bonding composite parts, it is important that all cured parts be thoroughly dried prior to bonding. In addition, it is important that prefit operations are conducted to make sure that the parts to be bonded will fit together without excessive high or low areas; otherwise, thick glue lines and unbonds may result.

Sandwich construction is very efficient structurally, particularly in stiffness critical applications. Unfortunately, honeycomb assemblies are difficult to build and have frequently experienced in-service durability problems. Aluminum honeycomb has been prone to severe corrosion problems and all honeycombs can be damaged from repeated freeze-thaw cycles if the honeycomb contains liquid water. In addition, if liquid water is present and the assembly is heated above 212 °F, there is the danger of steam pressure delamination.

Foam core sandwich structures are somewhat easier to build than honeycomb structures but the current foams do not have as good mechanical properties as the honeycombs. In addition, foam cores are subject to quite high moisture saturation levels. Both types of core materials can be fabricated into composite structures by either pre-curing the composite details and then bonding them to the core, or by cocuring the assembly with prepreg plies at one time.

Cocured unitized structure is another option for reducing assembly costs. Although it eliminates some of the potential durability costs associated with honeycomb, it is more difficult to design and more costly to tool than honeycomb assemblies. A variation of cocuring, called co-bonding can reduce some of the tooling costs for certain types of assemblies.

References

[1] Campbell F. C., "Secondary Adhesive Bonding of Polymer-Matrix Composites," in *ASM Handbook 21 Composites*, ASM International, 2001.

[2] "Redux Bonding Technology," Hexcel Composites, December 2001

[3] Heslehurst R. B., Hart-Smith L. J., "The Science and Art of Structural Adhesive Bonding," *SAMPE Journal* **38**(2), March/April 2002, pp. 60-71.

[4] Scardino W. M., "Adhesive Specifications," in *ASM Engineered Materials Handbook Volume 1 Composites*, ASM International, 1987, pp. 689-701.

[5] Hart-Smith L. J., Brown D., Wong S., "Surface Preparations for Ensuring that the Glue Will Stick in Bonded Composite Structures," 10th DOD/NASA/FAA Conference on Fibrous Composites in Structural Design, 1-4 November 1993, Hilton Head Island, SC.

[6] Hart-Smith L. J., Redmond G., Davis M. J., "The Curse of the Nylon Peel Ply", 41st SAMPE International Symposium and Exhibition, 25-28 March 1996, Anaheim, CA.

[7] Venables J. D., McNamara D. K., Chen J. M., Sun T. S., Hopping J. L., *Applied Surface Science* **3**, 1979, pp. 88.

[8] Blohowiak K.Y., Cadwell-Stancin L., Anderson R. A., Mazzitelli C., Preedy K., Grob J. W., Glidden M., "Factors Influencing Durability of Sol-Gel Surface Treatments in Metal Bonded Structures," SAMPE International Symposium and Exhibition, 12-16 May 2002, Long Beach, CA.

[9] McCray D. B., Huff J. M., Smith J. F., "An Ambient-Temperature Adhesive Bonded Repair Process for Aluminum Alloys," 46th SAMPE International Symposium and Exhibition, 6-10 May 2001, pp. 1135-1147.

[10] Voevodin N. N., Balbyshev V. N., Vreugdenhil A. J., Johnson J. A., Donley M. S., "Evaluation of Corrosion Protection Performance of Sol-Gel Surface Treatments on AA2024-T3," 33rd SAMPE Technical Conference, 5-8 November 2001.

[11] Krieger R. B., "A Chronology of 45 Years of Corrosion in Airframe Structural Bonds," 42nd International SAMPE Symposium, 4-8 May 1997, pp. 1236-1242.

[12] Hinrichs R . J., "Vacuum and Thermal Cycle Modifications to Improve Adhesive Bonding Quality Consistency," 34th International SAMPE Symposium, 8-11 May 1989, pp. 2520-2529.

[13] Gleich D. M., Tooren M. J., Beukers A., "Structural Adhesive Bonded Joint Review," 45th International SAMPE Symposium, 21-25 May 2000, pp. 818-832.

[14] "HexWeb Honeycomb Sandwich Design Technology," Hexcel Composites, 2000.

[15] Kindinger J., "Lightweight Structural Cores," in *ASM Handbook 21: Composites*, ASM International, 2001.

[16] Corden J., "Honeycomb Structures," in *ASM Engineered Materials Handbook, Volume 1, Composites*, ASM International, 1987.

[17] Bitzer T., *Honeycomb Technology-Materials, Design, Manufacturing, Applications and Testing*, Chapman and Hall, 1997.

[18] Danver D., "Advancements in the Manufacture of Honeycomb Cores," 42nd International SAMPE Symposium, 4-8 May 1997, pp. 1531-1542.

[19] "HexWeb Honeycomb Selector Guide," Hexcel Composites, 1999.

[20] Black S., "Improved Core Materials Lighten Helicopter Airframes," High-performance Composites, May 2002, p. 56-60.

[21] Gintert L., Singleton M., Powell W., "Corrosion Control For Aluminum Honeycomb Sandwich Structures," 33rd International SAMPE Technical Conference, 5-8 November 2001.

[22] Radtke T. C., Charon A., Vodicka R., "Hot/Wet Environmental Degradation of Honeycomb Sandwich Structure Representative of F/A-18: Flatwise Tension Strength," Australian Defence Science & Technology Organization (DSTO), Report DSTO-TR-0908.

[23] Whitehead S., McDonald M., Bartholomeusz R. A., "Loading, Degradation and Repair of F-111 Bonded Honeycomb Sandwich Panels-A Preliminary Study," Australian Defence Science & Technology Organization (DSTO), Report DSTO-TR-1041.

[24] Loken H.Y., Nollen D. A., Wardle M. W., Zahr G. E., "Water Ingression Resistant Thin Faced Honeycomb Cored Composite Systems with Facesheets Reinforced with Kevlar Aramid Fiber and Kevlar with Carbon Fibers," E. I. DuPont de Nemours & Company.

[25] Shafizadeh J. E., Seferis J. C., "The Cost of Water Ingression on Honeycomb Repair and Utilization," 45th International SAMPE Symposium, 21-25 May 2000, pp. 3-15.

[26] Stankunas T. P., Mazenko D. M., Jensen G. A., "Cocure Investigation of a Honeycomb Reinforced Spacecraft Structure," 21st International SAMPE Technical Conference, 25-28 September 1989, pp. 176-188.

[27] Brayden T. H., Darrow D. C., "Effect of Cure Cycle Parameters on 350 °F Cocured Epoxy Honeycomb Panels," 34th International SAMPE Symposium, 8-11 May 1989, pp. 861p-874.

[28] Zeng S., Seferis J. C., Ahn K. J., Pederson C. L., "Model Test Panel for Processing and Characterization Studies of Honeycomb Composite Structures," *Journal of Advanced Materials*, January 1994, pp. 9-21.

[29] Renn D. J., Tulleau T., Seferis J. C., Curran R. N., Ahn K. J., "Composite Honeycomb Core Crush in Relation to Internal Pressure Measurement," *Journal of Advanced Materials*, October 1995, pp. 31-40.
[30] Hsiao H. M., Lee S. M., Buyny R. A., Martin C. J., "Development of Core Crush Resistant Prepreg for Composite Sandwich Structures," 33rd International SAMPE Technical Conference, 5-8 November 2001.
[31] Harmon B., Boyd J., Thai B., "Advanced Products Designed to Simplify Cocure Over Honeycomb Core," 33rd International SAMPE Technical Conference, 5-8 November 2001.
[32] Weiser E., Baillif F., Grimsley B. W., Marchello J. M., "High Temperature Structural Foam," 43rd International SAMPE Symposium, 31 May–4 June 1998, pp. 730-740.
[33] Herbeck I. L., Kleinberg M., Schoppinger C., "Foam Cores in RTM Structures: Manufacturing Aid or High-performance Sandwich?" 23rd International Europe Conference of SAMPE, 9-11 April 2002, pp. 515-525.
[34] Carstensen T., Cournoyer D., Kunkel E., Magee C., "X-Cor™ Advanced Sandwich Core Material," 33rd International SAMPE Technical Conference, 5-8 November 2001.
[35] Burkes J. M., Griffen M. A., Parr C. H., "Performance of Aluminum Honeycomb Panels with Structural Defects and Core Anomalies: Part I-Test Methodology and General Results," *SAMPE Journal* **28**(2), March/April 1992, pp. 25-31.
[36] Burkes J. M., Griffen M. A., Parr C. H., "Performance of Aluminum Honeycomb Panels with Structural Defects and Core Anomalies: Part II-Specimen Description and Test Results," *SAMPE Journal* **28**(3), May/June 1992, pp. 35-42.
[37] Moors G. F., Arseneau A. A., Ashford L. W., Holly M. K., "AV-8B Composite Horizontal Stabilator Development," 5th Conference on Fibrous Composites in Structural Design, 27-29 January 1981.
[38] Watson J. C., Ostrodka D. L., "AV-8B Forward Fuselage Development," 5th Conference on Fibrous Composites in Structural Design, 27-29 January 1981.

Chapter 9

Liquid Molding: You Get a Good Preform and Tool ...You Get a Good Part

Liquid molding is a composite fabrication process that is capable of fabricating extremely complex and accurate dimensionally parts. One of the main advantages of liquid molding is part-count reduction, in which a number of parts that would normally be made individually and either fastened or bonded together are integrated into a single molded part. Another advantage is the ability to incorporate molded-in features, such as the incorporation of a sandwich core section in the interior of a liquid molded part. Resin transfer molding (RTM), the most widely used of the liquid molding processes, is a matched mold process that is well suited to fabricating three-dimensional (3D) structures requiring tight dimensional tolerances on several surfaces. Excellent surface finishes are possible, mirroring the surface finish of the tool. The major limitation of the RTM process is the relatively high initial investment in the matched-die tooling. Sufficient part quantities, usually in the 100–5,000 range, are necessary to justify the high non-recurring cost of the tooling. A summary of the advantages and disadvantages of the RTM process is given in Table 1.

Table 1.1 RTM Process Advantages and Disadvantages

Advantages	*Disadvantages*
• Best tolerance control- tooling controls dimensions • Class A surface finish possible • Surfaces my be gel coated for better surface finish • Cycle times can be very short • Molded-in inserts, fittings, ribs, bosses, and reinforcements possible • Low pressure operation (usually less than 100 psig) • Prototype tooling costs relatively low • Volatile emissions (e.g., styrene) controlled by close mold process • Lower labor intensity and skill levels • Considerable design flexibility: reinforcements, lay-up sequence, core materials, and mixed materials • Mechanical properties comparable to autoclave parts (void content < 1%) • Part size range and complexity makes RTM appealing • Smooth finish on both surfaces • Near net molded parts	• Mold and tool design critical to part quality • Tooling costs can be high for large production runs • Mold filling permeability based on limited permeability data base • Mold filling software still in development stages • Preform and reinforcement alignment in mold is critical • Production quantities typically range from 100-5000 parts • Requires matched, leakproof molds

Adapted from Reference 1

The RTM process consists of fabricating a dry fiber preform that is placed in a closed mold impregnated with a resin and then cured in the mold. The basic resin transfer molding process, shown in Fig. 1, consists of the following steps:

- fabricate a dry composite perform;
- place the preform in a closed mold;
- inject the preform with a low-viscosity liquid resin under pressure;
- cure the part at elevated temperature in the closed mold under pressure; and
- demould and clean up the cured part.

Over the past several years, there have been many variations of this process developed, including resin film infusion (RFI), vacuum-assisted resin transfer molding (VARTM) and Seeman's composite resin infusion molding process (SCRIMP) to name a few. The objective of all of these processes is to fabricate near-net molded composite parts at low cost. In this section, we will examine the basic RTM process and some of the major variants that have evolved over the past several years. Table 2 gives a more complete listing of some of the various RTM processes that are currently in development and use.

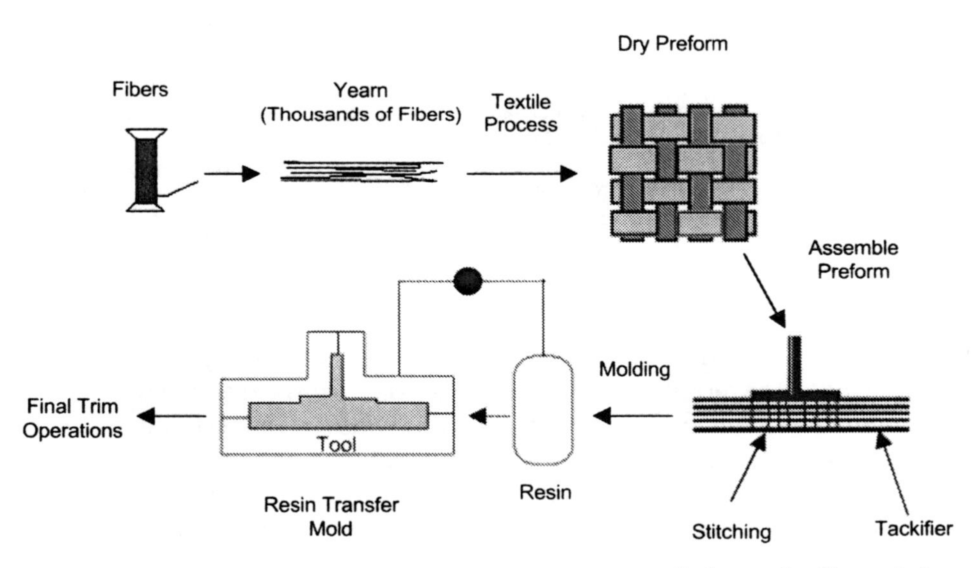

Fig. 1. Conceptuel Process Flow for Resin Transfer Molding (RTM)

Table 1.2 Comparaison of Various Liquid Molding Processes

Process Terminology	*Process Attributes*
Resin Transfer Molding (RTM)	• Resin injected into matched mold under presure • Vacuum assitance may or may not be used • Excellent surface finishes on both surfaces • Can obtain high fiber volumes (57-60 v/o) • Process called Co-Injection RTM (CIRTM) uses different resin systemes during injection process
Vacuum Assisted RTM (VARTM)	• Single sided tool normally used • Vecuum pulls liquid resin into preform (no applied pressure) • Requires low viscosity resins • Excellent surface finish on tool side • Tooling less expensive than RTM • Lower fiber volumes normally obtained (50-55 v/o) • Proprietary processes such as SCRIMP (Seeman's Composite Resin Infusion Molding Process) and FASTRAC used proprietary infusion media to speed resin infiltration
Resin Film Infusion Process (RFI)	• Resin film placed in bottom of tool and autoclave heat and pressure used To melt and force resin into preform • Normally requires matched die tools for complex parts • Variations include Resin Liquid Infusion (RLI) in which liquid resin replaces resin film and SPRINT in which resin layers are dispersed in the preform lay-up • Capable of producing high quality parts depending on the tooling
Expansion RTM	• Thermal-Expansion RTM (TERTM) uses matched die with internal core that expands to provide pressure during RTM • Rubber-Assisted RTM (RARTM) uses silicone rubber which expands on heating to provide pressure during RTM

Adapted from Reference 1

9.1 Preform Technology

The most important types of preforms for liquid molding processes are: (1) woven, (2) knitted, (3) stitched, (4) braided and (5) non-woven mats. In many cases, conventional textile machinery has been modified to handle the high-modulus fibers needed in structural applications and to reduce costs through automation. In addition, to meet the growing demand for 3D reinforced preforms, manufacturers have developed specialized machinery. A large variety of different advanced textile architectures are possible, several of which are shown in Fig. 2. NASA Langley Research Center has led much of the development of textile preform technology for aerospace applications. An excellent overview of their work can be found in Ref. 3.

9.2 Fibers

Textile machines have been adapted to handle most of the fibers[2] commonly used in structural composites, including glass, quartz, aramid and carbon. The main limitation is that most textile processes subject yarns to bending and abrasion. Although machines have been modified to minimize fiber damage, in many processes, exceptionally brittle or stiff fibers will suffer significant strength degradation. In general, the higher the modulus of the fiber, the harder it will be to process and the more prone it is to damage. Strength reductions can vary, depending upon the property

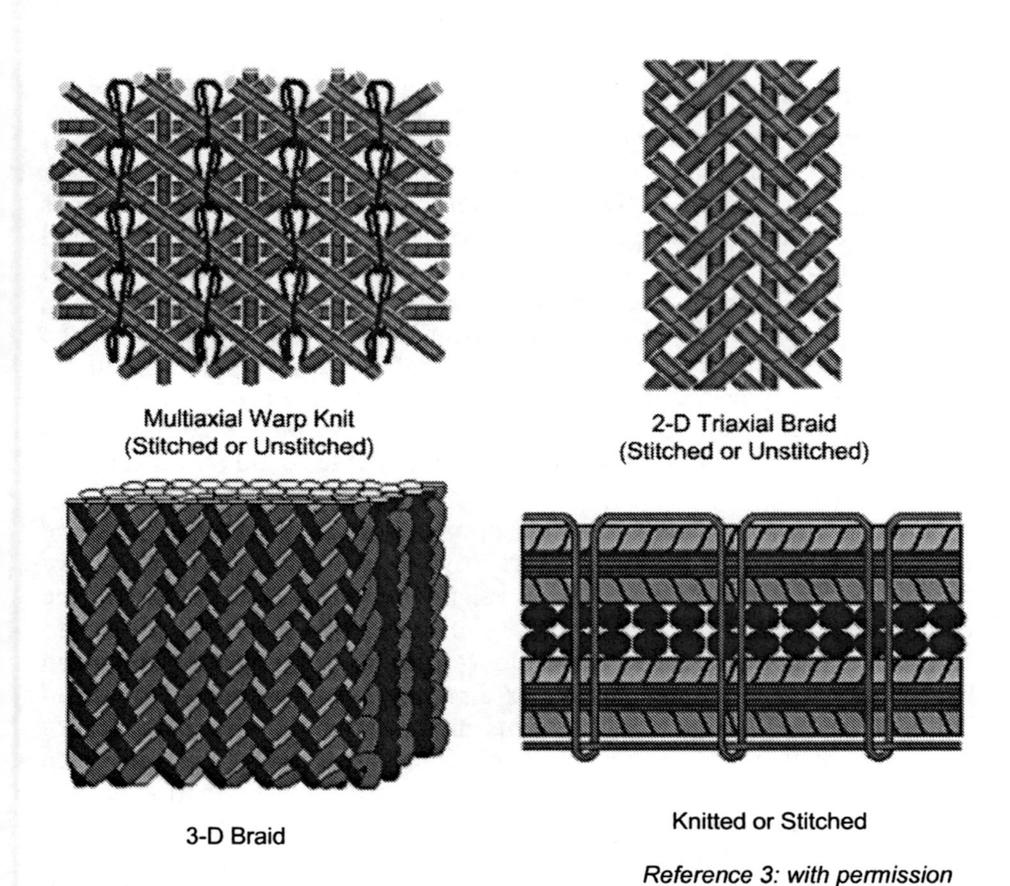

Fig. 2. Advanced textile Material Forms–Multiaxial Warp Knit, Triaxial braid, Three-Dimensional Braid and Knitted/Stitched

being measured and the textile process used to fabricate the preform. Polymeric sizings are usually applied to fibers to improve their handling characteristics and minimize strength degradation during processing. The sizings may be removed after processing or left on the fibers for the lamination process. If the sizing remains on the fibers, it is important that it be compatible with the matrix resin. A surface treatment is frequently used to improve the adhesion between the fibers and the matrix.

In traditional textile processes, yarns are usually twisted to improve handling, structural integrity and their ability to hold shape. However, twist reduces the axial strength and stiffness of the fibers, which is paramount in structural applications. Therefore, yarns with minimal or nominally zero twist (strands and tows) are preferred. Different processes and weaves require different strand or tow sizes. In general, the smaller the tow size, the more expensive the material will be on a per pound basis, particularly for carbon fiber.

9.3 Woven Fabrics

Woven fabrics[2] are available as 2D reinforcements (x- and y-directions) or 3D reinforcements (x-, y- and z-directions). When high in-plane stiffness and strength are required, 2D woven reinforcements are used. The various types of 2D weaves were previously covered in Chapter 2 on Fibers and Reinforcements; however, it should be noted that 2D woven products can be supplied either as a prepreg or as a dry cloth for either hand lay-up, preforming or repair applications. Two-dimensional weaves have the following advantages: (1) they can be accurately cut using automated ply cutters; (2) complicated lay-ups with ply drop-offs are possible; (3) there are a wide variety of fibers, tow sizes and weaves that are commercially available; and (4) 2D weaves are more amendable to thinner structures than 3D weaves.

Three-dimensional reinforced fabrics are normally used to (1) improve the handlability of the perform; (2) improve the delamination resistance of the composite structure; or (3) carry a significant portion of the load in the composite structure, such as a composite fitting that would be subject to complex load paths and major out-of-plane loading. If improved handlablity is the objective, usually z-direction fiber volumes as low as 1–2% will suffice. Major improvements in the delamination resistance of composite structures can be obtained with as little as 3–5% z-direction fiber; however, as the amount of z-directional fiber is increased, the delamination resistance and durability increases.[4] If the application calls for major out-of-plane loading, as much as 33% z-direction fiber reinforcement may be required. The fibers will then be arranged by with roughly equal load-bearing capacity along all three axes of a Cartesian coordinate system.

9.4 Three-dimensional Woven Fabrics[2]

Historically, composite designs have been restricted to structure that experiences primary in-plane loading such as fuselages or wing skins. One of the key reasons for the lack of composite structures in complex substructure, such as bulkheads or fittings, is the inability of 2D composites to handle complex, out-of-plane loads effectively. The planar load-carrying capability of composites is primarily a fiber-dominated property requiring well-defined load paths. Unfortunately, the planar loads must ultimately be transferred through a 3D joint into adjacent structure (e.g., a skin attached to a bulkhead). These 3D joints are subject to high shear, out-of-plane tension and out-of-plane bending loads, all of which are properties determined principally by the matrix properties in a traditional composite design. Since loading the matrix with large primary loads is a totally unacceptable design practice, metallic fittings are used to attach composite structure to metallic bulkheads with mechanical fasteners. With the introduction of high-performance 3D textiles, both woven and braided, this barrier to composite designs has the potential to be eliminated.

Three-dimensional preform technology enables composite structural concepts that effectively manage out-of-plane loads. This is done by weaving fibers in the x-, y- and z-directions. Other design benefits include stiffening concepts for monolithic composite aircraft skins, in which the stiffeners are woven integral with the skin structure. This eliminates the need for mechanical or bonded stiffener attachment and reduces the part count and costs for the assembly.

Three-dimensional woven fabrics are usually produced on a multiwarp loom as shown in Fig. 3. In a conventional 2D loom, harnesses alternately lift and lower the warp yarns to form the interlacing pattern. In a multiwarp loom, separate harnesses lift different groups of warp yarns to different heights, so that some are formed into layers while others weave the layers together to form net shape preforms. A 3D weave contains multiple planes of nominally straight warp and fill yarns that are connected together by warp weavers to form an integral structure. The most common classes are shown in Fig. 4. Within each class, there are several parameters that can be varied. Angle interlock weaves can be categorized by the number of layers that the warp weavers penetrate. A through-the-thickness interlock fabric, in which the warp weavers pass though the entire thickness, is shown in Fig. 4a, Figs. 4b and c show layer-to-layer interlock patterns, where a given weaver connects only two planes of fill yarns, but the weavers collectively bind the entire thickness. Various intermediate combinations can be fabricated with the weavers penetrating a specified number of layers. In orthogonal interlock weaves, the warp weavers pass through the thickness orthogonal to both in-plane directions, as shown in Fig. 4d. Interlock weaves are sometimes manufactured without straight warp yarns (stuffers)

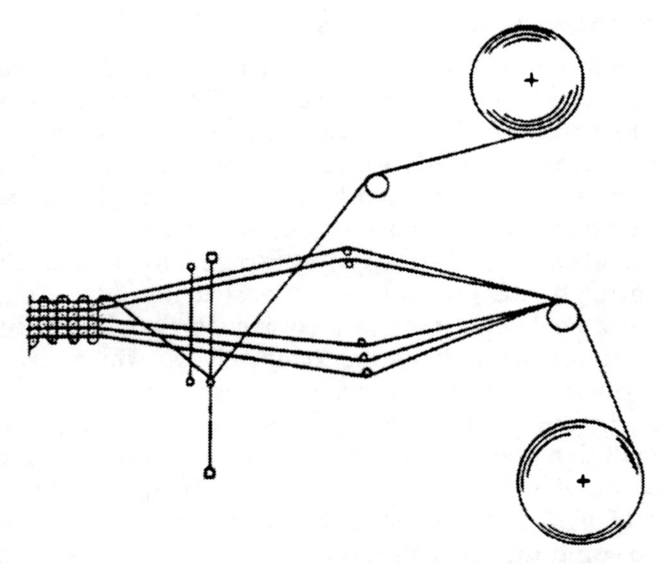

Fig. 3. Schematic of Multiwarp Weaving Loom

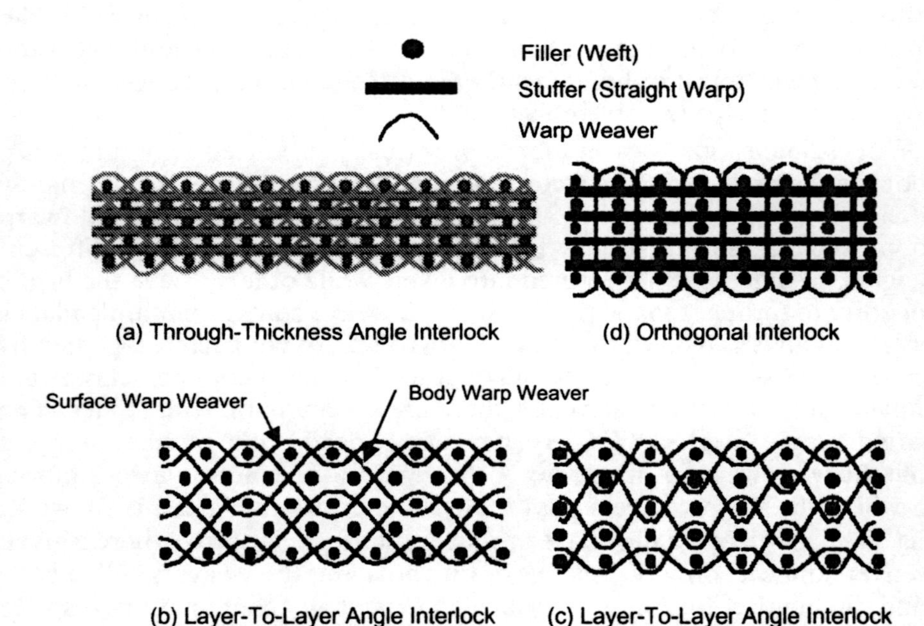

Fig. 4. Three Dimensional Weave Architectures

to produce a composite reinforced predominantly in one direction. They may also be fabricated with fill rather than warp yarns used for interlock. A major limitation of 3D weaves is the difficulty of introducing bias direction yarns to achieve in-plane isotropy. One solution is to stitch additional 2D fabric plies oriented at ±45° onto the woven preform. Three-dimensional weaving is capable of producing a wide variety of architectures, several of which are shown in Fig. 5. A relatively new approach to 3D weaving is shown in Fig. 6. In this process, called 3Weave, the *z*-directional fibers are introduced in a manner such that they are straight, which improves the mechanical properties of the final product.[7–9]

While 3D woven preforms show a potential for being able to fabricate complex net shaped preforms, the setup time is extensive and the weaving process is slow. In addition, small tow sizes that generally increase cost must be used to achieve high iber volume percents and eliminate large resin pockets that are susceptible to matrix microcracking during cure or later when the part is placed in-service.

9.5 Knitted Fabrics

Traditional knitted fabrics,[2] such as the weft and warp knits shown in Fig. 7, are highly flexible and conformable fabrics, but the extreme crimp of the fibers results in low structural properties. However, knitting can be effectively used to produce multiaxial warp knits (MWKs), also called stitch bonding, that combines the mechanical property advantages of unidirectional tape with the handling advantages and low-cost fabrication advantages of fabrics. MWKs, as shown in Fig. 8, consist of unidirectional tows of strong, stiff fibers woven together with fine yarns of glass or polyester thread. The glass or polyester threads, which normally amount to only 2% of the total weight, serve mainly to hold the unidirectional tows together during subsequent handling. An advantage of this process is that the *x*- and *y*-tows remain straight and do not suffer as much strength degradation as woven materials in which the tows are crimped during the weaving process.

The MWK process is used to tie tows of unidirectional fibers together in layers with 0°, 90° and ±θ° orientations. During knitting, the polyester threads are passed around the primary yarns and around one another in interpenetrating loops. Selecting the tow percentage in each of the orientations can be used to tailor the mechanical properties of the resulting stack. The MWK stacks form two-to-nine-layers thick building blocks that can be laminated to form the thickness desired the structure. Multiple layers of MWKs are often stitched together in a secondary operation to form stacks of any desired thickness and can be stacked, folded and stitched into net shapes. The stitching operation also greatly improves the durability and damage tolerance of the cured composite. MWK has the

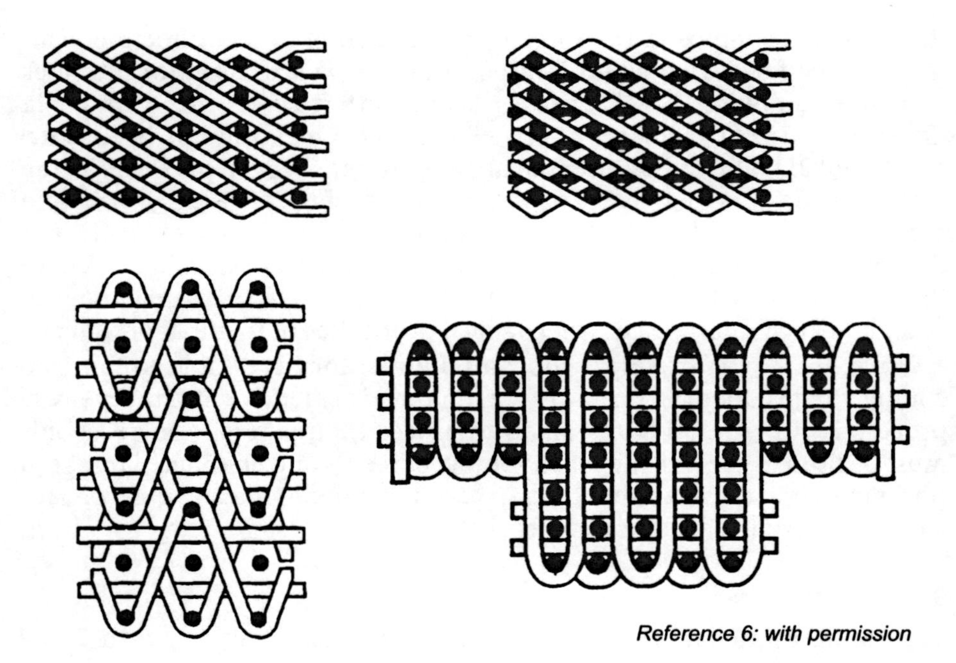

Fig. 5. Examples of 3D Woven Architectures

advantages of being fairly low cost, has uniform thicknesses, can be ordered in prefabricated blanket-like preforms, and is very amenable to gentile or no contour parts such as large skins.

9.6 Stitching

Stitching[2] has been used for more than 20 years to provide through the thickness reinforcement in composite structures, primarily to improve damage tolerance. The major manufacturing advancement in recent years has been the introduction of liquid molding processes which allows stitching of dry preforms rather than prepreg material. This enhances speed, allows stitching through thicker material, and greatly reduces damage to the in-plane fibers. Besides enhancing the damage tolerance, stitching also aids fabrication. Many textile processes generate preforms that cannot serve as the complete structure. Stitching provides a mechanical connection between the preform elements before the resin is introduced, allowing the completed preform to be handled without shifting or damage. In addition, stitching compacts (debulks) the fiber preform closer to the final desired thickness. Therefore, less mechanical compaction needs to be applied to the preform in the tool.

Two forms of stitching are normally employed for structural applications: the modified lock stitch and the chain stitch (Fig. 9). The chain stitch uses

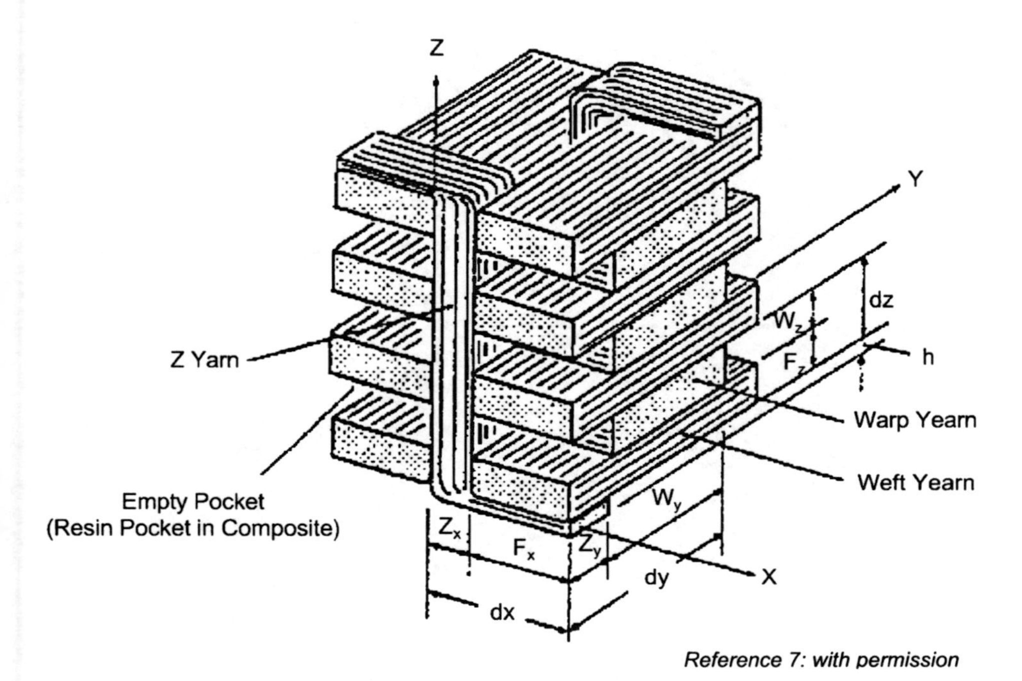

Reference 7: with permission

Fig. 6. schematic of 3Weave Three Dimensional Woven Structure

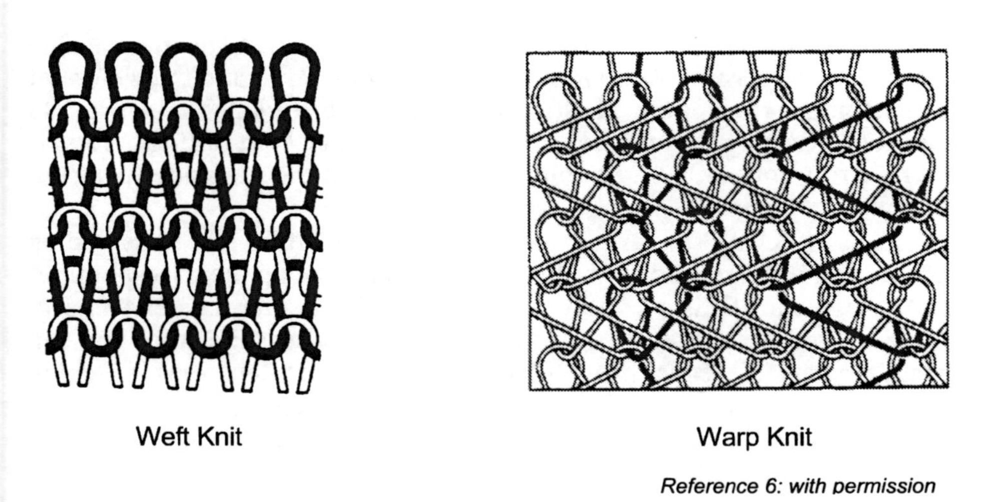

Reference 6: with permission

Fig. 7. Basic Knitted Structures

only one stitching thread, while the lock stitch requires a separate bobbin and needle threads. In the modified lock stitch, the thread tension is adjusted so that the knot forms on the outer surface of the laminate, rather

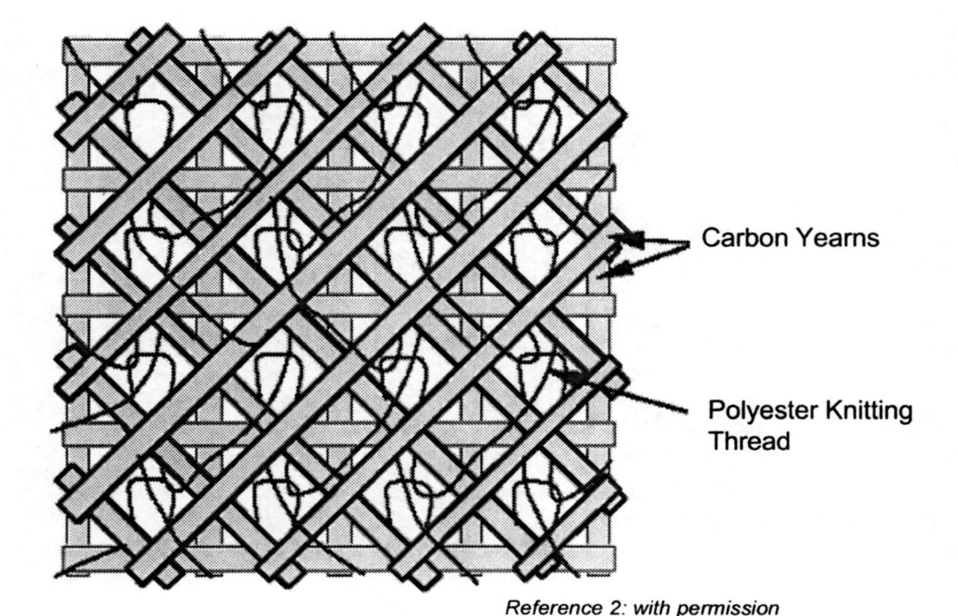

Fig. 8. Multi-Axial Warp Knit Architecture

than internally helping to minimize distortion of the laminate. Important stitching parameters include the pitch between penetrations, the spacing between parallel rows of stitching, the stitching material and the weight of the stitching yarn. Robotic 3D stitching machines are available that can do one-sided stitching using either a lockstitch or a tuft.[10] In tufting, a loop is inserted from one side and the loop folds over on the backside to secure the reinforcement.

Various stitching materials have been successfully used, including carbon, glass and aramid, with Kevlar 29 (aramid) being the most popular. Yarn weights for Kevlar of between 800 and 2,000 denier have been used. However, one disadvantage of aramid is that it absorbs moisture and can sometimes exhibit leaks through the skin at the stitch locations. Stitching that contributes around 3–5% to the total areal weight of the completed fabric has usually been found to impart satisfactory damage tolerance (compression strength after impact).

9.7 Braiding

Braiding[2] is a commercial textile process dating from the early 1800s. In braiding, a mandrel is fed through the center of the machine at a uniform rate, and fiber yarns from moving carriers on the machine braid over the mandrel at a controlled rate. The carriers work in pairs to accomplish an

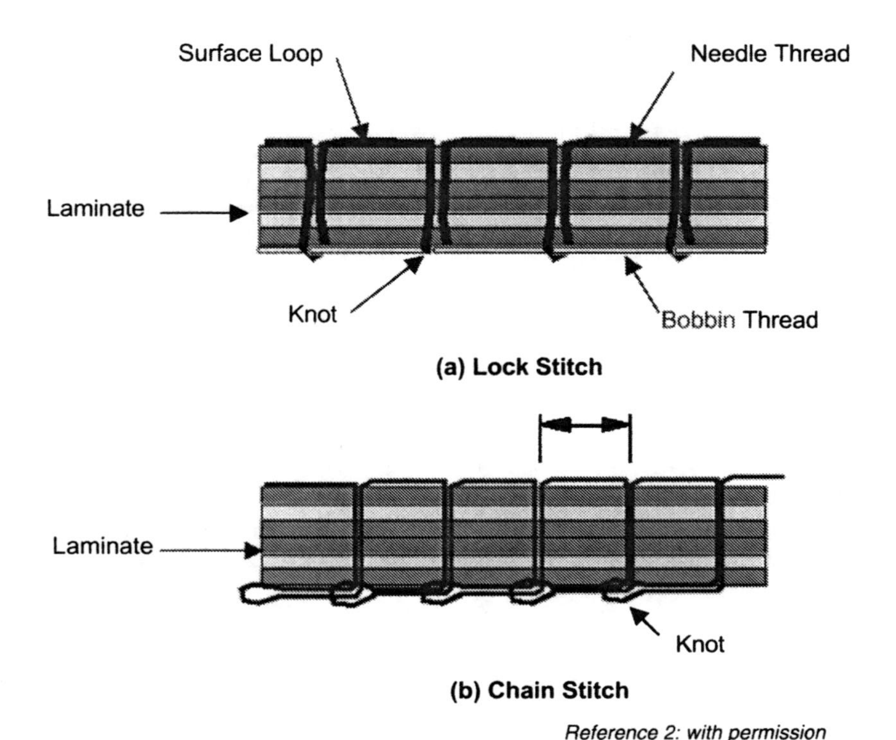

Fig. 9. Modified Lock and Chain Stitches for stitching Preforms

over/under braiding sequence. A schematic of the basic braiding principle is shown in Fig 10. Two or more systems of yarns are intertwined in the bias direction to form an integrated structure. Important parameters in braiding include yarn tension, mandrel feed rate, braider rotational speed, number of carriers, yarn width, perimeter being braided and reversing ring size.

Braided preforms are known for their high level of conformity, torsional stability and damage resistance. Either dry yarns or prepregged tows can be braided. Typical fibers include glass, aramid and carbon. Braiding normally produces parts with lower fiber volume fractions than filament winding but is much more amendable to intricate shapes. The rotational speed of the yarn carriers relative to the transversing speed of the mandrel controls the orientation of the yarns. The mandrel can vary in cross-section, with the braided fabric conforming to the mandrel shape.

Typical braiding machines contain anywhere from 3 to 144 carriers. A schematic of a simple flat braider is shown in Fig. 11. The main components are the track plate, the spool carriers, the former and the take-up device. The track plate supports the carriers and controls their path through a series of machined tracks. The spool carriers contain the composite yarn

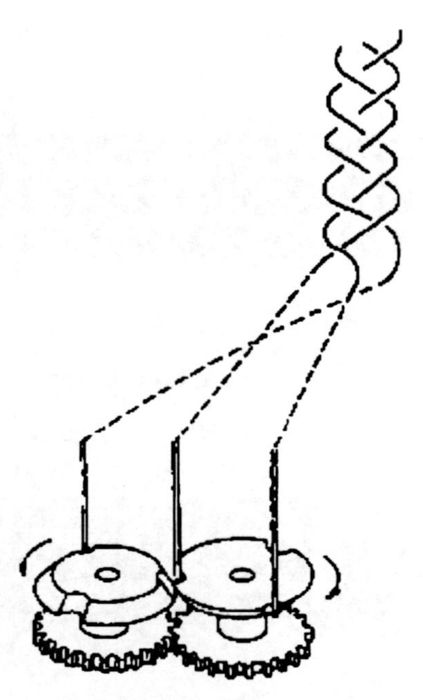

Reference 11: with permission

Fig. 10. Schematic of Braiding Principe

and are usually equipped with spring mechanisms to control the yarn tension. The details of a typical carrier mechanism are shown in Fig. 12 and a schematic of the braid formation is shown in Fig. 13. Guides or formers are used to control the braid dimensions as well as the shape of the braid. A take-up device takes up the braid as it is formed. A schematic of a large 144-carrier braider is shown in Fig 14 and an actual braider in operation is shown in Fig. 15.

Yarn width and the number of carriers determine the approximate braid angle for a given part perimeter in accordance with the relationship

$$\sin \theta = \frac{WN}{2P}$$

where θ = braid angle, W = yarn width, N = number of carriers, and P = part perimeter.

If the number of carriers and the braiding speed are controlled, the orientation of the braid angle and the diameter of the braid can be controlled. The total thickness of a braided part can be controlled by overbraiding, in which multiple passes of the mandrel are made through the braiding machine, laying down a series of nearly identical layers, similar

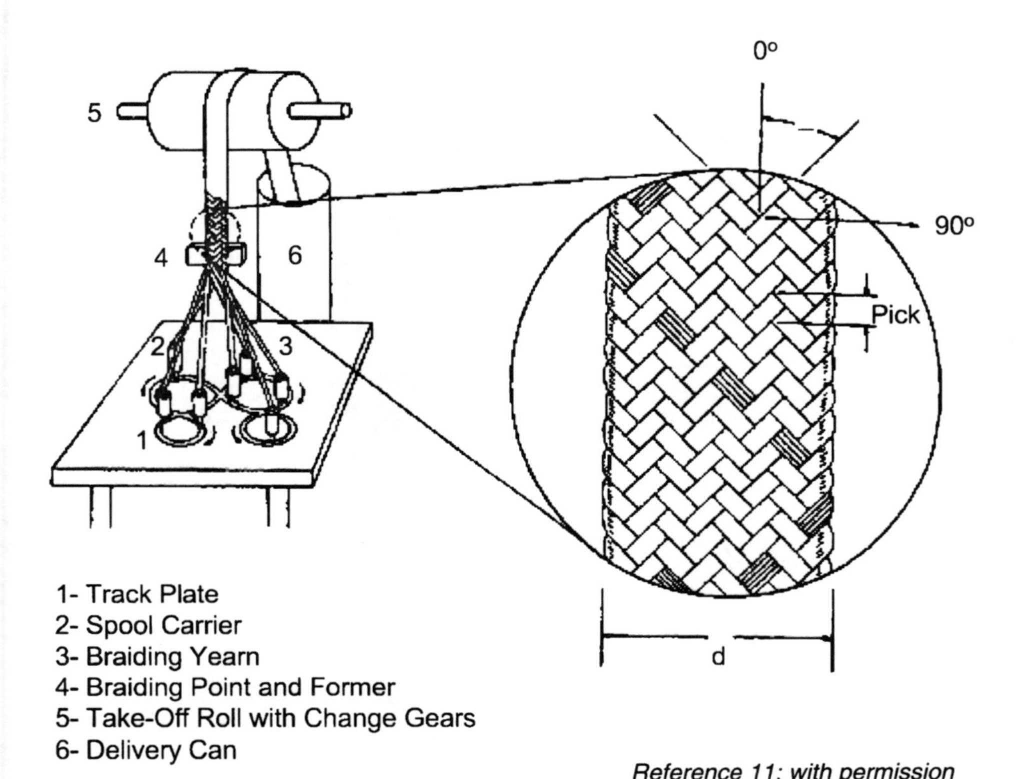

Fig. 11. Flat Braider and Braid Nomenclature

to a lamination. Possible fiber orientations are ±θ° or 0°/±θ° with no 90° layers unless the braider is fitted with filament winding capability.

Since the larger the number of carriers, the wider the range of diameters that can be produced, the trend has been to build larger machines with more carriers. A&P Technology developed a line of Megabraiders[13] with carriers ranging from 172 to 800. These large machines are capable of braiding parts up to 100 inches in diameter and 15 ft long. State-of-the-art braiding equipment provides full control over all of the braiding parameters, including translational and rotational control of the mandrel, vision systems for in-process inspection, laser projection systems to check braid accuracy and even integrated circumferential filament winding.

Due to the material conformity inherent in a braided product form, braided socks can be removed from the braiding mandrel and formed over a mandrel of a different shape for curing. In other situations, the braided part is cured directly on the mandrel. Both permanent and water-soluble or breakout mandrels can be used.

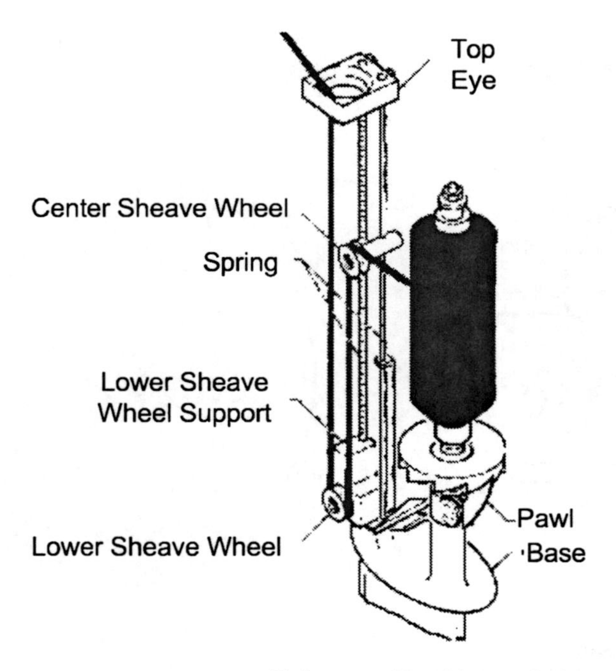

Fig. 12. Typical Braider Carrier Design

Fixed, straight axial yarns (0°) can also be introduced at the center of orbit of the braider yarn carriers. The braider yarns lock the axial yarns into the fabric, forming a triaxial braid, i.e., a braid reinforced in three in-plane directions. Cutting the cylindrical sheet from the mandrel and stretching it out flat can form a flat braided sheet.

Three-dimensional braiding can produce thick, net section preforms, in which the yarns are so intertwined that there may be no distinct layers. A schematic of a 3D process is shown in Fig. 16. In one process, the yarn carriers (bobbins) are arranged in a 2D grid, often in rectangular or annular patterns, sometimes in the cross-sectional shape of the final component. Multiple rectangles can be concatenated to form more complex cross-sections. Braiding proceeds by alternately exchanging rows and columns of yarn carriers. While the tracks that move yarn carriers return to their original positions after a small number of steps, the carriers themselves can follow complex paths by passing from one track to another. Three-dimensional braided socks can also be shaped into a preform suitable for use in joints and stiffeners. Due to the nature of braiding, a bias or 45° fiber orientation is inherent in the braided preform and theoretically allows the preform to carry high shear loads without the necessity of 45° hand layed-

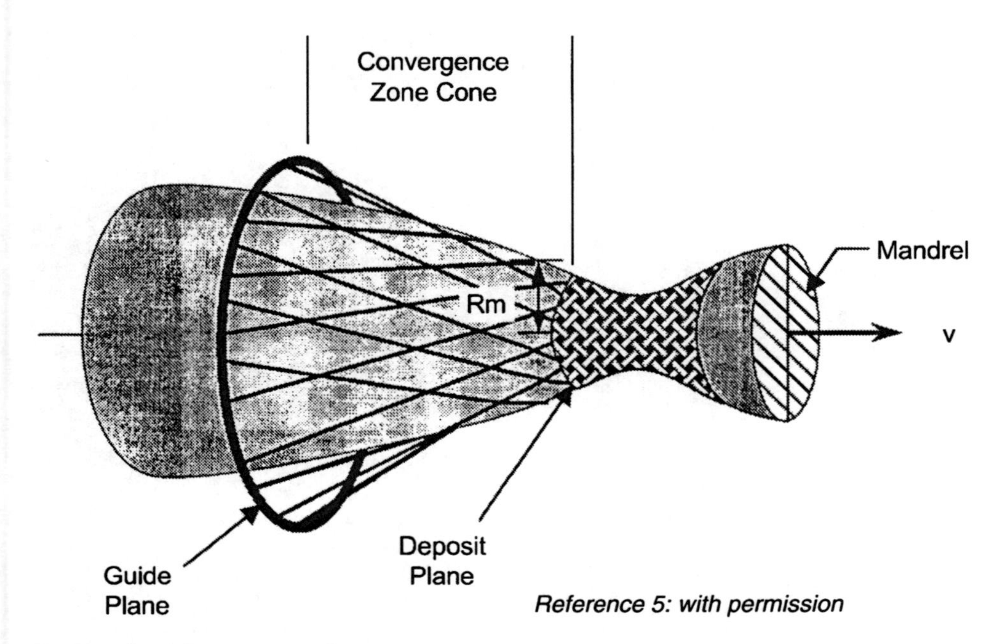

Fig. 13. Braid Formation over Mandrel

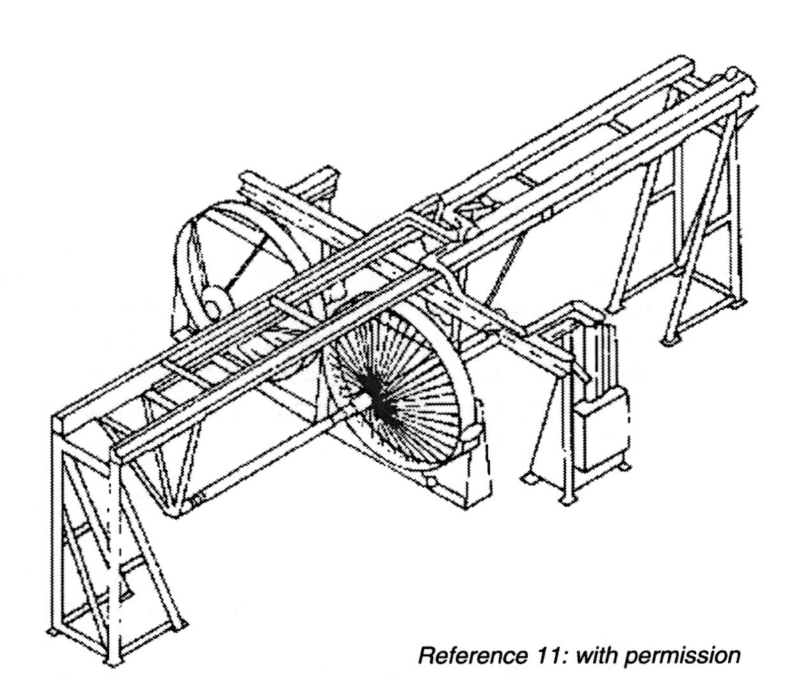

Fig. 14. 144 Carrier Horizontal Braiding Machine

Reference 3: with permission

Fig. 15. Braiding Machine Braiding Composite Preform

up overwrap plies. A 3D braiding machine can be set up to produce a near net-shape to the cross-section of the final part. An example of the complex fiber architecture that can be produced by 3D braiding is the structural I-beam shown in Fig. 17, and the range of potential structural shapes that can be produced by 3D braiding is shown in Fig. 18. The disadvantages of 3D braiding are similar to 3D weaving, i.e., complicated setups and slow throughput. Again, resin microcracking can be a problem with maximum fiber volumes of 45–50% obtainable.

9.8 P4A Process

Chopped fiberglass spray-up is a process that is used to make structural commercial products; however, since the fiber lengths are short and the orientation is random, the process is not competitive from a strength-to-weight ratio for structural applications. However, Owens–Corning developed a process for the automotive industry that uses a computer-

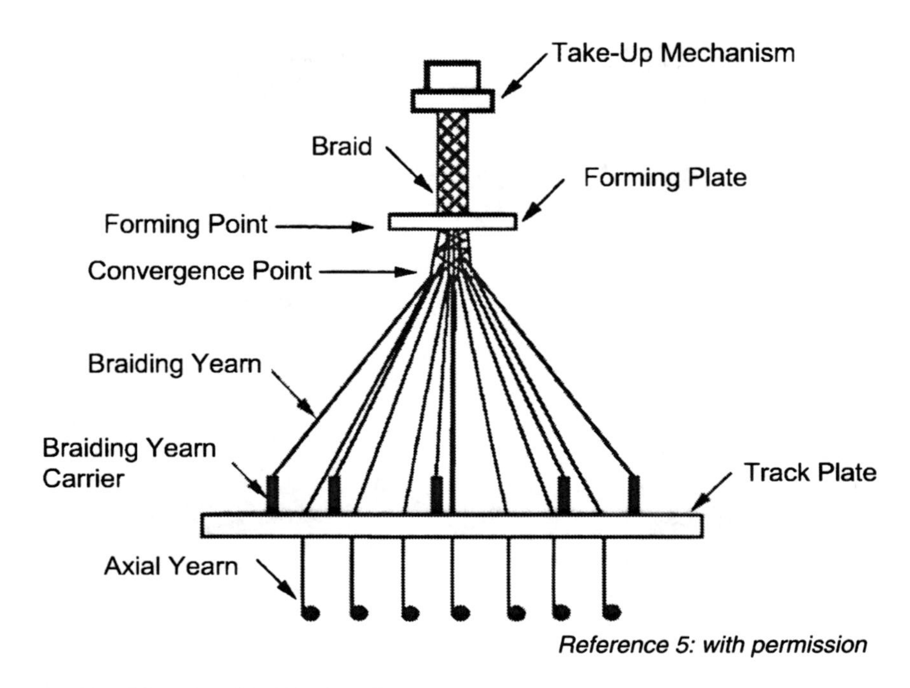

Fig. 16. Schematic of 3D Braiding Concept

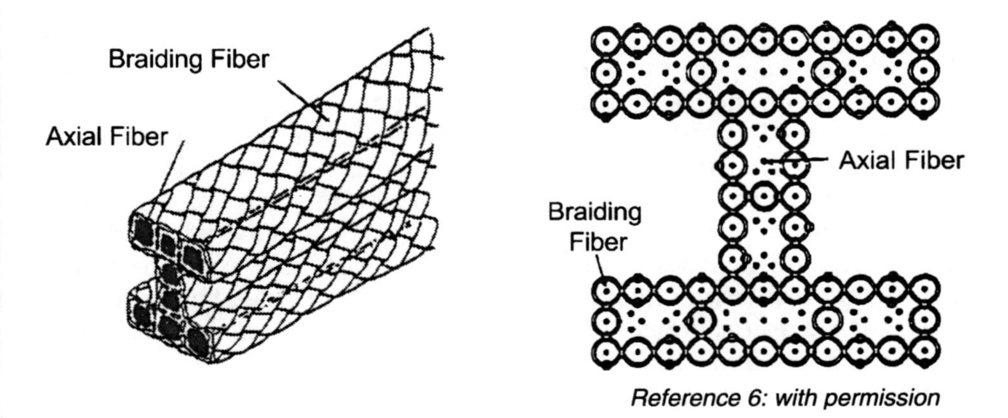

Fig. 17. 3D Braided I-Beam

controlled robot to spray chopped glass fibers in an oriented manner. This process, called the Programmable Powdered Preform Process (P4), is being adapted for carbon fibers for the aerospace industry (P4A). A computer-controlled robot and high-speed fiber chopper gun (Fig. 19) provides feedback to the computer that allows real time control of the process. The

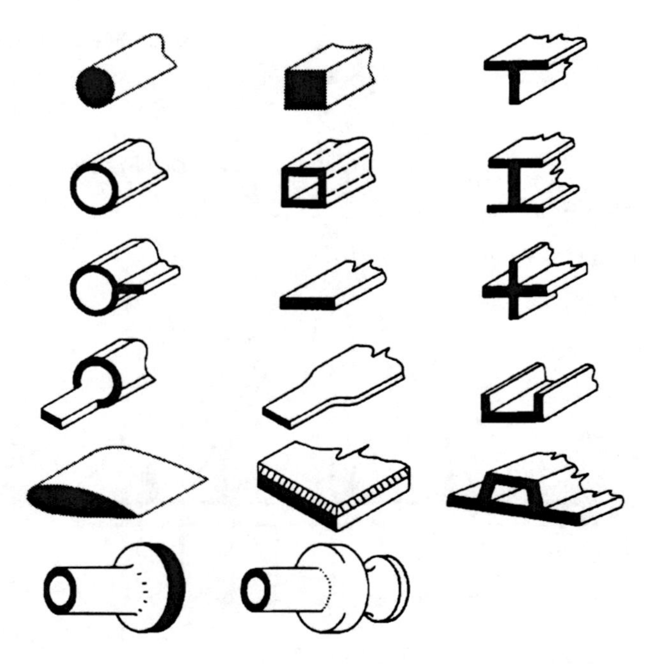

Reference 6: with permission

Fig. 18. Examples of Net Shaped 3D Braided Structures

chopper gun is designed to allow the lengths of the fibers (0.5–5.0 inches) to be varied "on the fly" during perform fabrication. By using a deflector plate at the end of the nozzle, up to 90% of the chopped fibers can be oriented in the desired direction. The fibers are directed along with a powder binder material onto a perforated preform screen of the desired shape and held in position by a vacuum system. A surface veil may be used on one or both surfaces to improve surface finish. The preform is then densified to the desired thickness by the application of heat and pressure. During the densification step, the powder binder melts and flows to give the perform-handling capability. While this process will not produce parts with properties equivalent to continuous fiber-reinforced parts, it has the potential to reduce costs while offering attractive mechanical properties. A notional trade-off of cost of processing versus performance is shown in Fig. 20.

9.9 Random Mat

Many preforms for commercial applications that do not require high mechanical properties can be made from random mat using either discontinuous chopped fiber or swirled fiber mat. This is a very economical

National Composite Center

Fig. 19. P4A Process Fabricating a Preform

approach that allows complex shapes with compound contours. Since the fiber volume is about 40% maximum, they are also easy to infuse by a variety of liquid molding concepts.

9.10 Preform Advantages[2]

Textile preforms will continue to play an integral role in the composites industry due to their ability to improve damage resistance and their potential to reduce composite part costs.

- Textile preforms have a handling advantage compared to unidirectional products forms. They are manufactured as dry fiber preforms that are held together without any polymer or matrix material. A textile preform can be shipped, stored, draped (within

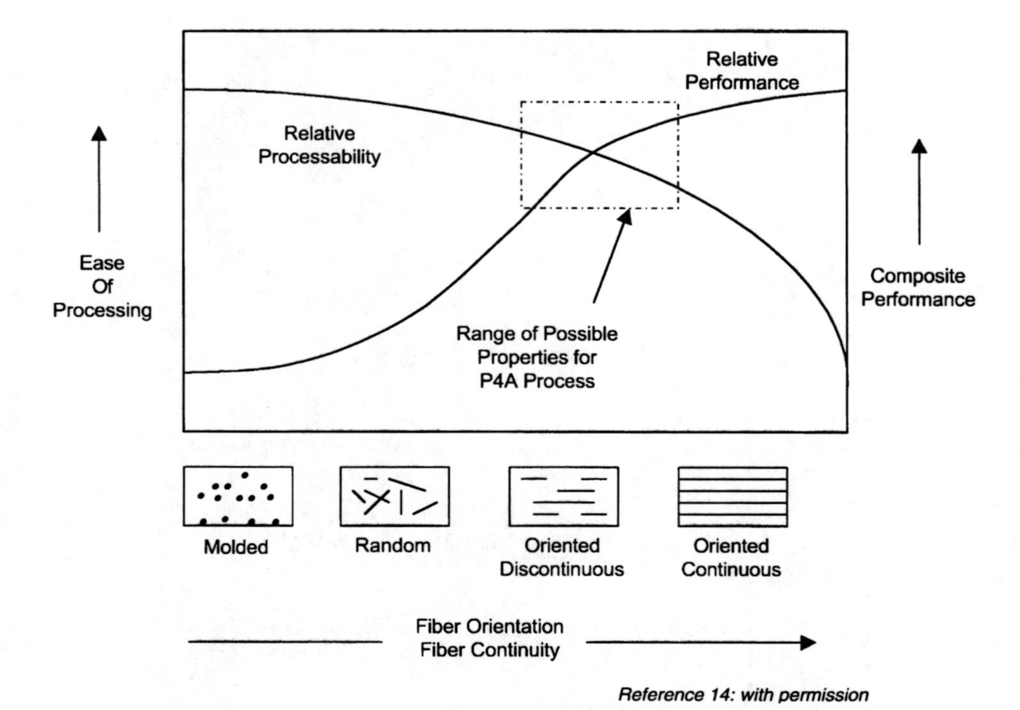

Fig. 20. Influence of Fiber Orientation on Processing and Performance

limits that depends on the kind of fabric) and pressed into shaped molds. The finished product can be cured in a mold by liquid molding.

- Separate preforms can easily be joined by co-curing if joints of moderate strength suffice, or by stitching if joints must be stronger. Textile product forms allow designers to step beyond conventional laminate concepts. For example, with conventional tape lay-up, a laminate skin is stiffened to prevent buckling by non-integral ribs, which must be attached in a separate process, such as co-curing, bonding or mechanical fastening. Textile preforms and liquid molding processes allow the manufacture of net shape integral parts. The skin and stiffeners can be manufactured as one piece. Net shape manufacturing of integral structures provides considerable potential cost savings over prepreg lay-up, because forming complex shapes through hand lay-up is difficult and integral structures eliminate joining steps. Integral structures are also superior in performance, because, given correct design, failure by delamination of the attached parts should be eliminated as a failure mechanism.
- Preforming frees the high-cost matched metal molding die in the RTM process from everything but loading, curing and unloading. It is

a common practice to use lower-cost preforming tools to lay up and heat-set the preform to shape.

9.11 Preform Disadvantages[2]

With the exception of improved damage tolerance and low cost, most preform product forms exhibit lower mechanical properties than unidirectional products. The lower properties are a combination of several factors.

- Anytime you handle or bend high-strength fibers, particularly glass and carbon, you degrade their properties. In many of the textile processes described above, there is considerable mechanical abrasion and bending of the tows or yarns.
- In weaving and braiding, the interlocking nature of the fiber architecture creates crimp or bending in the fibers that is not present in a laminate made from unidirectional material. There is also considerable pinching of the tows or yarns at the crossover points in the weave. Knitted or stitched unidirectional preforms generally behave more like unidirectional material.
- When a flat woven preform is formed to a tool with a complex compound contour, the fiber orientation will change and the original orthogonal construction will become distorted. In braided products, changes in the mandrel diameter will generally result in changes in the braid angle. For 3D woven products, during compaction, the *z*-direction reinforcement can become compressed and lose its straightness.
- In general, the more complex the preform becomes (e.g., 3D woven preforms), the more difficult it becomes to achieve as high a fiber volume in the cured composite as with unidirectional material. For example, unidirectional laminates usually contain about 60 vol.% fiber, while many 3D product forms are in the 50–55 vol.% range. This can also lead to large resin pockets that are prone to microcracking on cool-down from cure.[15]

Some of the advantages and disadvantages of the different textile processes used to make preforms are summarized in Table 3.

9.12 Integral Structures Made by Textile Processes[2]

The performance and manufacturing advantages of eliminating joints by making integral structures is an attractive approach. A weaving pattern does not need to be constant over the entire width or length of a fabric. By programming the loom, it is possible to have segments of fabric that are not locked through-the-thickness along a specified plane. After weaving, the flat fabric may be unfolded to create a branched structure. This approach

Table 1.3 Relative Advantages and Disadvantages of Various Textile Processes

textile Process	*Advantages*	*Limitations*
Low Crimp Uniweave	• High in-plane properties • Good tailorability • Highly automated preform fabrication process	• Low transverse and out-of-plane properties • Poor fabric stability • Labor intensive ply lay-up
2-D Woven Fabric	• Good in-plane properties • Good drapeablility • Highly automated preform fabrication process • Integrally woven shapes possible • Suited for large area coverage • Extensive data base	• Limited tailorability for off-axis properties • Low out-of-plane properties
3-D Woven Fabric	• Moderate in-plane and out-of-plane properties • Automated preform fabrication processes • Limited woven shapes possible	• Limited tailorability for off-axis properties • Poor drapeablility
2-D Braided Preform	• Good balance of off-axis properties • Automated preform fabrication process • Well suited for complex curved parts • Good drapeablility	• Size limitation due machine availability • Low out-of-Plane properties
3-D Braided Preform	• Good balance of in-plane and out-of-plane properties • Automated preform fabrication process • Well suited for complex shapes	• Slow preform fabrication process • Size limitation due to machine availability
Multiaxial Warp Knit	• Good tailorability for balanced in-plane properties • Highly automated preform fabrication process • Multi-layer high throughput material suited for large area coverage	• Low out-of-plane properties
Stitching	• Good in-plane properties • Highly automated process provides excellent damage tolerance and out-of-plane strength • Excellent assembly aid	• Small reduction in in-plane properties • Poor accessibility to complex curved shapes

Reference 16

can be used to fabricate crossing stiffeners with continuous reinforcements passing through the intersection.

An illustration of the use of weaving technology to produce an integral skin and stiffener assembly is shown in Fig. 21. The skin contains orthogonal warp and fill, as well as ±45° bias plies. These are all supported by through-the-thickness reinforcement in an interlock architecture. The stiffeners are formed by the bias yarns, which pass continuously from the skin up and over each stiffener in turn and back into the skin again. Therefore, the stiffeners and skin are formed in the same weaving process in an entirely integral manner. A schematic of a skin and stiffener formed integrally by stitching is shown in Fig. 22. Once again, a liquid molding process in a net shape tool can accomplish final curing.

9.13 Preform Lay-up

Since the stiffness and strength of polymeric composites are dominated by the reinforcing fibers, maintaining accurate positioning of the fibers during all steps of manufacturing process is paramount. Poor handling and processing after preforming can destroy fiber uniformity. Uncontrolled

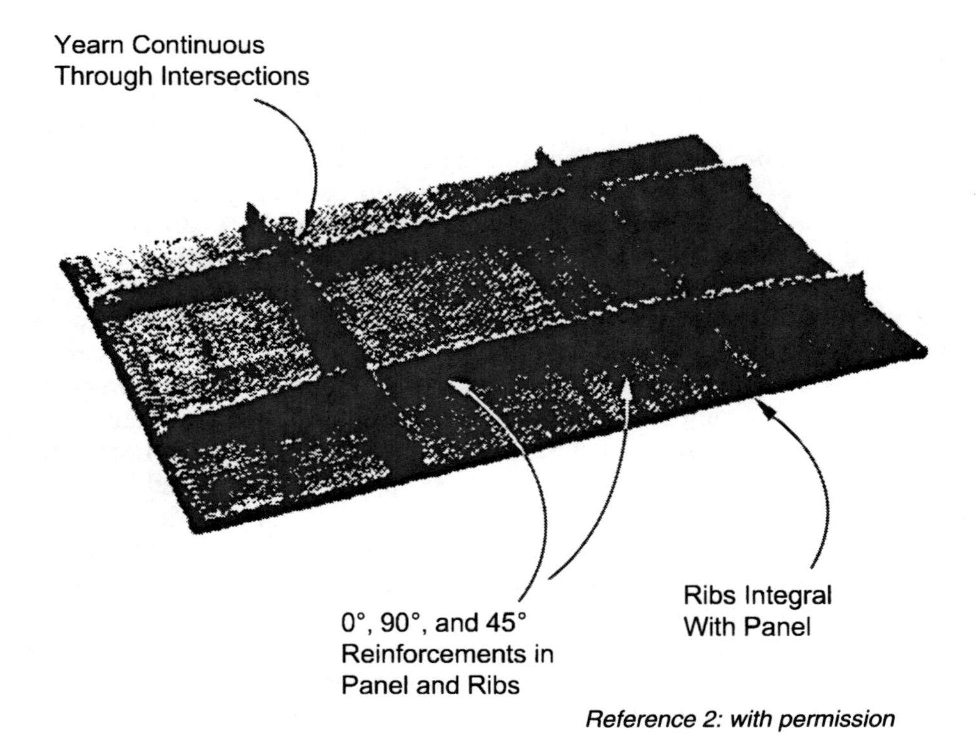

Fig. 21. Woven Preform with Integral Reinforcements

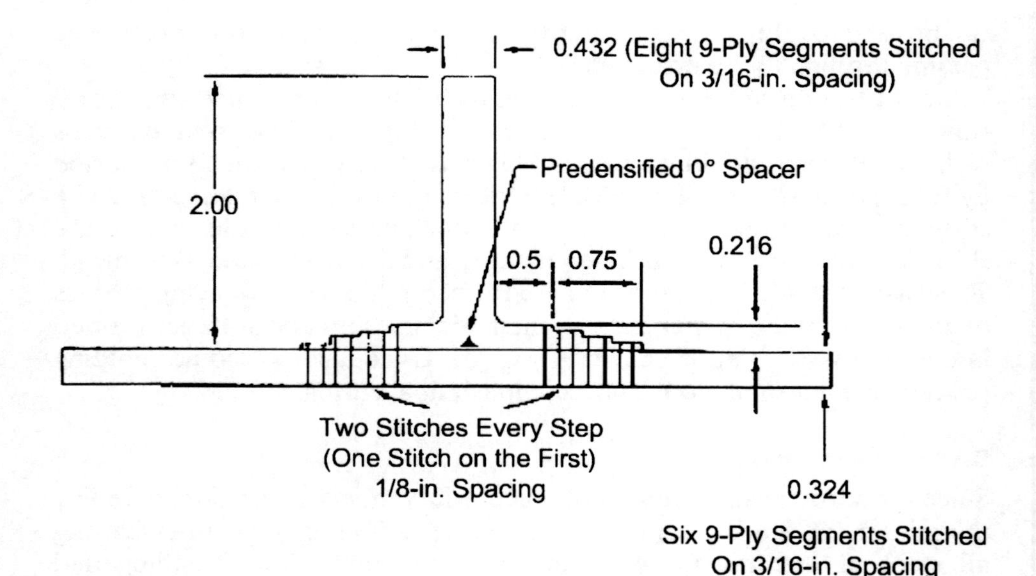

Fig. 22. Example of Attaching Stiffener to Skin using Stitching

material handling, draping the material over curved tools, debulking and tool closure can spread or distort the fibers. Manufacturing prove-out parts should be examined to establish that the minimum fiber volume fractions have been obtained, with particular attention paid to geometric details such as joints. The problem of maintaining the desired fiber content is most challenging when fabrics are draped. The draping characteristics of a fabric over a singly curved surface are a direct function of the shear flexibility of the weave. Satin weaves have fewer cross over points than plain weaves and have lower shear rigidity and are therefore more easily draped. Draping over a complex compound contour also depends on the in-plane extensibility and compressibility. This is difficult for fabrics containing high volume fractions of more or less straight in-plane fibers, as required for most structural applications. For these products, only mild double curvature can be accommodated by draping without a significant loss of fiber regularity. However, compound contours can be achieved through net shape processes, such as braiding onto a mandrel, thus avoiding the problems of draping.

There are several reasons that preforming is conducted prior to the injection process. First, preforming does not tie up the expensive matched die tool, i.e., the tool can be used to cure parts while the preforming operations are done ahead of time and offline. Second, a well-constructed preform (Fig. 23) will be rather stiff and rigid as opposed to laying up loose

Fiberglass Preform

Liquid Molded Part

Fig. 23. Example of Preform and Liquid Molded Part

fabric directly into the mold. Therefore, preforming improves the fiber alignment of the resultant part and reduces part-to-part variability.

Planar fabric preforms can be stitched together or held together with a tackifier. A tackifier is usually an uncatalyzed thermoset resin that is applied as a thin veil, a solvent spray or a powder. Veils can be placed between adjacent plies of fabric followed by fusing the ply stacks with heat and pressure to form the preform. Tackifiers can also be thinned with a solvent and then sprayed on the fabric plies. A third method is to apply powders to the surface followed by heating to melt the powder and allow it to impregnate the fabric. Tackified fabric can be thought of as a low resin content prepreg (usually in the range of 4–6%) that can be made into ply kits using conventional automated broadgoods cutting equipment. It is important to keep the tackifier content as low as possible because it reduces the permeability of the preform and makes resin filling more difficult. It is also important that the tackifier and the resin to be injected are chemically compatible, preferably the same base resin system.[17] Once the tackifier has been applied to the fabric layers, they are formed to the desired shape on a low-cost preforming tool and then heat-set by heating to approximately 200 °F for 30–60 s. The compaction behavior of a preform depends on the preform method used, the type of reinforcement, the tackifier used, the compaction pressure and the compaction temperature. The effects of compaction pressure and temperature are notionally shown in Fig. 24. A tackifier can act as a lubricant and increase compaction, but

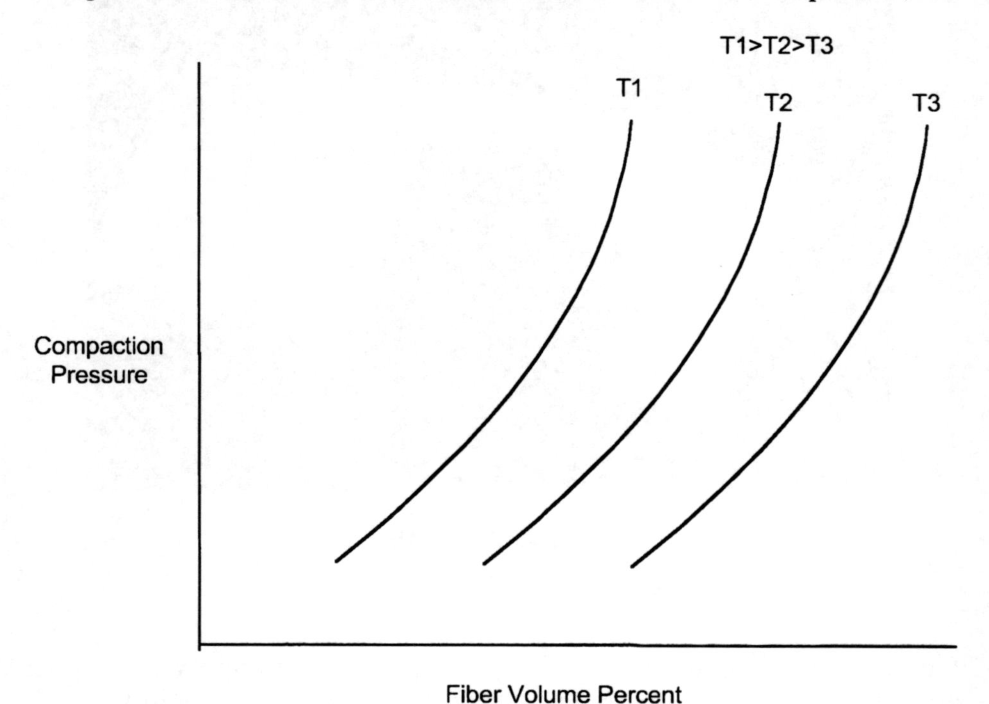

Fig. 24. influence of Compaction Pressure and Temperature on Preform Fiber Volume

this will also increase the preform permeability and make injection more difficult. For any preform construction, it is important that the preform be dried prior to resin injection to remove all surface moisture that may have condensed on the surface from the atmosphere.

9.14 Resin Injection

Resin injection follows Darcy's law of flow through a porous media that predicts that the flow rate per unit area (Q/A) is proportional to the preform permeability (k) and the pressure gradient ($\Delta P/L$), and inversely proportional to the viscosity (η) of the resin and the flow length (L):

$$\frac{Q}{A} = \frac{k\,\Delta P}{\eta L}$$

Therefore, for a short injection time (high Q/A), one would want a preform with a high permeability (k), a high pressure (ΔP), a low resin viscosity (η) and a short flow length (L). Using this equation can provide useful guidelines for RTM: (1) use resins with low viscosity; (2) use higher pressures for faster injections; and (3) use multiple injection ports and vents for faster injections.

The ideal resin for RTM will have (1) a low viscosity to allow flow through the mold and complete impregnation of the fiber preform; (2) a sufficient pot life where the viscosity is low enough to allow complete injection at reasonable pressures; (3) a low volatile content to minimize the occurrence of voids and porosity; and (4) a reasonable cure time and temperature to produce a fully cured part. A more complete list of the variables that can affect the RTM process is given in Table 4.

Resin viscosity is a major consideration when selecting a resin system for RTM. Low-viscosity resins are desirable with an ideal range being in the 100–300 centipoise (cps) range with about 500 cps being the upper limit. Although resins with a higher viscosity have been successfully injected, high injection pressures or temperature are required which results in more massive tools to prevent tool deflection. Normally, the resin is mixed and catalyzed before it is injected into the mold, or, if the resin is a solid at room temperature with a latent curing agent, it must be melted by heating. Vacuum degassing in the injection pot (Fig. 25) is a good practice to remove entrained air from mixing and low-boiling-point volatiles. Both epoxies and bismaleimides are amenable to RTM, with preformulated resins available from a large number of suppliers. Similar to prepreg resins, it is important to understand the resin viscosity and cure kinetics of any resin used for RTM.

Although resin injection pressures can range from vacuum only up to 400–500 psig applied pressure, applied pressures are normally 100 psig or lower. Although high pressures are often needed to fully impregnate the

Table 1.4 RTM Processing Variables and Their Effects

RTM Process Parameter	*Potential Effects on Processing or Structure*
Resin Viscosity	• 100-1000 cps typical flow processing range • Processing at 10-100 cps at high temperatures also typical • Higher viscosity- preform fails to wet out • Lower viscosity- rapid infusion may leave dry areas and viods
Resin Pot Life	• Too short- resin fails to fill preform • Too long- process cycle lengthened unnecessarily
Resin Injection Pressure	• Helps drive resin into mold ans preform • Too fast or too high- may move preform within mold • Too high- may damage mold or tooling • Too high- may 'blow' out seals and cause leakage • Too low- cycle times very long • Too low- resin may gel during fill period
Resin Injection Vacuum Level	• 10-28 inch Hg is typical processing range • Helps pull resin into mold and preform • Aids in reducing void content • Assists in holding mold halves together • Aids in removing moisture and volatiles
Multiple Injection Ports	• Commonly used to assure complete wet out • Sometimes used sequentially to fill very long parts
Internal Rubber/Elastomeric Tooling	• Rubber inserts used to provide very high compaction • High fiber volumes (> 65%) achievable • Very low void contents typical • Tooling must be robust to wthstand high pressures
Closed Mold Pressurization	• Pressure increased to 100-200 psig after resin wet out • Decreases microvoids by collapse of bubble cavities
Fiber Sizing or Coupling Agents	• Sizing chemistry must be compatible with resin selection • Sizing level reduces resin flow (lower permeability)
Fiber Volume	• Resin flow permeability inversely proportional to volume • High fiber volumes (> 60%) require more work to wet out • Commercial market- usually 25-55% by volume • Aerospace market- usually 50-70% by volume
Mold-In Inserts and Fittings	• Very possible with RTM process • Resin flow around fittings can leave dry areas and voids

Reference 1

preform, the higher the injection pressure, the greater the chance of preform migration, i.e., the pressure front can actually cause the dry preform to migrate and move out of its desired location. Injection molding dies are normally either designed so that they are stiff enough to react the

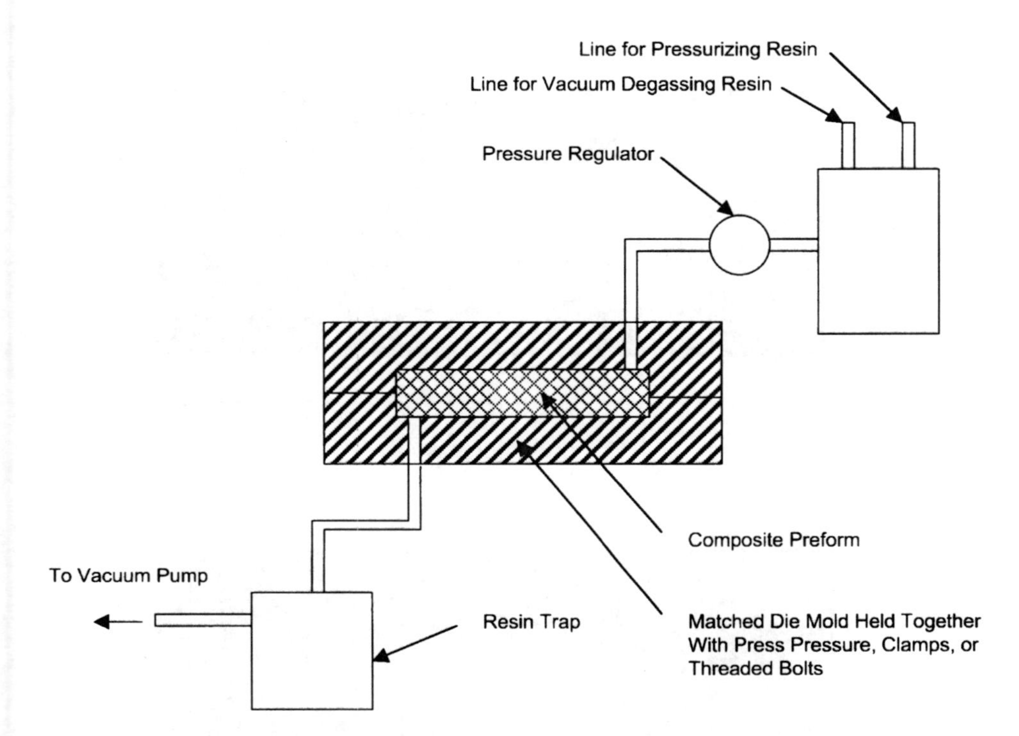

Fig. 25. Schematic of a Typical RTM Process

injection pressures or they may be placed in a platen press under pressure to react the injection pressure. As a rule of thumb, the higher the injection pressure, the higher the tooling cost. Heating the resin or tool prior to or during injection can be used to reduce the viscosity but will also reduce the working or pot life of the resin. It is normal practice to vacuum-degas the resin prior to injection to remove as many volatiles as possible and thereby reduce the chance of voids and porosity in the cured part. A vacuum is also frequently used during the injection process to remove entrapped air from the preform and mold. The vacuum pressure also helps to pull resin into the mold and preform, helps to remove moisture and volatiles and aids in reducing voids and porosity. It has been reported that the use of a vacuum is a significant variable in improving product quality by reducing the occurrence of voids and porosity.[19]

The time it takes for the resin to fill the mold is a function of the resin viscosity, the permeability of the fiber preform, the injection pressure, the number and location of the injection ports and the size of the part. The injection strategy usually consists of one of three main types shown in Fig. 26: (1) point injection, (2) edge injection or (3) peripheral injection. Point injection is usually done by injecting at the center of the part and

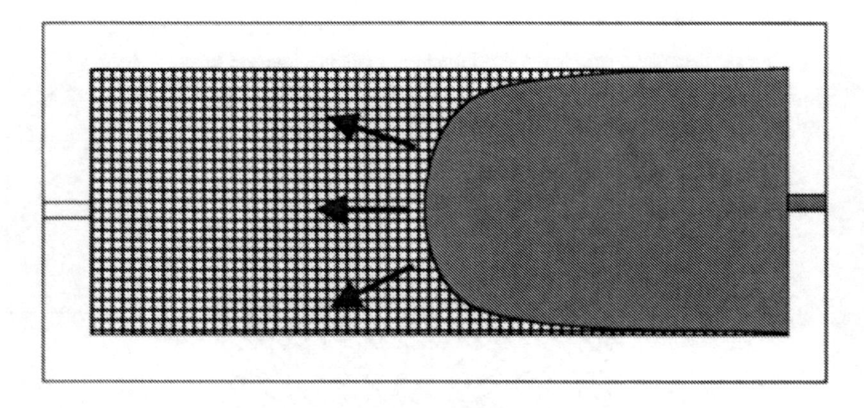

Edge Injection

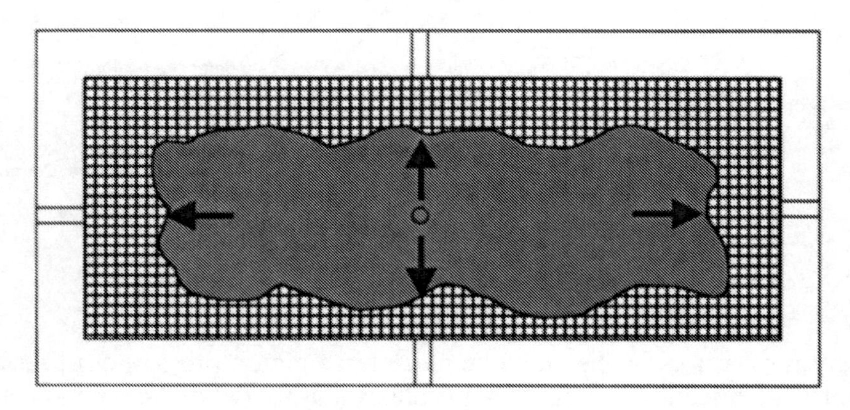

Point Injection

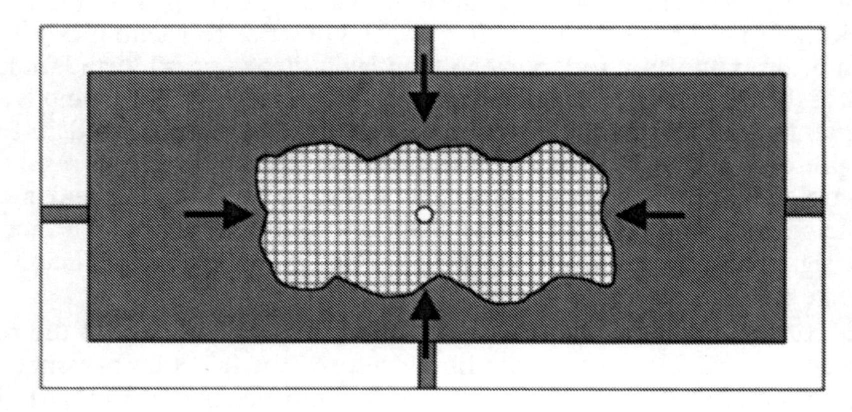

Peripheral Injection

Fig. 26. Resin Injection Strategies

allowing the resin to flow radially into the reinforcement as air is vented along the part periphery. Edge injection consists of injecting the resin at one end of the part and allowing the resin to flow unidirectionally down the length as air is vented at the opposite end. Finally, in peripheral injection, the resin is injected into a channel around the part and the flow is radially inward as air is vented at the center of the part. Peripheral injection is usually the fastest of the three, although it is not uncommon to use more than one of these methods on the same part. Also, the locations of the injection and venting ports are important considerations in the ability to effectively achieve complete filling without entrapped air pockets or unimpregnated dry spots. Although there are several ways that the time to fill the mold can be reduced, such as using lower viscosity resins or higher injection pressures, the most effective method is to design an injection and porting system that minimizes the distance the resin has to flow. However, in designing an injection and porting system, the most important consideration is to have a system that will minimize any entrapped air pockets as these will result in dry unimpregnated areas in the cured part. In peripheral injection, a phenomenon known as "race tracking" can occur in which the resin runs around the peripheral injection channel and then migrates inward but traps air pockets resulting in dry spots. This can usually be avoided by the judicious selection of the location and number of the porting vents.

Vacuum assistance during injection will usually help to reduce the void content significantly. However, it is important that the mold be vacuum tight (sealed) if vacuum assistance is going to be used. If the mold leaks, air will actually be sucked into the mold causing a potentially higher void content. Besides entrapped air in the preform and mold, entrained air, moisture and volatiles within the resin itself can also be a source of voids and porosity. The normal practice is to thoroughly vacuum-degas the mixed resin at room or slightly elevated temperature to remove these gases prior to resin injection. During the injection process, when the mold is almost full, resin will start flowing out through the porting system. If there is evidence of bubbles in the exiting resin, allow the resin to continue to bleed out until the bubbles disappear. To further reduce the possibility of voids and porosity, once the injection is complete, the ports can be sealed while the pumping system is allowed to build up hydrostatic resin pressure within the mold.

9.15 Curing

Curing can be accomplished using several methods:

- Matched-die molds with integral heaters – electric, hot water or hot oil.
- Matched-die molds placed in an oven.

- Matched-die molds placed between a heated platen press that provides the heat and reaction pressure on the mold.
- For liquid molding processes that use vacuum injection only, such as VARTM and SCRIMP, a single-sided tool with only a vacuum bag is used for pressure application. In this case, heat can be provided by integral heaters, ovens or even heat lamps.

As opposed to autoclave curing, where the operator can control the variables time, temperature, and pressure (t, T, P), in RTM the P variable is often predetermined by the pressure applied to the resin during the injection process, or in VARTM it is limited to the pressure that can be developed by a vacuum (≤14.7 psia). In some match mold applications the vent ports can be sealed off and pressure can continue to be applied by the pump. To improve productivity, RTM parts are frequently cured in their molds, demolded, and then given free-standing post-cures in ovens. Several examples of completed RTM composite parts are shown in Fig. 27.

9.16 RTM Tooling

Tooling is probably the single most important variable in the RTM process. A properly designed and built mold will normally yield a good part, while a poorly designed or fabricated mold will almost certainly produce a deficient part. There are several characteristics that must be considered when designing a tool for RTM:

- The tool must be stiff enough not to deflect when the preform is initially loaded into the tool and it must be able to react with the resin

GKN Aerospace Services

Fig. 27. Examples of RTM Carbon Composite Parts

injection pressures during molding. If the mold is going to be used free standing (i.e., not in a platen press), it may be necessary to incorporate external stiffening structure to prevent the mold from deflecting during the injection process.

- It is extremely important that the tool is capable of being vacuum-sealed to prevent air from being sucked into the tool when vacuum assistance is used during injection. Although a number of different approaches are used, they usually involve some type of rubber O-ring that fits within a groove placed around the tool periphery. If possible, it is best to keep the seal on a single plane. It is more difficult to effectively seal a tool along a curved surface.
- The mold clamping system must be sufficient to hold the mold together during preform compression and resin injection. Large threaded bolts are frequently used although hydraulic systems have been used for high-volume applications. Note that injection pressures can exert tremendous separation forces on a tool. For example, an injection pressure of 60 psig on a mold with an area of 20 square ft will exert a force of 80 tons on the mold.
- The mold must either be integrally heated or placed in an oven or press for curing.
- The mold must have an injection and porting system that will allow complete preform impregnation and mold filling during injection. The location and number of injection and venting ports is not a science. They are frequently determined by experience and trial and error.

Conventional RTM tooling consists of matched molds usually machined from tool steel. Steel dies yield long lives for large production runs and are resistant to handling damage. The dies are usually blended and buffed to a fine surface finish that will yield good surface finishes on the RTM part. Many matched metal molds are built with sufficient rigidity that they do not need to be placed in a platen press during injection and cured to react with the resin injection pressures. Since these molds necessarily become extremely heavy, attachment fittings are built into the mold to provide hoisting capability for cranes. They are held together with a system of heavy bolts and are often designed with internal ports for heating with hot water or oil. Hot-water heaters are effective to about 280 °F. Above that temperature, hot oil must be used. Electric heaters can also be placed within the mold, but are generally less reliable than hot oil because of the maintenance problems of replacing burned-out heaters. RTM molds can also be placed in convectively heated ovens but for large tools, the heat-up rates will be extremely slow.

Steel matched metal molds have two disadvantages: (1) they are expensive and (2) the heat-up and cool-down rates are slow. Matched metal molds have also been fabricated from Invar 42 to match the coefficient of thermal expansion of carbon composites, and from aluminum because it is easier to machine (less costly), has a high coefficient of thermal expansion which can be useful in some applications, but is much more prone to wear and damage than steel or Invar. For prototype and short production runs, matched molds can be made of high-temperature resins that are frequently reinforced with glass or carbon fibers. Prototype dies can be NC-machined directly from mass-casted blocks, laminated on a master model or laminated on a master model and finished by NC-machining the surface.

Much lighter-weight and less expensive tooling is a distinct advantage of processes that use only vacuum pressure for injection and cure, such as the VARTM and SCRIMP processes. In fact, most of these processes use single-sided hard tooling on the one side and a vacuum bag on the other side. A porous media is almost always used on top of the fiber preform to aid in resin filling during injection.

9.17 Resin Transfer Molding Effects

If the radii in a part are designed too tight, it is not uncommon that some bridging of the preform in the radius can occur, often resulting in a resin-rich area on the outside of the radius as illustrated in Fig. 28. Since the resin is not reinforced with fiber, it will often craze or break off. If the bridging condition at the radius is extremely severe, there can also be a problem

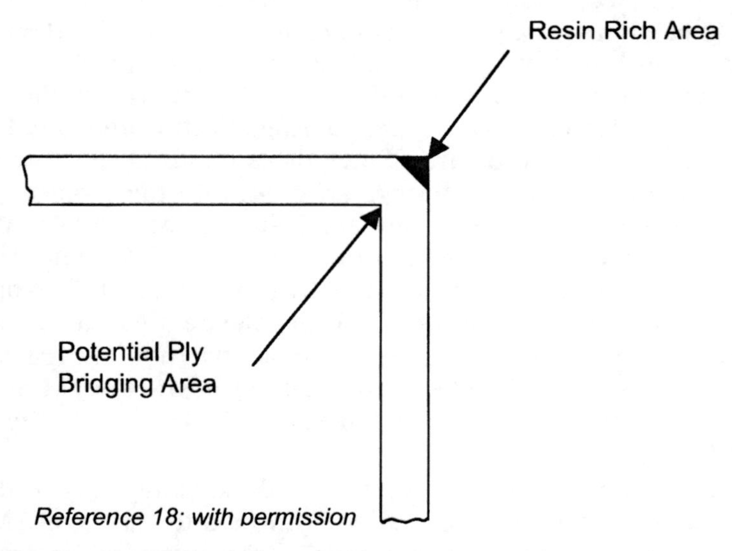

Fig. 28. Resin Rich Area Formed Between Preform and Tool

with inadequate ply compaction and delaminated plies may result. As a general rule of thumb, the radius should be at least 3 times larger than the part thickness. Other potential problems with too small a radii[18] are: (1) they can restrict resin impregnation; (2) they make it more difficult to fit the preform in the tool; (3) fiber damage can occur when the mold is closed; and (4) if the mold is made of a soft material, such a composite laminate, the tool itself can be damaged.

During injection the resin is very fluid and will follow the path of least resistance. If there are gaps between the preform and the tool or locations in the preform where the permeability is higher than average, the resin can "race" ahead of the flow front and isolate areas in perform where air pressure cannot escape and dry or unimpregnated spots result. This "race tracking" phenomenon is shown schematically in Fig. 29. Proper design of

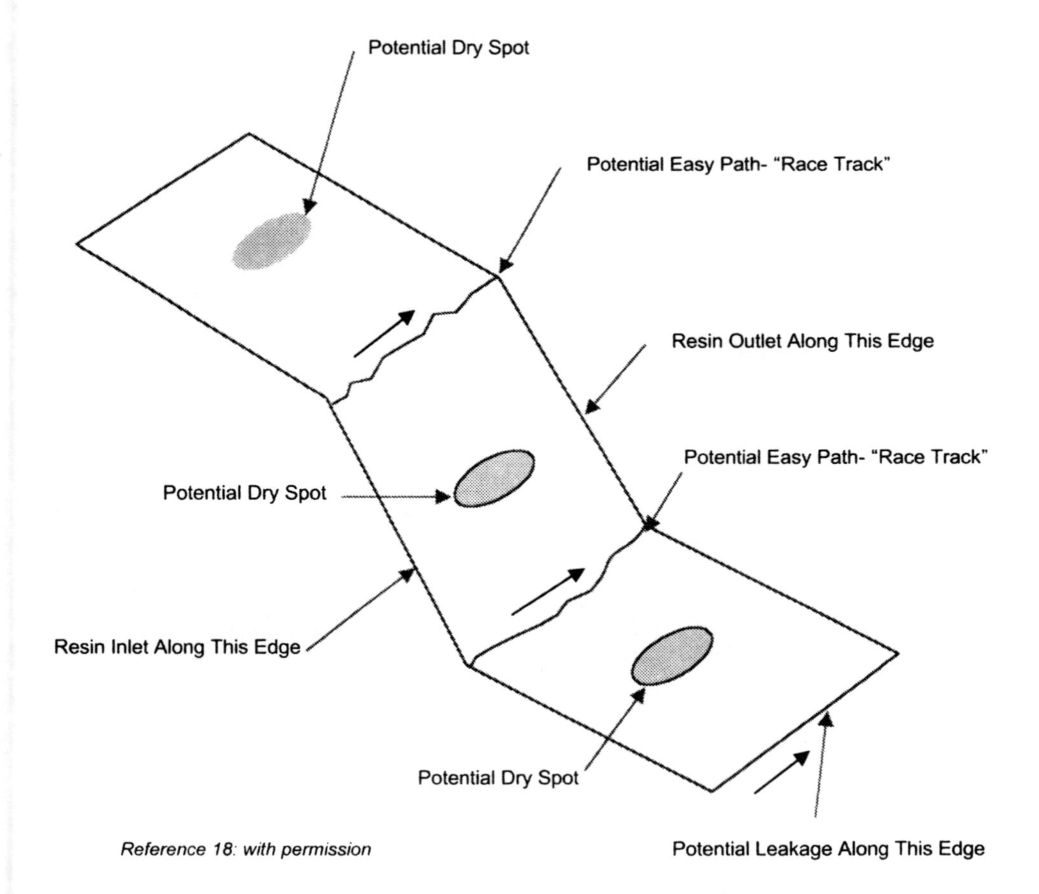

Fig. 29. Resin "Race Tracking" Phenomena During Injection

the inlet and venting systems is critical in preventing "race tracking." It is a more prevalent problem with peripheral injection systems than point or edge injection designs. A somewhat similar phenomenon can also occur in sandwich construction (Fig. 30) in which the resin flows faster along one surface than the other, resulting in non-uniform skin thicknesses. The core is actually pushed toward the thinner skin by the advancing flow front. To circumvent some of the infusion problems with foam cores, core manufacturers can supply core with scoured intersecting lines on the surface to provide infusion paths and even holes through the foam that allows the resin to flow through the core and equalize the pressure on both sides. This approach is frequently used with VARTM to reduce infusion time and improve part quality.

If the injection pressure is too high, the advancing flow front can actually displace portions of the preform resulting in "fiber wash." This is a more prevalent problem in parts with lower fiber volume percents. With higher fiber volume loadings, there are usually higher compaction pressures from

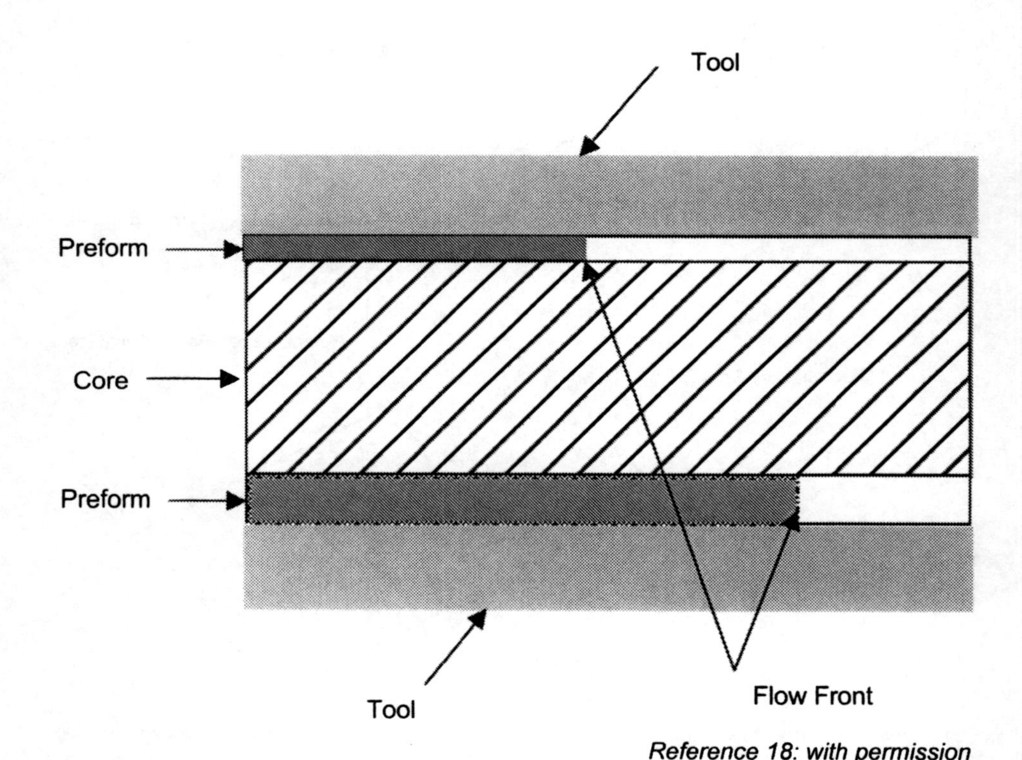

Fig. 30. Nonuniform Skin Thicknesses in Sandwich Panel

the tool that react the forces causing fiber wash. However, higher compaction pressures also lower the permeability of the preform and make injection more difficult.

Voids in resin-transfer-molded parts are usually not as big a problem as in conventional autoclave cured parts. However, voids can occur if vacuum assistance is used to remove air from the mold during the injection process. This is usually a result of a leak in the mold sealing system that allows the vacuum to actually pull additional air into the mold. There are several steps that can be taken to reduce voids: (1) make sure the mold is vacuum tight, particularly if vacuum is going to be used during the injection process; (2) vacuum-degas the resin in the pot to remove volatiles and entrained air just prior to injection; (3) let resin exiting the vent tubes flow until all air bubbles disappear; and (4) continue to apply pressure after the mold is sealed off to insure that there is adequate hydrostatic pressure on the resin during the cure cycle.

All liquid molding processes, including RTM, can benefit from mathematical modeling of both the preforming and the injection process. An example of a potential sequence is shown in Fig. 31. Using the product geometry, there are finite element models that can be used to predict both fiber movement and distortions during the performing process and the flow of resin into the tool and preform during injection. The proper use of these advanced modeling capabilities can significantly reduce the time and cost to develop a new part configuration. The proper application of these models can also be a significant aid in developing injection and venting strategies for new tool configurations. A review of the current state of liquid molding simulation technology can be found in Ref. 20.

9.18 Resin Film Infusion

RFI is a process developed by NASA and the Long Beach division of McDonnell-Douglas (now Boeing). There were two drives for the development of this process: (1) the desire to have 3D reinforcement for damage-tolerant commercial wing design and (2) the desire to use a qualified prepreg resin system for the matrix resin. The problem with using prepreg resin systems for conventional RTM is that the minimum viscosities (>500 cps) are too high to be able to successfully inject and fill the stitched preform during injection. The process they developed, shown schematically in Fig. 32, consists ofplacing a controlled layer of matrix resin (Hexcel's 3501-6), which is a solid at room temperature, in the bottom of the mold. A stitched preform is then loaded into the mold on top of the resin layer. During autoclave cure, the resin melts and vacuum and autoclave pressure is used to draw the liquid resin up through the tool to impregnate the preform. Once the infiltration cycle is complete, the temperature is raised to the cure temperature and the part is cured using

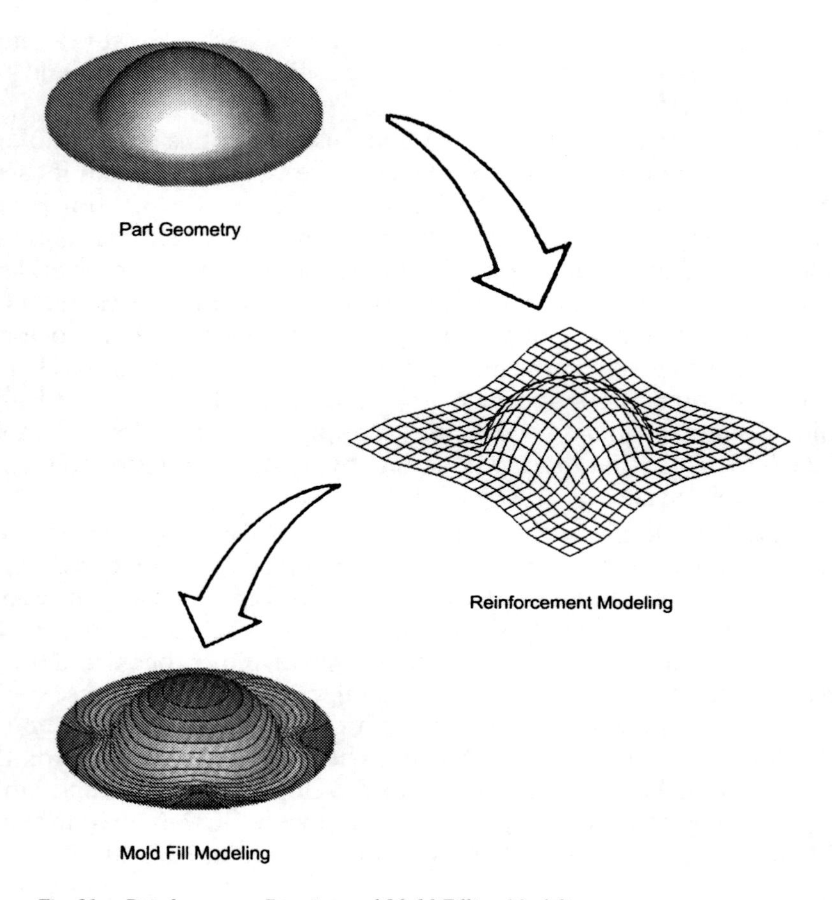

Fig. 31. Reinforcement Draping and Mold Filling Modeling

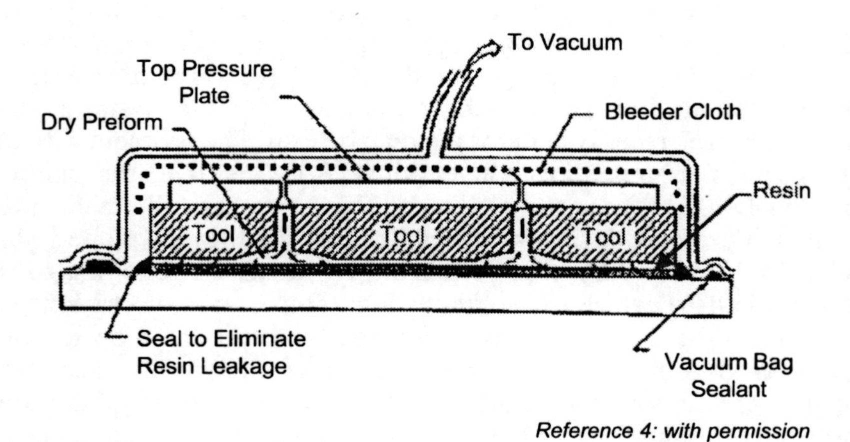

Reference 4: with permission

Fig. 32. Schematic of Resin Film Infusion (RFI) Process

autoclave heat and pressure. The keys to the RFI process are understanding the compaction and permeability of the preform and the viscosity and kinetics of the resin system. For example, to infuse a large preform may require a resin with a viscosity of lower than 250 cps at 250 °F for 1–4 h. In addition, preform design and placement within the tool, and tool design and dimensional control are critical for this process. A special cure cycle must be developed to achieve the correct time–temperature–viscosity profile to insure complete preform saturation. A variant of this process is resin liquid infusion (RLI), in which a liquid resin instead of a solid resin is placed or injected into the bottom of the tool prior to loading the preform.

Boeing-Long Beach, under the NASA Composite Wing Program,[21] successfully designed, built, assembled, and tested a 42 ft long by 8 ft wide composite wing using the RFI process. For this program, they used the matrix resin 3501-6 with stitched performs supplied by the Saertex company in Germany. A completed stitched, infused and cured wing cover is shown in Fig. 33.

Reference 21: with permission

Fig. 33. Integral Wing Cover Panel Fabricated by RFI Process

The Saertex material can typically be supplied as either seven or nine layer thick material with typical orientations of 0°, ±45° and 90° orientations. This material is made on a Libra warp knitting machine (Fig. 34) that uses a 72-denier polyester thread to hold the stack-ups together. One of the main advantages of this process is that the individual layers remain straight without the crimp associated with normal weaving practice.[22]

A schematic of the stitching process is shown in Figs. 35 and 36. The skin, stringers and intercostals were stitched together individually and then the skin, stringers and intercostal clips were stitched together. A typical stitching pattern is shown in Fig. 37, where the Saertex stack-ups were stitched together using 1,600-denier Kevlar 29 thread. A modified lock

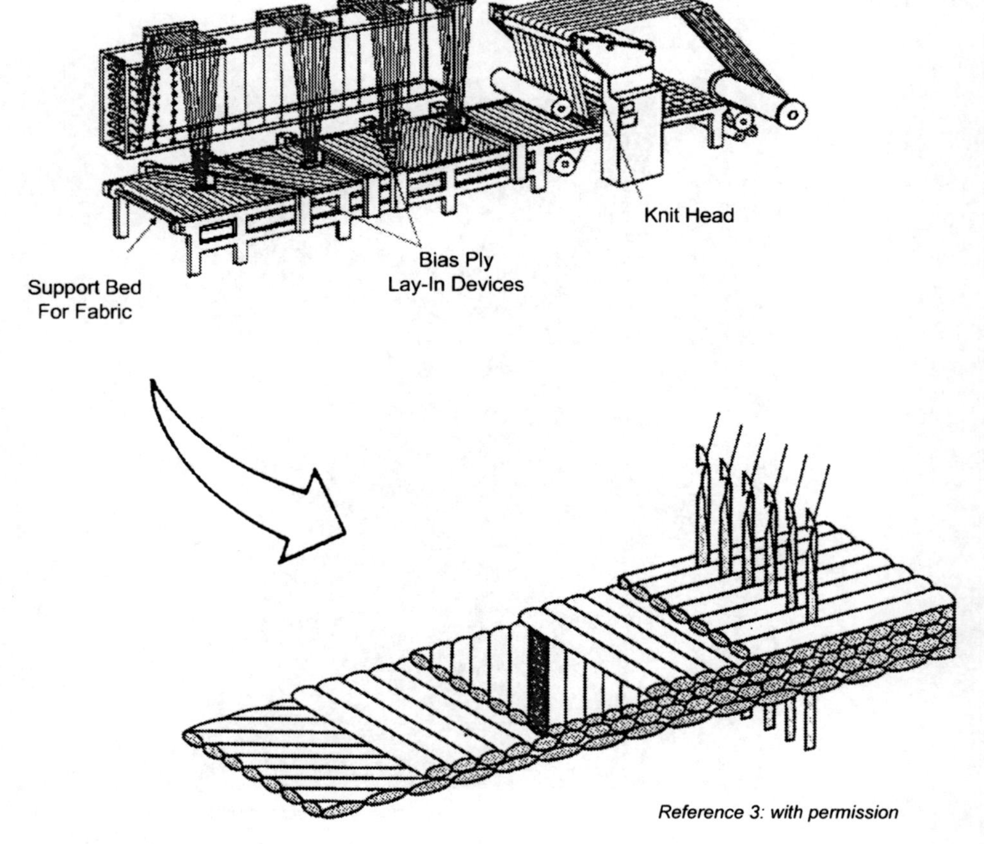

Fig. 34. Libra Warp Knitting machine and Typical Product Form Produced

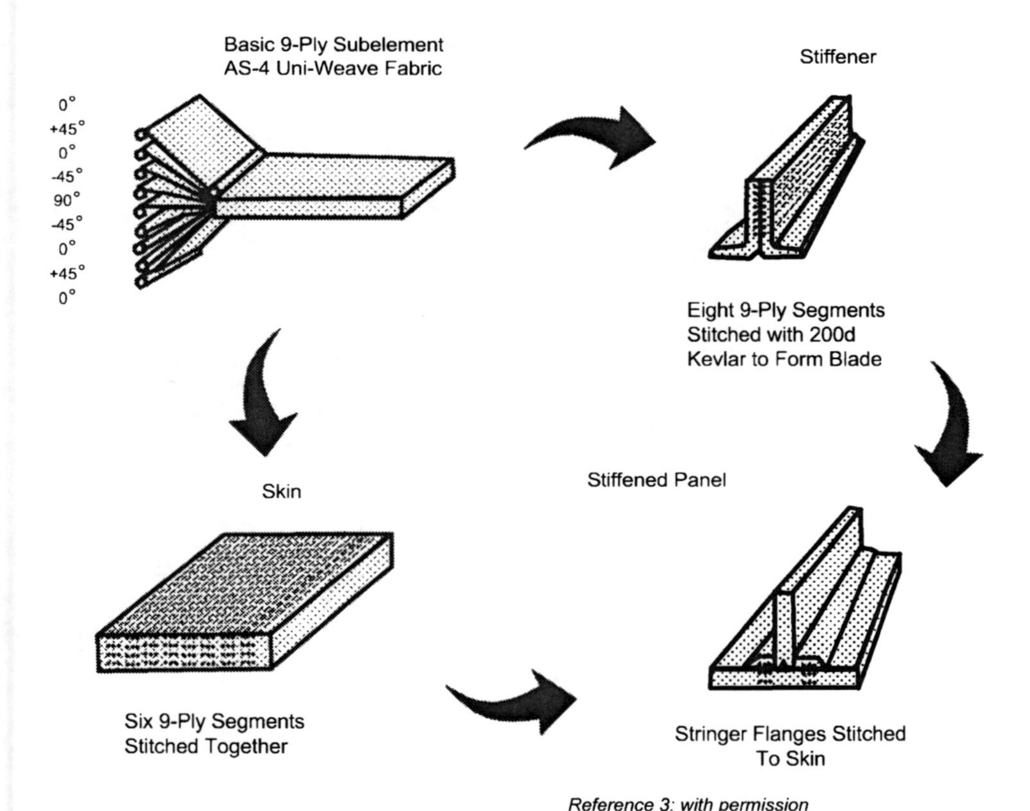

Fig. 35. Process Flow for Stitching RFI Wing Cover

stitch was used to stitch the stack-ups on a 0.20 inch wide spacing with a pitch of 8 stitches per inch. The large wing covers normally contained 5–11 stacks of the Seartex material with buildup areas as thick as 17 stacks (0.94 inches). The large stitching machine used to fabricate the wing covers is shown in Figs. 38 and 39. Details of stitching an intercostal substructure member are shown in Fig. 40.

To extend the time at low viscosity, a reduced catalyst version of 3501-6 was used that maintained a low viscosity (100–300 cps) for up to 120 min to allow time for the resin to flow through up the tool and impregnate the preform. They also had the resin vacuum degassed to reduce voids and surface porosity. A typical cure cycle is shown in Fig. 41. It is important to allow sufficient time during the initial heat-up to 250 °F and hold at 250 °F for the resin to melt and thoroughly infuse the preform.

A variant of the RFI process involves using thin layers of resin film between the dry preform layers instead of putting a large mass of resin in

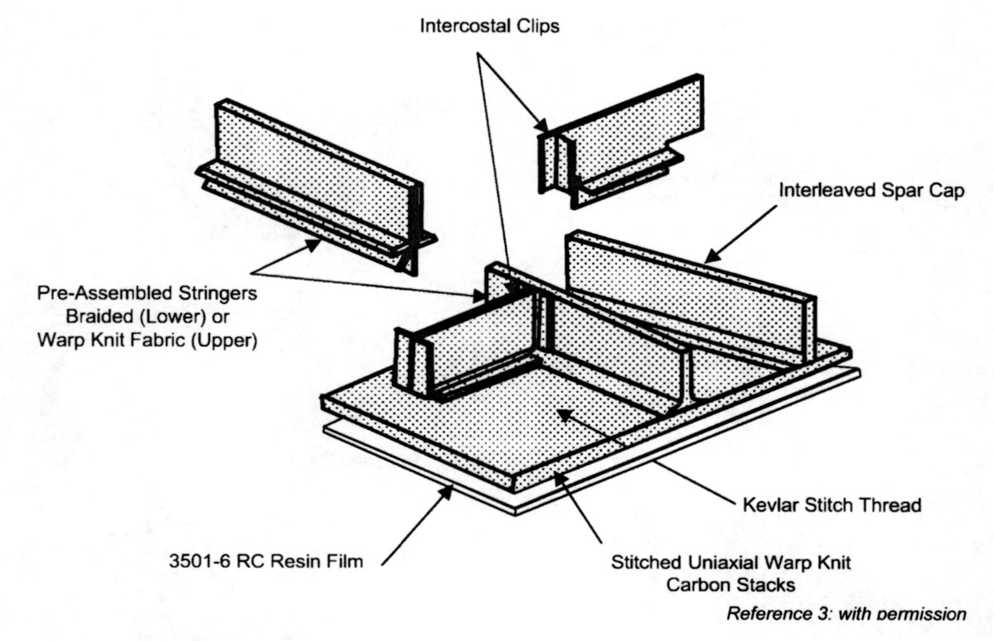

Fig. 36. Basic Construction of Integral RFI Wing Cover

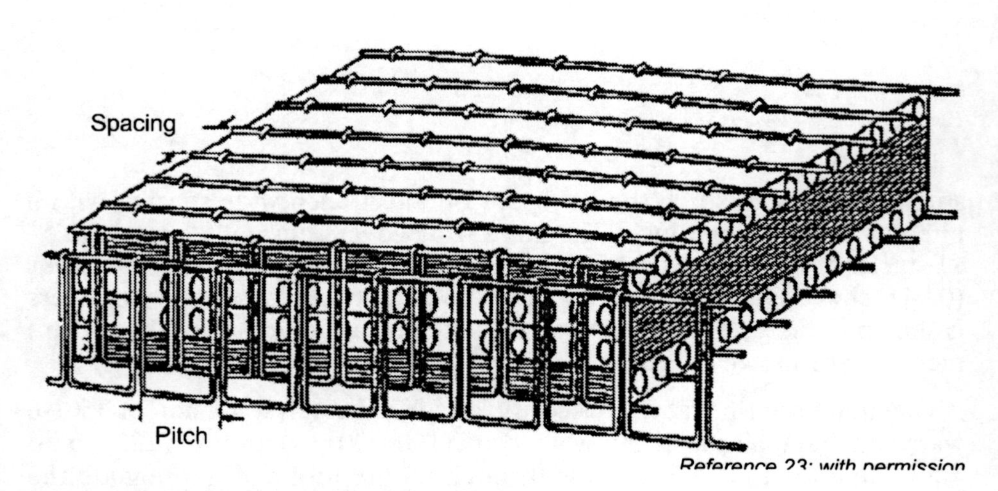

Fig. 37. Schematic of Stitcing Pattern for Stack-Ups

the bottom of the tool. The advantage of this process, sometimes referred to as SPRINT, is that the resin does not have to flow very far to impregnate each layer and vacuum bag pressure is often sufficient for full impregnation.

Reference 21: with permission

Fig. 38. Front View of Large Multi-Axis Stitching Machine

Reference 21: with permission

Fig. 39. Large Multi-Axis Stitching Machine for Wing Covers

Reference 21: with permission

Fig. 40. Stitching of Intecostal Substructure to Wing Cover

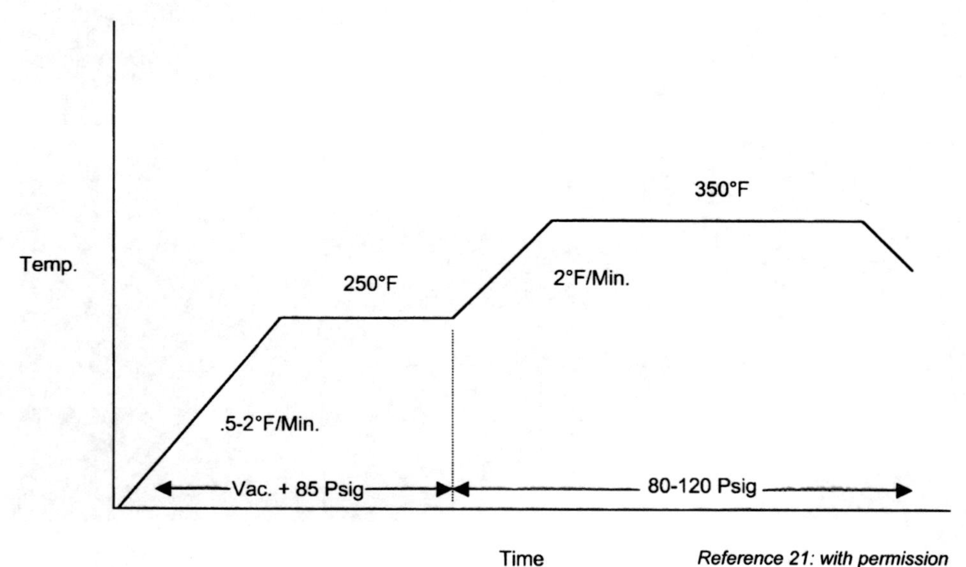

Reference 21: with permission

Fig. 41. Typical Autoclave Cure Cycle used for RFI Process

9.19 Vacuum-assisted Resin Transfer Molding

Since VARTM processes use only vacuum pressure for both injection and cure, the single biggest advantage of VARTM is that the tooling is much less expensive and simpler to design than for conventional RTM processes. In addition, since an autoclave is not required for curing, the potential exists to make very large structures using the VARTM process. Also, since much lower pressures are used in VARTM processes, lightweight foam cores can easily be incorporated into the lay-ups. VARTM-type processes have been used for many years to build fiberglass boat hulls, but have only recently attracted the attention of the aerospace industry.

A typical VARTM process, shown in Fig. 42, consists of single-sided tooling with a vacuum bag. VARTM processes normally use some type of porous media on top of the preform to facilitate resin flow across the preform and then saturation through the preform. This porous distribution media should be a highly permeable material that allows resin to flow through the material with ease. When a porous distribution media is used, the resin typically flows through the distribution media and then migrates down into the preform. Typical distribution media include nylon screens and knitted polypropylene. Since resin infiltration is in the through-the-thickness direction, race tracking and resin leakage around the preform are largely eliminated.[24]

One of the distinct advantages of VARTM-type processes is the ability to build very large parts (Fig. 43) on relatively inexpensive tooling. This

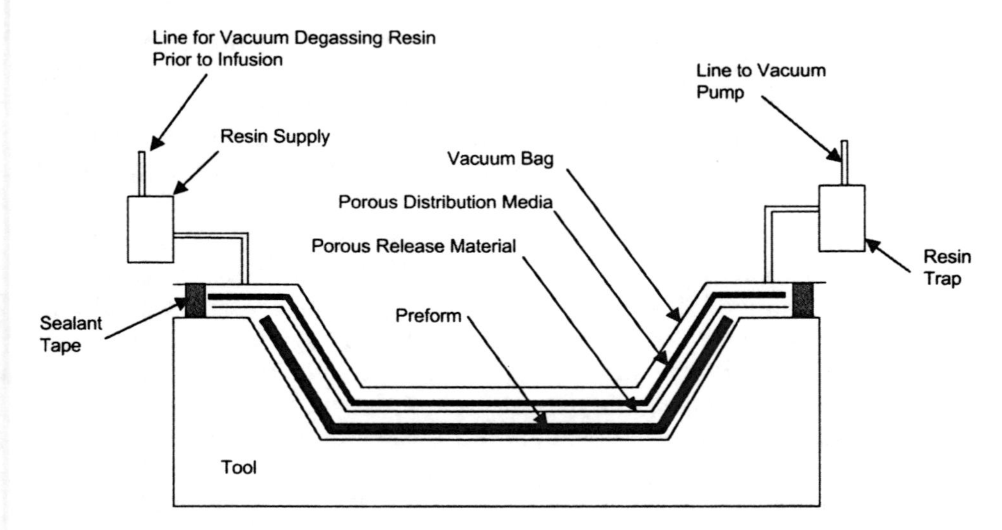

Fig. 42. Typical VARTM Process Setup

Fig. 43. Large Part Being Infused Using SCRIMP

part was fabricated using the patented SCRIMP process that utilizes a special resin distribution media to allow short infusion times over large areas. Note the locations where the resin has wetted the preform indicating the use of multiple injection ports. Another example is the complex helicopter fuselage (Fig. 44) in which the frames and bulkheads were fabricated using RTM and then co-bonded to the skin during the SCRIMP process.

A rather innovative method of infusing the resin quickly without the need for a porous media is given in Ref. 25. This method, called FAST Remotely Actuated Channeling (FASTRAC), is shown schematically in Fig. 45. A vacuum is drawn between the FASTRAC outer bag, which contains ridges, and the inner bag, which is drawn up into the ridges allowing channels for the resin to be infused. Once infusion is complete, the outer FASTRAC bag is removed, and the vacuum on the inner bag then provides the compaction pressure during cure. Since the FASTRAC outer bag never contacts the resin directly, it can be reused multiple times. Extremely fast infusion times have been reported with this process.

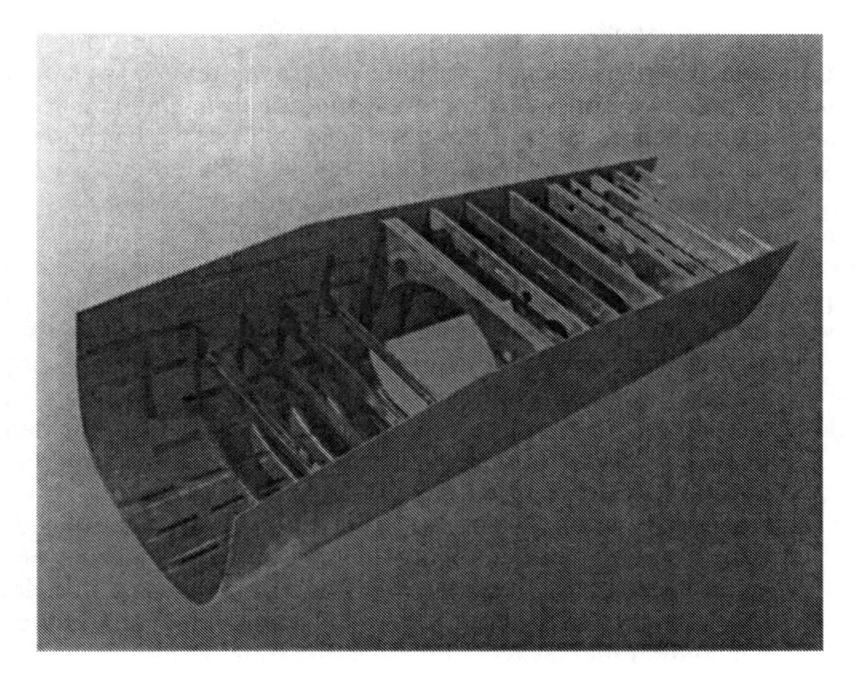

Fig. 44. Complex Helicopter Fuselage Made by Liquid Molding

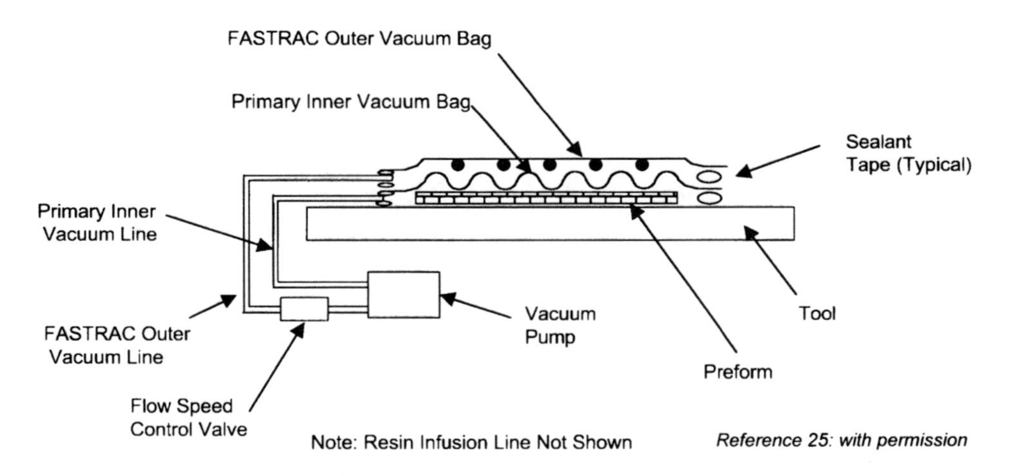

Fig. 45. Schematic Showing Principle of FASTRAC Process

Since the VARTM process uses only vacuum pressure for both injection and cure, autoclaves are not required and very large part sizes can be made. Ovens and integrally heated tools are normally used, and since the pressures are low (i.e., ≤14.7 psia), low-cost lightweight tools can be used. Some manufacturers use double vacuum bags to minimize variations in compaction pressure and guard against potential vacuum leaks in the primary vacuum bag. A layer of breather between the two bags increases the ability to remove any air from leak locations. Reusable vacuum bags can also be used to reduce the cost of bagging complex shapes.

The resins used for VARTM processing should have even a lower viscosity than those used for traditional RTM. Resin viscosities less than 100 cps are desirable to give the flow needed to impregnate the preform at only vacuum pressure. Vacuum degassing prior to infusion is often used to help remove entrained air from the mixing operation. Some resins may be infused at room temperature, whereas others require heating. It is desirable to keep the resin source and vacuum trap away from the heated tool. This makes it easier to control the temperature of the resin at the source and minimizes the chance of an exotherm at the trap.

For large part sizes, multiple injection and venting ports are utilized. As a rule of thumb, resin feed lines and vacuum sources should be placed about 18 inches apart. As the resin moves away from its source, its velocity decreases in accordance with Darcy's law and the final thickness decreases and the resin content increases as shown by the trend lines in Fig. 46. It is

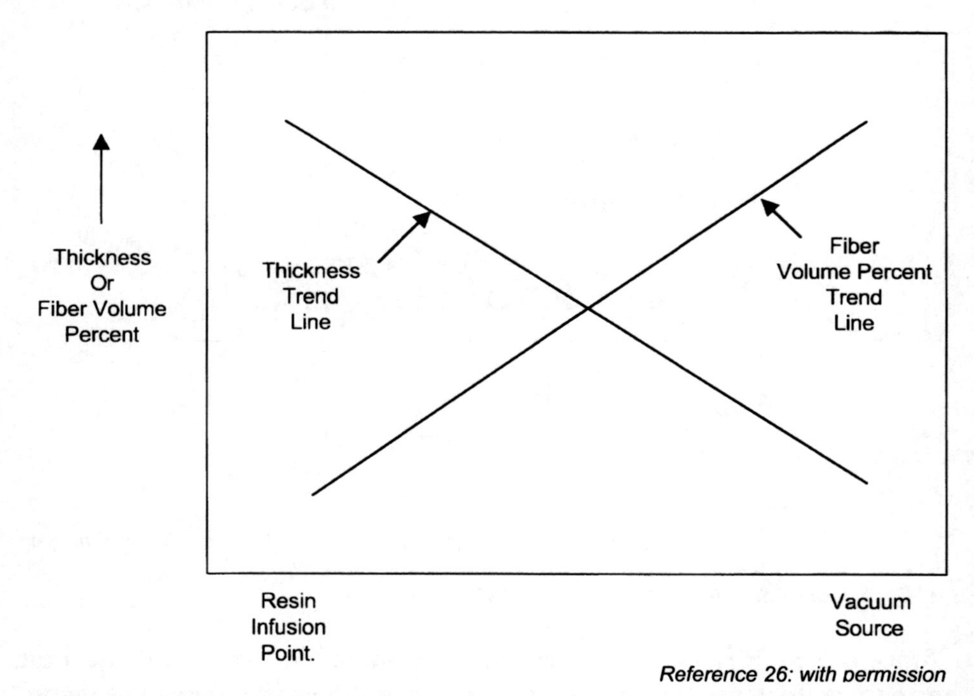

Fig. 46. Cured Part Thickness and Fiber Volume Fraction Changes vs. Distance Relative to Vacuum Source

also more difficult to obtain high fiber volume contents in thick preforms. Since perfect fiber bundle nesting does not occur, there is an increase in free volume with every additional layer, which results in lower fiber volume contents in thick parts.

Since the pressure is much lower than that normally used in the conventional RTM or autoclave processes, it is more difficult to obtain as high a fiber volume percent as with the higher-pressure processes; however, this process disadvantage of this process is being overcome with near net preforms. In addition, the VARTM processes cannot hold as tight dimensional tolerances as conventional RTM and the bag side surface finish will not be as good as a hard tooled surface. Thickness control is generally a function of the perform lay-up, the number of plies, the fiber volume percent and the amount of vacuum applied during the process. During the NASA Advanced Composite Wing program,[21] they evaluated VARTM in addition to RFI. Their initial wing panel structures contained low fiber volume percents (≤54), so they VARTM infused a wing panel and then placed it in an autoclave for a pressurized cure. The resulting panel (Fig. 47) contained an acceptable fiber volume percent (57+). They called this process VARTM-PB (pressure bleed).

Reference 21: with permission

Fig. 47. Integral Wing Cover Panel Fabricated by VARTM-PB Process

9.20 Summary

RTM is an excellent process for making highly dimensionally controlled parts that have complex geometries that would be difficult to hand lay-up on a consistent basis. However, the major disadvantage is the cost of the tooling. A well-designed and fabricated tool is a prerequisite for obtaining a good part. When combined with 3D textile geometries, liquid molding processes offer the potential to introduce composites into new areas that were previously restricted to metals, such as fittings and bulkheads that have multi-directional complex load paths not amendable to 2D reinforced designs.

Preforms can be fabricated from a number of processes including weaving, knitting, stitching and braiding. Woven materials can be supplied either as conventional 2D woven fabrics or as more complex 3D constructions. Knitting for structural composites is the MWK material in which a number of unidirectional plies are knitted or stitch-bonded together with minimal 3D reinforcement to provide handling capabilities. MWK materials, along with 2D woven or braided constructions, are often given true 3D strength improvements by stitching operations. Braided 2D and 3D reinforced parts are ideal for bodies of revolution or they can be cut off the mandrel and formed into substructure parts. For less structurally demanding parts, the P4A and random mat materials are attractive methods for preforming.

The main advantages of preforms are the reduced labor required for manual ply-by-ply collation of complex shapes and the potential for 3D reinforcement. These advantages must be weighed against the lower in-plane mechanical properties compared to unidirectional tapes caused by the somewhat lower fiber volume contents for preforms and the damage done to the fibers by mechanical twisting and abrasion.

RTM is the most mature of the high-performance composite liquid molding processes. In RTM a preform is placed in a rigid matched-die mold and injected with a liquid resin under pressure. The part is then cured in the mold. While RTM can produce extremely accurate parts with good mechanical properties, the tooling is usually expensive.

RFI is another liquid molding process in which the resin is placed in the bottom of a matched-die mold. Autoclave pressure is used to drive the resin up through the preform during cure. Other variants of this process use thin resin film layers between the individual preform layers.

VARTM uses a vacuum to draw the resin through the preform rather than injection pressure. Normally single-sided tooling is used along with a vacuum bag and a porous media material to aid in resin infiltration. The major advantage of VARTM-type processes is the potential for lower tooling costs and the ability to make large parts because an autoclave is not required for curing. While VARTM processes have been used for years in the commercial boat building industry, the technology for high-performance composites is still evolving but shows great potential for cost savings.

References

[1] Beckwith S.W., Hyland C.R., "Resin Transfer Moulding: A Decade of Technology Advances," *SAMPE Journal* **34**(6), November/December 1998, pp. 7–19.

[2] Cox B.N., Flanagan G., "Handbook of Analytical Methods for Textile Composites," NASA Contractor Report 4750, March 1997.

[3] Dow M.B., Dexter H.B., "Development of Stitched, Braided and Woven Composite Structures in the ACT Program and at Langley Research Center (1985 to 1997)," Summary and Bibliography, NASA/TP-97-206234, November 1997.

[4] Palmer R., "Techno-Economic Requirements for Composite Aircraft Components," Fiber-Tex 1992 Conference, NASA Conference Publication 3211, 1992.

[5] Ko F.K., Du G.W., "Processing of Textile Preforms," in *Advanced Composites Manufacturing*, Wiley, 1997.

[6] Gutowski T.G., "Cost, Automation, and Design," in *Advanced Composites Manufacturing*, Wiley, 1997.

[7] Singletary J.N., Bogdanovich A.E., "Processing and Characterization of Novel 3-D Woven Composites," 46th International SAMPE Symposium, May 2001, pp. 835–845.

[8] Dickinson L., Salama M., Stobbe D., "Design Approach for 3D Woven Composites: Cost vs. Performance," 46th International SAMPE Symposium, May 2001, pp. 765–777.

[9] Mohamed M.H., Bogdanovich A.E., Dickinson L.C., Singletary J.N., Lienhart R.B., "A New Generation of 3D Woven Fabric Preforms and Composites," *SAMPE Journal* **37**(3), May/June 2001, pp. 8–17.

[10] Wittig J., "Robotic Three-dimensional Stitching Technology," 46th International SAMPE Symposium, May, 6–10, 2001, pp. 2433–2444.

[11] Ko F.K., "Braiding," *Volume 1 Engineered Materials Handbook – Composites*, ASM International, 1987, pp. 519–528.

[12] Sanders L.R., "Braiding – A Mechanical Means of Composite Fabrication," 8th National SAMPE Conference, October 1976.

[13] Braley M., Dingeldein M., "Advancements in Braided Materials Technology," 46th International SAMPE Symposium, May 2001, pp. 2445–2454

[14] Reeve S., Robinson W., Cordell T., "Carbon Fiber Evaluation for Directed Fiber Preforms," 46th International SAMPE Symposium, May 2001, pp. 790–802.

[15] Shim S.B., Ahn K., Seferis J.C., "Cracks and Microcracks in Stitched Structural Composites Manufactured with Resin Film Infusion Process," *Journal of Advanced Materials*, July 1995, pp. 48–62.

[16] Poe C.C., Dexter H.B., Raju I.S., "A Review of the NASA Textile Composites Research," AIAA, 1997

[17] Kittleson J.L., Hackett S.C., "Tackifier/Resin Compatibility is Essential for Aerospace Grade Resin Transfer Moulding," 39th International SAMPE Symposium, April 1994, pp. 83–96.

[18] Gebart B.R., Strombeck L.A., "Principles of Liquid Moulding," in *Processing of Composites*, Hanser, 2000, pp. 359–386.

[19] Hayward J.S., Harris B., "Effect of Process Variables on the Quality of RTM Mouldings," *SAMPE Journal* **26**(3), May/June 1990, pp. 39–46.

[20] Simacek P., Lawrence J., Advani S., "Numerical Mould Filling Simulations of Liquid Composite Moulding Processes – Applications and Current Issues," 2002 European SAMPE, pp. 137–148.

[21] Karal M., "AST Composite Wing Program – Executive Summary," NASA/CR-20001-210650, March 2001.

[22] Palmer R., "Manufacture of Multi-axial Stitched Bonded Non-crimp Fabrics," 46th International SAMPE Symposium, May 2001, pp. 779–788.

[23] Hinrichs S., Palmer R., Ghumman A., "Mechanical Property Evaluation of Stitched/ RFI Composites," 5th NASA/DoD Advanced Composites Technical Conference, NASA CP-3294, vol. 1, 1995, pp. 697–716.

[24] Loos A.C., Sayre J., McGrane R., Grimsley B., "VARTM Process Model Development," 46th International SAMPE Symposium, May 2001, pp. 1049–1060.

[25] Rigas E.I., Walsh S.M., Spurgeon W.A., "Development of Novel Processing Technique for Vacuum Assisted Resin Transfer Moulding," 46th International SAMPE Symposium, May 2001, pp. 1086–1093.

[26] Rigas E.I., Mulkern T.J., Walsh S.M., Nguyen S.P., "Effects of Processing Conditions on Vacuum Assisted Resin Transfer Moulding Process (VARTM)," Army Research Laboratory, Report ARL-TR-2480, May 2001.

Chapter 10

Thermoplastic Composites: An Unfulfilled Promise

During the 1980s and early 1990s, government agencies, aerospace contractors and material suppliers invested hundreds of millions of dollars in developing thermoplastic composites to replace thermosets. In spite of all of this investment and effort, continuous fiber thermoplastic composites account for only a handful of production applications on commercial and military aircraft. In this chapter, we will examine the potential advantages of thermoplastics and their processing characteristics and point out why they failed to replace thermoset composites in the aerospace industry.

10.1 The Case for Thermoplastic Composites

Before considering the potential advantages of thermoplastic composite materials, it is necessary to understand the difference between a thermoset and thermoplastic. As shown in Fig. 1, a thermoset cross-links during cure to form a rigid intractable solid. Prior to cure the resin is a relatively low-molecular-weight semi-solid that melts and flows during the initial part of the cure process. As the molecular weight builds during cure, the viscosity increases until the resin gels and then strong covalent bond cross-links form during cure. Due to the high cross-link densities obtained for high-performance thermoset systems, they are inherently brittle unless steps are

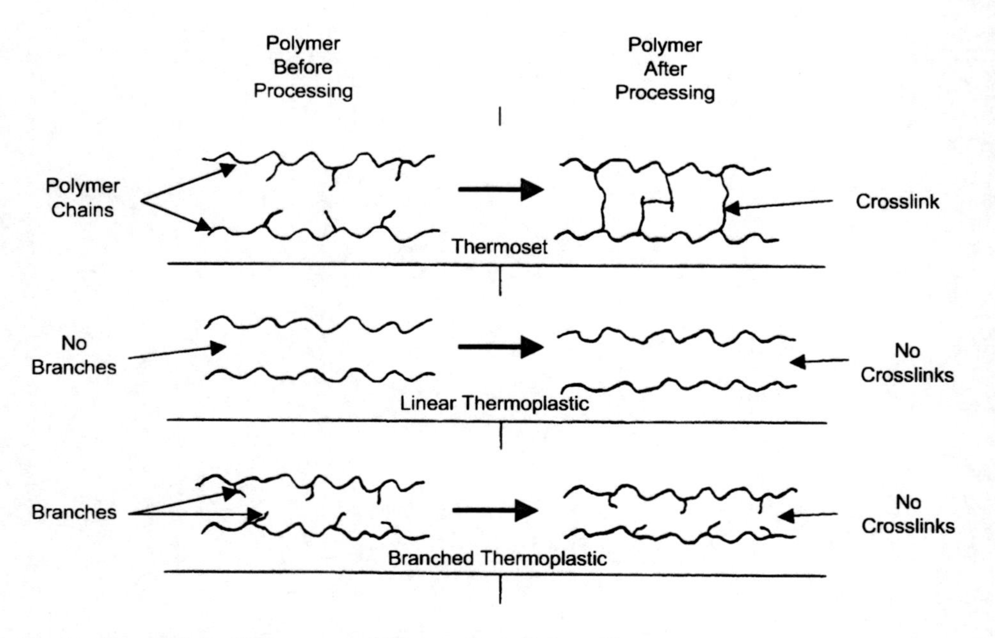

Fig. 1. Comparison of Thermoset and Thermoplastic Polymer Structures

taken to enhance toughness. On the other hand, thermoplastics are high-molecular-weight resins that are fully reacted prior to processing. They melt and flow during processing but do not form cross-linking reactions. Their main chains are held together by relatively weak secondary bonds. However, being high-molecular-weight resins, the viscosity of thermoplastics during processing is orders of magnitude higher than that of thermosets (e.g., 10^4–10^7 poise for thermoplastics versus 10 poise for thermosets[1]). Since thermoplastics do not cross-link during processing, they can be reprocessed; for example, they can be thermoformed into structural shapes by simply reheating to the processing temperature. On the other hand, thermosets, due to their highly cross-linked structures, cannot be reprocessed and will thermally degrade and eventually char if heated to high enough temperatures. However, there is a limit to the number of times a thermoplastic can be reprocessed. Since the processing temperatures are close to the polymer degradation temperatures, multiple reprocessing will eventually degrade the resin and in some cases it may cross-link.

The structural difference between thermosets and thermoplastics yields some insight to the potential advantages of thermoplastics. Since thermoplastics are not cross-linked, they are inherently much tougher than thermosets. Therefore, they are much more damage tolerant and resistant to low-velocity impact damage than the untoughened thermoset resins used in the early to mid-1980s. However, as a result of improved toughening approaches for thermoset resins, primarily with thermoplastic additions to the resin, the thermosets available today exhibit toughness comparable to thermoplastic systems.

Since thermoplastics are fully reacted high-molecular-weight resins that do not undergo chemical reactions during cure, the processing for these materials is theoretically simpler and faster. Thermoplastics can be consolidated and thermoformed in minutes (or even seconds), while thermosets require long cures (hours) to build molecular weight and cross-link through chemical reactions. However, since thermoplastics are fully reacted, they contain no tack, and the prepreg is stiff and boardy. In addition, competing thermoset epoxies are usually processed at 250–350 °F, whereas high-performance thermoplastics require temperatures in the range of 500–800 °F. This greatly complicates the processing operations requiring high-temperature autoclaves or presses, and bagging materials that can withstand the higher processing temperatures. Another advantage of thermoplastic composites involves health and safety issues. Since these materials are fully reacted, there is no danger to the worker from low-molecular-weight unreacted resin components. In addition, thermoplastic composite prepregs do not require refrigeration, as do thermoset prepregs.

They have essentially an infinite shelf life but may require drying to remove surface moisture prior to processing.

Another potential advantage of thermoplastics is low moisture absorption. Cured thermoset composite parts absorb moisture from the atmosphere that lowers their elevated temperature (hot-wet) performance. Since many thermoplastics absorb only very little moisture, the design does not have to take as severe a structural "knock-down" for lower hot-wet properties. However, since thermosets are highly cross-linked, they are resistant to most fluids and solvents encountered in service. Some amorphous thermoplastics are very susceptible to solvents and may even dissolve in methylene chloride, a common base for many paint strippers, while others, primarily semi-crystalline thermoplastics, are quite resistant to solvents and other fluids.

Since thermoplastics can be reprocessed by simply heating above their melting temperature, they offer advantages in forming and joining applications. For example, large flat sheets of thermoplastic composite can be autoclave or press-consolidated, cut into smaller blanks and then thermoformed into structural shapes. Unfortunately, this has proved to be much more difficult in practice than originally anticipated. Press-forming processes are limited to relatively simple geometric shapes because of the inextensible nature of the continuous fiber reinforcement. If a defect (e.g., an unbond) is discovered, the part can often be reprocessed to heal the defect, but in practice such repairs are rarely practical without undesirable fiber distortion and the associated structural property degradation. The melt fusible nature of thermoplastics also offers a number of attractive joining options such as melt fusion, resistance welding, ultrasonic welding and induction welding in addition to conventional adhesive bonding and mechanical fastening.

10.2 Thermoplastic Composite Matrices

During the 1980s dozens of thermoplastic matrices and product forms were available to industry. The number commercially available today is much more modest. The four most important materials are shown in Fig. 2. Polyetheretherketone (PEEK), polyphenylene sulfide (PPS) and polypropylene (PP) are semi-crystalline thermoplastics while polyetherimide (PEI) is an amorphous thermoplastic. PEEK, PPS and PEI are normally used for continuous fiber-reinforced thermoplastic composites, while PP is a lower temperature resin that is used quite extensively in the automotive industry as a discontinuous glass fiber stampable sheet product form called glass mat reinforced thermoplastic (GMT). High-performance thermoplastics, such as PEEK, PPS and PEI, have high T_g's with good mechanical properties, much higher than conventional thermoplastics but are also more costly. High-performance

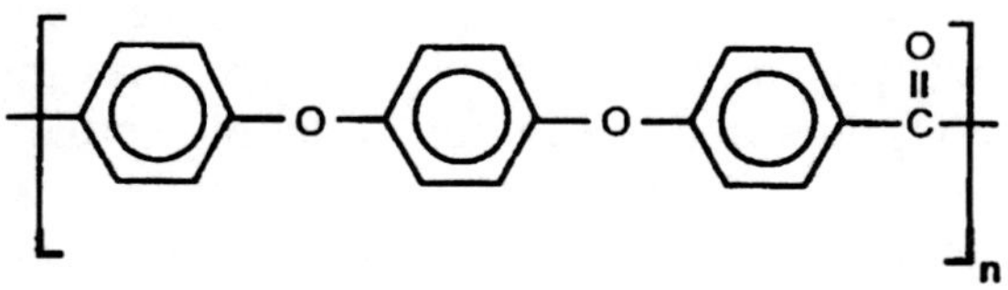

Polyetheretherketone

Tm= 633-650°F
Tg= 284-293°F
Processing Temperature= 680-750°F

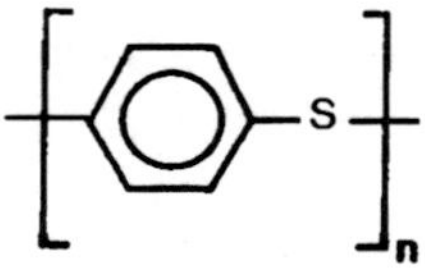

Polyphenylene Sulfide

Tm= 527-555°F
Tg= 185-194°F
Processing Temperature= 600-650°F

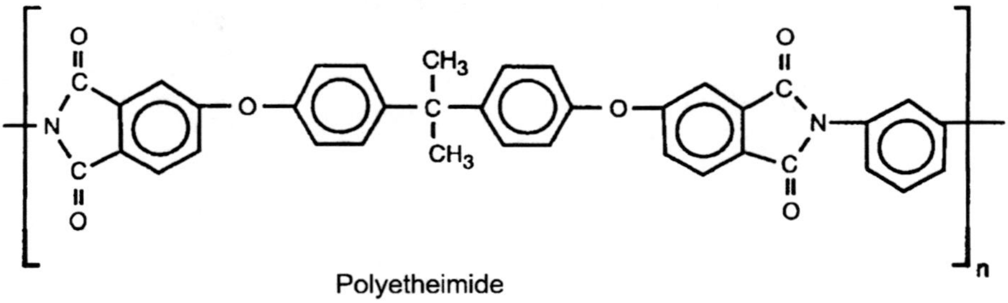

Polyetheimide

Tm= Not Applicable (Amorphous)
Tg= 420-423°F
Processing Temperature= 645-750°F

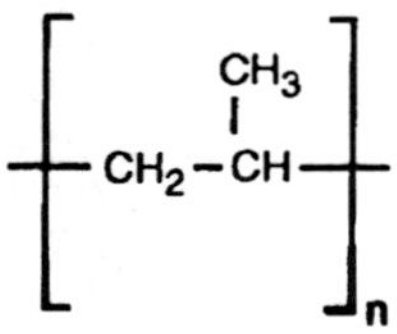

Polyproplyene

Tm= 325-349°F
Tg= -17°F
Processing Temperature= 160-175°F

Fig. 2. Chemical structure of Several Thermoplastic Resins

thermoplastics are usually aromatic, containing the benzene ring (actually the phenylene ring) that increases the T_g and provides thermal stability. Also, when "*n*" (i.e., the number of units in the molecular chain) is large, there is a high degree of orientation in the liquid state that helps promote crystallinity during freezing.[1] Highly aromatic thermoplastics exhibit good flame retardance because of their tendency to char and form a protective surface layer.

The differences between an amorphous and a semi-crystalline thermoplastic are shown in Fig. 3. An amorphous thermoplastic contains a massive random array of entangled molecular chains. The chains themselves are held together by strong covalent bonds while the bonds

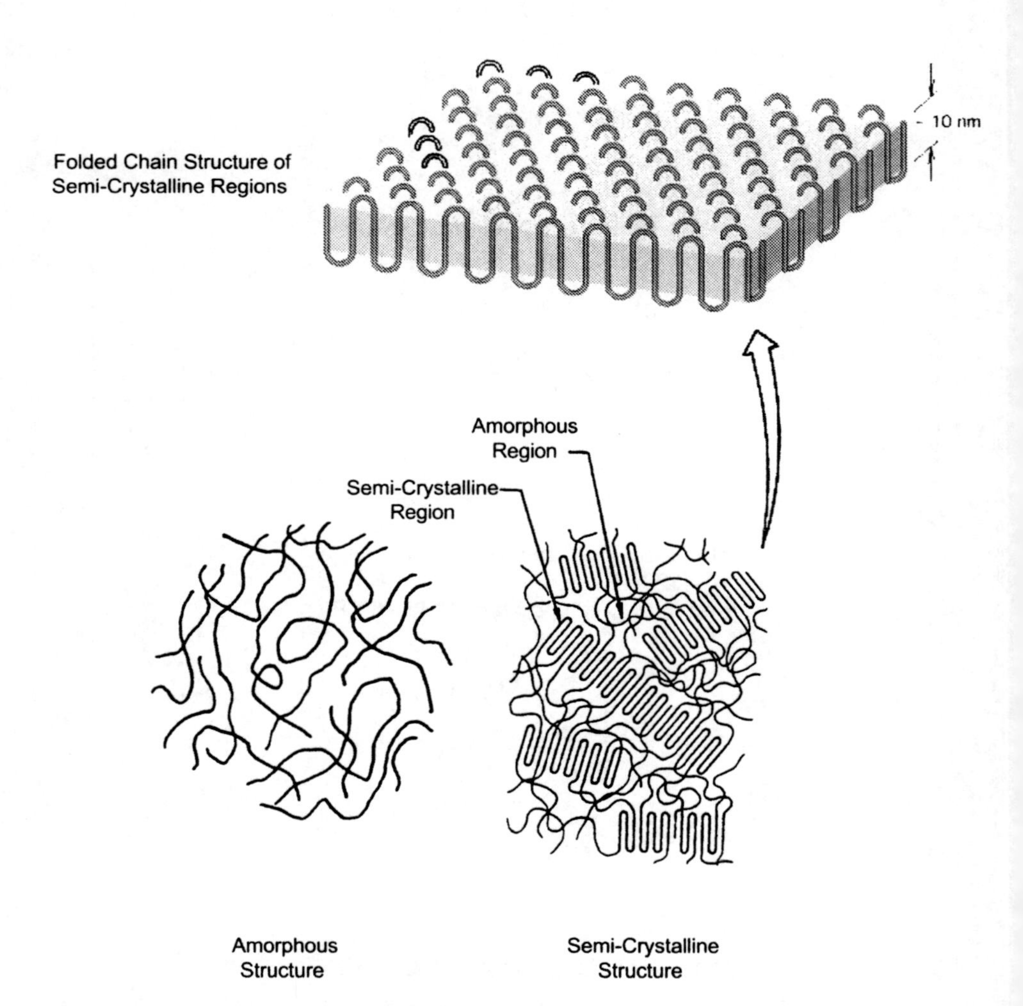

Fig. 3. Comparaison of Amorphous and Semi-Crystalline Thermoplastic Structures

between the chains are much weaker secondary bonds. When the material is heated to its processing temperature, it is these weak secondary bonds that breakdown and allow the chains to move and slide past one another. Amorphous thermoplastics exhibit good elongation, toughness and impact resistance.[1] As the chains get longer, the molecular weight increases resulting in higher viscosities, higher melting points and greater chain entanglement, all leading to higher mechanical properties.[2] Semi-crystalline thermoplastics contain areas of tightly folded chains (crystallites) that are connected together with amorphous regions. As shown in Fig. 4, amorphous thermoplastics exhibit a gradual softening on heating while semi-crystalline thermoplastics exhibit a sharp melting point when the crystalline regions start dissolving. As the polymer approaches its melting point, the crystalline lattice breaks down and the molecules are free to rotate and translate, while non-crystalline amorphous thermoplastics exhibit a more gradual transition from a solid to a liquid. In general, the melting point T_m increases with increasing chain length, greater attractive forces between the chains, greater chain stiffness and increasing crystallinity, while the glass transition temperature T_g increases with lower free volume, greater attractive forces between the molecules,

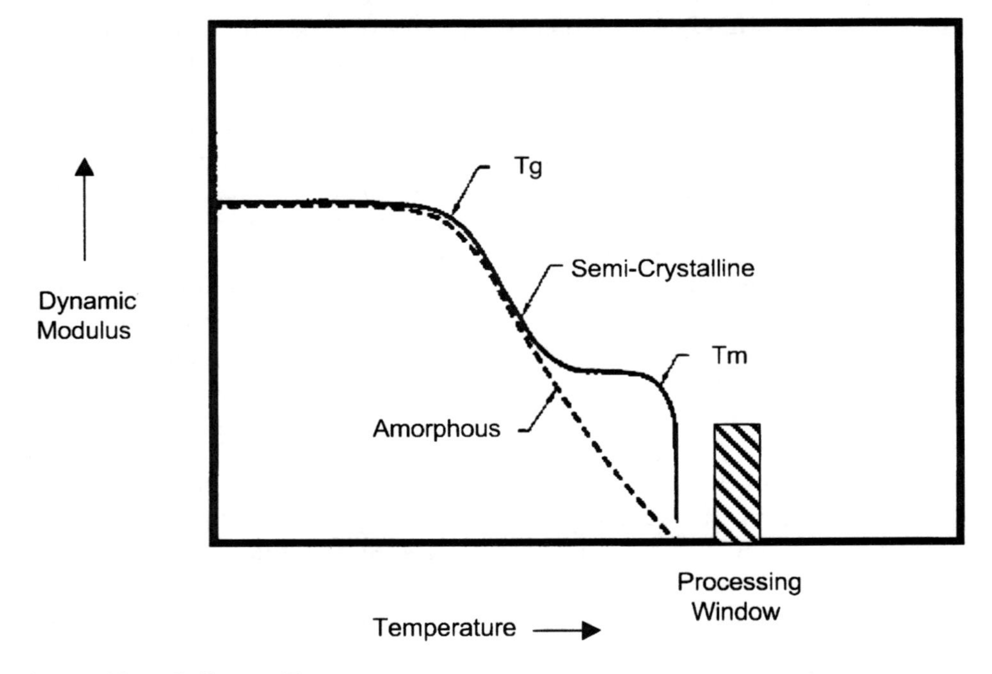

Fig. 4. Matrix Stiffness vs. Temperature

decreasing chain mobility, increasing chain stiffness, increasing chain length and, for thermosets, increasing crosslink density.[3] Crystallinity provides the following attributes to a thermoplastic resin:[3]

- Crystalline regions are held together by amorphous regions. The maximum crystallinity obtainable is about 98%, whereas metallic structures are usually 100% crystalline and exhibit much more ordered structures.
- Crystallinity increases density. The density increase helps to explain the improved solvent resistance since it becomes more difficult for the solvent molecules to penetrate the tightly packed crystallites.
- Crystallinity increases strength, stiffness, creep resistance and temperature resistance but usually decreases toughness. The tightly packed crystalline structure behaves somewhat like cross-linking in thermosets by decreasing and restricting chain mobility.
- Crystalline polymers are either opaque or translucent, while transparent polymers are always amorphous.
- Crystallinity can be increased by mechanical stretching.
- Crystallinity is an exothermic process where heat is given off to obtain the lowest free energy state.

In general, thermoplastics used for composite matrices contain 20–35% crystallinity.[4] It should also be noted that all thermoset resins are amorphous but are cross-linked to provide strength, stiffness and temperature stability. As a general class of polymers, thermoplastics are much more widely used than thermosets, accounting for about 80% of the polymers produced.[2]

Crystallites form from the melt as spherulites during cooling by a nucleation and growth process, as shown in Fig. 5. Spherulites are families of crystallites radiating from a single nucleation point.[4] If carbon fibers are present, they will frequently nucleate on the surface of the fibers and grow outward until they impinge on another spherulite. The degree of crystallinity is dependent on the cooling rate. As shown in Fig. 6, it is possible to quench the material from the melt at very high cooling rates and form primarily an amorphous structure. Slow cooling rates are required to provide the time necessary for the nucleation and growth process. The optimum cooling rate for PEEK is in the range of 0.2–20 °F min^{-1}, which will yield a crystalline content of 25–35%.[4] If the material is quenched to produce an amorphous structure, the proper amount of crystallinity can be established by a short (1 min) annealing cycle at 430–520 °F.[4] The rate of crystallization is also dependent on the specific annealing temperature. As shown in Fig. 7, the peak rate lies at about the mid-point between the glass transition temperature (T_g) and the melt temperature (T_m). The specific volume versus temperature curve of Fig. 8 illustrates the reduction in free volume that occurs during crystallization

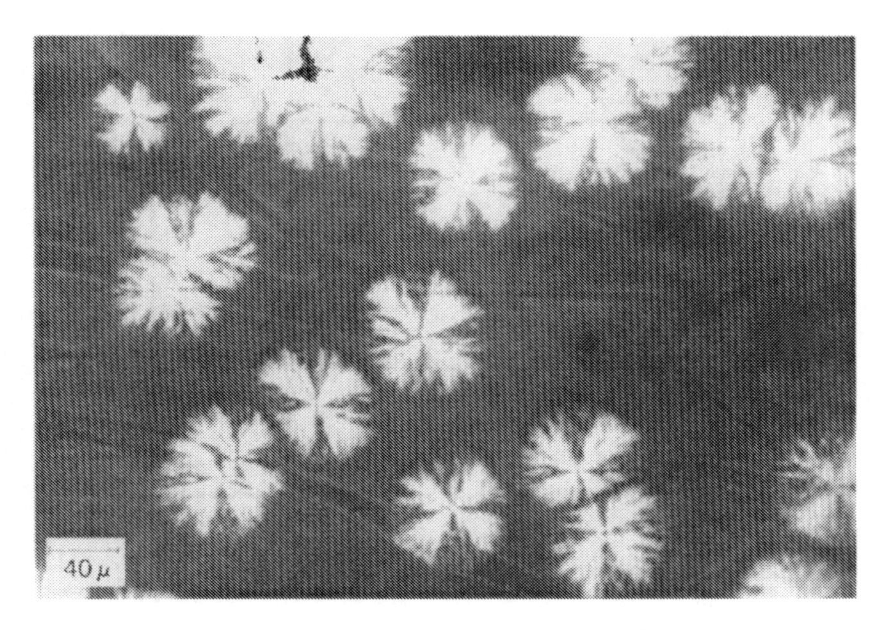

Nucleation of Spherulites in Pure PEEK Resin

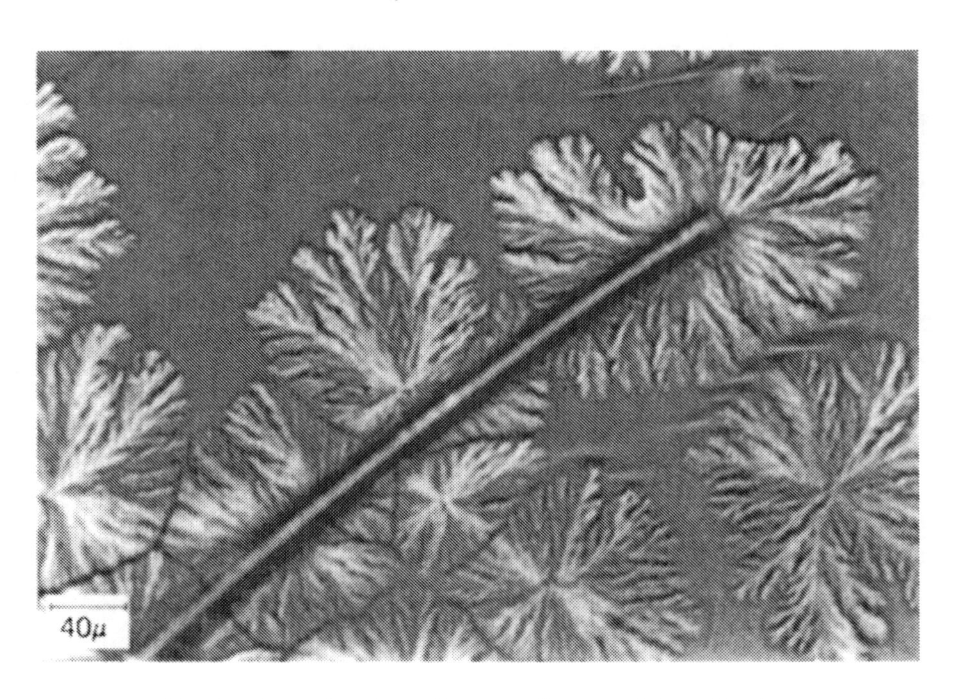

Nucleation of Spherulites on Carbon Fiber Surface

Fig. 5. Nucleation and Growth of Spherulites in PEEK

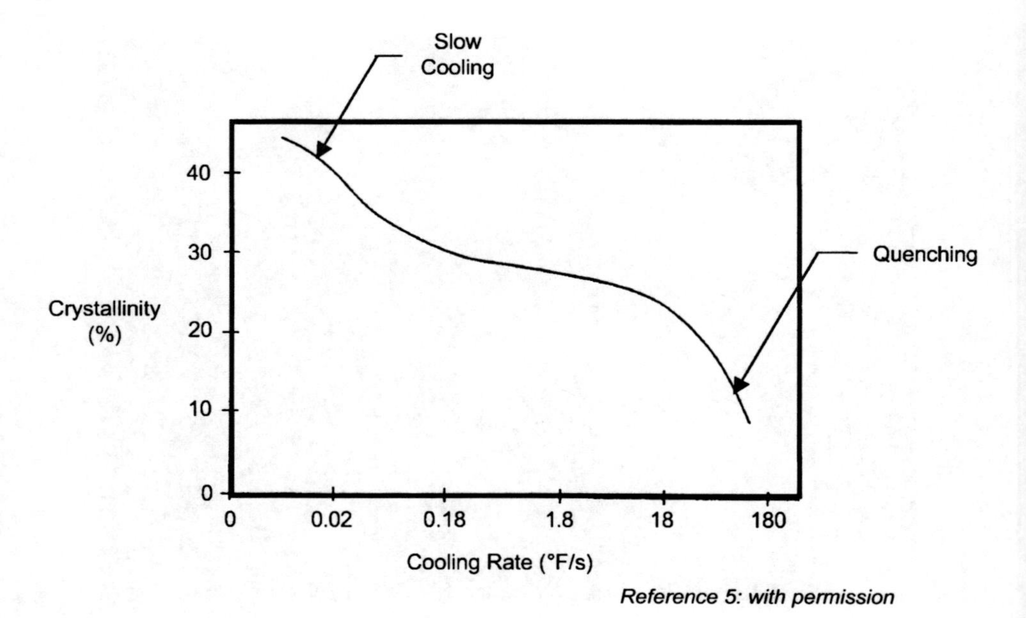

Fig. 6. Crystallization vs. Cooling Rate for Carbon/PEEK Laminate

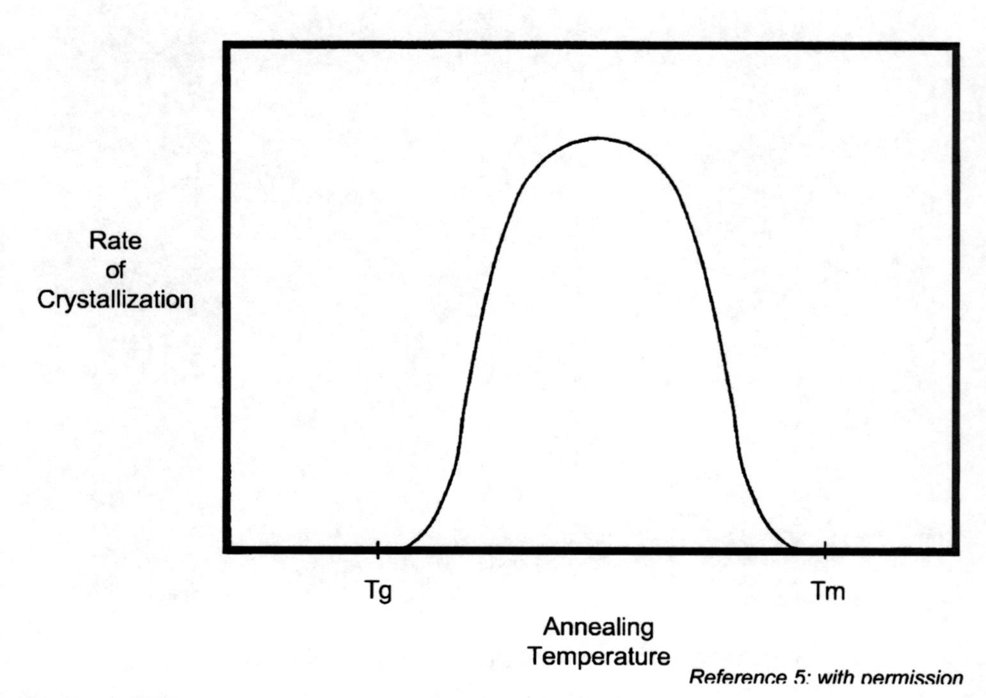

Fig. 7. Crystallization Rate Dependency on Annealing Temperature

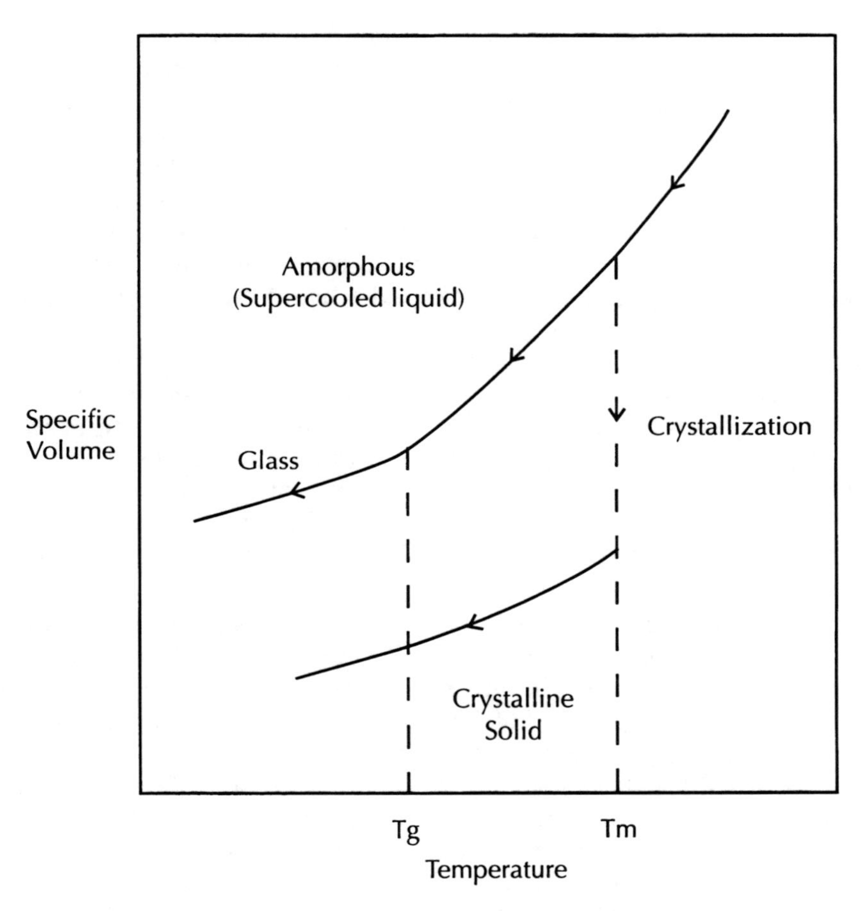

Fig. 8. Cooling Behavior of Amorphous and Semi-Crystalline Thermoplastics

and helps to explain the superior solvent resistance of semi-crystalline thermoplastics. Since there is no step change in density for amorphous thermoplastics, they have less tendency to warp or distort during rapid cooling since residual stresses are typically lower.

It should be pointed out that many condensation polyimides are also thermoplastics. Some, like Avimid K-III, are often classified as pseudo-thermoplastics because they can undergo light cross-linking during processing. These materials can be supplied either in low- or high-molecular-weight forms. When supplied as a low-molecular-weight prepreg, they process in a manner similar to thermoset prepregs, i.e., long processing cycles are required to achieve the molecular-weight buildup. Low-molecular-weight thermoplastic polyimide prepregs are normally produced using conventional thermoset prepregging equipment. Thus, both unidirectional and woven data product forms are readily available. However, due to the chemical inertness of some of the resin components, they must be dissolved in high-boiling-point solvents (e.g., NMP) to facilitate prepregging. Although some of the solvent is removed in a subsequent drying operation, an appreciable amount of solvent remains (as much as 12–18% in Avimid K-III). This residual solvent can cause problems with volatile evolution resulting in voids in the form of microporosity during elevated temperature processing. In addition, the condensation reactions give off water and ethanol, further contributing to the void problem. Extensive use of breather and vacuum porting arrangements can help in volatile removal. Although these materials do possess tack and drapeablility, their handling properties are inferior to production grades of carbon/epoxy prepreg. High-temperature heat guns and respirators are required when forming the prepreg to complex contours. The high-molecular-weight forms are somewhat easier to process; however, voids can also be formed in some of these materials through additional high-temperature reactions evolving carbon dioxide.

The low-molecular-weight thermoplastic polyimide prepregs are processed in a manner similar to thermoset composite parts but at much higher pressures and temperatures. The prepregs can be layed up, bagged and autoclave-processed. The processing cycles are necessarily long, and high temperatures (e.g., 650–700 °F) are required to build up the required final molecular weight and allow time for volatile removal. The processing is complicated by water, alcohol and solvent evolution. These volatile by-products can cause voids in the final laminate; therefore, they are very difficult to process. After initial volatile removal and consolidation, they can be reheated and formed in a press; however, the viscosity remains high and high pressures must be used to form even simple structural shapes. K-polymers based on the polyamideimide family are actually prepolymers with tack that first polymerizes to a powdery solid that allows removal of volatiles before finally melting and fusing.[6] In addition, since many of them have a tendency to lightly cross-link (~10–15%), their formability is much less than the true melt fusible thermoplastics. As such, they are not true thermoplastics. Even this small amount of cross-linking prevents makes them much more difficult to reprocess and thermally form. The cross-linking does, however, impart added solvent resistance. The remainder of this chapter will deal with true melt fusible thermoplastics.

10.3 Product Forms

Thermoplastic composite materials can be supplied in a number of different product forms, several of which are shown in Fig. 9. Unidirectional tape, tow and woven cloth prepregs are boardy and contain no tack. If the resin is amorphous (e.g., PEI), the resin can be dissolved in solvents and prepregged similar to thermoset prepregs. A distinct disadvantage of this process is that the solvent must be removed prior to processing. Even trace amounts of solvent have been shown to reduce the T_g and properties of the composite due to the plastisizing effect of the solvent. One investigation[7] reported that T_g decreased from 400 to 250 °F as the residual solvent increased from 0 to 3.7%. Hot-melt impregnation of thermoplastics, a requirement for the semi-crystalline materials that will not dissolve in solvents, is much more difficult than the equivalent process for thermosets. Much higher temperatures are required to melt the high-molecular-weight resins, and even at their melt temperature, their viscosities are orders of magnitude higher than for thermoset resins. This makes uniform fiber impregnation very difficult, and those who have mastered the process consider it a closely guarded trade secret. References to flowing the resin through a series of porous-heated spreader plates combined with nip rollers to push the resin into the fiber tows followed by an extrusion die are reported in the literature.[4] The extrusion die produces shear thinning (i.e., reduces viscosity by several orders of magnitude) by forcing the resin through a heated die that creates large shear stresses.[2] Hot-melt prepregging with viscous polymers requires high pressures, slow speeds and thin fiber beds.[8] A well-impregnated hot-melt tape usually contains resin-rich surface layers.

An alternate method developed for impregnating carbon tows is called powder coating. As shown in Fig. 10, a fluidized bed is used to introduce powder onto the carbon fiber surface, which is then fused to the fibers in an oven. The powder is charged and fluidized, while the tow is spread and grounded in order to pick up the charged powder.[8] Other processes, such as electrostatic dry powder and slurry coatings, have also been developed. Comingled fiber prepreg is another method of making a drapable product form. Fine fibers of the thermoplastic resin are extruded and then comingled with the carbon fiber tows. Both the powder coated and the comingled tows are then normally woven into a tackless but very drapable product form. It should be pointed out that the fiber distribution for powder coated and comingled product forms will not be as uniform as the fiber distribution achievable by prepregging; therefore, it is difficult to achieve consistent fiber volume contents during consolidation. In addition, there is considerable bulk in these product forms. This bulk, combined with their extensive drapeablility, can also result in wrinkling and buckling of fibers during material placement and consolidation.

10.4 Consolidation

Consolidation of melt fusible thermoplastics consists of heating, consolidation and cooling, as depicted schematically in Fig. 11. As with thermoset composites, the main processing variables are time (t),

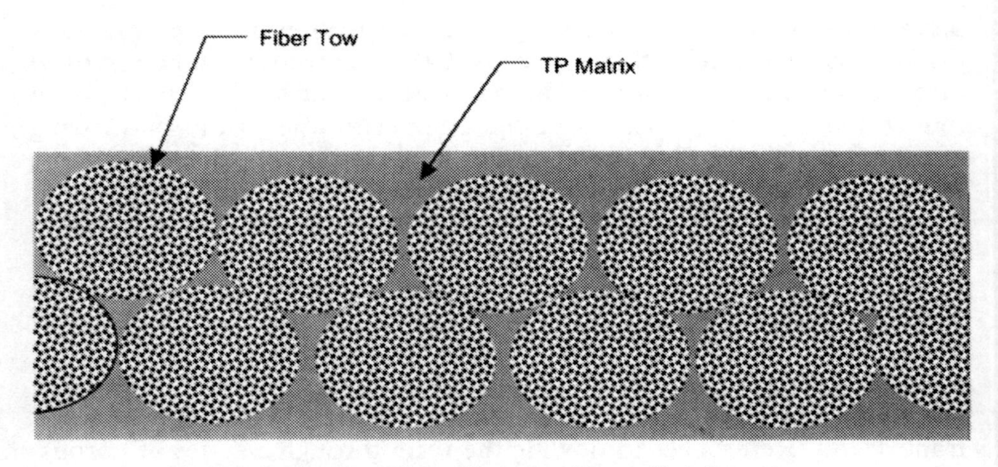

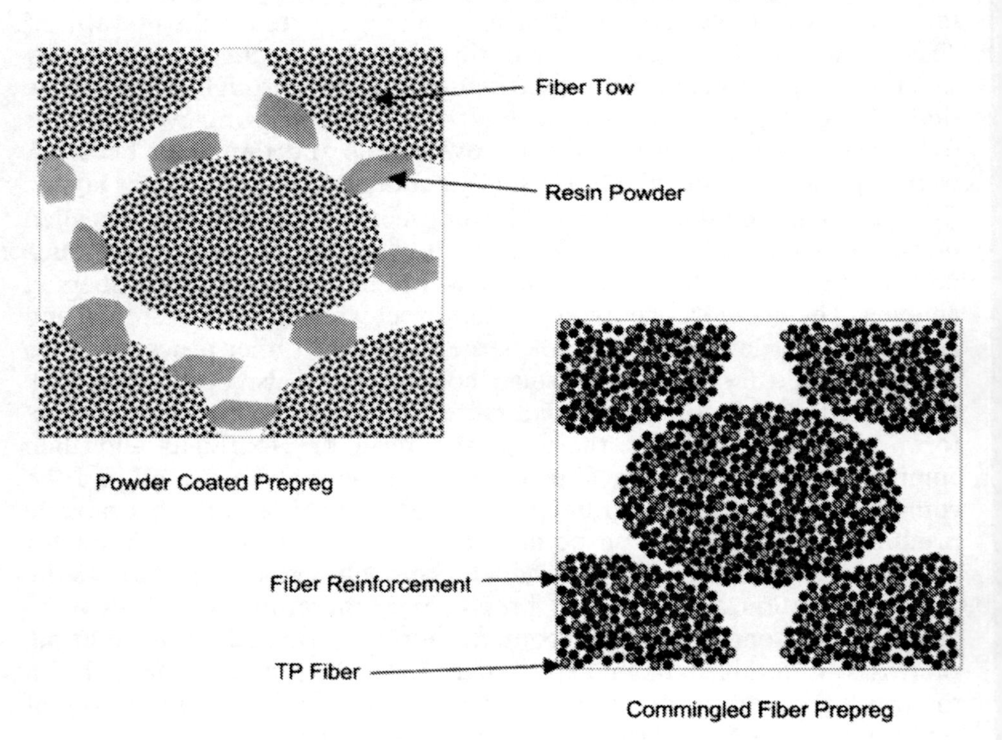

Fig. 9. Forms of Thermoplastic Composite Prepreg

temperature (T) and pressure (P). Heating can be accomplished with infrared heaters, convection ovens, heated platen presses or autoclaves. The time required to reach consolidation temperature is a function of the

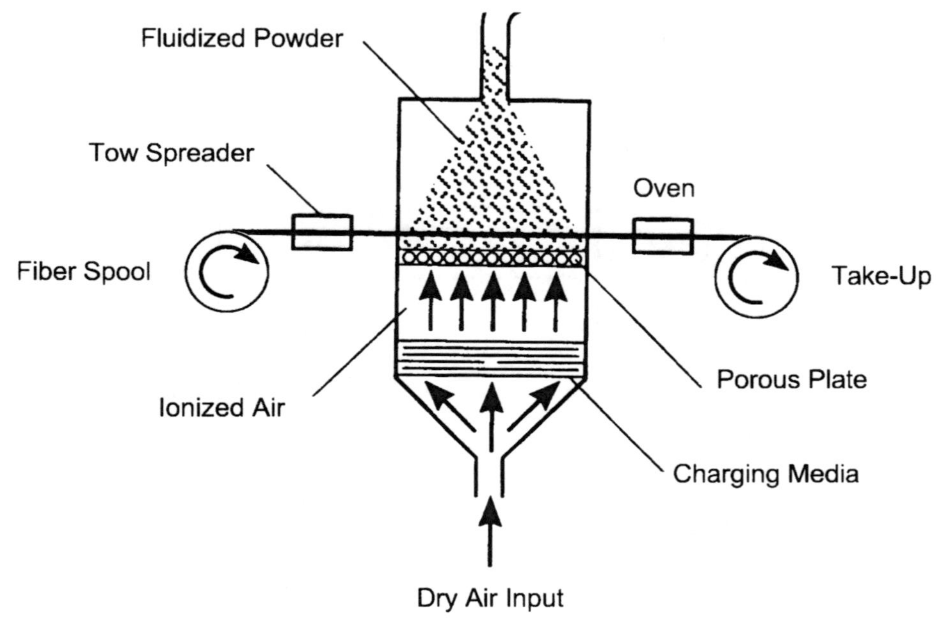

Fig. 10. Powder Coating Process for Thermoplastic Tow

heating method or mass of the tooling since time for chemical reactions is not required. The consolidation temperature depends on the specific thermoplastic resin, but should be above the T_g for amorphous resins or above the T_m for semi-crystalline materials. As a general rule of thumb, the processing temperature for an amorphous thermoplastic composite should be 400 °F above its T_g, and for a semi-crystalline material, it should be 200 °F or less above its T_m.[11] However, heating most thermoplastics above 800 °F will result in degradation. The time at temperature for consolidation is primarily a function of the product form used. For example, well-consolidated hot-melt impregnated tape can be successfully consolidated in very short times (minutes if not seconds), while woven powder coated or comingled prepregs require longer times for the resin to flow and impregnate the fibers. Occasionally a process called film stacking is used in which alternating layers of thermoplastic film and dry woven cloth are layed-up and consolidated. The time for successful consolidation for film-stacked lay-ups becomes even longer since the high-viscosity resin has even longer distances to flow. A typical processing cycle to achieve fiber wet-out and full consolidation for a film-stacked laminate would be 1 h at 150 psig applied pressure. Like heat-up, the cool-down rate from consolidation is a function of the processing method used and the mass of the tooling. The

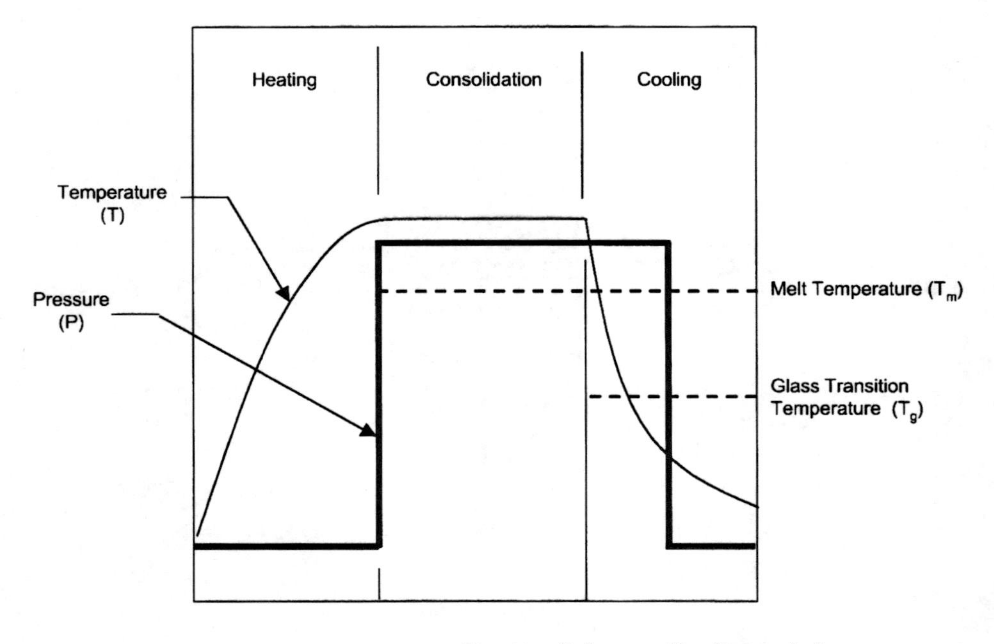

Fig. 11. Typical Thermoplastic Composite Process Cycle

only caveat on cooling is that semi-crystalline thermoplastics should not be cooled so quickly (i.e., quenched) that they fail to form the desired semi-crystalline structure that provides optimal elevated temperature performance and solvent resistance. During cooling the pressure should be maintained until the temperature falls well below the T_g of the resin. This restricts the nucleation of voids, suppresses the elastic recovery of the fiber bed and helps to maintain the desired dimensions.[8] Finally, pressure during the process provides the driving force to put the layers in intimate contact, push them together and help further impregnate the fiber bed. It should be noted that the properties of solvent-impregnated prepreg, powder-coated, comingled and film-stacked laminates are not as good as those made from hot-melt impregnated prepreg due to the superior fiber-to-matrix bond formed during the hot-melt impregnation process.[2]

Thermoplastic consolidation occurs by a process called autohesion depicted in Fig. 12. When two interfaces come together, they must obtain intimate contact before the polymer chains can diffuse across the interface and obtain full consolidation. Due to the low flow and tow height non-uniformity of thermoplastic prepregs, the surfaces must be physically deformed under heat and pressure to provide the intimate contact required for chain migration at the ply interfaces. To obtain intimate contact and

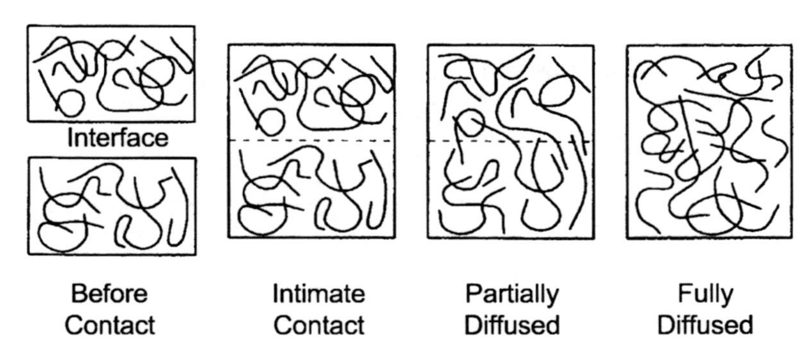

Fig. 12. Autohesion at Thermoplastic Interfaces

autohesion, the material must be heated above the T_g if it is amorphous, and above the T_m if it is semi-crystalline. In general, higher pressures and higher temperatures lead to shorter consolidation times. Autohesion is a diffusion-controlled process in which the polymer chains move across the interface and entangle with neighboring chains. As the contact time increases, the extent of polymer entanglement increases and results in the formation of a strong bond at the ply interfaces.[8] Consolidation times are usually longer for amorphous thermoplastics since they do not melt and generally maintain higher viscosities at the processing temperature;[12] however, shorter times can be used if higher pressures are employed. The time required for autohesion is directly proportional to the polymer viscosity.[4] Therefore, a certain amount of bulk consolidation must occur at the interfaces prior to the initiation of autohesion. As shown in Fig. 13, consolidation is also aided by resin flow due to the applied pressure that aids in ply contact and eventually leads to 100% autohesion. The process is essentially complete when the fiber bed is compressed to the point it reacts to the applied processing pressure.

There are several methods employed to consolidate thermoplastic composites. Flat sheet stock can be pre-consolidated for subsequent forming in a platen presses. Two press processes are shown in Fig. 14. In the platen press, pre-collated ply packs are preheated in an oven and then rapidly shuttled into the pressure application zone for consolidation. The press may require heating if the material requires time for resin flow for full consolidation or crystallinity control. If a well-consolidated prepreg is used, then rapid cooling in a cold platen press may suffice. It should be pointed out that this process still requires collation of the ply packs or layers, usually a hand lay-up operation. Since the material contains no tack, soldering irons heated to 800–1,200 °F are frequently used to tack the edges to prevent the material from slipping. Hand held ultrasonic guns have also

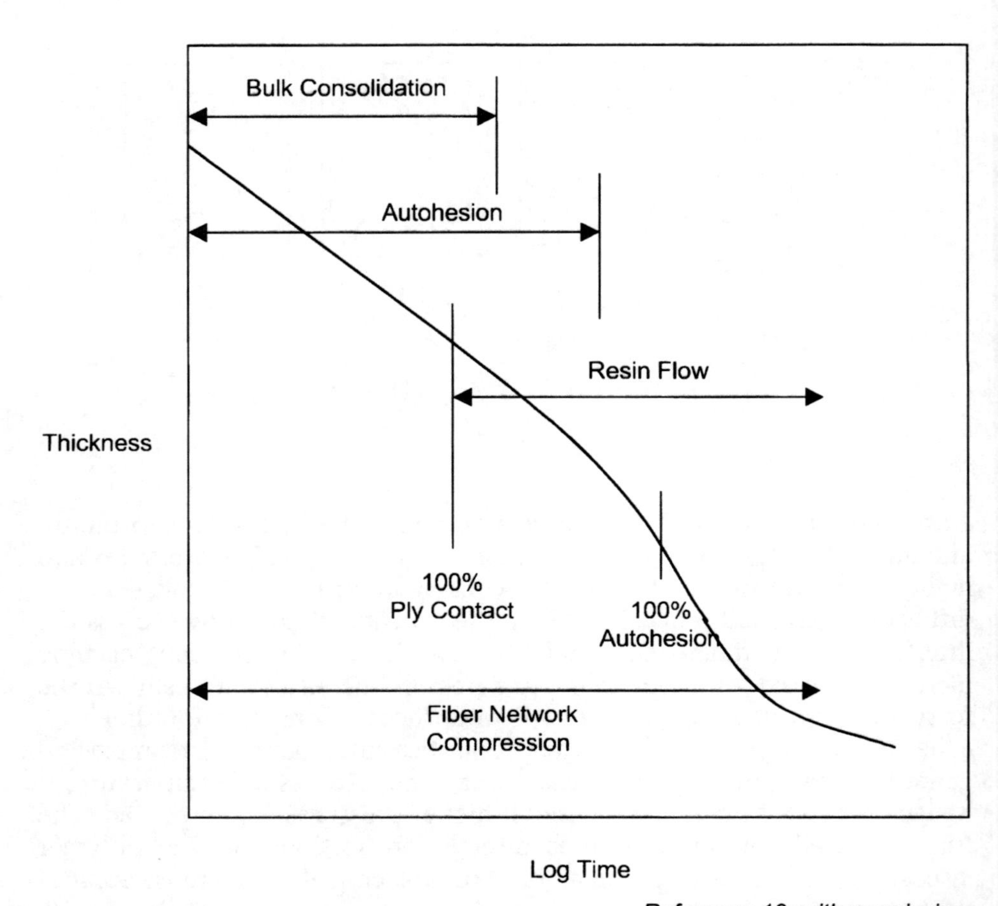

Fig. 13. Decrease in Thiskness During Consolidation

been used for ply tacking. A continuous consolidation process is the double belt press that contains both pressurized heating and cooling zones. This process is widely used in making glass mat thermoplastic (GMT) prepreg for the automotive industry with PP as the resin and random glass mat as the reinforcement. The processing of GMT is discussed in Chapter 11, Commercial Processes.

If the part configuration is complex, an autoclave is certainly an option for part consolidation. However, there are several disadvantages to autoclave consolidation. First, it may prove difficult even finding an autoclave that is capable of attaining the 650–750 °F temperatures and 100–200 psig pressures required for some advanced thermoplastics. Second, at these temperatures, the tooling is going to be expensive and may be

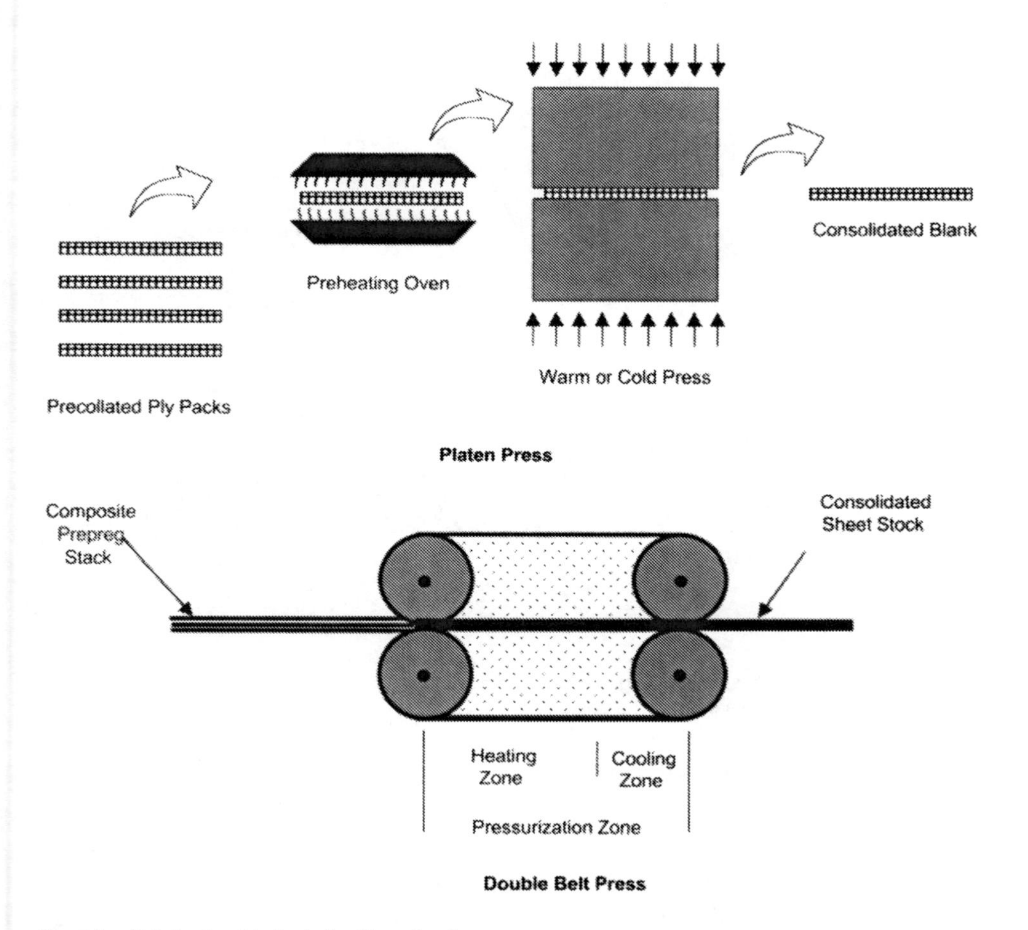

Fig. 14. Fabrication Methods for Sheet Stock

massive, dictating slow heat-up and cool-down rates. Third, since high processing temperatures are required, it is very important that the coefficient of thermal expansion of the tool matches that of the part. For carbon fiber thermoplastics, monolithic graphite, cast ceramic and Invar 42 are normally used. Fourth, the bagging materials must be capable of withstanding the high temperatures and pressures. In a typical bagging operation (Fig. 15), the materials required include high-temperature polyimide bagging material, glass bleeder cloth and silicone bag sealant. The polyimide bagging materials (e.g., Kapton or Uplilex) are more brittle and harder to work with than the nylon materials used for 250–350 °F curing thermosets. In addition, the high-temperature silicone rubber sealants have minimal tack and tend not to seal very effectively at room temperature. Clamped bars are often placed around the periphery to help

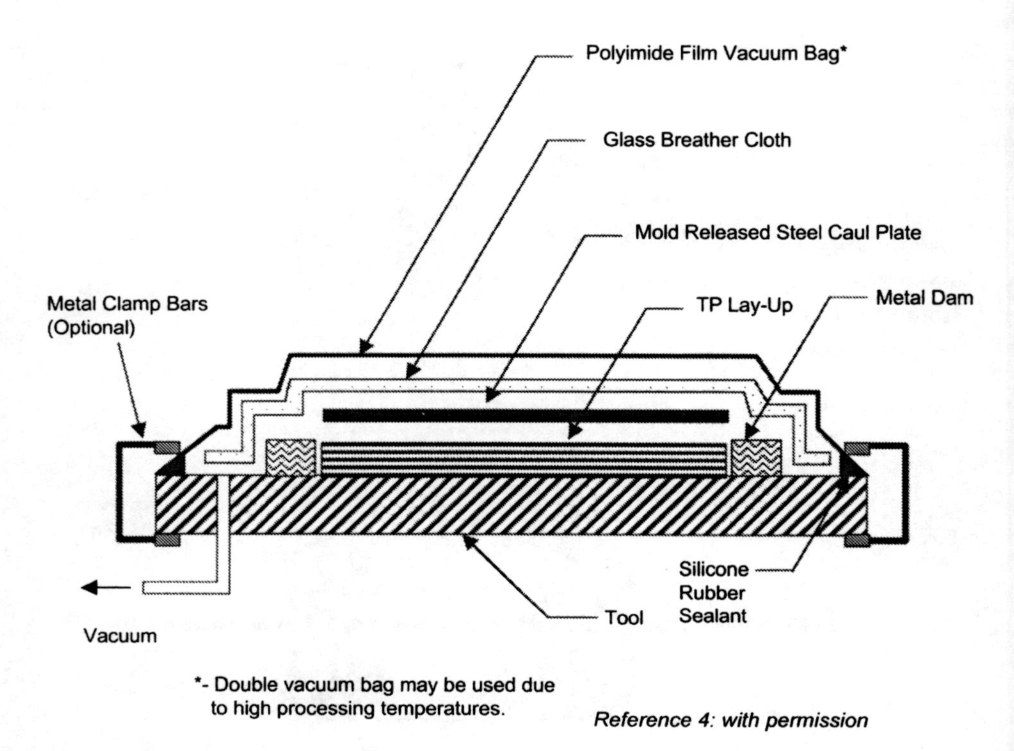

Fig. 15. Typical Vacuum Bag Configuration

get the seal to take at room temperature. As the temperature is increased, the sealant develops tack under pressure and the seal becomes much more effective. A typical autoclave consolidation cycle for carbon/PEEK prepreg would be 680–750 °F at 50–100 psig pressure for 5–30 min; however, the actual cycle time to heat and cool large tools is normally in the range of 5–15 h. In spite of all of these disadvantages, autoclaves nevertheless have a place in thermoplastic composite part fabrication for parts that are just too complex to make by other methods.

Auto-consolidation or *in situ* placement of melt fusible thermoplastics are comprised of a series of processes that include hot tape laying, filament winding and fiber placement. In the auto-consolidation process, only the area that is being immediately consolidated is heated above the melt temperature, the remainder of the part is held at temperatures well below the melt temperature. Two processes are shown in Fig. 16: a hot tape laying process that relies on conduction heating and cooling from hot shoes, and a fiber placement process that uses a focused laser beam at the nip point for heating. Other forms of heating include hot gas torches, quartz lamps and infrared heaters.

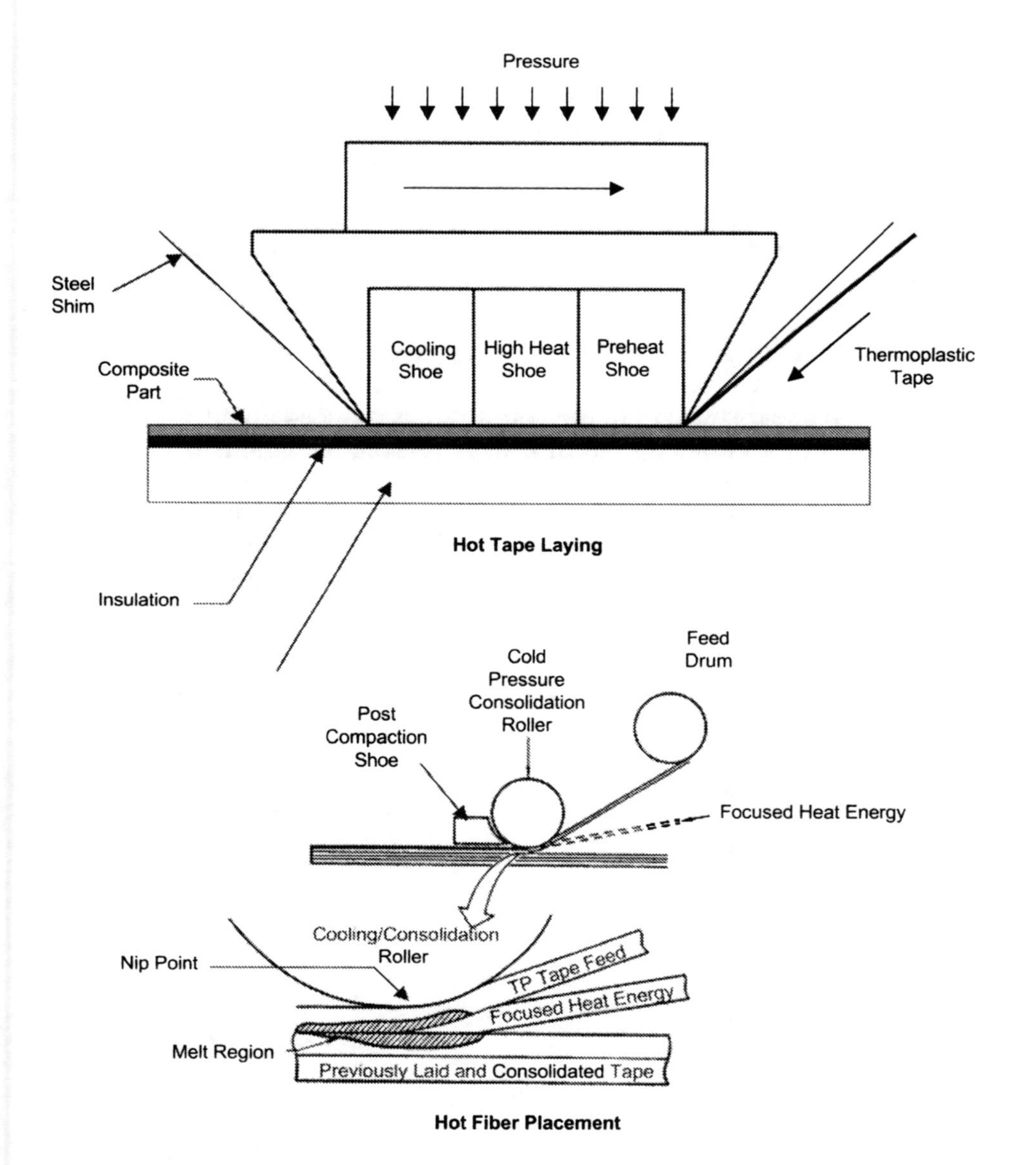

Fig. 16. Principle of Autoconsolidation

The mere fact that auto-consolidation is possible illustrates that the contact times for many thermoplastic polymers at normal processing temperatures can be quite short. Provided full contact pressure is made at the ply interfaces, auto-consolidation can occur in less than 0.5 of a second.[2] As shown in Fig.17, thermoplastics will degrade if held at long times at high temperatures, particularly in an open air environment. For example,

PEEK tends to cross-link when exposed to oxygen at high temperatures, increasing its viscosity and decreasing its ability to crystallize.[4] While this is not generally a problem for fast processes such as auto-consolidation or thermoforming, it should be taken into account when consolidating in a heated press.

A potential problem with auto-consolidation is lack of consolidation due to insufficient diffusion time. If a well-impregnated prepreg is used, then only the ply interfaces need be consolidated. However, if there are intraply voids, then the process time is so short that there is insufficient time to heal and consolidate these voids and a post-consolidation cycle will be required to achieve full consolidation. Previous studies have shown that the interlaminar shear strength of the composite is reduced about 7% for each 1% of voids up to a maximum of 4%. A reasonable goal is 0.5% or less porosity.[13] It has been reported[2] that hot taping laying operations usually result in 80–90% consolidation, indicating the necessity of secondary processing to obtain full consolidation. However, productivity gains for processes such as hot tape laying of 200–300% have been sited compared to traditional hand lay-up methods.[14]

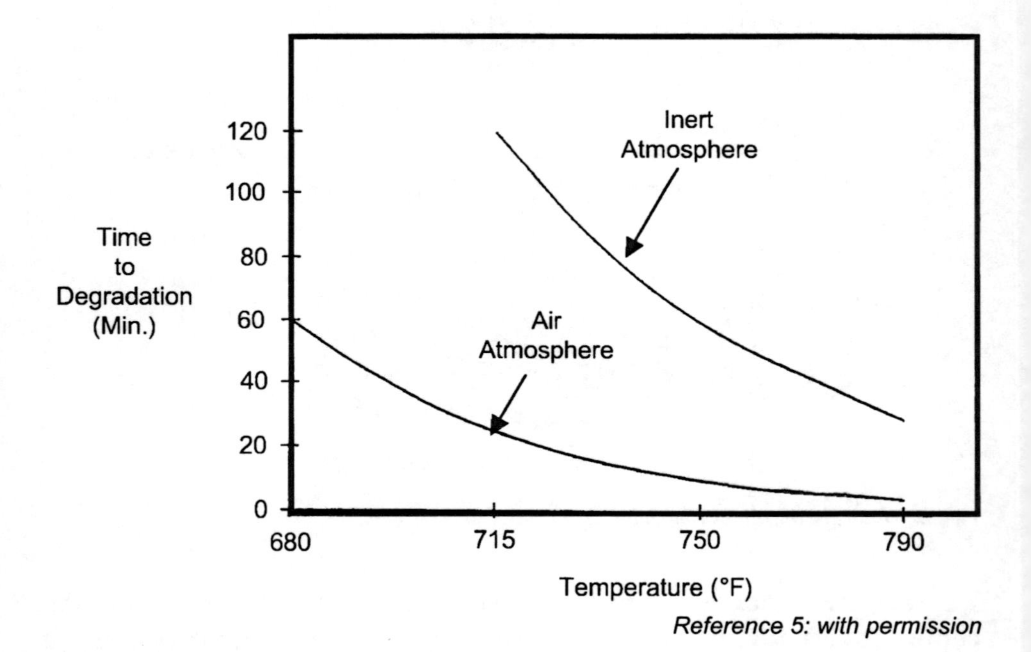

Fig. 17. Surface Degradation of PEEK as a Function of Time and Temperature

10.5 Thermoforming

One of the main advantages of thermoplastic composites is their ability to be rapidly processed into structural shapes by thermoforming. The term thermoforming encompasses quite a broad range of manufacturing methods but is essentially a process that uses heat and pressure to form a flat sheet or ply stack into a structural shape. A typical thermoforming process for a melt fusible thermoplastic part, shown schematically in Fig. 18, consists of (1) collating the plies, (2) press consolidating a flat blank, (3) placing the blank in a second press for cooling, (4) trimming the blank to shape if required, (5) reheating the blank to above its melt temperature, and then quickly transferring to a press containing dies of the desired shape. The part must be held under pressure until it cools below its T_g to avoid inducing residual stresses and part warpage.

The primary preheating methods used for press thermoforming are infrared (IR) heater banks, convection ovens and heated platen presses. In IR heating, the heating time is typically short (i.e., 1–2 min) but temperature gradients can form within thick ply stacks. Since the surface heats considerably faster than the center, there is the danger of overheating unless the temperature is carefully controlled. In addition, it is difficult to obtain uniform heating on complex contours. Still, IR heating is a good choice for thin pre-consolidated blanks of moderate contour. On the other hand, convection heating takes longer (i.e., 5–10 min.) but is generally more uniform through the thickness.[15] It is the preferred method for unconsolidated blanks and blanks containing high contour. Impingement heating is a variation of convection heating that uses a multitude of high-velocity jets of heated gas that impinge on the surfaces, greatly enhancing the heat flow and reducing the time required to heat the part.[16]

Although matched metal dies can be used for thermoforming, they are expensive and unforgiving, i.e., if the dies are not precisely made, there will be high- and low-pressure points that will result in defective parts. The dies can be made with internal heating and/or cooling capability. Facing one of the die halves with a heat resistant rubber, typically a silicone rubber, can help in equalizing the pressure. Similarly, one of the die halves can be made entirely from rubber, either as a flat block (Fig. 19) or as a block that is cast to the shape of the part (Fig. 20). Although the flat block is simpler and cheaper to fabricate, the shaped block provides a more uniform pressure distribution and better part definition.[2] Silicone rubber of 60–70 Shore-A hardness is commonly used. For deep draws, it is usually better to make the male tool half metal and the female half rubber. If the female tool is metal, as is customary for moderate draws, it should incorporate draft angles of 2–3° to facilitate part removal.[2] Another method of applying pressure during the forming process is hydroforming (Fig. 21), in which an elastomeric bladder is forced down around the part and lower die half using

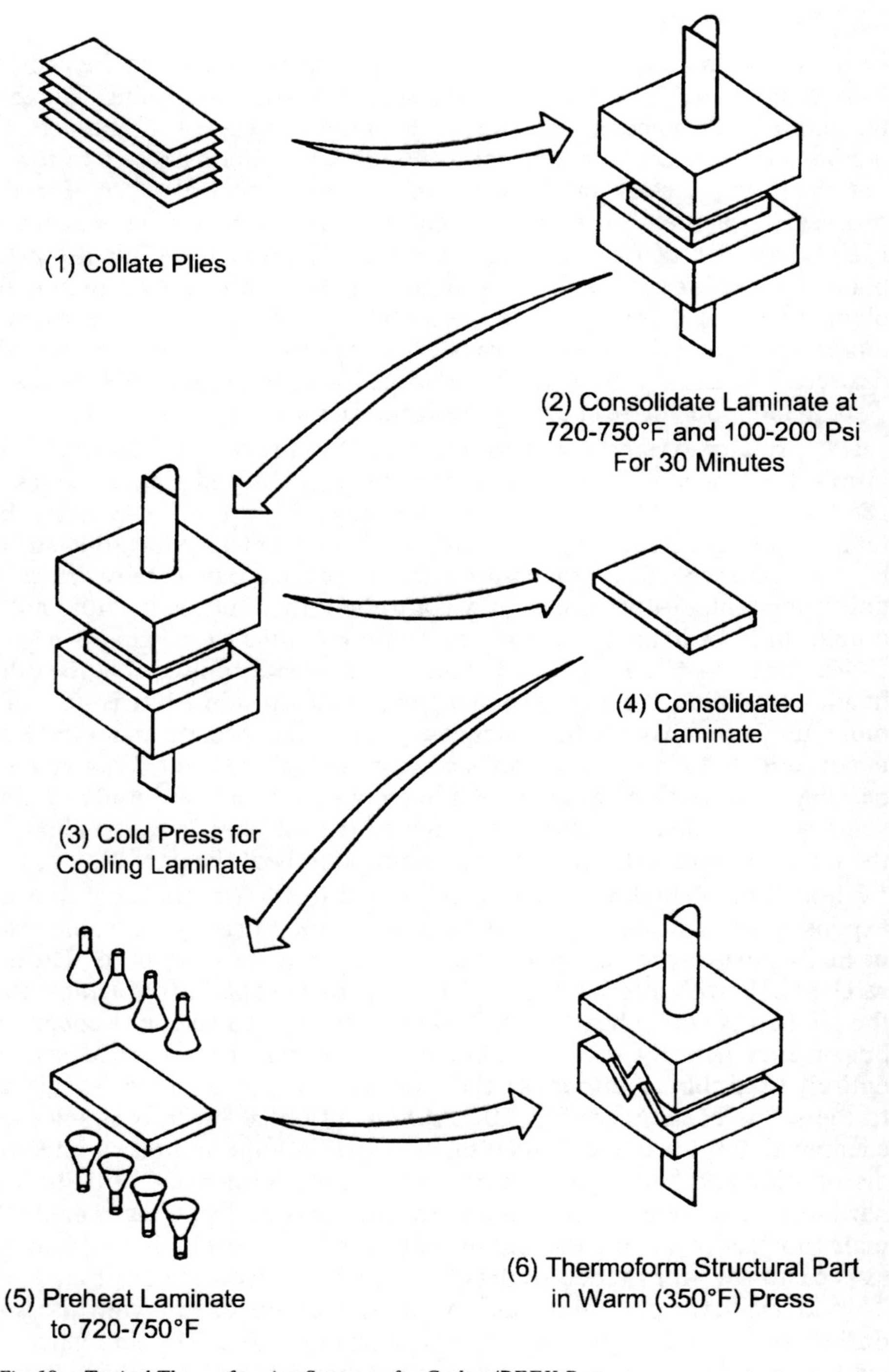

Fig. 18. Typical Thermoforming Sequence for Carbon/PEEK Part

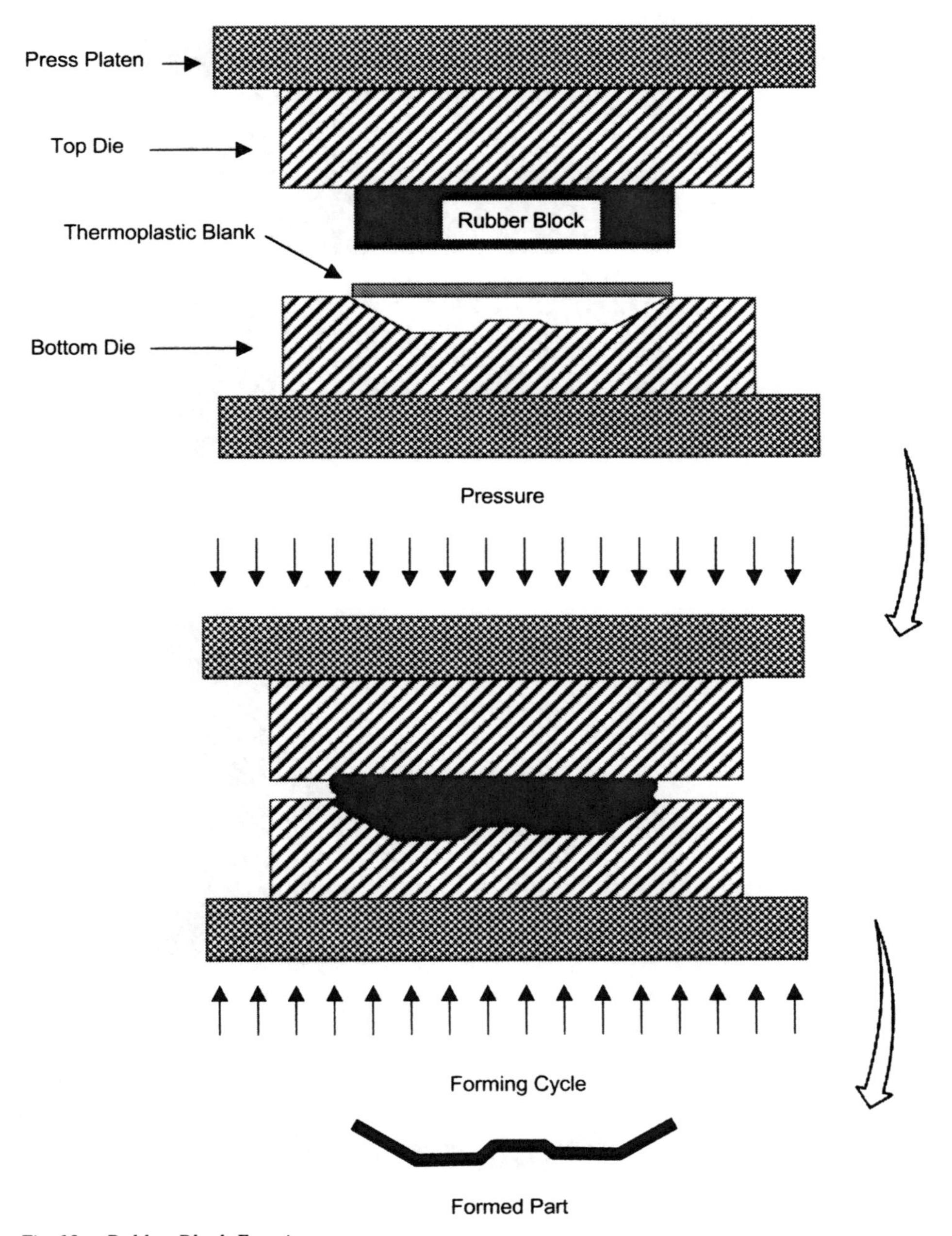

Fig. 19. Rubber Block Forming

fluid pressure. Typical thermoforming pressures are 100–500 psig; however, some hydroforming presses are capable of pressures as high as 10,000 psig.[15]

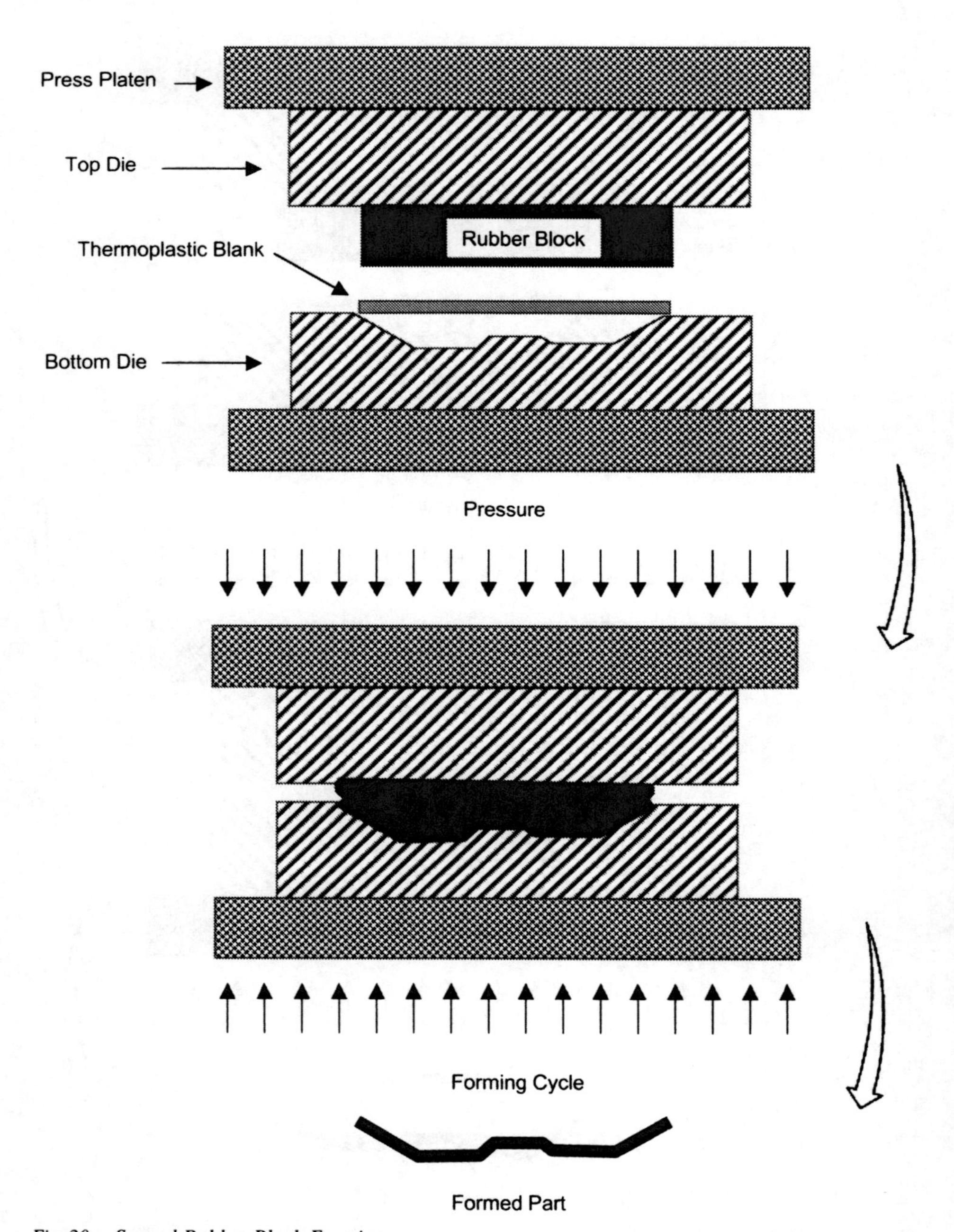

Fig. 20. Sgaped Rubber Block Forming

In any thermoforming operation, transfer time from the heating station to the press is critical. The part must be transferred or shuttled to the press and formed before it cools below its T_g for amorphous resins or below its

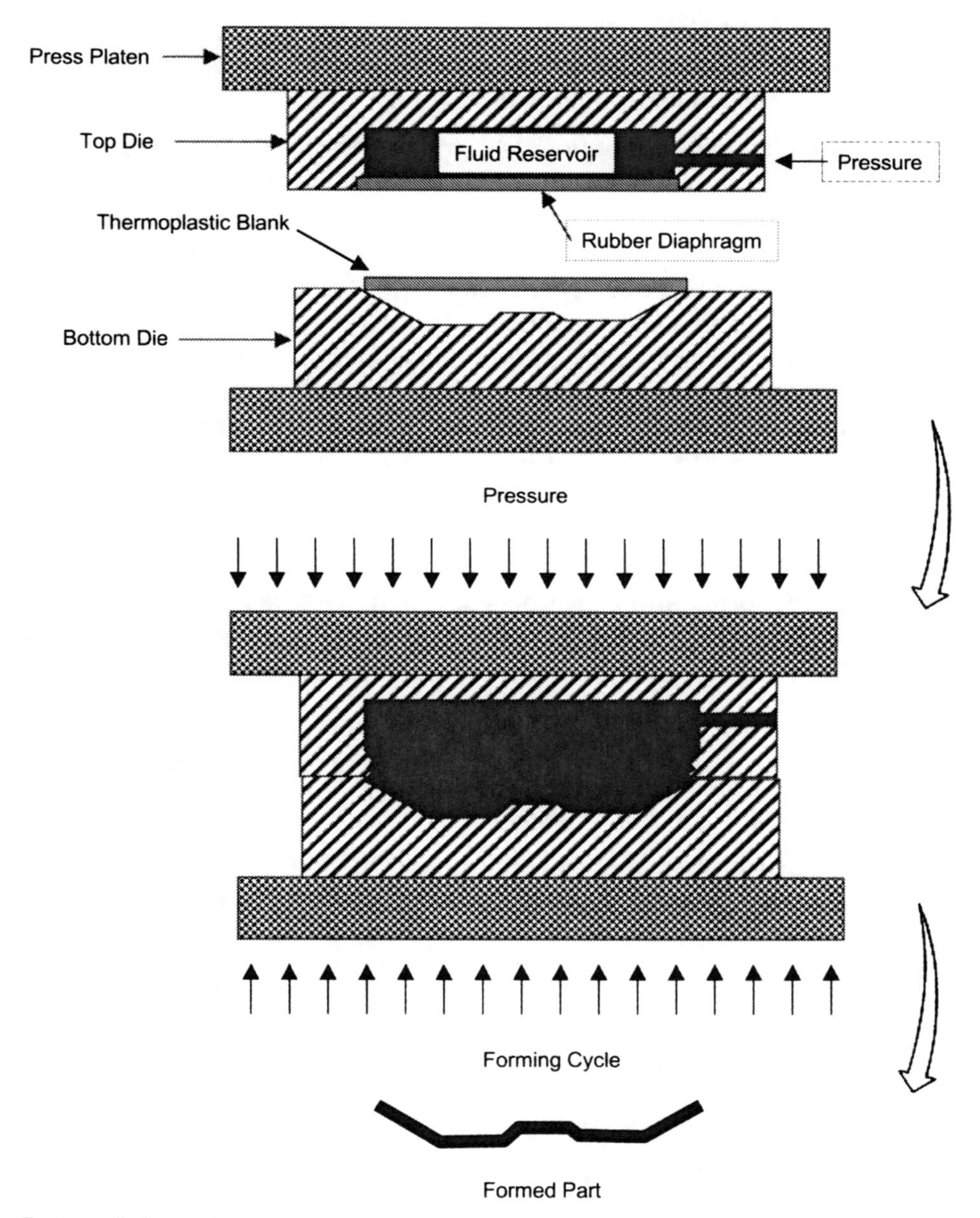

Fig. 21. Hydroforming

T_m for semi-crystalline resins. This usually dictates a transfer time of 15 s or less. For optimum results presses with fast closing speeds (e.g., 200–500 inch min^{-1}) are preferred that are capable of producing pressures of 200–500 psig. There is some disagreement as to whether it is better to use pre-consolidated blanks or lose unconsolidated ply packs for thermoforming.

Pre-consolidated blanks offer the advantage of being well consolidated with no voids or porosity but do not slip as well during the forming operation as loose unconsolidated ply packs.

While these approaches seem fairly straightforward on the surface, they are actually quite complicated because of the inextensibility of the continuous fiber reinforcement. There are four primary resin flow phenomena that must be dealt with when forming thermoplastic composite parts (Fig. 22): (1) resin percolation, (2) transverse squeeze flow, (3) interply slip and (4) intraply slip. Resin percolation and transverse squeeze flow normally occur during consolidation but are also factors during forming. Resin percolation is the flow of the viscous polymer through or along the fiber bed that allows the plies to bond together, while transverse squeeze flow eliminates slight variations in prepreg thickness by allowing the prepreg layers to spread latterly due to applied pressure. The polymer matrix tends to flow parallel to the fiber axis, while flow through the fiber bed is much more difficult. When flow occurs in a direction off-axis to the fiber orientation, the fibers tend to move with the resin.[8] Interply slip, or slip between the plies at the resin-rich ply interfaces, and intraply slip, or slip within the plies themselves by a combination of transverse and

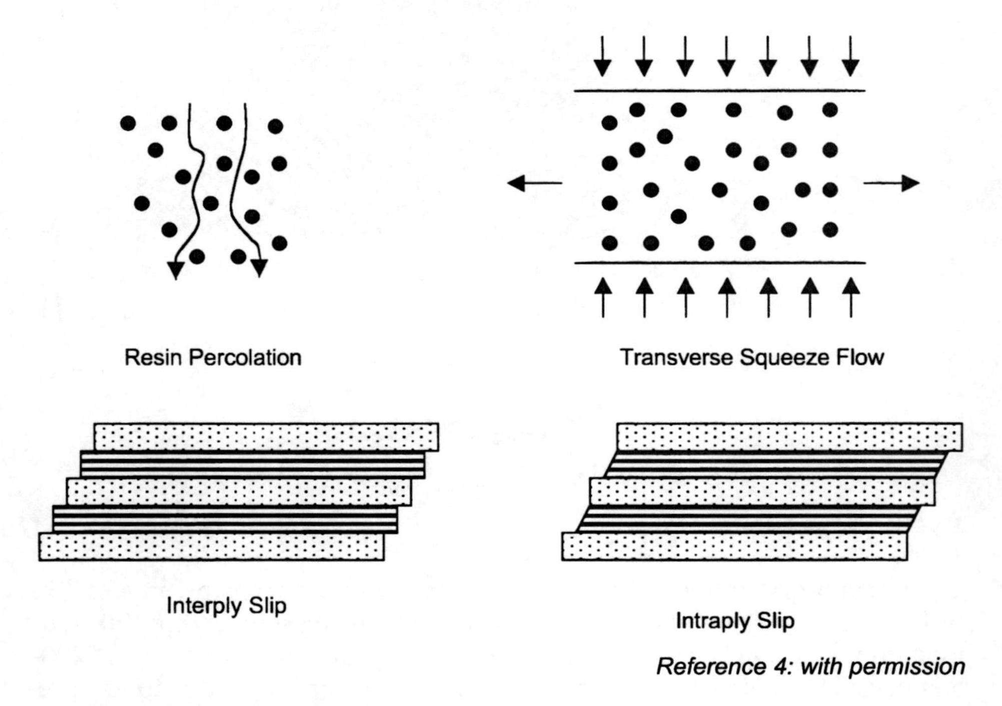

Fig. 22. Resin Flow Patterns During Thermoplastic Composite Processing

axial shear, are normally encountered in thermoforming.[17] If you did not get these slip mechanisms during forming, you would either break or buckle the reinforcing fibers or fail to form the part. Another way of looking at these flow mechanisms is shown in Fig. 23. To form a double contoured part requires all four mechanisms, while a single curvature part requires only three. Consolidation of flat or mildly contoured skins requires only the first two. A couple of examples of these mechanisms are shown in Fig. 24. In the first, transverse flow of a single curvature part often results in ply thickening and ply thinning, particularly at or near radii. In the second, interply slip prevents buckling or wrinkling of the plies at the radius.

Carbon fibers in a viscous or near liquid thermoplastic resin are still extremely strong in tension but will buckle and wrinkle readily when

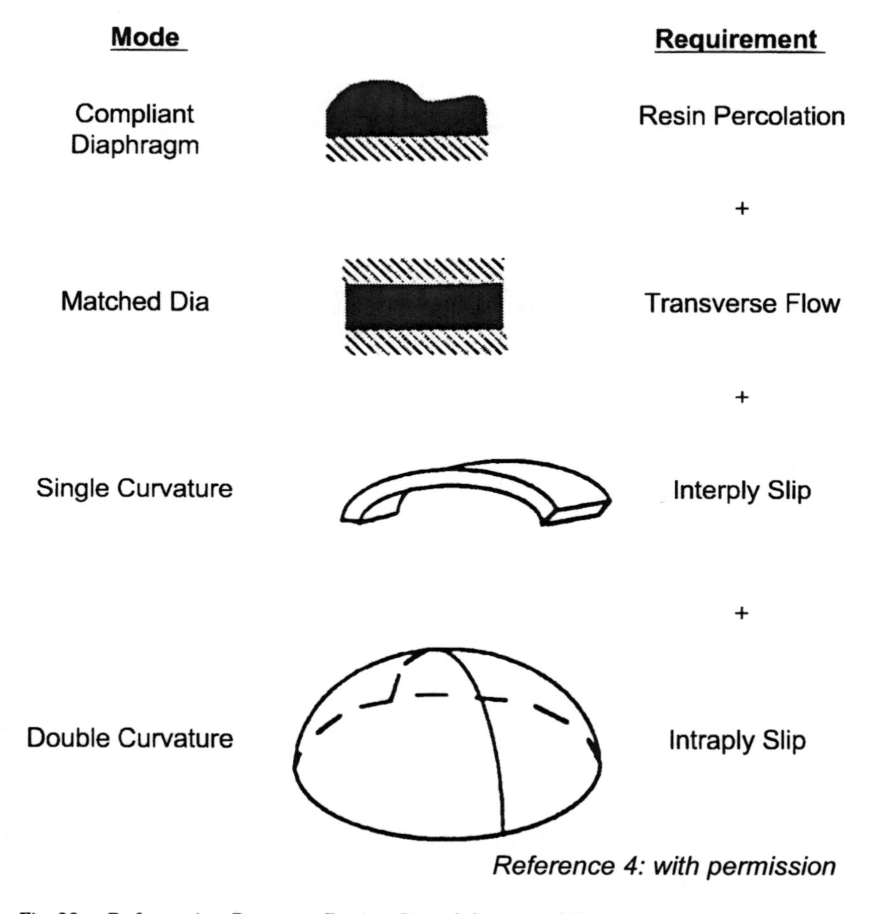

Fig. 23. Deformation Processes During Consolidation and Forming

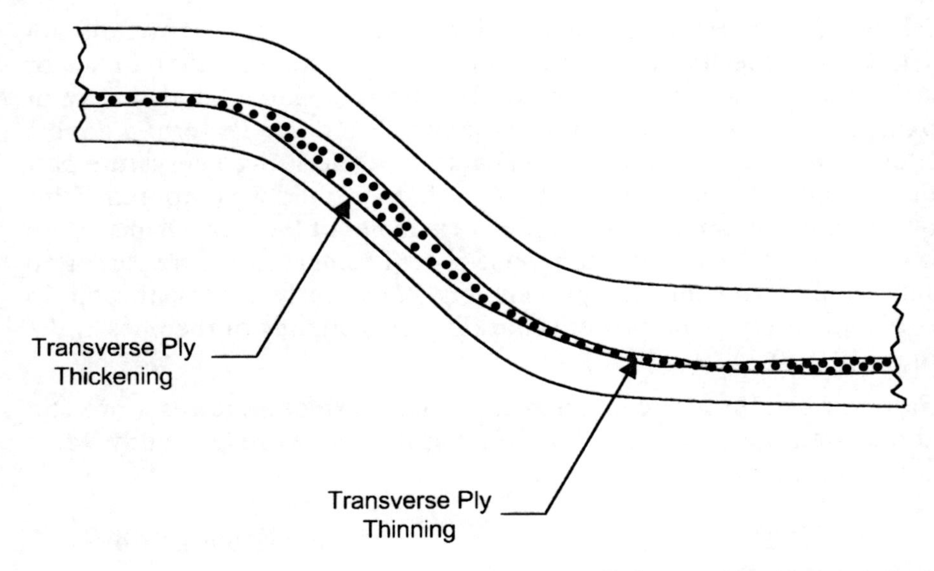

Reference 11: with permission

Transverse Flow in Reversed Single Curvature

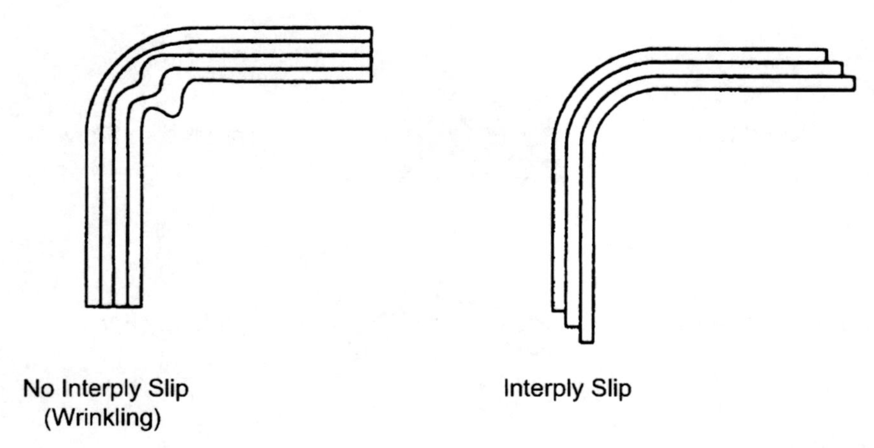

Reference 18: with permission

Importance of Interply Slip During Forming

Fig. 24. Examples of Slip Mechanisms During Thermoplastic Forming

placed in compression. Therefore, either the part shape or the die has to be designed to keep the fibers in tension throughout the forming process, but at the same time allow them to move through either interply and/or intraply slip. If neither the part shape nor the die design is amendable to preventing compression buckling of the fibers, a special holding/clamping fixture can be used during the forming operation. These fixtures can be as simple as peripheral clamping fixtures for the blank that allow the material to slip as necessary during forming, or they can be rather sophisticated mechanisms involving springs located at strategic positions to provide variable tension. Properly designed, the springs allow the part to rotate out-of-plane yielding improved force to fiber directional alignment and allowing greater variations in draw depth.[2] The type of holding fixture and the location of its springs is usually determined by previous experience and by considerable trial and error. Slower forming speeds also help to reduce wrinkling and buckling.[19] In the same study, it was shown that fiber buckling and waviness reduced part strength by up to 50% and that tension pressures of 40–100 psig were often sufficient to suppress fiber buckling.

Diaphragm forming is a rather unique process that is capable of making a wider range of part configurations and severe contours than can be made by press forming. A typical diaphragm forming cycle for a PEEK thermoplastic part is shown in Fig. 25. Diaphragm forming can be done in

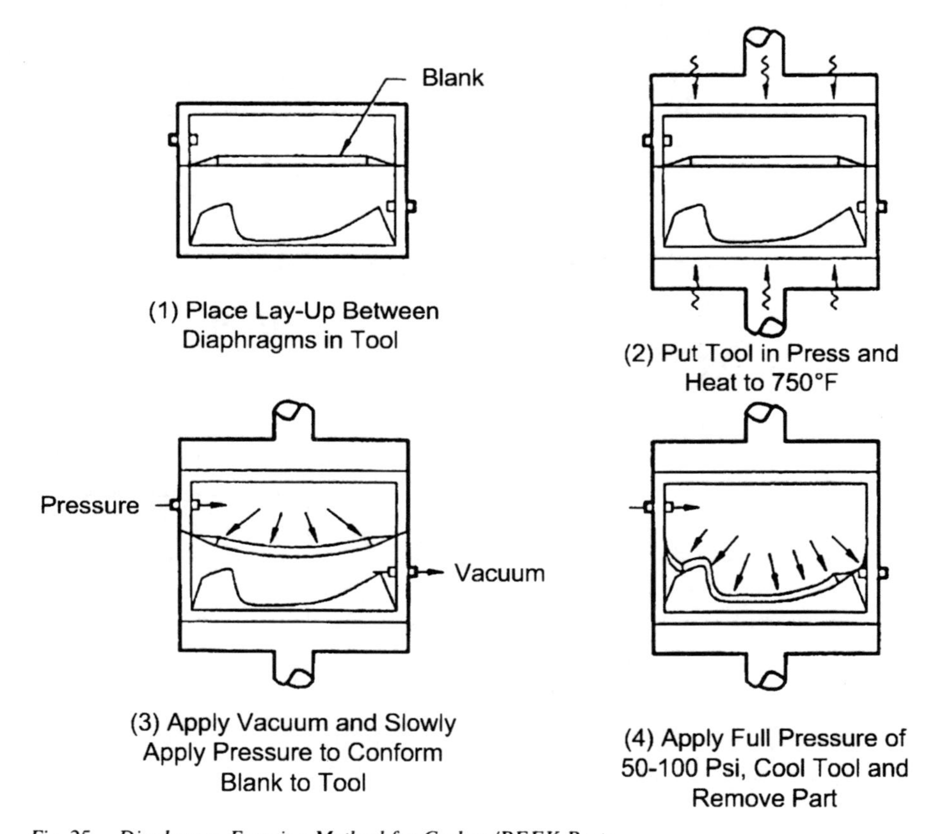

Fig. 25. Diaphragm Forming Method for Carbon/PEEK Parts

either a press or an autoclave. In this process, unconsolidated ply packs, to more readily promote ply slippage, rather than pre-consolidated blanks are placed between two flexible diaphragms. A vacuum is drawn between the diaphragms to remove air and provide tension on the lay-up. The part is then placed in a press and heated above the melt temperature. Gas pressure is used to form the pack down over the tool surface. During forming, the plies slide within the diaphragms creating tensile stresses that reduce the tendency for wrinkling. The gas pressure both forms and consolidates the part to the tool contour. Pressures usually range from 50 to 150 psig with cycle times of 20–100 min; however, for more massive tools, cycle times of 4–6 h are not unusual. Slow pressurization rates are recommended to avoid out-of-plane buckling.

Diaphragm materials include Supral superplastic aluminum and high-temperature polyimide films (Upilex-R and -S). Supral aluminum sheet is more expensive than the polyimide films but is less susceptible to rupturing during the forming cycle. Polyimide films work well for thin parts with moderate draws, while Supral sheet is preferred for thicker parts with complex geometries. Typical diaphragm forming temperatures are 750 °F for Supral and 570–750 °F for the polyimide films.[15] A comparison of thickness variations in a part as a function of the forming pressure and diaphragm materials used is shown in Fig. 26. Note that even with

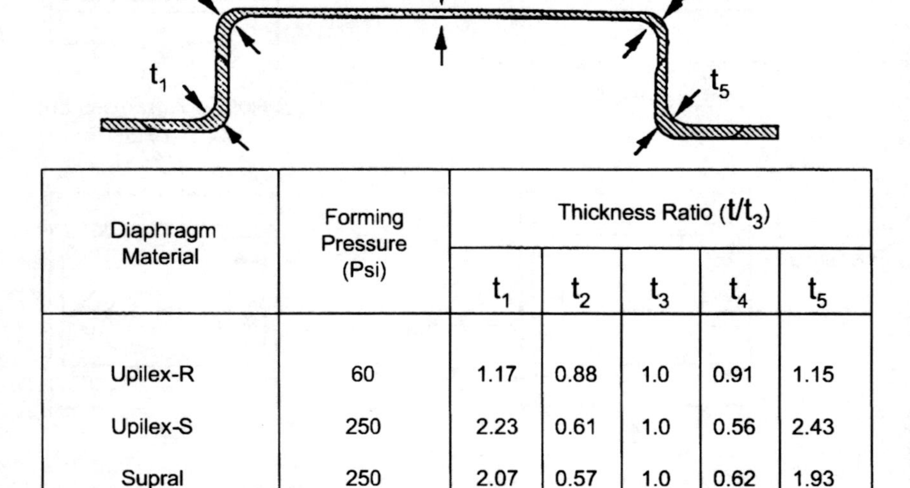

Diaphragm Material	Forming Pressure (Psi)	Thickness Ratio (t/t_3)				
		t_1	t_2	t_3	t_4	t_5
Upilex-R	60	1.17	0.88	1.0	0.91	1.15
Upilex-S	250	2.23	0.61	1.0	0.56	2.43
Supral	250	2.07	0.57	1.0	0.62	1.93

Reference 18: with permission

Fig. 26. Thickness Variations Observed in Diaphragm Formed Parts

diaphragm forming, material thinning occurs in the male radii and thickening occurs in the female radii. One disadvantage of this process is that the materials that can be formed must comply with the forming temperatures of the available diaphragm materials. In addition, the diaphragm materials are expensive and can be used only once.

Many other processes have been evaluated for fabricating thermoplastic composite structural shapes, including roll forming, pultrusion and even resin transfer molding. A typical roll-forming process is shown in Fig. 27. A pre-consolidated blank is heated above its melt temperature and then fed through a series of rollers to gradually form the desired shape. In some

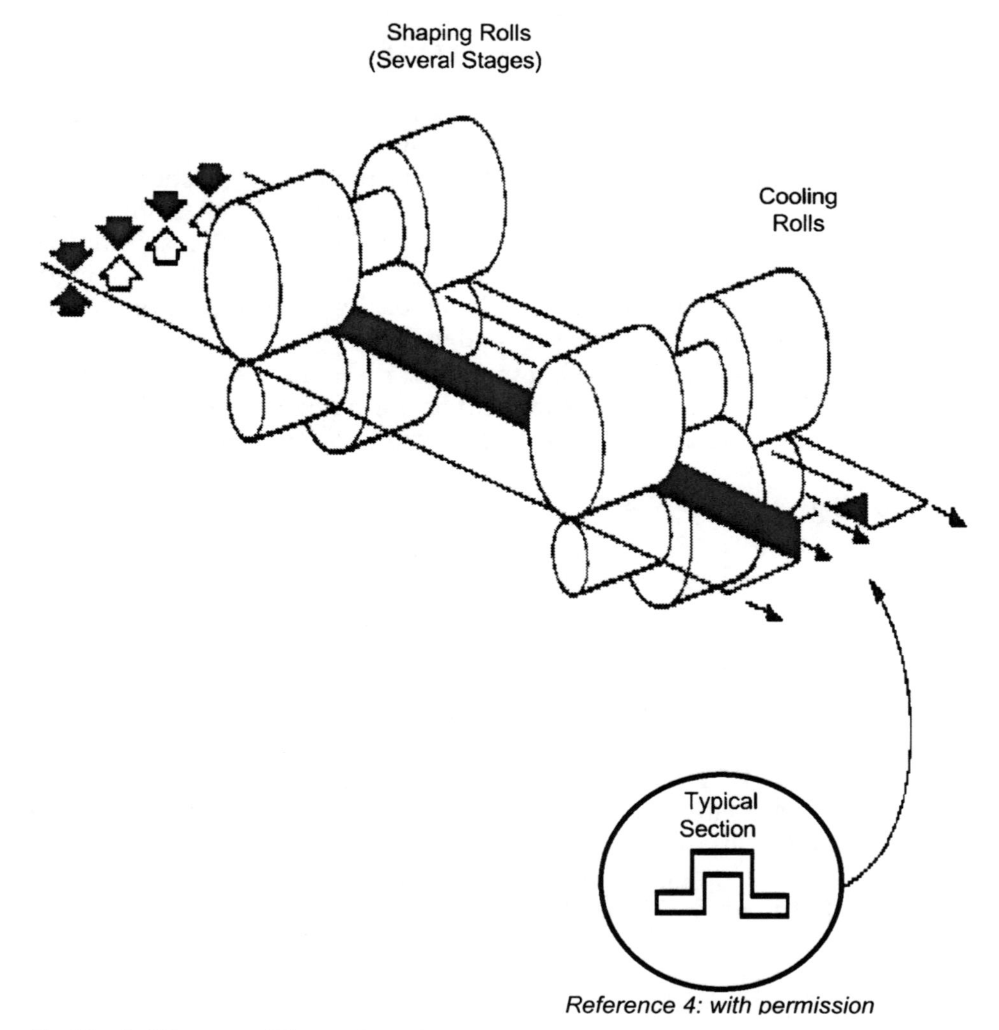

Fig. 27. Roll Forming of Thermoplastic Composite Parts

operations, only the portion of the blank that is to be formed is heated. The trick to roll forming is to maintain uniform pressure on all portions of the part throughout heat-up, forming and cool-down. If uniform pressure is not maintained on the molten portions of the part being formed, deconsolidation will occur due to relaxation of the fiber bed. Thermoplastic composites have also been successfully pultruded but the process is much more difficult and expensive than that for thermosets due to the high melt temperatures and viscosities. Resin transfer molding of thermoplastics is not a feasible process with the materials we have discussed so far. The viscosity is just too high for the long flow paths required for RTM and total wet-out of a dry reinforced fiber bed is rarely achieved. However, a relatively new class of materials called "cyclics" offers great potential for RTM. These materials initially melt and flow like thermosets and then undergo a ring opening mechanism to form a linear thermoplastic on further heating.[20] The molecular weight increases during heating in the presence of an anionic catalyst.[2] At the present time, the technology is applicable only to low-temperature thermoplastics, such as nylon and polybutylene terephathalate (PBT). Currently, these materials show great promise for commercial industries such as the automotive industry. In the future, if this technology can be extended to high-temperature thermoplastics, it could drastically alter the approach to thermoplastic composite processing and usage.

"It is virtually impossible to control fiber placement as exactly during thermoforming of a thermoplastic sheet as during hand or machine layup of prepreg tapes, which is a disadvantage in highly strength-critical parts.[16]"

10.6 Joining

Another unique advantage of thermoplastic composites is the rather extensive joining options available. While thermosets are restricted to either co-curing, adhesive bonding or mechanical fastening, thermoplastic composites can be joined by melt fusion, dual resin bonding, resistance welding, ultrasonic welding or induction welding, as well as by conventional adhesive bonding and mechanical fastening.

Adhesive bonding. In general, structural bonds using thermoset (e.g., epoxy) adhesives produce lower bond strengths with thermoplastic composites than with thermoset composites. This is believed to be due primarily to the differences in surface chemistry between thermosets and thermoplastics. Thermoplastics contain rather inert, non-polar surfaces that impede the ability of the adhesive to wet the surface. A number of different surface preparations have been evaluated including the following:[2]

- sodium hydroxide etch,
- grit blasting,

- acid etching,
- plasma treatment,
- silane coupling agents,
- corona discharge, and
- kevlar (aramid) peel plies.

While a number of these surface preparations, or combinations of them, give acceptable bond strengths, the long-term service durability of thermoplastic adhesively bonded joints has not been established.

Mechanical fastening. Thermoplastic composites can be mechanically fastened in the same manner as thermoset composites. Initially, there was concern that thermoplastics would creep excessively when fastened, resulting in a loss of fastener torque and thus lower joint strengths. Extensive testing has shown that this was an unfounded fear and mechanically fastened thermoplastic composite joints behave very similar to thermoset composite joints.

Melt fusion. Since thermoplastics can be processed multiple times by heating above their T_g for amorphous or T_m for semi-crystalline resins with minimal degradation, melt fusion essentially produces joints as strong as the parent resin. An extra layer of neat resin film can be placed in the bondline for gap-filling purposes and to insure that there is adequate resin to facilitate a good bond. However, if the joint is produced in a local area, adequate pressure must be provided over the heat-affected zone to prevent elasticity of the fiber bed from producing delaminations at the ply interfaces.

Dual resin bonding. In this method, a lower melting temperature thermoplastic film is placed at the interfaces of the joint to be bonded. As shown in Fig. 28, in a process called amorphous bonding or the Thermabond process, a layer of amorphous PEI is used to bond two PEEK composite laminates together. To provide the best bond strengths, a layer of PEI is fused to both PEEK laminate surfaces prior to bonding to enhance resin mixing. In addition, an extra layer of film may be used at the interfaces for gap filling purposes. Since the processing temperature for PEI is below the melt temperature of the PEEK laminates, the danger of ply delamination within the PEEK substrates is avoided. Like the melt fusion process, dual resin bonding would normally be used to join large sections together, such as bonding stringers to skins.

Resistance welding. Two approaches have been used to join thermoplastic composite parts using resistance heating. As shown in Fig. 29, a carbon ply can be used as a resistance heater or a separate metallic heater can be embedded in the bondline. The advantages of the carbon-ply method are that there is no foreign object left in the bondline after bonding, and the thermoplastic resin may adhere better to the carbon fibers than to a metallic material. In general, the resin generally is removed

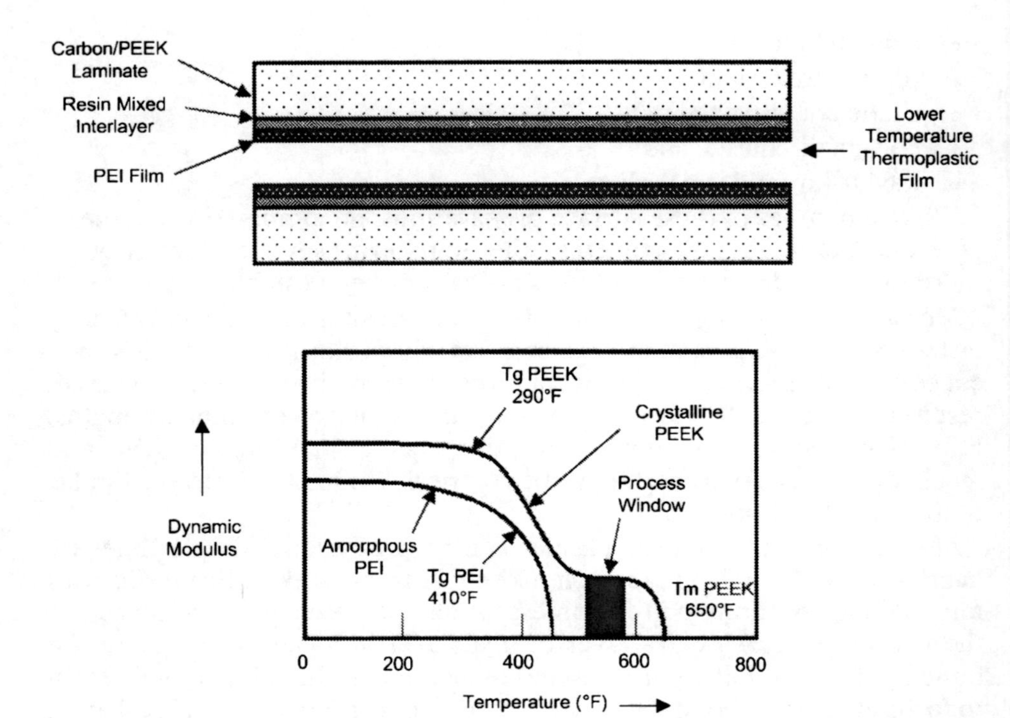

Fig. 28. Amorphous or Dual-Resin Bonding

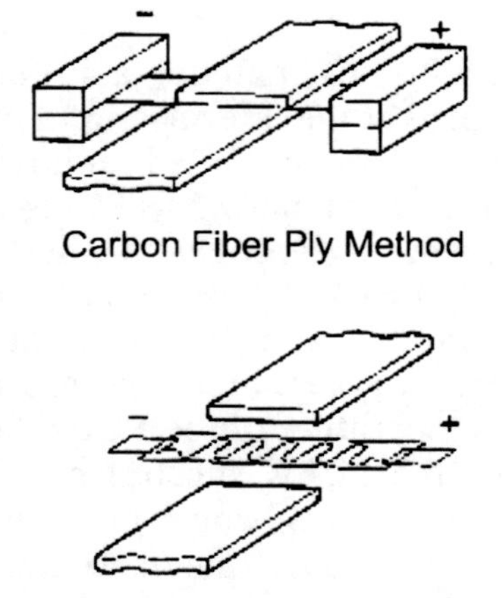

Fig. 29. Resistance Heating Methods for Joining Thermoplastic Composites

from the ply ends where it is clamped to the electrical bus bars. In addition, the carbon-ply method does not supply heat to the joint as effectively as an embedded metallic heater and is more prone to electrical shorts during the bonding process. Thin layers of polymer film are usually added to both sides of the resistance heater to provide electrical insulation from the carbon fibers and additional matrix material to fill any gaps. For all fusion welding operations with thermoplastic composites, it is necessary to maintain adequate pressure at all locations that are heated above the melt temperature. If pressure is not maintained at all locations that exceed the melt temperature, de-consolidation due to fiber bed relaxation will likely occur. The pressure should be maintained until the part is cooled below its T_g. Typical processing times for resistance welding are 30 s to 5 min at 100–200 psig pressure.[21]

Ultrasonic welding. It is used extensively in commercial process to join lower-temperature unreinforced thermoplastics and can also be used for advanced thermoplastic composites. As shown in Fig. 30, an ultrasonic

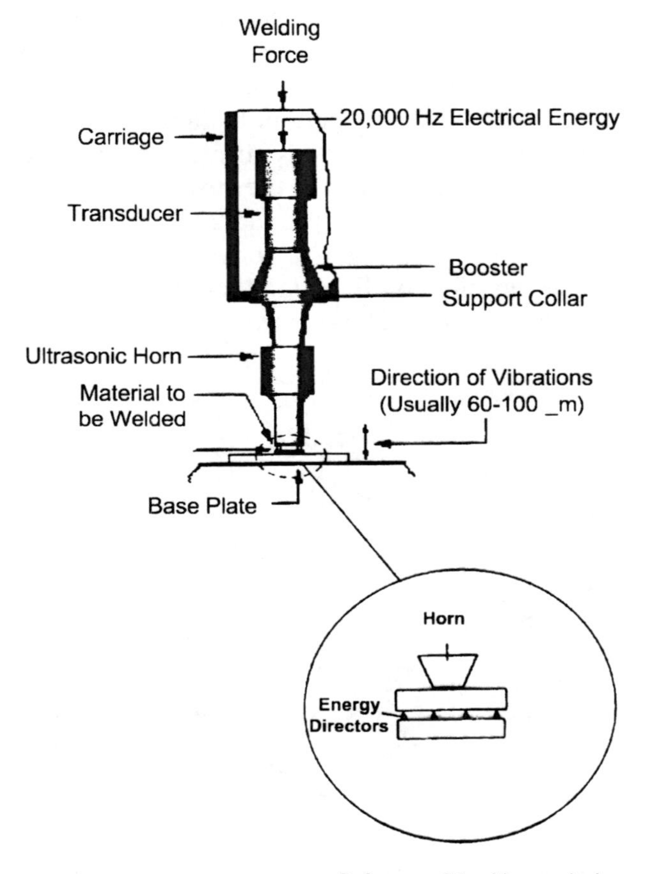

Fig. 30. Ultrasonic Joining Method for Thermoplastic Composites

horn, also known as a sonotrode, is used to produce ultrasonic energy at the composite interfaces. Electrical energy is converted into mechanical energy and the sonotrode is placed in contact with one of the pieces to be joined. The second piece is held stationary while the vibrating piece creates frictional heating at the interface. Ultrasonic energies of 20–40 kHz are normally used. The process works best if one of the surfaces has small asperities that act as energy directors or intensifiers. The asperities have a high energy per unit volume and melt before the surrounding material. The quality of the bond is increased with increasing time, pressure and the amplitude of the signal.[2] Again, it is common practice to incorporate a thin layer of neat resin film to provide gap filling. Typical weld parameters are less than 10 s at 70–200 psig pressure.[21] This process is somewhat similar to spot welding of metals but is difficult to scale up to large area bondlines.

Induction welding. Similar to resistance welding, induction welding techniques have been developed in which a metallic susceptor may or may not be placed in the bondline. It is generally accepted that the use of a metallic susceptor produces superior joint strengths. A typical induction setup, shown in Fig. 31, uses an induction coil to generate an electromagnetic field that results in eddy current heating in the conductive susceptor and/or by hysteresis losses in the susceptor. Susceptor materials that have been evaluated include iron, nickel, carbon fibers and copper meshes. As with resistance heating, it is normal practice to place a layer of polymer film on each side of the metallic susceptor. Typical welding parameters are 5–30 min at 50–200 psig pressure.[21]

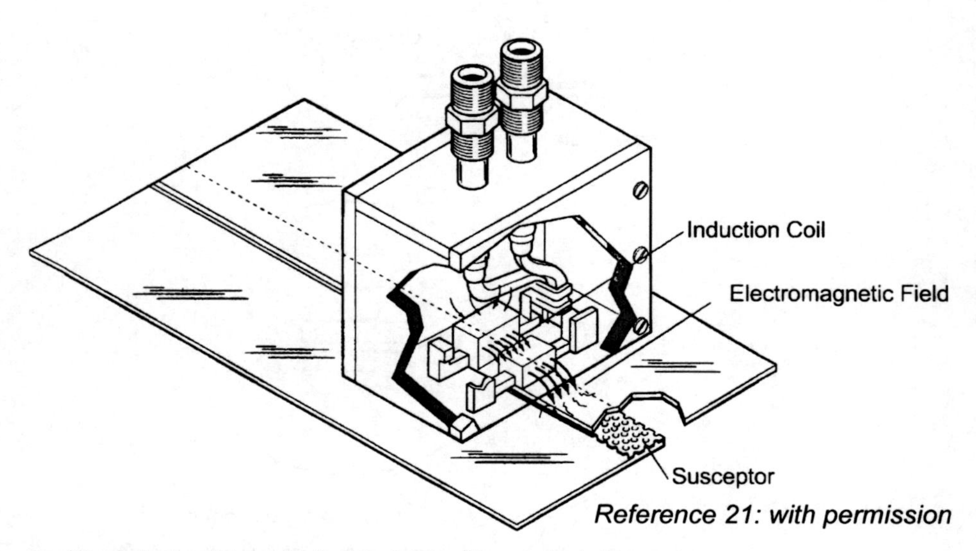

Fig. 31. Induction Joining Method for Joining Thermoplastic Composites

A comparison of single lap shear strengths produced in thermoplastic composites using the various techniques described above is given in Fig. 32. Note that adhesive bonding yields lower joint strengths than the fusion bonding techniques and is very dependent on the surface preparation method used. Autoclave co-consolidated (melt fusion) joint strengths approach virgin autoclave molded strengths. Typically, resistance and induction welding strengths are similar and exhibit similar properties, both of which are superior to those of ultrasonic welding.

10.7 Summary

Thermoplastic composites offer some definite advantages compared to thermoset composites; however, in spite of large investments over the last 20 years, very few continuous fiber thermoplastic composites have made it into production applications.

Compared to thermoset composites, thermoplastic composites offer the potential for short processing times, but their inherent characteristics have prevented them from replacing thermoset composites in the aerospace industry, namely:

- high processing temperatures (500–800 °F) increase the cost of the prepreg and complicate the use of conventional processing equipment;

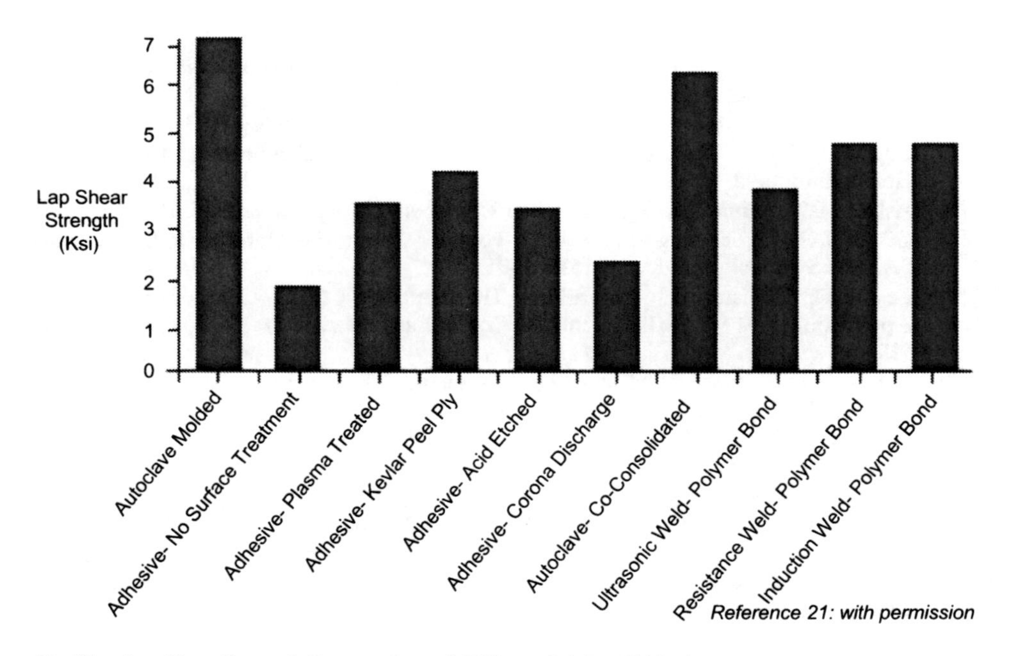

Fig. 32. Lap Shear Strength Comparaison of Different Joining Méthods

- the lack of tack and boardiness of the prepreg results in expensive manual handling operations;
- thermoforming of continuous fiber-reinforced thermoplastics has proven to be much more difficult than first anticipated due to the tendency of the fibers to wrinkle and buckle if not maintained under tension during the forming operation;
- the early claims of superior toughness and damage tolerance have largely been negated by the development of much tougher thermoset resins; and
- solvent and fluid resistance properties remain major barriers to the use of amorphous thermoplastic composites.

In the author's opinion, two criteria must be satisfied to take advantage of continuous fiber thermoplastic composites: (1) the demand for a large quantity of parts, and (2) the process must be automated to remove almost all manual operations. Unfortunately, neither of these criteria has been met in the aerospace industry where lot sizes are small and production rates cannot justify the investment in highly sophisticated automated equipment. It is interesting and insightful that GMTs have made significant inroads in the automotive industry where the demand for parts is large and the process has been almost totally automated.

References

[1] Strong,A.B., *Fundamentals of Composite Manufacturing: Materials, Methods, and Applications*, Society of Manufacturing Engineers, 1989.

[2] Strong A.B., *High Performance and Engineering Thermoplastic Composites*, Technomic Publishing , 1993.

[3] Rosen S.L., *Fundamental Principles of Polymeric Materials*, Wiley, 1971.

[4] Cogswell F.N., *Thermoplastic Aromatic Polymer Composites*, Butterworth-Heinemann, 1992.

[5] Astrom B.T., *Manufacturing of Polymer Composites*, Chapman & Hall, 1997.

[6] Gibbs H.H., "Processing Studies on K-Polymer Composite Materials," 30th National SAMPE Symposium, 1985, pp. 1585–1601.

[7] Lesser D., Banister B., "Amorphous Thermoplastic Matrix Composites for New Applications," 21st SAMPE Technical Conference, September 25–28, 1989, pp. 507–513.

[8] Muzzy J.D., Colton, J.S., "The Processing Science of Thermoplastic Composites," in *Advanced Composites Manufacturing*, Wiley, 1997.

[9] Jang B.J., *Advanced Polymer Composites: Principles and Applications*, ASM International, 1994.

[10] Muzzy J., Norpoth L., Varughese B., "Characterization of Thermoplastic Composites for Processing," *SAMPE Journal* **25**(1), January/February 1989, pp. 23–29.

[11] Leach D.C., Cogswell F.N., Nield E., "High Temperature Performance of Thermoplastic Aromatic Polymer Composites," 31st National SAMPE Symposium, 1986, pp. 434–448.

[12] Loos A.C., Min-Chung L., "Consolidation during Thermoplastic Composite Processing," in *Processing of Composites*, Hanser/Gardner Publications, 2000.

[13] Strong A.B., "Manufacturing," in *International Encyclopedia of Composites* (ed. Stuart Lee), VCH Publishers, 1990, pp. 102–126.
[14] Harper R.C., "Thermoforming of Thermoplastic Matrix Composites--Part II," *SAMPE Journal* **28**(3), May/June 1992, pp. 9–17.
[15] Okine R.L., "Analysis of Forming Parts from Advanced Thermoplastic Sheet Materials," *SAMPE Journal* **25**(3), May/June 1989, pp. 9–19.
[16] Harper R.C., "Thermoforming of Thermoplastic Matrix Composites--Part I," *SAMPE Journal* **28**(2), March/April 1992, pp. 9–18.
[17] Cogswell F.N., Leach D.C., "Processing Science of Continuous Fibre Reinforced Thermoplastic Composites," *SAMPE Journal* May/June 1988, pp. 11–14.
[18] Dillon G., Mallon P., Monaghan M., "The Autoclave Processing of Composites," in *Advanced Composites Manufacturing*, Wiley, 1997, pp. 207–258.
[19] Soll W., Gutowski T.G., "Forming Thermoplastic Composite Parts," *SAMPE Journal* May/June 1988, pp. 15–19.
[20] Dave R.S., Udipi K., Kruse R.L., "Chemistry, Kinetics, and Rheology of Thermoplastic Resins Made by Ring Opening Polymerization," in *Processing of Composites*, Hanser/Gardner Publications, 2000.
[21] McCarville D.A., Schaefer H.A., "Processing and Joining of Thermoplastic Composites," in *ASM Handbook Volume 21 Composites*, ASM International, 2001, pp. 633–645.

Chapter 11

Commercial Composite Processes: These Commercial Processes Produce Far More Parts than the High-performance Processes

There are a number of important composite fabrication processes that are used more widely in the commercial marketplace than for high-performance aerospace products. In this chapter, five of the most important of these processes will be covered: lay-up, compression molding, injection molding, reaction injection molding (RIM) and pultrusion. Some of these processes are capable of producing over a million parts per year while others are more amenable to low production rates. Some are restricted to thermoset resins while others can accommodate both thermoset and thermoplastic resins. Some use continuous fiber reinforcement while others are restricted to short fiber reinforcements. Although a number of different types of reinforcement can be used, glass fibers are the most predominate due to their relatively low cost; the availability of a large number of product forms; and their overall good combination of physical and mechanical properties.

11.1 Lay-up Processes

Lay-up processes are ideally suited for the manufacture of low-volume medium-to-large parts. These processes are capable of making very large parts with minimal tooling costs, such as custom-built yacht hulls. However, manual lay-up processes, similar to the ones covered in Chapter 5 on Collation, are labor intensive and part quality is very dependent on worker skill. Three lay-up processes will be covered: (1) wet lay-up, (2) spray-up and (3) low-temperature curing/vacuum bag (LTVB) prepreg lay-up.

Wet lay-up. In the wet lay-up process, shown schematically in Fig. 1, a dry reinforcement, usually a woven glass roving or cloth, is manually placed on the mold. A low-viscosity liquid resin is then applied to the reinforcement by pouring, brushing or spraying. Squeegees or rollers are used to densify the lay-up, thoroughly wetting the reinforcement with the resin and removing excess resin and entrapped air. The laminate is built up layer-by-layer until the required thickness is obtained. E-glass is the most prevalent material but S-2 glass, carbon and aramid can be used where the improved properties justifies their higher costs. Heavy glass woven rovings (500 g m^{-2}) can be used to build up thickness quickly and reduce labor costs, yielding a part with approximately a 40% glass content. Although heavy woven rovings reduce lay-up times, the heavy weaves are more difficult to impregnate than the lighter weight glass clothes. Where the high strength of woven roving or glass cloth is not required, glass mats can be used to save costs. Glass mats can be either continuous strand mats, in which continuous strands of glass are swirled onto a moving carrier and then tacked with a binder, or chopped strand mats, in which chopped fibers (~1–2 inches) are sprayed onto a moving carrier and again heat tacked with a liquid, spray or powder binder. Frequently, to save weight

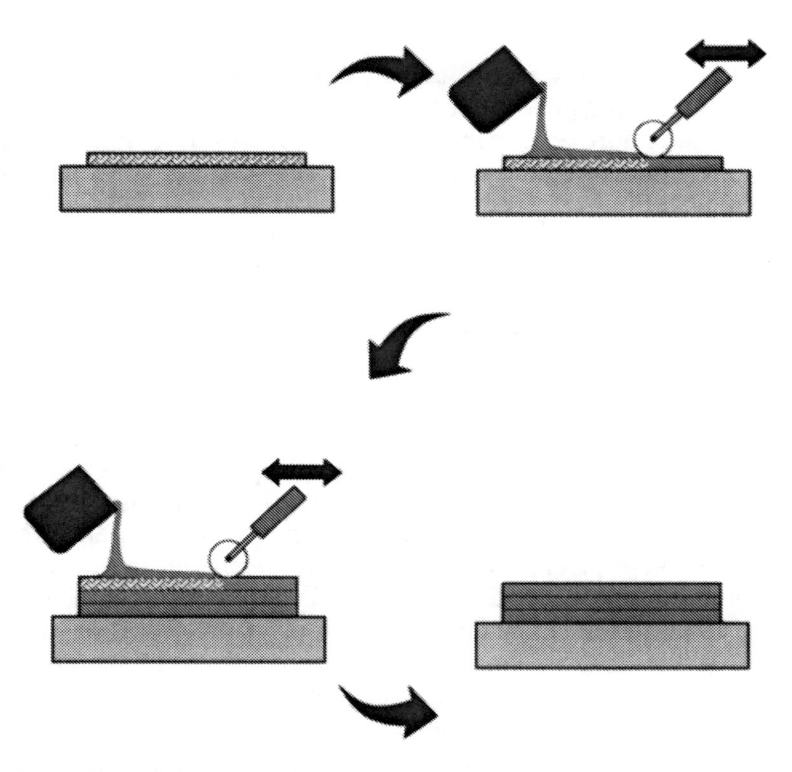

Fig. 1. Typical Wet Lau-Up Operation

and labor costs, core materials such as honeycomb, balsa, or foams are added to produce a sandwich construction, which in the case of a boat hull can also help provide flotation. In general, foam cores should be sealed prior to lay up to reduce excessive resin absorption. Since impregnation is done by hand, voids, resin-rich and resin-starved areas can be a problem. An improvement in the impregnation consistency can be obtained by pre-impregnating the reinforcement before it is placed on the tool. This can be accomplished by placing a layer of Mylar (clear plastic sheet) on a flat bench; covering the Mylar with the dry reinforcement; applying a predetermined amount of resin; covering with another layer of Mylar; and then thoroughly rolling the resin into the reinforcement. The Mylar sheets can then be used to support the pre-impregnated ply as it is moved to the lay-up. Some manufacturers have built their own impregnation machines to improve quality and productivity.

To provide a smooth surface finish on the tool side, a gel coat is often applied to the mold released tool prior to the start of lay-up. Gel coats, which are specially formulated resins that provide a resin-rich surface on the cured laminate, can be applied by either brushing or spraying. The normal thickness for a gel coat should be 0.020–0.040 inches. If the gel coat is too thick, resin crazing and cracking can develop in-service. Usually lay-up can begin after the gel coat cures to a tacky condition. Gel coats can be formulated to improve flexibility, blister resistance, stain resistance, weatherability and toughness. Tough and resilient gel coats can provide impact and abrasion resistance to the laminate surface. Gel coats can also be pigmented to provide the cured part with a variety of colors. Some manufacturers also use a fine mat or woven cloth (veil) as the first ply to further enhance surface finish. Veils, which are used primarily for surface plies, consist of thin weaves with very fine fibers to enhance the surface finish.

If the part is cured at room or low temperatures, extremely inexpensive tooling can be fabricated from wood, plaster, sheet metal or glass laminates. These tooling approaches make this process attractive for large parts where the size and expense of autoclave curing would not be practical. It is also a good process for making prototype parts where the design may change prior to production. Typically, wet lay-up molding is done on an open single-sided tool. The tool configuration can be designed to control the internal shape and surface finish (male tool) or the outside shape and surface finish (female tool). Cured parts will have one finished surface (i.e., the tool surface) that is essentially as smooth as the tool itself, while the untooled surface will be somewhat rougher.

Wet layed-up parts are usually cured at room temperature without a vacuum bag. Although a vacuum bag increases costs, it provides better consolidation and more uniform laminates. A vacuum bag cure also produces laminates with higher reinforcement contents, more uniform thicknesses and better surface finishes. Vacuum bags can also be used for intermediate debulks during the lay-up process. If the part is cured at slightly elevated temperature (i.e., <200 °F), then heat lamps are often used or a simple forced air convection oven can be built around the part. The ovens are frequently constructed of plywood with foam insulation and heated with hot air blowers. If this process is used, it is a good idea to conduct a trial heat cycle on the tool prior to part fabrication to identify any hot or cold areas. Thermocouples can be attached to the tool to monitor the cure cycle.

Polyesters and vinyl esters are the predominant resins used for wet layed-up glass fiber reinforced parts. In fact, polyesters are the most commonly used thermoset resin for all commercial composite parts. These resins provide a balance of good mechanical, chemical, electrical properties, dimensional stability, ease of handling and low cost. They can be formulated for low- or high-temperature usage, for room or high-temperature cure, and for flexible or rigid products. Additives can be incorporated to provide flame retardant properties, superior surface finishes, pigmentation, low shrinkage, weather resistance and other properties. As discussed in Chapter 3, vinyl esters, although somewhat most costly than regular polyesters, offer some advantages in toughness and weathering resistance (i.e., lower moisture absorption). Vinyl esters can also be formulated for higher-temperature resistance.

Polyester resins are usually supplied in liquid form as a mixture of resin and a liquid monomer, usually styrene. The amount of monomer is the major determinant of resin viscosity. The addition of a catalyst and its subsequent activation (usually by heat) causes the cross-linking reaction. Completion of the reaction is dependent on both the formulation and the cure cycle for the selected formulation. In room temperature curing systems, an accelerator can be used to promote a catalytic reaction. Inhibitors can also be added to provide slower cures and a longer working life (i.e., pot life), an important consideration when laying up large parts. Since polyester resins are more susceptible to exotherms than epoxies, the cure must be properly controlled.

Polyester resins can be formulated to provide special processing characteristics such as:[1]

(i) Hot strength allows hot parts to be removed from the tool or die without losing their dimensional stability or shape.

(ii) Low exothermic heat is used for thick laminates to minimize the heat given off during cure, an important consideration for parts with extremely thick sections.

(iii) Extended pot life is necessary for large, complex parts, where resin flow is needed for some time during the lay-up and cure process.

(iv) Air drying provides a tack free cure at room temperature, again useful when fabricating very large parts such as boat hulls and pool liners.

(v) Thixotropy, a property of the resin causing it to resist flowing or sagging on a vertical surface, is important when laying up boat hulls or pool liners.

(vi) Additives for special end use requirements can be added to the resin formulation to provide the finished part with special properties as dictated by end use requirements. These include the following additives:

(a) Pigments are available that can provide almost any color and shade to the finished part. Pigments can also be added to gel coats.
(b) Fillers are usually inorganic or inert materials that can improve surface appearance, processability, some mechanical properties, and reduce cost.
(c) Flame retardants are often used when interior parts are being manufactured and toxic fumes from a fire are a concern.
(d) Ultraviolet absorbers can be added to the resin to improve resistance to extended sunlight exposure.

(e) Mold release agents can either be applied directly to the mold or blended with the resin to facilitate part removal.

(f) Low shrink and low profile additives, usually thermoplastic additives, that give the cured part minimum surface waviness and low part shrinkage.

Epoxies exhibit better temperature resistance than polyesters. They have good-to-excellent mechanical strength-to-weight ratios and better dimensional stability than polyesters. Epoxies are ideal for applications requiring elevated temperatures; however, they are not as economical as polyesters, but their extended range of properties can make them cost effective in certain applications. Flame retardants, pigments, and other additives can also be added to epoxies. They can be formulated for room temperature cure, but more commonly heat is used to cure the resin if high mechanical properties are required.

Spray-up. Spray-up is another low-to-medium volume, open mold method similar to hand lay-up in its suitability for simple medium-to-large part sizes. However, greater shape complexity is possible with spray-up than with hand lay-up. Continuous strand glass roving is fed through a combination chopper and spray gun. The gun (Fig. 2) simultaneously deposits chopped roving (1-3 inches long) and catalyzed resin onto the tool. The laminate is then densified with rollers or squeegees to remove air and thoroughly work the resin into the reinforcing strands. Additional layers of chopped roving and resin are added as required for thickness. Cure is usually done at room temperature or can be accelerated by the moderate

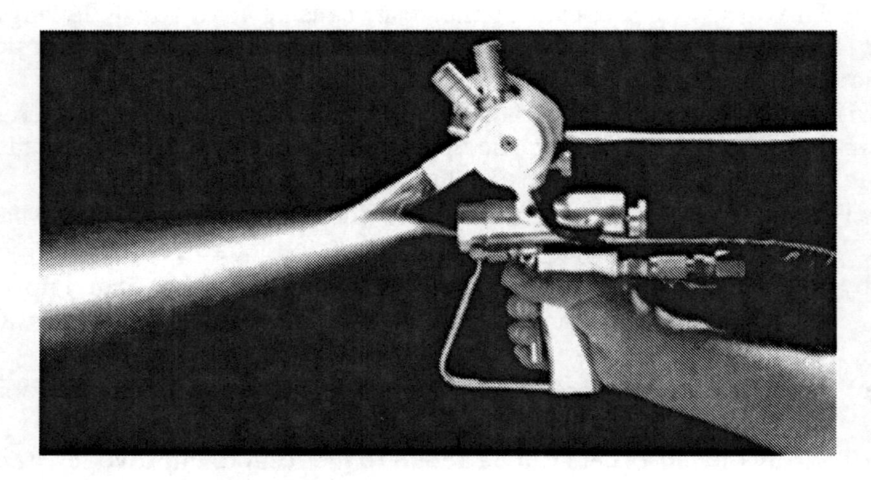

Fig. 2. Chopped Fiber Spray-Up Gun

application of heat. As with hand lay-up, superior surface finishes can be achieved by first spraying a gel coat onto the tool surface prior to spray-up. Woven roving or woven cloth is occasionally added to the laminate to provide higher strengths in certain locations. Also, core materials are easily incorporated. Again, general purpose, room temperature curing or low heat curing polyesters are used with single-sided tools. If the part complexity requires it, the tool can be assembled in sections and then disassembled when removing the part. As with the wet lay-up process, the major advantages are low cost tooling, simple processing, portable equipment that permits on-site fabrication and virtually unlimited part sizes. An additional advantage of spray-up is it is very amendable to automation, thereby reducing labor costs and the exposure of workers to potentially hazardous fumes. Sprayed-up glass parts are limited to a maximum of about 35% reinforcement.

If higher mechanical properties are required, the following methods may be used to reduce voids and porosity and insure consistent glass contents:

- *Vacuum bag.* A flexible film (nylon) is placed over the completed lay-up or spray-up, its joints sealed and a vacuum drawn. Bleeders can be used to remove excess resin and help with the evacuation of air. The vacuum bag pressure helps minimize voids in the laminate and forces excess resin and air from the lay-up. The addition of pressure also yields higher glass concentrations and provides better adhesion between the layers of a sandwich construction.
- *Pressure bag.* A tailored rubber sheet is placed against the finished lay-up or spray-up, and air pressure is applied between the rubber sheet and a pressure plate. Steam may be applied to heat the resin to accelerate cure. Pressure eliminates voids and drives excess resin and air out of the laminate, densifying it and improving the surface finish.
- *Autoclave.* Either a vacuum bag or pressure bag process can be further improved by using an autoclave, which provides additional heat and pressure capabilities, producing greater laminate densification. This process is usually employed in the production of high-performance laminates using epoxy resin systems in aircraft and aerospace applications. However, either a pressure bag or an autoclave usually increases the costs substantially and negates many of the cost advantages of the wet lay-up and spray lay-up processes.

The main advantage of both the wet lay-up and spray-up processes is that they are simple methods offering low-cost tooling, simple processing and can produce a wide range of part sizes. There is a minimum investment in equipment; however, skilled operators are needed to produce consistent part quality and the processes are somewhat messy and labor intensive.

Low-temperature curing/vacuum bag prepreg. LTVB prepregs were originally developed for building composite tooling.[2] However, over the last 10 years they have evolved to the point where they are useful for making composite structural parts. They are normally supplied as carbon fabric prepregged with a low-temperature curing epoxy resin; however, other reinforcements and even unidirectional material forms are available. The advantages of these materials over wet lay-up are: (1) being a prepreg the resin content is much more tightly controlled; (2) there is no chance for mixing errors that can occur with liquid resins; (3) much higher fiber volumes are obtainable, i.e., 55–60% versus 30–50% for a typical wet lay-up; (4) these materials are net resin content prepregs that do not require bleeding excess resin during cure; and (5) they are available in both unidirectional tape and woven cloth product forms. The disadvantage is that they cost as much as conventional 250–350 °F curing prepregs; however, much of the additional material cost is offset by the labor involved in wet lay-up to mix the resin, impregnate the dry plies and then roll out air and excess resin. As for wet lay-up and spray lay-up, the main driver for using LTVB materials is the ability to make large parts with minimal tooling investment. These prepregs have been successfully used on a number of low-rate prototype aerospace programs.[3–5]

Since these materials are formulated to cure at temperatures in the range of 100–150 °F, they contain very reactive curing agents and their working life is normally short compared to the prepregs previously discussed in Chapter 5. In general, the lower the cure temperature, the shorter the out-time.[6] For example, there are materials that initially cure at temperatures as low as 100 °F for 14 h under vacuum bag pressure but the total out-time for the material is only 2–3 d. On the other hand, if tooling can be designed to tolerate temperatures of 150 °F for 12 h, the out-time increases to 6–7 d at room temperature. A typical cure and post-cure cycle is shown in Fig. 3. The manufacturing approach is to initially cure the part at low temperatures (100–160 °F) for 12–14 h under a vacuum bag on an inexpensive tool. Low-cost tooling can be made from plywood, plaster, syntactic core, tooling dough or other materials. Heat for cure can be provided by placing the part in an oven or by building a low-cost foam oven with heat blowers around the part. The initial low-temperature cure produces a part that is approximately 40–50% cured. After the initial cure, the part is removed from the tool and post-cured at 350 °F to optimize mechanical properties and elevated-temperature resistance. Lower-temperature post-cures can be used if the service temperature is lower. During the initial cure, the part develops enough green strength to prevent it from sagging or warping during post-cure. As the part heats up to the post-cure temperature, the glass transition temperature T_g steps ahead of the post-cure temperature to prevent part distortion. Since the initial gellation and cure of these materials is done at a comparatively low

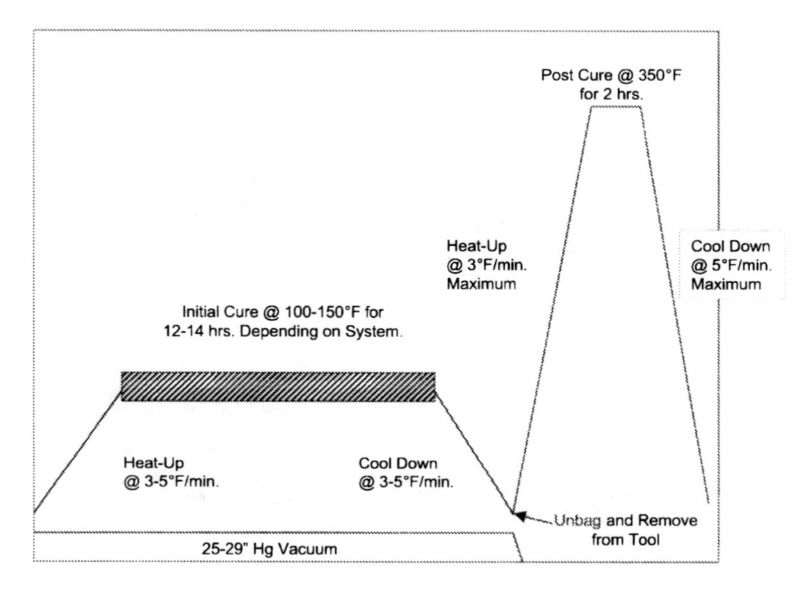

Fig. 3. Typical Cure Cycle for Low Temperature Curing/Vacuum Bag Prepreg Parts

temperature, they normally exhibit less spring-in or distortion than conventional 250–350 °F cured materials.[7]

When the materials were first developed for use on aerospace prototype vehicles, excessive porosity was a problem. Depending on the material (e.g., fabric style) and lay-up orientation, porosity levels as high as 3–5% were occasionally experienced. This is really not too surprising, considering the only consolidation pressure during cure is vacuum pressure (<15 psig). Since the initial cure is done at low temperatures (100–160 °F), water and volatile vapor pressure should not be a problem as could be the case for the 250–350 °F curing systems discussed in Chapter 6. It should be noted that a conventional 350 °F curing prepreg cured under only vacuum pressure will contain 5% or more porosity due to the combination of entrapped air and volatiles (i.e., water) coming off during cure.[8] However, air entrapped during lay-up can be a major problem for the low-temperature curing materials. Attempts were made to reduce porosity by reducing the viscosity of these resins. While this approach was partially successful, porosity remained a problem since the viscosity is constantly changing as the material ages during lay-up. To reduce and eliminate the porosity problem, manufacturers have developed three methods6,9 (Fig. 4) to allow the air to escape during the initial portion of cure. The first, partially impregnated prepreg was originally developed for 350 °F curing prepregs to help eliminate porosity. The partially impregnated prepreg contains some "dry fiber" in the middle to provide a path for the air to escape. On further

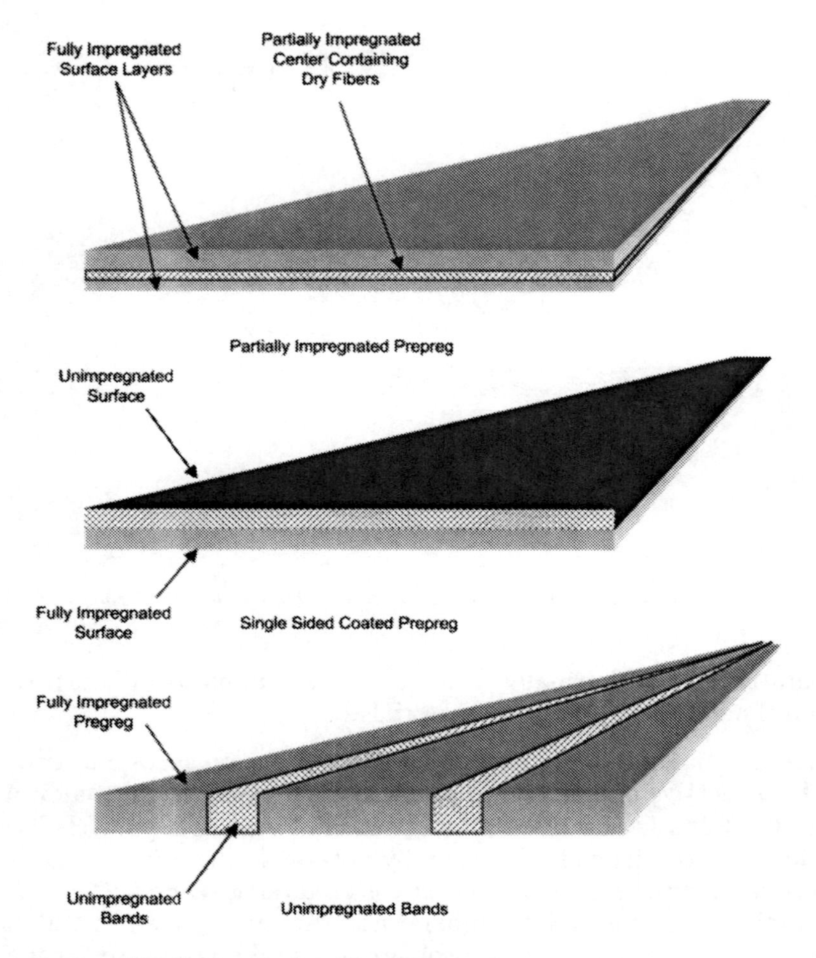

Fig. 4. LTVB Prepregs Designed for Air Removal

heating during the initial cure, the resin flows in behind the evacuating air to fully impregnate the plies. A photomicrograph of a carbon/epoxy tape laminate made using this technique (Fig. 5) shows essentially no significant porosity. The second method is to coat only one surface of the prepreg with resin that allows entrapped air to evacuate from the uncoated surface with the resin again subsequently flowing through the fiber bed to impregnate the plies. Finally, the third method consists of leaving narrow unimpregnated bands (0.50–1.50 inches wide) to allow the air to evacuate. These methods have been successful in reducing the porosity contents to less than 1%, essentially equivalent to autoclave processed parts.

Since the success of making good parts with these materials is providing an escape route for air during the initial evacuation and cure, several

Reference 9: with permission

Fig. 5. LTVB Cured Laminate

precautions need to be taken during lay-up and bagging. During lay-up it is important to work as much air out of the lay-up as possible. When laying, plies tack them in the middle first and work outward towards the edges. Extreme care should be taken to make sure there is no "bridging" or gaps between the plies, especially at the inside radii of stiffeners. Minimize the number of vacuum debulks during lay-up as this can often seal off the edges making it harder to remove entrapped air. If debulks are required, place edge breather around the part and make sure the vacuum bag does not pinch off at the edges. A recommended bagging schematic for debulking and cure is shown in Fig. 6 illustrating the importance of edge bleeding and avoiding vacuum bag pinching during evacuation.[9] Since the only source of pressure for consolidation of the plies is vacuum, it is important to have a vacuum bag with no leaks and to pull as high a vacuum as possible during cure. The minimum acceptable vacuum is in the range of 25 inches of Hg. Some manufacturers recommend that, once the final bag is applied, the part remain under vacuum for 4–8 h to allow time for all air to be evacuated.[9] During cure, it is important not to exceed the manufacturers recommended heat-up rates. These resins contain extremely reactive curing agents and there could be the potential for an exothermic reaction if they are heated too rapidly.

It should be noted that these materials can also be processed in an autoclave. Since the initial cure is still low (100–160 °F), many of the low-cost tooling approaches can still be used; however, the tool must be capable of withstanding autoclave pressures of 50–100 psig. While an autoclave may not always be available, it does provide positive pressure and allows

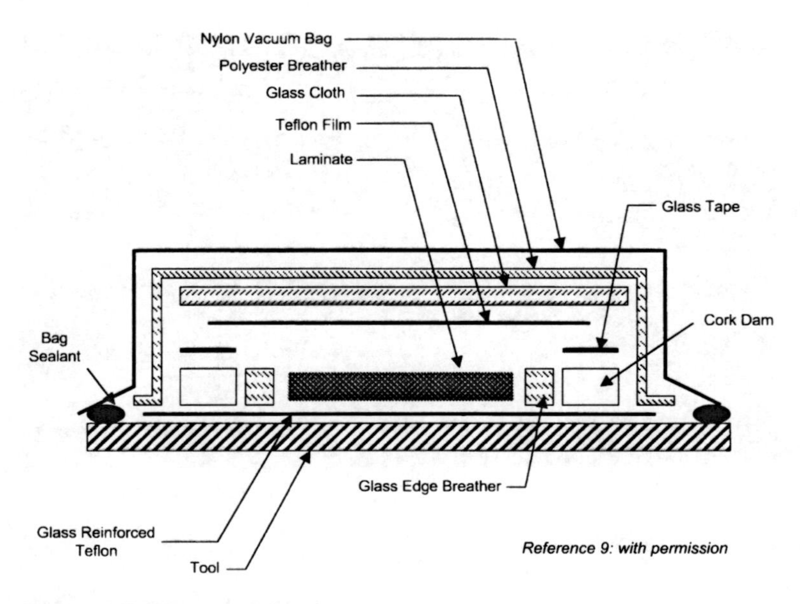

Fig. 6. LTVB Bagging Schematic

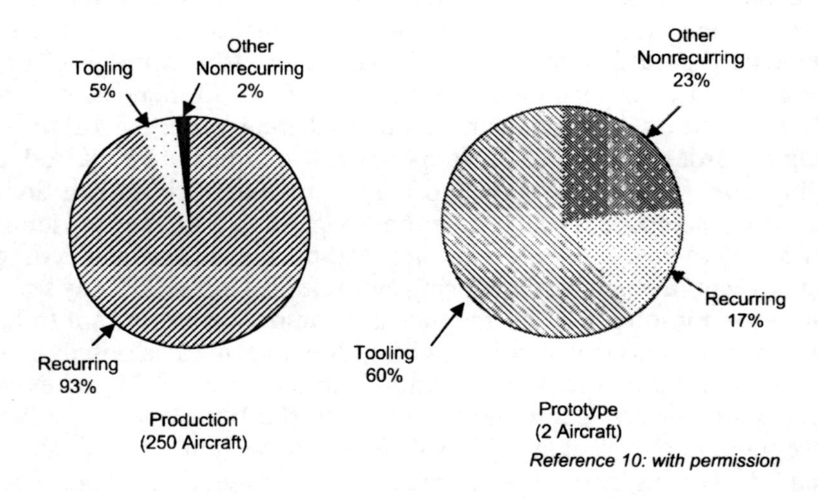

Fig. 7. Tooling Cost Comparaison - Production vs. Prototype Program

more complex configurations to be made than a vacuum bag alone. Also, the 12–14 h initial cure cycles can be shortened considerably if higher cure temperatures can be tolerated.

Prototype tooling. As shown in Fig. 7, the tooling cost for a large production program is minimal when it is amortized over a large number of parts; however, if only a few to several parts are going to be fabricated, the tooling cost can be a substantial cost of the total program, up to 60–70% if conventional autoclavable tooling (Chapter 4) is produced.[10] These tooling approaches can be used for wet lay-up methods and LTVB prepregs. A number of them would also be applicable to spray-up methods.

Low-cost molds can be fabricated by a number of approaches, several of which are shown in Fig. 8. A number of facesheet materials can be used with a plywood substructure, including thin aluminum, drapeable plywood or wet laid-up or sprayed-up fiberglass if a master model is available. Many parts can be fabricated directly from a master model constructed of plaster, or NC-machined polyurethane or syntactic foams. Simple aluminum headers can be used if a full computer-aided design (CAD) model of the part is not available. For very large parts, the mold can be built in sections (Fig. 9) and then bonded together, splined smooth and then surface-sealed.

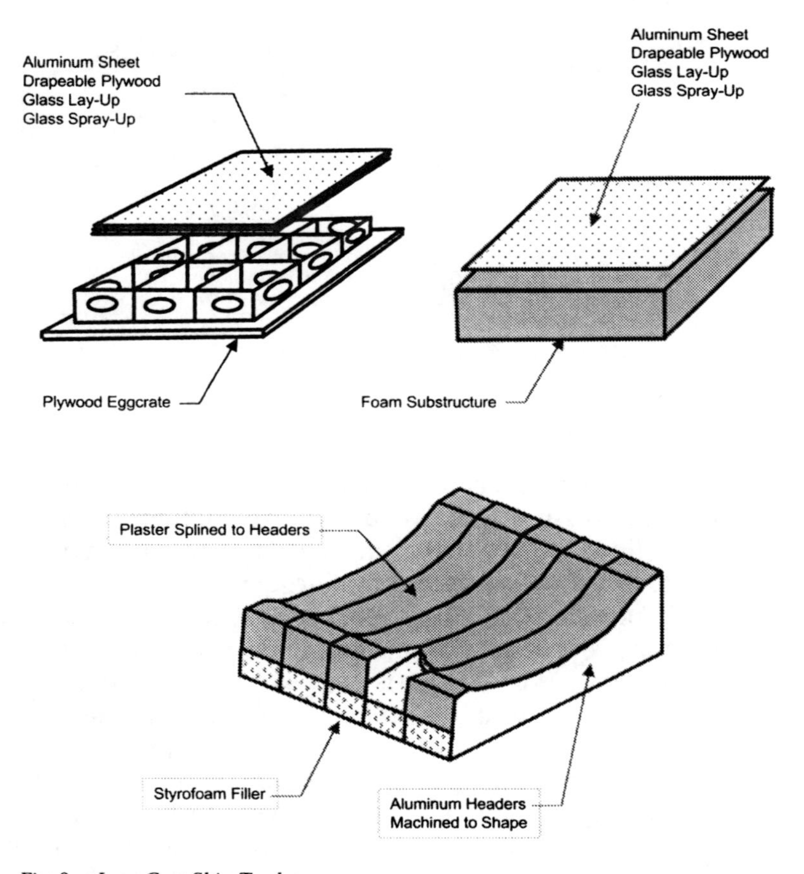

Fig. 8. Low Cost Skin Tools

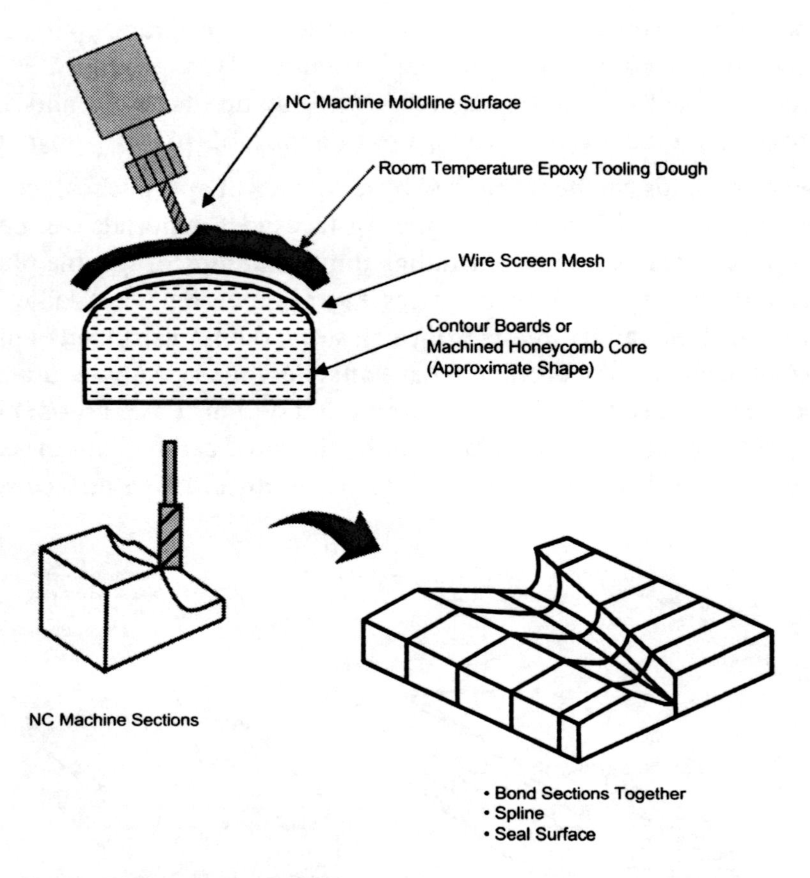

Fig. 9. Fabrication of Large Prototype Lay-Up Tools

11.2 Compression Molding

At the other end of the manufacturing spectrum, compression molding is a high-volume, high-pressure process suitable for molding complex high-strength glass fiber reinforced parts using either thermoset or thermoplastic resins. Compression molding (Fig. 10) is a matched-die process that can produce fairly large parts with excellent overall surface finishes, good dimensional control and a high degree of complexity. High volume means that at least 1,000 parts per year, but more typically in the range of 100,000 parts per year, would be needed to justify the investment in equipment and tooling. In thermoset compression molding, precombined glass fiber chopped strands and resin are available as compression molding compound, as a sheet form (sheet molding compound (SMC)) or a bulk form (bulk molding compound (BMC)). A "charge" of these compounds is placed in matched metal molds and cured

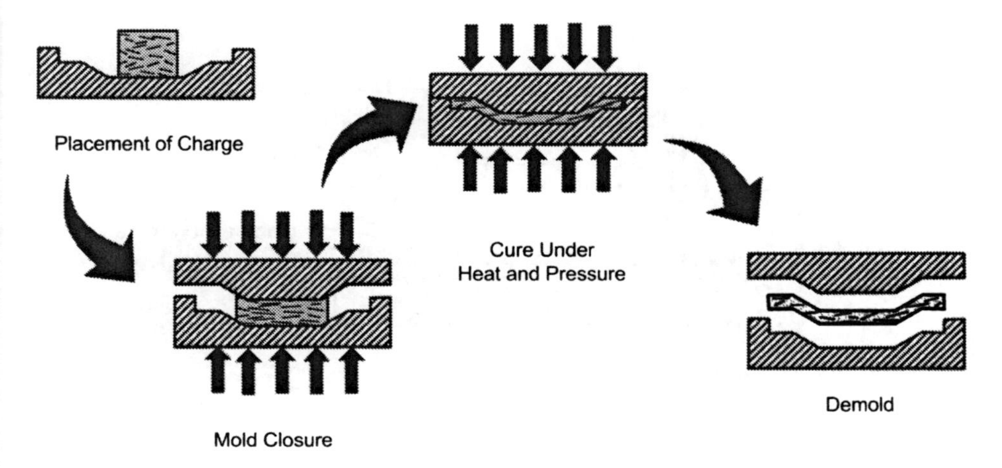

Fig. 10. Compression Molding of Thermosets

under heat and pressure. During cure the material flows to fill the mold. For more complex parts, preforms of glass fiber can be fabricated ahead of time and then placed in the tool for cure. In general, since the fiber reinforcement content is usually low (20–30%), the fibers are short (<2 inches), and the orientation is random, the mechanical properties of compression molded parts are much lower than the continually reinforced parts containing a high volume percentage of reinforcement.

Compression molding compound. Compression molding compounds usually consist of phenolics but alkyds and epoxies are also available. A typical compound might contain 40% glass fibers 0.040 inches in length or shorter. Prior to molding the compound is palletized into cylinders 0.5 to 2.5 inches in diameter by 0.25 to 1 inch long. A typical compression molding cycle for phenolics would be 340–375 °F and 700–3,000 psig for 1–10 min. The parts are frequently post-cured when elevated temperature resistance is required.

Sheet molding compound. A typical process schematic for making thermoset SMC is shown in Fig. 11. Continuous strand glass fiber roving is chopped to the desired length (1–2 inches) and deposited onto a coat of filled polyester resin paste traveling on a polyethylene film web. After fiber deposition is complete, a second web also carrying resin paste joins the first web forming a continuous sandwich of glass and resin. This is compacted and rolled under controlled tension onto standard package-sized rolls. Typical SMC is 0.25 inches thick and comes in rolls 40–80 inches wide. The SMC roll is then aged at 85–90 °F for about 1–7 d to thicken it by increasing the viscosity from about 100 poise to 10,000–1,000,000 poise.[11]

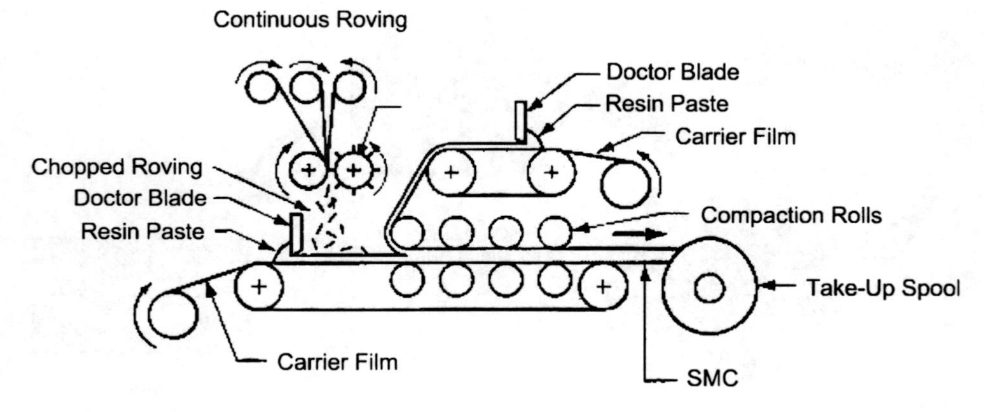

Fig. 11. Process for Making SMC

A typical composition would consist of 25% polyester resin, 25% glass fiber and up to 50% filler;[12] however, molding compounds with glass contents as high as 60% are available. Although polyesters dominate the market for compression molding compounds, vinyl esters, phenolics, ureas, melamines and epoxies are commonly used for compression molding compounds. Chopped strand mats can be used when higher properties are required and occasionally longitudinal glass strands will be added. The resin can be a standard polyester, a low shrink formulation or a low profile formulation. Standard polyesters are used when the highest mechanical properties are required but they also shrink the most during molding (0.3%). Low shrinkage formulations shrink less (0.05–0.3%), while low profile resins exhibit almost no shrinkage (<0.05%) resulting in better surface finishes and less tendency for cracking. Fillers, such as calcium carbonate, aluminum trihydrates or kaolins (clays), reduce cost and shrinkage. Mold release agents, such as zinc or calcium sterates, are often added to the formulation to provide self- releasing properties after molding. Thickeners (MgO, CaO, $Mg(OH)_2$ or $Ca(OH)_2$) can be added to increase viscosity and improve the flow characteristics during molding.

As shown in Fig.12, SMCs can be classified as either SMC-R (random), SMC-C (continuous) or SMC-R/C (random/continuous). SMC-R contains 1–2 inches long glass fibers oriented in a random order, while SMC-C consists of an oriented array of unidirectional strands. SMC-R/C is a hybrid containing a mixture of random fibers placed on top of unidirectional strands. Due to the unidirectional strands, SMC-C and SMC-R/C are stronger but are also harder to mold. SMC-R is used more frequently due to its lower cost, better flow and moldability.

Layers of SMC are stacked into "charges" for compression molding. The charge size is typically 40–70% smaller than the projected mold surface.[12]

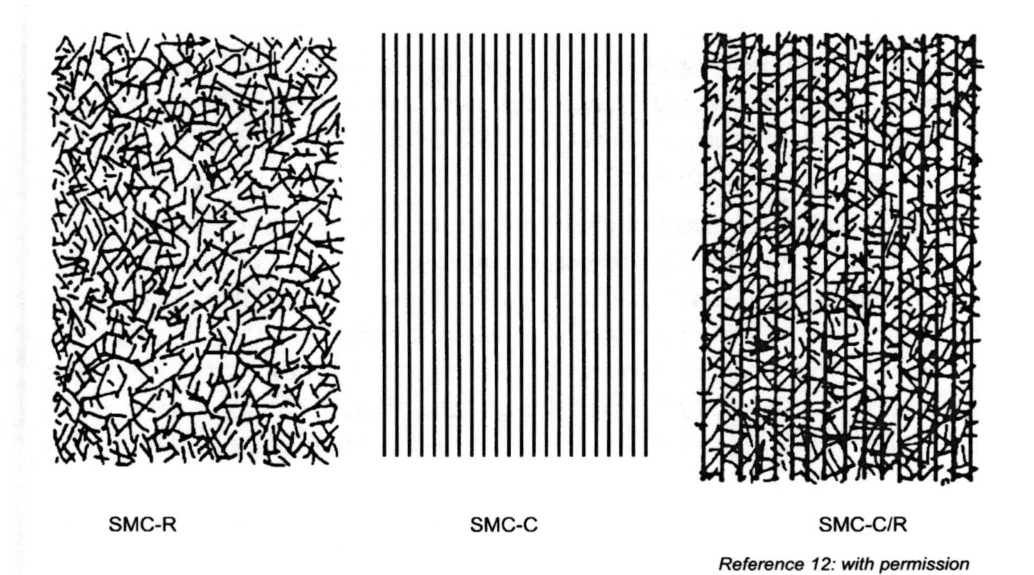

Fig. 12. Types of Sheet Molding Compounds

The weight of the charge is determined by the weight of the part to be molded plus some additional material for edge trimming. Different thicknesses of charge can be placed in different portions of the mold to account for thickness or geometry changes in the part. Close control of charge weight and size allows the production of molded parts with very little excess material on the edge that has to be trimmed after cure. Charge placement is critical to minimizing knit lines where two flow fronts merge and create a location of lower strength. The charges are heated in a continuous oven with infrared (IR) or force air impingement heaters. They are then placed in heated matched-die molds and press cured.

Bulk molding compound. BMC is a mixture of shorter (1/8 to 1-1/4 inches) glass fibers and resin containing filler, catalyst, pigment and other additives. The fiber reinforcement level is usually 10–20%. BMCs are usually made in an extruder. This premixed material, with the consistency of modeling clay, can be provided in bulk form or can be extruded into rope (1–2 inches in diameter) or log-like forms for easier handling. Like SMC, the weighed charge of BMC is placed into a heated matched metal die, the die is closed and pressure is applied. The completed, cured part is removed from the die after an interval of several seconds to a few minutes depending on part size and thickness. BMC is commercially available in various combinations of resin, additives and glass reinforcements, to satisfy a variety of end use requirements in high-volume applications where good finishes, good dimensional stability, part complexity and good overall mechanical properties are important. The longer fiber lengths obtainable in SMC provide higher mechanical properties compared to parts made with BMC, especially those having relatively thin cross-sections. Actually, more BMC is used for thermoset injection molding than for the traditional compression molding process.

Preforming. A third approach to preparation of a ready-to-mold form of preblended glass fiber reinforcement and resin is the preform. This is a mat of chopped glass fiber strands with a binder that allows the preform to hold its shape. The preform is made to the approximate shape of the product to be molded. There are two common methods of producing preforms, the directed fiber method and the plenum chamber method. In the directed fiber method (Fig. 13), continuous strand glass fiber roving is cut into 1–2 inches lengths and blown with a binder onto a rotating metal screen that approximates the shape of the final part. Suction is used to hold the fibers in place. After heat setting the binder in an oven, the preform is ready for the die. The plenum chamber method is similar but involves distribution of chopped fibers and binder into an air chamber where they are sucked onto a rotating preform screen. Preform methods are used for applications requiring medium to large parts having relatively

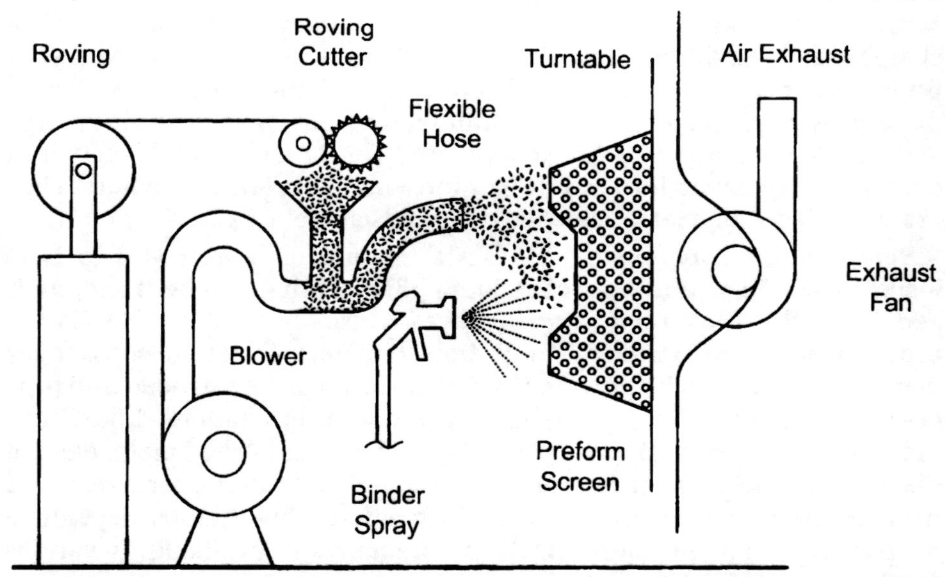

Fig. 13. Directed Fiber Spray-Up Process for Preforms

constant cross-sections and high concentrations of glass reinforcement. In preform molding, the major portion of the resin is added just before molding or after the preform is placed within the mold.

While most applications use either SMC or BMC, for flat sheets and simple shapes, glass mats can be placed directly into the mold; resin poured over the mat; and the mold closed under pressure to consolidate the resin and glass. For complex parts, preforms are normally made before placing them in the tool. The matched-die molds are mounted in a hydraulic or modified mechanical molding press such as the one shown in Fig. 14. The two halves of the die are closed and heat and pressure applied (225–320 °F, 150–2,000 psig). Parts with higher glass contents or more complex geometries require the higher pressures. Depending on thickness, size and shape of the part, cure cycles range from less than a minute to about five minutes. Press closing occurs in two steps: a fast initial step followed by a slower step to allow the material time to flow. Higher molding temperatures

Fig. 14. Compression Molding Machine

and faster curing times require faster closing speeds to prevent premature gel. If a condensation curing resin, such as a phenolic, is being molded, it is a common practice to momentarily open the mold to let the volatiles escape during the curing process. After cure, the mold is then opened and the finished part is removed. If hollow parts are required, low melting eutectic metals can be inserted into the mold and then melted out after molding. The matched metal dies can be single- or multiple-cavity hardened and chrome-plated molds, usually cored for steam or hot oil heating. Electrical and hot water heating systems can also be used. Side cores, provisions for inserts and other features are often employed. The dies are also equipped with part ejectors that can either be mechanical pins or air pressure. Mold materials include cast or forged steel, cast iron and cast aluminum. Compression molding is an ideal process when the volume is high enough to justify the investment in precision matched metal dies; however, with the advent of high-speed machining of aluminum dies, the process is becoming more economical for lower volume production runs. It is capable of producing extremely uniform parts with superior surface finishes. Inserts and attachments can be molded-in, and final trimming and machining operations are minimal.

Transfer molding. Transfer molding (Fig. 15) is similar to compression molding, except that the charge of molding compound is heated to 300–350 °F in a separate chamber; transferred by a ram under heat and pressure into a closed die where the shape of the part is determined and cure takes place in usually 45–90 s. The process is used for small intricate parts requiring tight dimensional tolerances. Although good resin flow is required, the process is not sensitive to precise charge weight. Typical reinforcement levels are 10–35%.

Glass mat thermoplastic. Glass mat thermoplastic (GMT) consists of glass-reinforced thermoplastic, usually polypropylene, that flows similar to thermoset resins to fill the mold during the compression molding cycle. A typical GMT fabrication line is shown in Fig. 16. Several reinforcements are available including continuous fiber mat, chopped fiber mat and unidirectional fiber mat, normally at a glass content of 40 vol.%. This process has gained a foothold in the automotive market due to the automation employed and the short processing cycles, some as short as 30 s, although longer times are sometimes required to obtain the proper amount of crystallinity. GMT material is made using the double-belt process described in Chapter 10. The material is then heated in continuous moving furnaces using IR heater banks to 600 °F in 2–3 min. It is then inserted into a high-speed hydraulic press with a mold heated to 275 °F and consolidated under 1,000–4,000 psig pressure. A two-stage pressurization cycle is frequently used; in the first stage, the pressure is applied quickly (200–500 inch min^{-1}) followed by a slower stage at 1–10 inch min^{-1} to allow

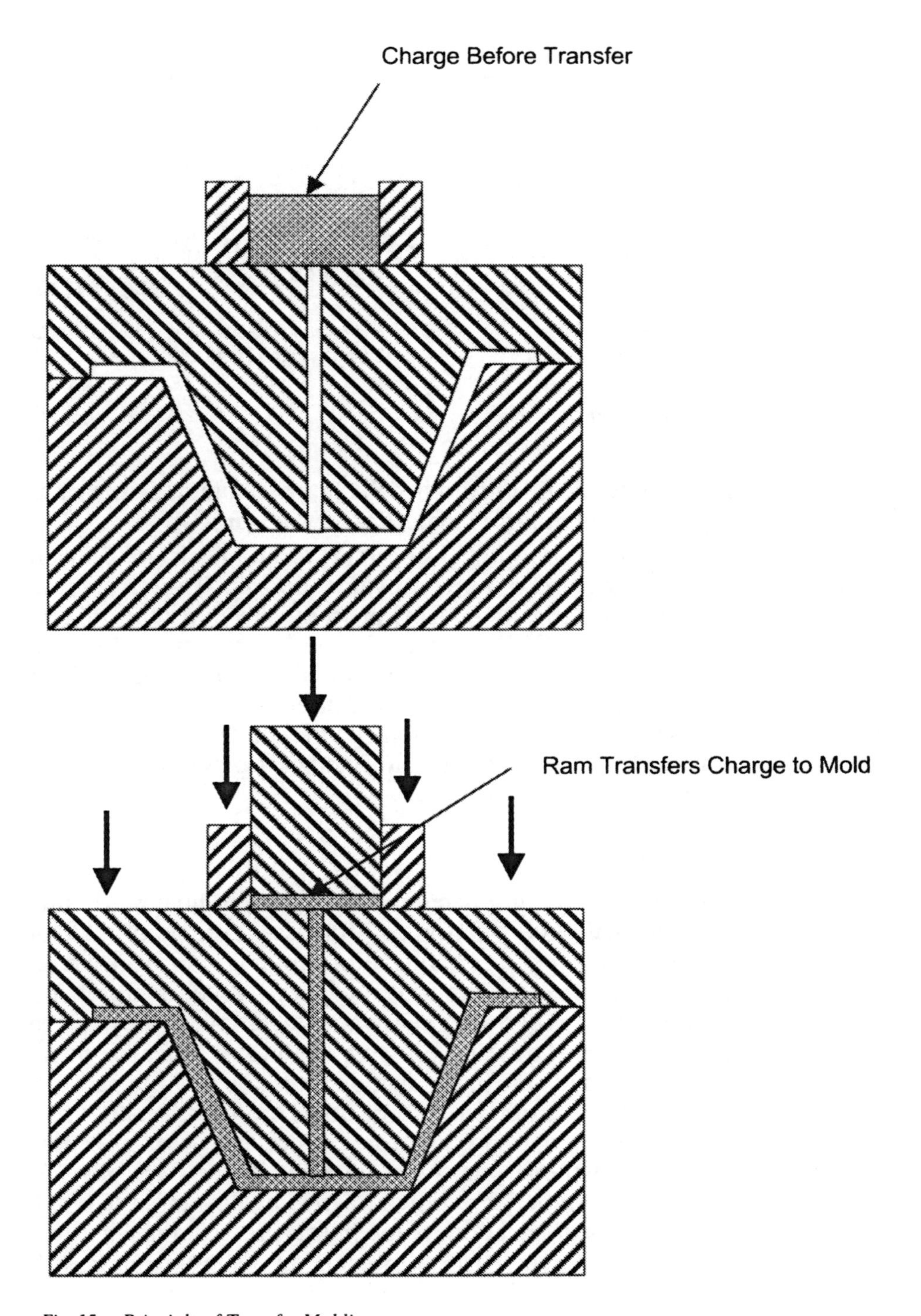

Fig. 15. Principle of Transfer Molding

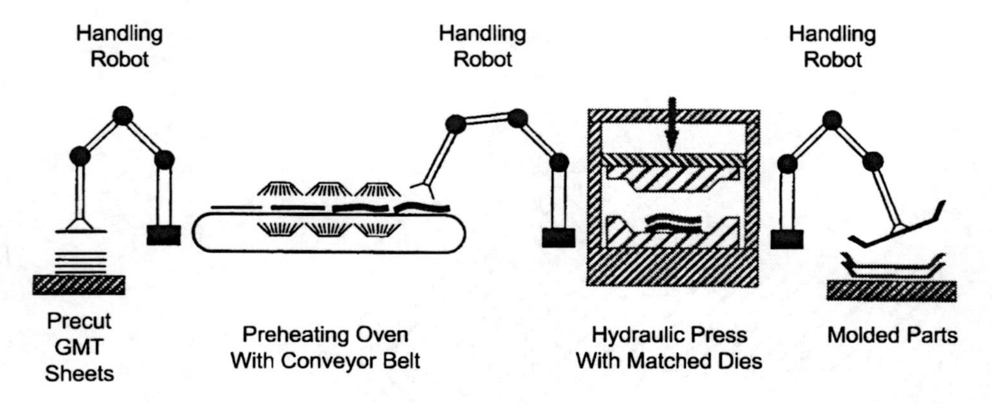

Fig. 16. Automated Fabrication of GMT Parts

the material time to flow and fill the mold.[14] Sometimes holding fixtures are required during cooling to minimize distortion or warping.

Long discontinuous thermoplastic. Long fiber thermoplastic (LFT) again uses glass reinforcements in a polypropylene matrix. The fibers are normally 0.5, 1.0, or 2.0 inches long. This process is even a more highly automated process than the GMT process. The polypropylene is fed into an extruder that heats and melts the material and then adds the glass fiber near the end. This feedstock is then transferred directly to the compression molding press without having to reheat it as required in the GMT process, thus further reducing costs. For both the GMT and LFT processes, recycled material is frequently used to reduce costs.

11.3 Injection Molding

Thermoplastic injection molding. Injection molding is the highest volume method of any of the glass fiber reinforced processes, capable of making more than a million parts per year. Injection molding can be used to fabricate either unreinforced or reinforced parts. Like compression molding, injection molding requires expensive matched metal dies due to the high processing temperatures and pressures. However, because many complex parts can be molded at high production rates, the cost per part can often be quite low. Although other reinforcements are occasionally used, glass fibers are the predominate product form. Since the fibers are short (0.03–0.125 inches), the volume percentage of fiber is low (typically 30–40%), and the fiber orientation is random or aligned by flow into the mold, the mechanical properties of injection-molded parts are again much lower than continuous fiber reinforced parts. However, very complex parts with all surfaces controlled can be achieved with injection molding. Either single- or multi-cavity dies are used to produce very large volumes of complex parts at very high production rates, with a variety of properties provided by the wide range of thermoplastic resins that are injection moldable. Injection-molded glass-fiber-reinforced thermoplastics are used extensively in the automotive and appliance industries. Design flexibility, the ability of glass fiber reinforced thermoplastics to form intricate shapes, very high production rates and low per piece cost account for its extensive application.

In injection molding, the molding compound, in pellet form, compound concentrate, or dry blend, is heated in the injection chamber of the molding machine. A typical injection-molding cycle is illustrated in Fig. 17. The material is melted as it progresses down the screw by a combination of heat from the barrel and the shearing action of the screw. The material accumulates in the barrel and is then injected under high pressure and in a hot fluid form into a relatively cold closed die. After the desired "shot size" is accumulated, the screw stops rotating and moves forward at a controlled rate acting as a ram or plunger so that the plastic melt in front of the screw is forced in the mold. The molten material flows through the nozzle into a water cooled die where it flows through a sprue, a runner system and finally into the mold cavity through the gate. When the mold is completely filled, the screw remains stationary to keep the thermoplastic in the mold under pressure. During this hold or dwell time, additional melt is injected into the mold to compensate for the shrinkage due to cooling. After the gate freezes,

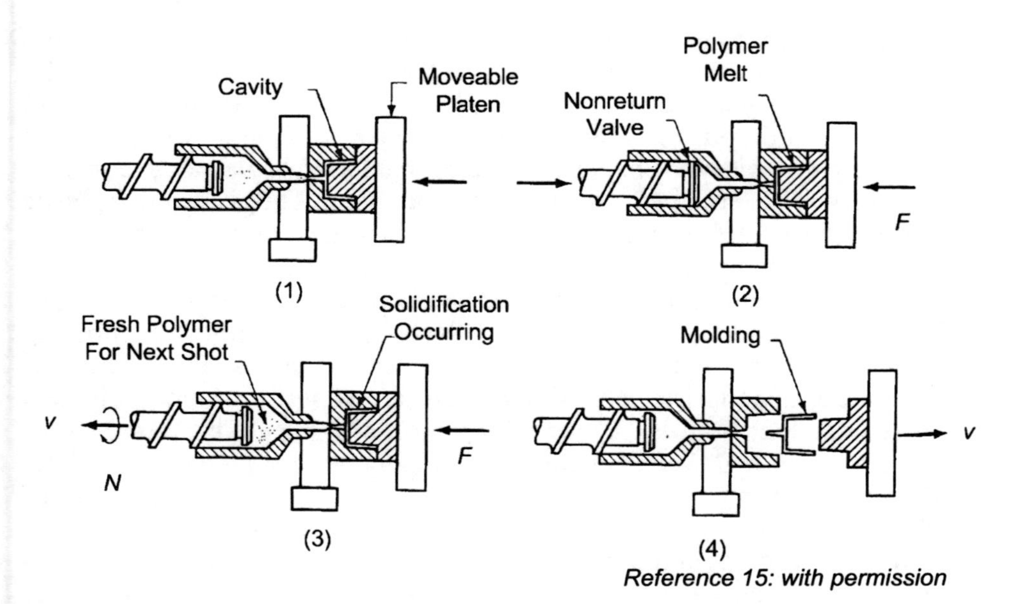

Fig. 17. Injection Molding Machine Sequence

the mold becomes isolated from the injection unit. The melt that accumulates at the end of the screw pushes the screw backward, i.e., the screw is rotating and moving backward at the same time. The rate at which the plastic melt accumulates in front of the screw for the next shot is controlled by the backpressure, which is the hydraulic pressure exerted on the screw. When a sufficient melt has accumulated in front of the screw, the screw stops rotating while the part in the mold cools and solidifies. After a short cooling cycle, usually 20–120 s, the part has solidified to a degree sufficient to enable it to be removed from the mold without distortion. The cycle time for the process is usually controlled by the cooling rate. Cooling rate is usually dependent on the die design, part thickness and reinforcement content. After cooling, knockout pins eject the part from the mold. The sprue, runners and gates are separated from the part and are usually recycled by grinding into powder and mixing with fresh material.

Virtually all thermoplastics are injection moldable including nylon, acetal, vinyl, polycarbonate, polyethylene, polystyrene, polypropylene, polysulfone, modified polyphenylene oxide, fluorocarbons, polyetherimide, acrylonitrile–butadiene–styrene (ABS) and styrene–acrylonitrile (SAN). For injection molding, glass fiber chopped strands can be blended with thermoplastic resin molding powders or pellets. Carbon fibers can be used when higher properties are required and they also provide enhanced electrical and thermal conductivity. Carbon fibers are frequently used as electrically conductive fillers in parts requiring electromagnetic interference (EMI) shielding, anti-static, electrostatic discharge (ESD) and other electronic applications.[16]

A lower cost thermoplastic resin can often be strengthened by the addition of glass fiber reinforcement to the point where it will offer the superior performance characteristics of a more costly unreinforced resin. Also, the addition of glass fibers provides sufficient strength and dimensional stability to enable many reinforced thermoplastics to compete effectively with stamped sheet metal and die castings. Thermoplastic molding compounds usually include pigments, fillers, mold release agents, lubricants and occasionally other additives for special purposes. The corrosion resistance and molded-in color also provides design advantages. The list of thermoplastics reinforced with glass fibers for injection molding is a long one, and includes resins offering an extremely broad range of mechanical, chemical, electrical and thermal properties. Some polymers can require that both the speed of injection and injection pressure vary during the molding cycle. A heat sensitive polymer can be degraded if the fill rate is too rapid. Forcing the polymer through the small restrictions of the sprue, runner and gate systems at high speed will increase the internal shear and raise the temperature to the point that polymer degradation occurs. Programming different injection speeds and pressures during the injection process may be required to prevent resin degradation.

Injection-molding compounds are usually supplied in a form that can be fed directly into the injection-molding machine. They are usually made by mixing short chopped fibers with a powdered or pelletized resin in an extruder. In extrusion processing, fiber breakage is a major problem. To minimize breakage, deep flighted screws or twin screws are used and the fibers are fed into the barrel after the resin has melted. Pre-pelletized blends are available in controlled ratios of glass fiber reinforcement to resin and with specified additives such as pigments, flame retardants, stabilizers and lubricants. These pellets are available in nearly all of the common injection-molding thermoplastics. Concentrates are similar to the glass fiber/resin prepelletized blends but have very high percentages of glass-to-resin, up to 80% by weight, as compared to the 10–40% customary in pellets. The concentrated pellets are blended with non-reinforced thermoplastic pellets to achieve the desired glass-to-resin ratio in the molded part. With these formulations, the injection molder can select a glass-reinforced compound to satisfy their end-use needs with respect to mechanical properties, chemical properties, impact strength, surface finish and color. Prior to injection molding, it is important that the material be thoroughly dried. Drying can be accomplished offline as a batch-drying process or in-line in a hopper dryer on the machine. It is important to note that some polymers absorb moisture primarily on the surface (nonhydroscopic) while others absorb moisture into the pellets or granules (hygroscopic). Drying of non-hydroscopic materials, such as polypropylene and polyethylene, can be accomplished by simply evaporating the moisture from the surface. Hydroscopic materials, such as nylon and polycarbonate, must be dried using dehumidified hot air to remove the moisture and generally require longer drying cycles. It is common practice to use up to 20% regrind for injected molded parts; however, since the properties of the part will often decrease with increasing regrind content, it is important to test prior to incorporating regrind. The reinforcing effects of glass and carbon fibers decrease with repeated processing due to fiber breakdown and thermal degradation of the resin after multiple heat exposures.

The advantages of glass-reinforced injection-molded parts over unreinforced parts include higher tensile strength and modulus, greater impact resistance, reduced shrinkage, improved dimensional stability and higher temperature capability. While normal polymers shrink about 5%, fiber-reinforced parts will normally shrink 1% or less. A disadvantage of fiber reinforcement is lower ductility. In general, slightly higher injection pressures and barrel temperatures are required for reinforced parts due to their higher viscosities. The mixing and shearing action of the injection molder reduces the fiber length; therefore, the gates and runners should be as large as possible. Injection pressures are usually 10–15,000 psig. Minimal back pressure (25–50 psig) and low screw speeds (30–60 rpm) should be utilized to avoid excessive fiber breakage. Cavities should be filled as rapidly as possible to minimize fiber orientation and enhance knit weld line integrity, especially for thin wall parts. Higher barrel temperatures (e.g., 30–60 °F higher than for unfilled resin) are also used with injection-molding reinforced parts to minimize fiber breakage by melting the polymer rather than by heat provided by the shearing action of the screw. After the cavity is filled, a longer hold time is needed to help maintain the dimensional tolerances. One problem with reinforced parts is that the fiber orientation is largely controlled by the flow forces and the part geometry and can result in parts with variable strengths in different sections of the part. Also, knit lines form where the flow fronts meet and result in lower-strength areas. This problem is accentuated for parts containing longer fibers and higher fiber contents.[14]

The main parts of an injection-molding machine (Fig. 18) are the clamping unit and the injection unit. The clamping unit holds the die. It is capable of closing, clamping and opening the die. It contains fixed and moving plates, tie bars, and the mechanism for opening, closing and clamping. The injection or plasticizing unit melts the thermoplastic and injects it into the mold, while a drive unit provides power for both the plasitcizing and clamping units. While both plunger and screw machines are available, screw machines dominate the injection-molding industry. Reciprocating plunger machines are inferior to screw machines because all of the heat has to be supplied by conduction from the barrel heaters and mixing is very poor, resulting in low plasticizing (melting) rates and thermally non-uniform melts. Screw machines generate heat within the material and provides mixing, thus producing much more uniform melts. Generally, a screw

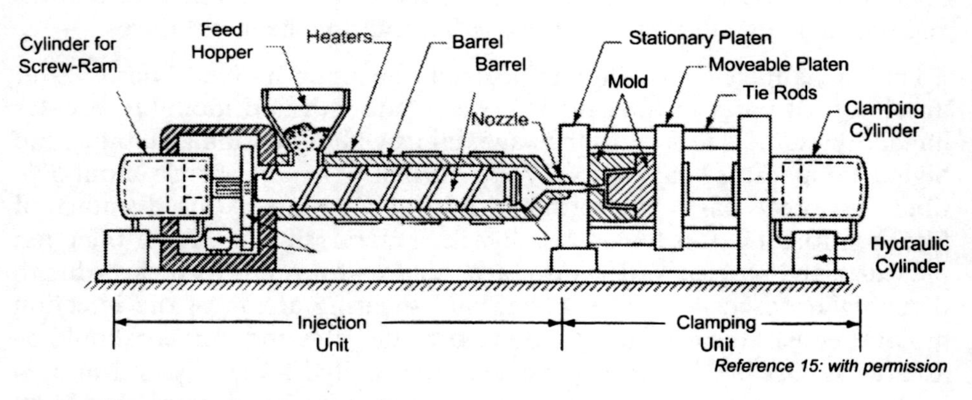

Reference 15: with permission

Fig. 18. Injection Molding Machine Schematic

consists of three sections: (1) the feed section, (2) a compression section and (3) a metering section. The feed section transports pellets from the hopper to the heated section of the barrel. Within the feed section, the screw flights are the same dimension. The next section, compression, is where the polymer is melted due to the combined heat provided by the barrel heaters and the shearing action of the screw. In this section, the volume of the screw flights decreases to compensate for changes in the material density. Finally, in the metering section, the flight dimensions are constant where the final melting and mixing of the molten polymer occurs. Screw designs vary with the material being molded.[17] Three screw designs are shown in Fig. 19. A design for highly crystalline polymers is shown in Fig. 19a. Since these materials have very sharp melting points, a very short compression section is used. For semi-crystalline thermoplastics (Fig. 19b), the compression section is longer to accommodate their more gradual melting behavior. Finally, for an amorphous thermoplastic (Fig. 19c) that does not have a true melting point, the screw is designed with a gradual increasing compression over its length. The barrel heaters are divided into zones to provide better control of the heating process.

Injection-molding machines are rated in terms of tons of mold-clamping capacity and ounces of shot size and can range from laboratory units

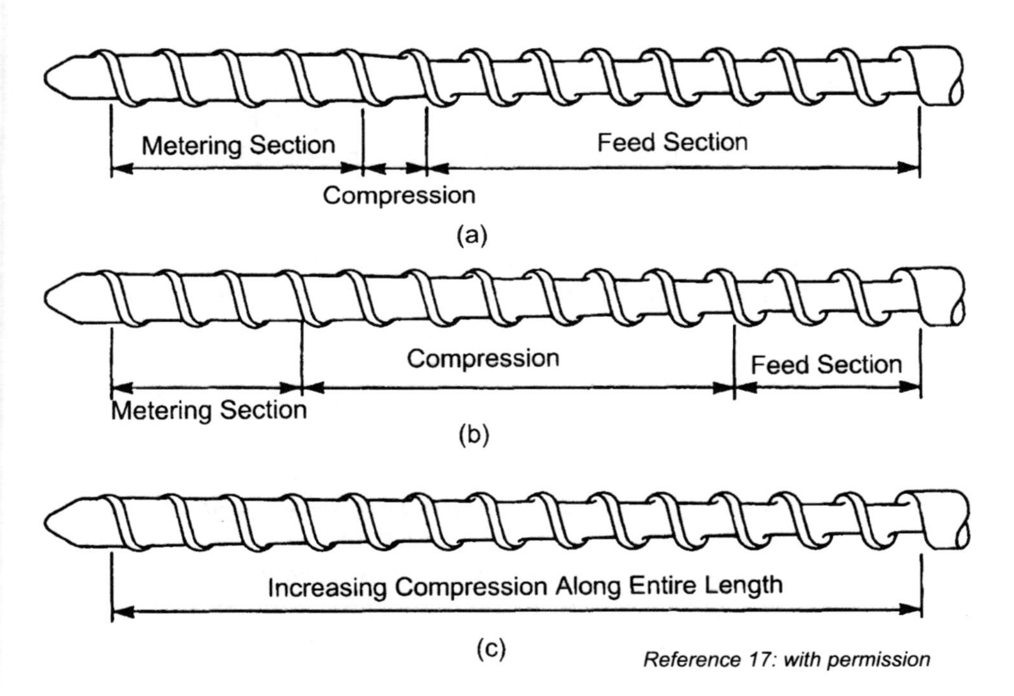

Fig. 19. Injection Molding Screw Designs

Fig. 20. Injection Molding Machine

(2 tons, 0.25 oz) to large industrial units (3,500 tons, 1,500 oz).[18] A typical lab size injection-molding machine is shown in Fig. 20. The clamping mechanism must have sufficient locking force to resist the tendency of the molten polymer moving at high pressures to force the mold halves apart. Since reinforced thermoplastics have higher viscosities than unreinforced resins, they are injected at higher pressures, i.e., the clamping mechanism must be even stronger. Mold clamping pressure is usually provided by either a toggle lock or a hydraulic cylinder, or a combination of the two. Modern injection-molding machines are equipped with automatic feedback control systems that monitor the process parameters to produce consistently high-quality parts. Important processing parameters that need to be controlled include individual control of the injection and holding pressures, ram position and velocity, back pressure and screw speed. Barrel and nozzle temperature control is important for achieving low thermal stresses in the molded part. Lower molding temperatures will promote rapid cooling and higher production rates but lower-quality parts. Rapid cooling affects internal stresses, degree of fiber orientation, post-mold shrinkage and warpage.

Matched metal dies for thermoplastic production runs are normally made of high-strength tool steels that are often plated for additional abrasion resistance. Typical coatings include nickel phosphorous impregnated with polymers, hard chrome, electroless nickel, titanium nitride and diamond black (boron carbide thin film with tungsten disulfide). In a two-plate cold runner configuration (Fig. 21a), the material is injected through the sprue bushing, runner system and into the die cavities through the gates. A typical eight-mold sprue, runner and gate system is illustrated in Fig. 22. After cooling, the parts are removed along with the sprue, runner and gates. The sprue, runner system and gates are then manually removed for recycling. In the three-plate cold runner system shown in the bottom portion of Fig. 21b, the middle plate separates the sprue, runner system and

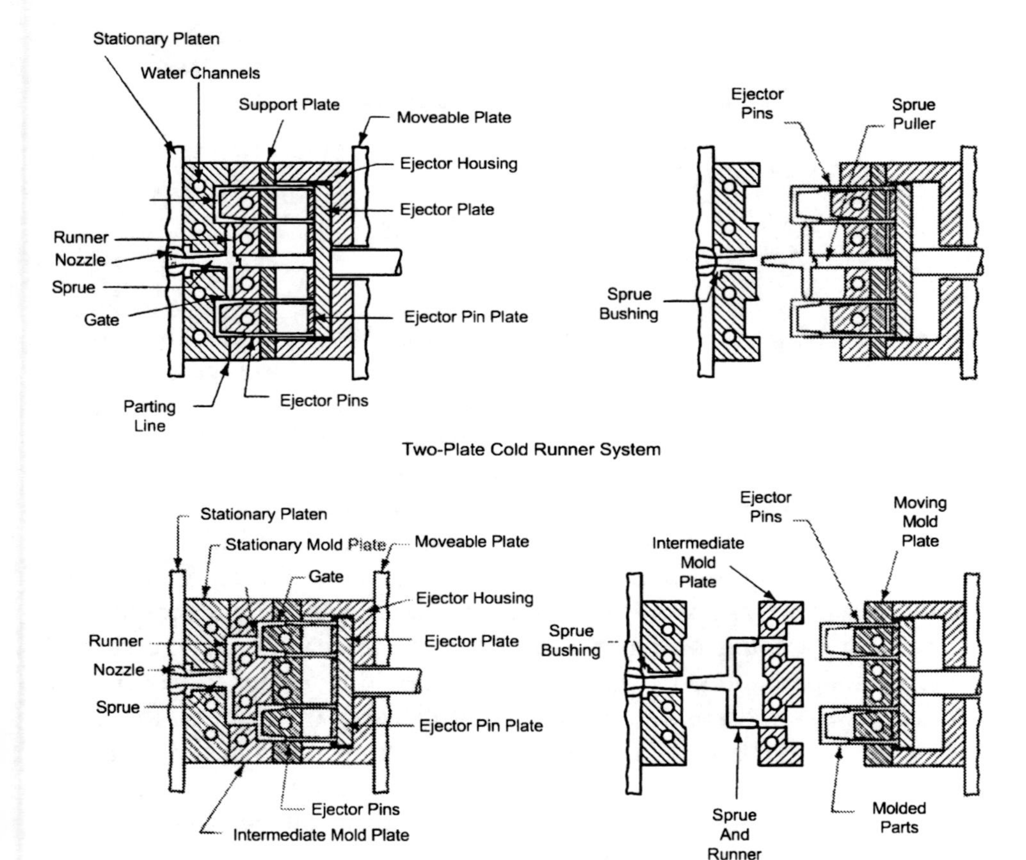

Fig. 21. Injection Molding Screw Designs

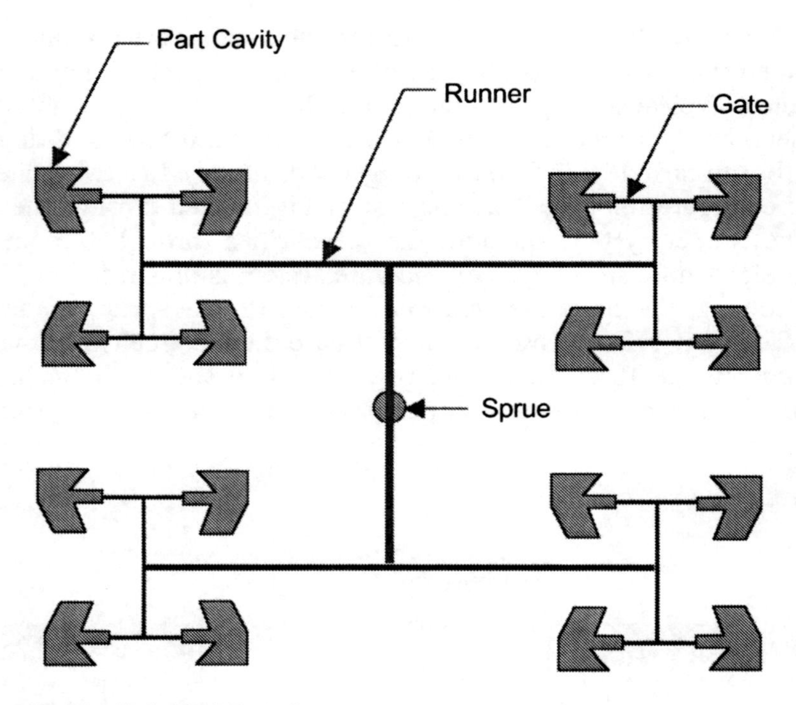

Fig. 22. Eight Mold Cavity System

gates from the parts during the ejection cycle. There are also three-plate hot runner systems that keep the sprue, runner and gates molten throughout the process. Although these systems are more expensive, there is no labor or material waste associated with recycling the sprue, runner and gate system. Injection-molding dies normally contain vents to let the air in the die escape as the hot polymer flows into the die cavity. Vents are usually placed in the areas that fill last, near knit lines and even in the runner system. Sprues and runners should have gentle bends. Runners should be as short as possible to avoid pressure drops. The gates control the fill rate, the amount of material flowing into the cavity and the rate of material solidification. The gating system should be designed to avoid or minimize weak weld or knit lines. If multiple gates are required to fill the mold, then weld lines will result and they should be located in lower stressed locations in the part. Since part cooling is often the longest part of the injection-molding cycle, special attention during die design is required to provide efficient cooling channels. In general, (1) since cooling rate is thickness dependent, put cooling fluid close to the cavity surface; (2) put cooling lines exactly where they are needed to uniformly cool the part; (3) insure turbulent flow of the cooling media for maximum heat transfer (sharp turns and high speeds promote turbulent flow); and (4) design the cooling system so cooling is uniform in each mold half.[19]

One of the main challenges in injection molding is producing parts with correct dimensional accuracy and minimal warpage. Warpage occurs when the internal stresses in the part exceed the inherent part stiffness and cause a permanent deformation. Cooling gradients, which lead to differential shrinkage, are the primary cause of part warpage. Shrinkage is dependent on the material, the part design, the mold design and the processing parameters. Due to their lower inherent shrinkages, amorphous resins are frequently selected over semi-crystalline resins when tight tolerances are required. In general, the addition of fibrous fillers increases part warpage due to the anistropic nature of their flow patterns as the fibers align with the flow. Due to differential cooling, part designs with non-uniform wall thicknesses will tend to warp more than those with uniform wall thicknesses. Stiffeners or gussets are often incorporated during design to reduce or minimize warpage. Gate location and cooling passages are important considerations during die design. Ideally, the mold should be filled uniformly during the injection process and then cool uniformly until it is ready for ejection. If the gates and cooling system are not properly designed, regions in the part cavity will fill prematurely and start to solidify before the remainder of the mold is filled. Likewise, an improperly designed cooling system will result in thermal gradients during the cooling process. During processing, cavity fill time is important. If the time is too short, high shear rates can cause substantial melt shearing and high residual stresses leading to part warpage. If the time is long, a large melt temperature drop occurs as cavity filling is completed resulting in thermal gradients leading to part warpage.

Gas-assist molding (GAM) is used to mold hollow injection-molded parts by the controlled injection of an inert gas (N_2) into the hot polymer melt. The gas forms a continuous channel through the hotter, less viscous thicker sections of the melt. Gas injection pressures of 400–800 psig are commonly used. Injection compression molding (ICM) is another variant where the melted polymer is injected into a partially opened mold. The mold closes, compresses and distributes the melt throughout the cavity. The ICM process helps to prevent fiber breakage and improve the properties of the molded part. Thin wall injection molding (TWIM) is a process capable of making parts with 0.020–0.080 inches walls and a flow length-to-thickness (L/t) ratio of greater than 75. Since thinner walls cool rapidly, equipment capable of extremely high fill pressures (15–35,000 psig) is required along with very short fill times (<0.75 s) to produce the part. This process usually requires thicker dies to resist deflections with specially designed runner and vent systems. In structural foam molding (SFM), chemical blowing agents are used to produce parts with solid skins and foamed cores. This process allows the fabrication of large parts with high strength-to-weight ratios and good dimensional control. Tooling costs can be lower with aluminum dies acceptable for some part configurations.[19]

Thermoset injection molding. Thermoset-molding compounds can be injection molded when the injection screw or plunger and chamber of the molding machine are maintained at low temperatures and the mold itself is heated (250–400 °F) rather than cooled as for thermoplastics. This causes the thermoset material to cure after several minutes in the mold under heat and pressure. Precise control of temperature and cycle time must be maintained to prevent the resin from gelling in the barrel. Low-viscosity resins are required that maintain low viscosity for a period of time but cure quickly after gel. Particulate fillers, chopped glass fibers and short milled fibers (<0.80 inches) can be used for reinforcement; however, chopped and milled glass fibers, which increase the viscosity of the resin more than particulate fillers, are harder to mold. For thermoset injection molding, the screw is usually shorter and has a lower compression ratio. The barrel is set at rather low temperatures (160–212 °F) and cure is conducted with heated molds with injection pressures of 7,500–15,000 psig. Adequate venting of volatiles is critical, especially for condensation curing phenolics, and venting passages are built into the mold at parting lines, ejector pins and core pins. The maximum weight of parts made using this are about 10 lb. BMCs are also frequently used for thermoset injection-molded parts. Polyesters, vinyl esters and epoxies are used with glass fibers 0.25–1 inches in length. When using BMC, larger machines are frequently used to produce larger moldings at lower pressures (750–1,500 psig) than for conventional thermoset injection molding. Typical cures occur in 2–5 min.

11.4 Structural Reaction Injection Molding

Reaction injection molding (RIM), as shown in Fig. 23, is a process for rapidly making unreinforced thermoset parts. A two-component highly reactive resin system is injected into a closed mold where the resin quickly reacts and cures. RIM resins must have low viscosities (500–2,000 cps) and fast cure cycles. Polyurethanes are the most prevalent but nylons, polyureas, acrylics, polyesters and epoxies have also been used. The two components (isocyanate and polyol in the case of polyurethanes) are kept separate and are constantly recirculated under high pressure. They are mixed in a dynamic mix head under high pressures (1,500–3,000 psig) and high speeds (4,000–8,000 ins^{-1}).[11] The cross-linking reaction is initiated by mixing rather than by heat and gel can occur as quickly as 2–10 s. The actual injection pressure is quite low and the die is heated to about 120–150 °F to accelerate cure and then remove the exothermic heat of reaction during the latter part of cure. RIM resin systems have very low viscosities and can be injected at low pressures (<100 psig). These low pressures allow the use of

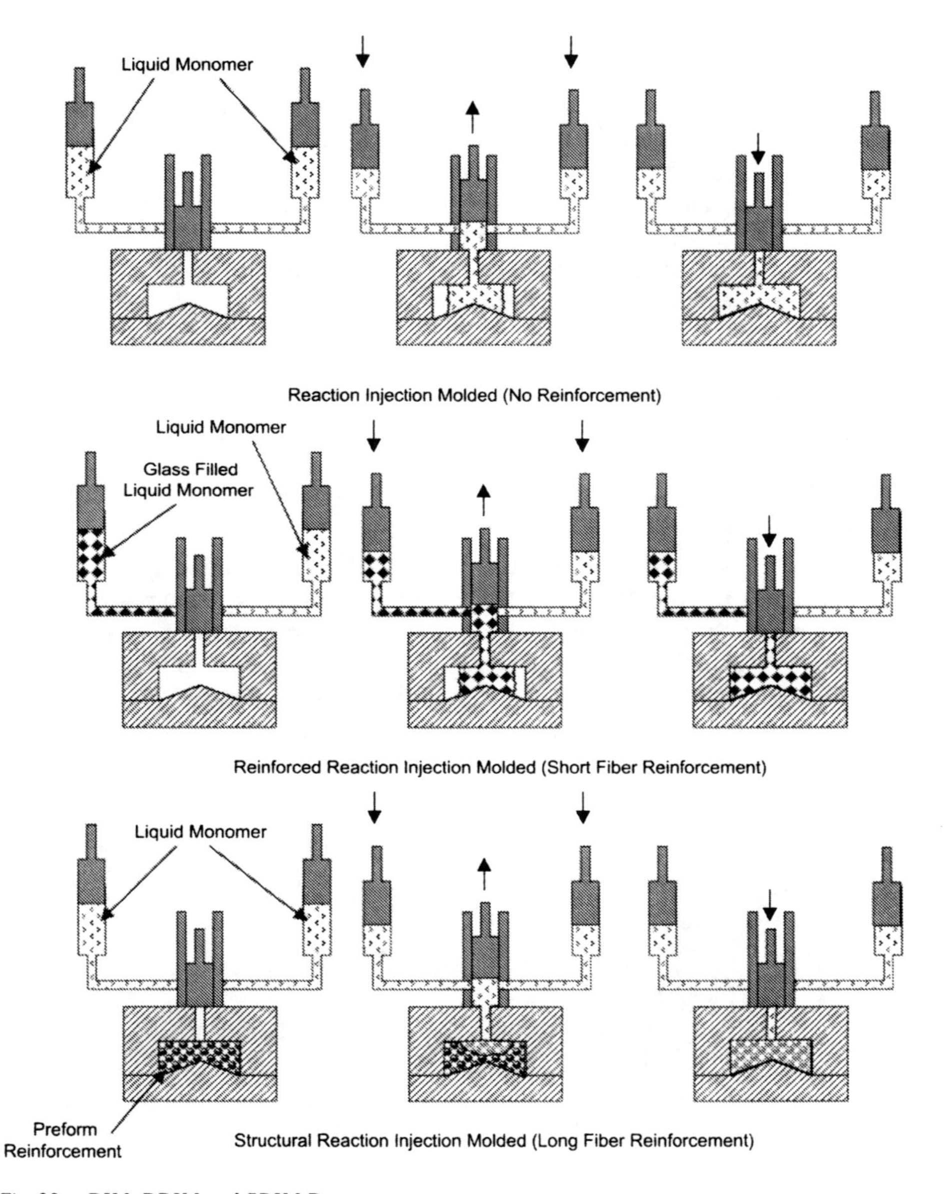

Fig. 23. RIM, RRIM and SRIM Processes

inexpensive molds and low-force clamping systems. Typical tooling materials include steel, cast aluminum, electroformed nickel and composites. Cycle times as short as 2 min for large automotive bumpers have been reported.[18] Since isocyanate vapors are harmful, RIM work cells should be provided with exhaust ventilation.

Reinforced reaction injection molding (RRIM) is similar to RIM except that short glass fibers are added to one of the resin components. The fibers must be extremely short (e.g., 0.03 inches) or the resin viscosity will be too great. Short fibers, milled fibers and flakes are commonly used. Again, polyurethanes are the predominant resin system. The addition of the fibers improves modulus, impact resistance, dimensional tolerances, and lowers the coefficient of thermal expansion.

Structural reaction injection molding (SRIM) is similar to the previous two processes except that a continuous glass preform is placed in the die prior to injection. Again, this process is used almost exclusively with polyurethanes. Due to the highly reactive resins and short cycle times, SRIM cannot produce as large a part size as with RTM (Chapter 9). Also SRIM parts have lower fiber volumes and generally more porosity than RTM parts.

11.5 Pultrusion

Pultrusion[20] is a rather mature process that has been used in commercial applications since the 1950s. In the pultrusion process, continuous fibrous reinforcement is impregnated with a matrix and then is continuously consolidated into a solid composite. While there are several different variations of the pultrusion process, the basic process for thermoset composites is shown in Fig. 24. The reinforcement, usually glass rovings, is

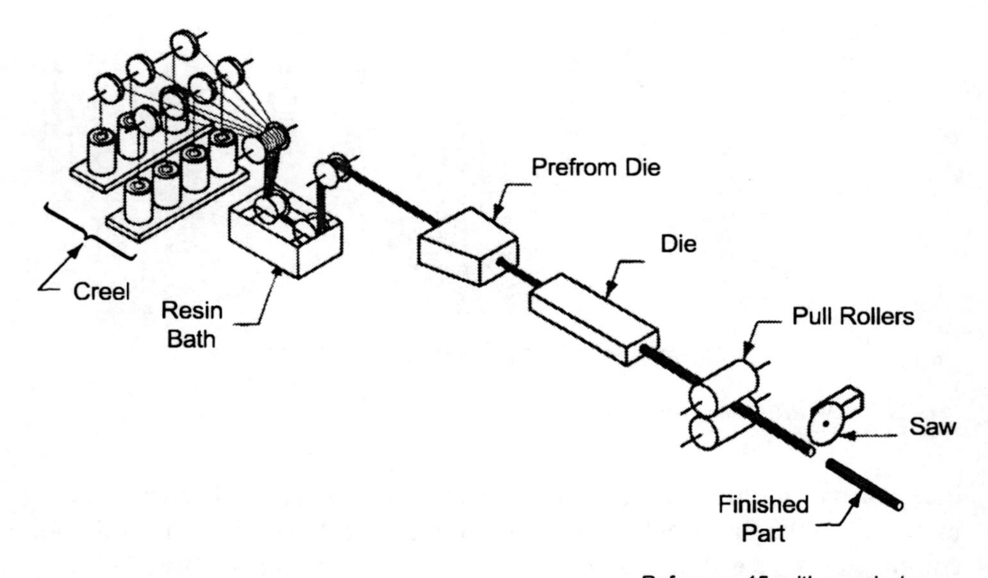

Fig. 24. Pultruded Process

pulled from packages in a creel stand and gradually brought together and pulled into an open resin bath where the reinforcement is impregnated with liquid resin. After emerging from the resin bath, the reinforcement is first directed through a preform die that aligns the rovings to the part shape and then guided into a heated constant cross-section die where it cures as it progresses through the die. Curing takes place from the outside of the part toward the interior. Although the die initially heats the resin, the exotherm resulting from the curing resin can also provide a significant amount of the heat required for cure. The temperature peak caused by the exotherm should occur within the confines of the die and allow the composite to shrink away from the die at the exit. The composite part emerges from the die as a fully cured part that cools as it is being pulled by the puller mechanism. Finally, the part is cut to the required length by a cut-off saw. While pultrusion has the advantage of being an extremely cost-effective process for making long constant cross-section composite parts, it is definitely a high volume process, as the setup time for a production run can be rather costly. In addition, there are limitations in that the part must be of constant cross-section, and the flexibility in defining reinforcement orientation is somewhat limited. While glass fiber/polyester materials dominate the market, a considerable amount of work has been done to develop the process for the aerospace industry with higher-performance carbon/epoxy materials. Floor beams in commercial aircraft are a potential application. Pultrusion is capable of making a wide variety of structural shapes as shown in Fig. 25, including hollow sections when a mandrel is used.

The major advantages of the pultrusion process are low production cost due to the continuous nature of the process; low raw material costs and minimal scrap; uncomplicated machinery; and a high degree of automation. Disadvantages include: the process is limited to constant cross-section shapes; setup times and initial process startup is labor intensive; parts may have higher void contents than allowed for some structural applications; the majority of the reinforcement is oriented in the longitudinal direction; the resin used must have a low viscosity and a long pot life; and in the case of polyesters, styrene emissions can create worker health concerns. The critical processing variables are die design, resin formulation, material guidance before and after impregnation and temperature control in the die.

Due to the nature of the pultrusion process, continuous reinforcement must be used, either in the form of rovings or rolls of fabric; however, discontinuous mats and veils can be incorporated. To facilitate setup, creel stands are often placed on wheels so that a majority of the setup can be done offline, reducing the down time for the pultruder. A consideration for the pre-impregnation guide mechanism is that the reinforcement is usually fragile, and in the case of glass and carbon, abrasive. Dry rovings are often

Fig. 25. Pultruded Parts

guided by ceramic eyelets to reduce wear on both the fibers and guidance mechanism. Fabrics, mats and veils can be guided with plastic or steel sheets with machined slots or holes. Chopped strand mat is often used at an areal weight of 1.5 oz yd^{-2} in rolls up to 300 ft with a minimum width of 4 inches. Several sets of guidance mechanisms may be required to gradually shape the reinforcements prior to impregnation. In a process called pull-winding, moving winding units are used to overwrap the primarily unidirectional reinforcement, thereby providing additional torsional stiffness.

There are several different methods that can be used for the impregnation process. The first, and most common, is to guide the reinforcements down into an open resin bath. Impregnation results from capillary action and by sets of rods in the bath that the reinforcements pass over and under. This method produces good impregnation and is simple; however, in the case of polyester resins, styrene emissions can be a problem. To reduce the styrene emission problem, another approach uses an enclosed bath in which the reinforcement travels horizontally and enters and exits the bath through slots cut in the ends of the bath. The main advantage of this method is that the reinforcement is not bent and the

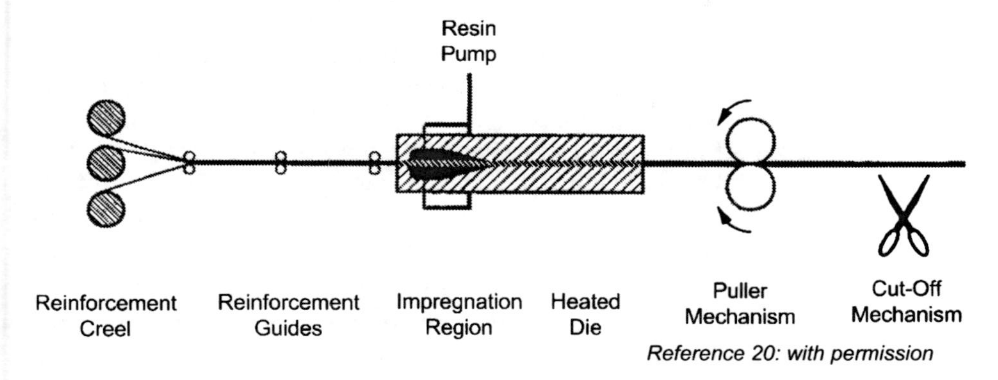

Fig. 26. Injection Pultrusion Die Process

styrene emissions are somewhat reduced. Vented hoods are frequently placed over the resin bath to help remove the styrene fumes from the immediate work area. A third method, called injection or reaction injection pultrusion (Fig. 26), injects the resin under pressure after the reinforcement has already entered the die. While this method virtually eliminates the styrene emission problem, the temperature of the die at the injection point must be closely controlled to prevent premature resin gellation. It should also be noted that this method significantly complicates die design. The dies are usually longer, more complex and more expensive than conventional dies. A fourth method, although rarely used to date, is to use preimpregnated reinforcements. Although more costly than the inline impregnation methods, pre-impregnation allows better control of the resin content and fiber areal weight.

For the open bath impregnation process, the resin must have a low viscosity (approximately 10 poise) and a long pot life so that it will readily impregnate the fibers and not gel in the bath. Resin impregnation baths are usually 3–6 ft long. The resin can be heated to reduce the viscosity but this will generally shorten the pot life considerably. For the injection pultrusion process, more reactive resins with shorter pot lives can be used. The resin should cross-link throughout the part thickness before it exits the die. The faster it cross-links, while keeping the exotherm under control so that thermally induced residual stresses are minimized and resin cracking is prevented, the faster and more productive the process. However, for epoxy resins that do not react as quickly as polyesters and vinyl esters, a free-standing post-cure in an oven after pultrusion may be required to complete the cure reaction. After the reinforcements have been impregnated, they pass through another set of guidance devices to further shape the reinforcements prior to entering the die cavity. These guidance devices aid in further gradually shaping the reinforcements prior to curing.

The pultrusion die itself is usually machined from tool steel and typically has a length of 24–60 inches. With the exception of a gradual taper at the entrance end of the die, the die cross-section is usually constant with extremely smooth surfaces that have been chrome-plated for durability and reduced friction. Almost all dies are segmented so that they can be taken apart for inspection and cleaning. Pultrusion dies generally have multiple heating zones that allow the temperature to be varied along the length of the die. For high exothermic resin systems such as polyesters, they may even have cooling zones near the exit to help maintain temperature control.

After the part is cured in the die, it is gripped and pulled over a rather long distance to allow the part to cool sufficiently before being cut to length. A number of different puller mechanisms are used in industry. The simplest method is to use a series of rubber wheels that grip the part in pairs and pull it. Although this method is simple and cheap, it is limited to small cross-sections where the pulling forces are low. Conventional belt pullers may also be used, frequently caterpillar belt pullers, where successive rubber pads are mounted on the belt. The most common pulling method is hydraulic clamp pullers with rubber pads. The pulling motion may be intermittent or continuous. It should be noted that the rubber pads must be tailored to fit the part geometry or excessive lateral pressure may be exerted on the part. A typical puller mechanism may have a pulling capacity of 10,000–20,000 lb, although larger machines may have a capacity of several hundred tons. A water-cooled abrasive cut-off saw is used to cut the pultruded part to the required length. The saw is mounted to the pultrusion unit and travels with the puller mechanism. Depending on the material to be cut, the cutting may be done dry or with flood coolant. Diamond impregnated or carbide circular saw blades are normally used to produce a clean cut.

Glass/polyester is the predominant material used for pultruded parts. Other matrices used include vinyl esters, acrylics, phenolics and epoxies. A processing advantage of polyesters and vinyl esters is the high shrinkage (7–9%) they undergo during cure. This helps the part to shrink away from the die reducing the friction and pulling forces required. On the other hand, epoxies exhibit much less shrinkage during cure (1–4%) and the friction and pulling forces are substantially higher. In general, epoxies do not produce as good a surface finish and the process must be run at higher temperatures and slower speeds. Typical processing speeds for polyesters are 24–48 in min^{-1}; however, under certain circumstances, speeds may approach 200 in min^{-1}. As the pulling speed increases, the pulling force increases. Too fast a pulling speed may move the peak exotherm temperature outside of the die and result in defects such as scaly and cracked surfaces, porosity and internal cracks, warpage and discoloration. Although a longer die may insure that the peak exotherm remains inside of the die, the trade-off is that greater pulling forces will be required. In fact, the pulling force required during pultrusion is often used as an indicator of the health of the process. If the pulling force starts to increase above the normal force required, it might be an indication that the process is going out of control.

Continuous rovings are the most prevalent reinforcement, although it is standard practice to interleave chopped or continuous strand mats to improve transverse properties, and surfacing mats, veils and non-wovens to improve the surface finish. For applications that require higher transverse strength or torsional stiffness, fabrics and braids that are stitched together can be used.

11.6 Summary

The commercial processes described in this chapter are capable of making anywhere from 1 to over a 1,000,000 parts per year. On both a dollar and volume basis, their importance far outweighs the high-performance composite processes dealt with in the majority of this book. Glass reinforcements, primarily E-glass, are the fibers of choice for most of these processes due to their low cost, abundance and varied product forms. Polyesters, and more recently vinyl esters, have demonstrated an excellent record of low cost combined with outstanding performance. Likewise, thermoplastic resins, such as polypropylene and many others, offer cost-effective solutions for many consumer products.

Lay-up processes include wet lay-up, spray lay-up and LTVB prepregs. These processes are amenable to low-to-medium production rates and are capable of making extremely large part sizes with minimal equipment or tooling investment.

Molding processes, such as compression molding, injection molding and the family of reaction injection-molding processes, while not offering the strength and stiffness advantages of continuous fiber composites, are capable of making extremely complex and tight tolerance parts in large production volumes.

Pultrusion is a rather specialized fabrication process that is ideal for long constant cross-section structural parts. Once a setup is completed, the process runs almost continuously with little intervention required.

It should be pointed out that there are many other processes for making commercial products from both unreinforced and reinforced polymers.

References

[1] "FRP – An Introduction to Fiberglas-Reinforced Plastics/Composites," Owens/Corning Fiberglas Corporation, 1976.

[2] Ridgard C., "Composite Tooling – Design and Manufacture, Getting It Right," The Advanced Composites Group.

[3] Ridgard C., "Affordable Production of Composite Parts Using Low Temperature Curing Prepregs," 42nd International SAMPE Symposium, May 4-8, 1997, pp. 147–161.
[4] Dragone T.L., Hipp P.A., "Materials Characterization and Joint Testing on the X-34 Reusable Launch Vehicle," *SAMPE Journal* **34**(5), September/October 1998, pp. 7–20.
[5] Niitsu M., Uzawa K., Kamita T., "HOPE-X: Development of the Japanese All-composite Prototype Re-entry Vehicle Structure," *SAMPE Journal* **38**(4), July/August 2002, pp. 34–39.
[6] Jackson K., "Low Temperature Curing Materials – The Next Generation," 43rd International SAMPE Symposium, May 31-June 4, 1998, pp. 1–8.
[7] Ridgard C., "Low Temperature Moulding (LTM) Tooling Prepreg with High Temperature Performance Characteristics," Reinforced Plastics, March 1990, pp. 28–33.
[8] Xu G.F., Repecka L., Boyd J., "Cycom X5215 – An Epoxy Pregreg that Cures Void Free Out of Autoclave at Low Temperature," 43rd International SAMPE Symposium, May 31-June 4 1998, pp. 9–19.
[9] Repecka L., Boyd J., "Vacuum-bag-only Prepregs that Produce Void-free Parts," SAMPE 2002, May 12-16, 2002.
[10] Ridgard C., "Advances in Low Temperature Curing Prepregs for Aerospace Structures," 45th International SAMPE Symposium, May 21-25, 2000, pp. 1353–1367.
[11] Astrom B.T., *Manufacturing of Polymer Composites*, Chapman & Hall, 1997.
[12] Peterson C.W., Ehnert G., Liebold K., Horsting K., Kuhfusz R., "Compression Molding," in *ASM Handbook Volume 21 Composites*, ASM International, 2001, pp. 515–535.
[13] Jang B.Z., *Advanced Polymer Composites: Principles and Applications*, ASM International, 1994.
[14] Strong A.B., *High Performance and Engineering Thermoplastics*, Technomic Publishing, 1993.
[15] Groover M.P., *Fundamentals of Modern Manufacturing – Materials, Processes, and Systems*, Prentice-Hall, 1996
[16] "User's Guide For Short Carbon Fiber Composites," Zoltek Companies, 2000
[17] John V., *Introduction to Engineering Materials*, Industrial Press, 1992, pp. 342–343
[18] Rosen S.L., *Fundamental Principles of Polymeric Materials*, Wiley, 1982
[19] "Injection Molding Processing Guide," LNP Engineering Plastics, 1998
[20] Astrom B.T., "Pultrusion" in *Processing of Composites*, Hanser, 2000, pp. 318–357.

Chapter 12

Assembly: The Best Assembly Is No Assembly Required

One of the main advantages of composites is the ability to make large integral structures that eliminates parts and the mechanical fasteners required to assemble them. Many structures can either be cocured or bonded together to integrate a significant number of detail parts into a single assembly. In spite of these technical advancements, assembly still represents a significant portion of the total manufacturing cost. As shown in Fig. 1, assembly costs can represent as much as 50% of the total delivered part cost.[1] Assembly operations are labor intensive and involve many steps. For example, the composite wing shown in Fig. 2 requires: (1) a framing operation in which all of the spars and ribs must be located in their proper location and connected together with shear ties; (2) each skin must then be located on the substructure, shimmed, holes drilled and fasteners installed. During and after skin installation there are various sealing operations that must be performed; and (3) the final wing torque box must have the leading edges, wing tips and control surfaces assembled. This brief description is a gross over simplification of the complexity involved in assembling a large structural component. In this chapter, the basic machining and assembly operations will be covered with an emphasis on hole preparation and the types of mechanical fasteners used in composite structures.

12.1 Trimming and Machining Operations

Composites are more prone to damage during trimming and drilling than conventional metals. Composites contain strong and very abrasive fibers held together by a relatively weak and brittle matrix. During machining,

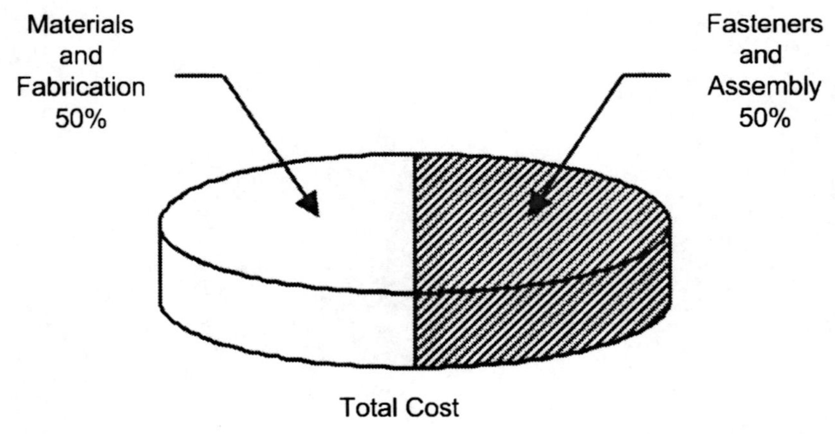

Fig. 1. The High Cost of Assembly Operations

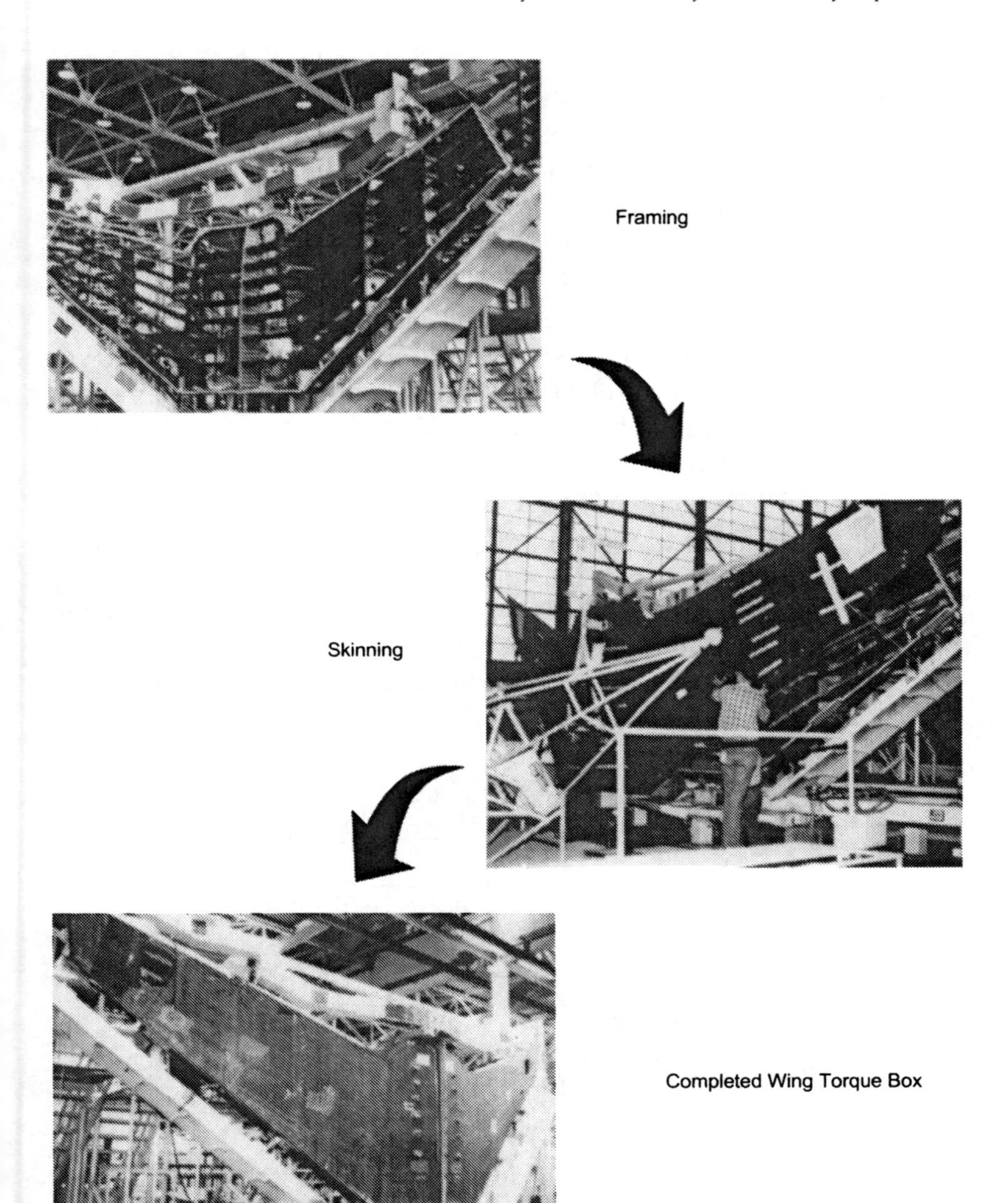

Fig. 2. Assembly Complexity

they are prone to delaminations, cracking, fiber pullout, fiber fuzzing (aramid fibers), matrix chipping and heat damage. It is important to minimize forces and heat generation during machining. During metallic machining, the chips help to remove much of the heat generated during the cutting operation. Due to the much lower thermal conductivity of the fibers (especially glass and aramid), heat build up can occur rapidly and degrade the matrix, resulting in matrix cracking and even delaminations. When machining composites, generally high speeds, low feed rates and small depths of cuts are used to minimize damage.[2]

Most composite parts require peripheral edge trimming after cure. Edge trimming is usually done either manually with high-speed cut-off saws or automatically with NC controlled abrasive water jet machines. Lasers have been often proposed for trimming and hole drilling of cured composites, but the surfaces become charred due to the intense heat and are unacceptable for most structural applications.

Carbon fibers are very abrasive and quickly wear out conventional steel cutting blades; therefore, trimming operations should be conducted using either diamond coated circular saw blades, carbide router bits or diamond coated router bits. A typical manual edge trimming operation, shown in Fig. 3, can be conducted with a high speed air motor (e.g., 20,000 rpm) with

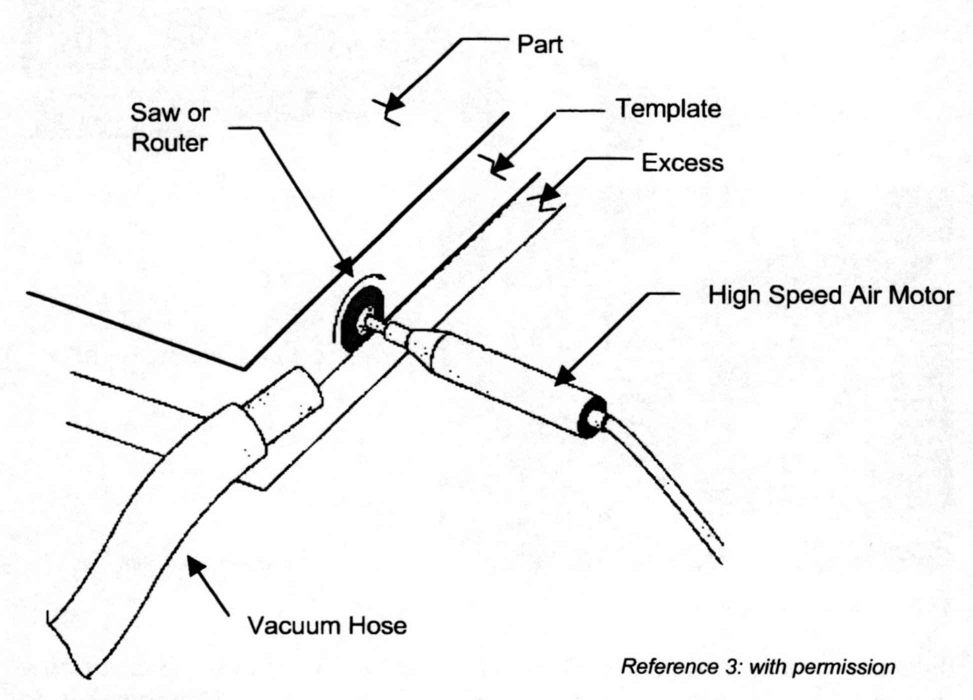

Fig. 3. Hand Trimming

either a diamond impregnated cut-off wheel or more typically a carbide router bit. Fiberglass laminate trim templates are often clamped to the part to insure that the true trim path is followed and provide edge support to help prevent delaminations. Typical feed rates are 10–14 inch min^{-1}. Hand trimming is a dirty job. The operator should wear a respirator, have eye and ear protection, and wear heavy duty gloves. Many facilities have installed ventilated trim booths to help control the noise and fine dust generated by the operation. Being a hand operation, the quality of the cut is very dependent on the skill of the operator. Too fast a feed rate can cause excessive heat leading to matrix overheating and ply delaminations.

Abrasive water jet trimming has emerged as probably the most accepted method for trimming cured composites; however, these are large NC-controlled machine tools (Fig. 4) that are expensive. The advantages of abrasive water jet cutting are that consistent delamination-free edges are produced, and since the cutting path is NC-controlled, the requirement for tooling is much simpler. Abrasive water jet cutting is primarily an erosion

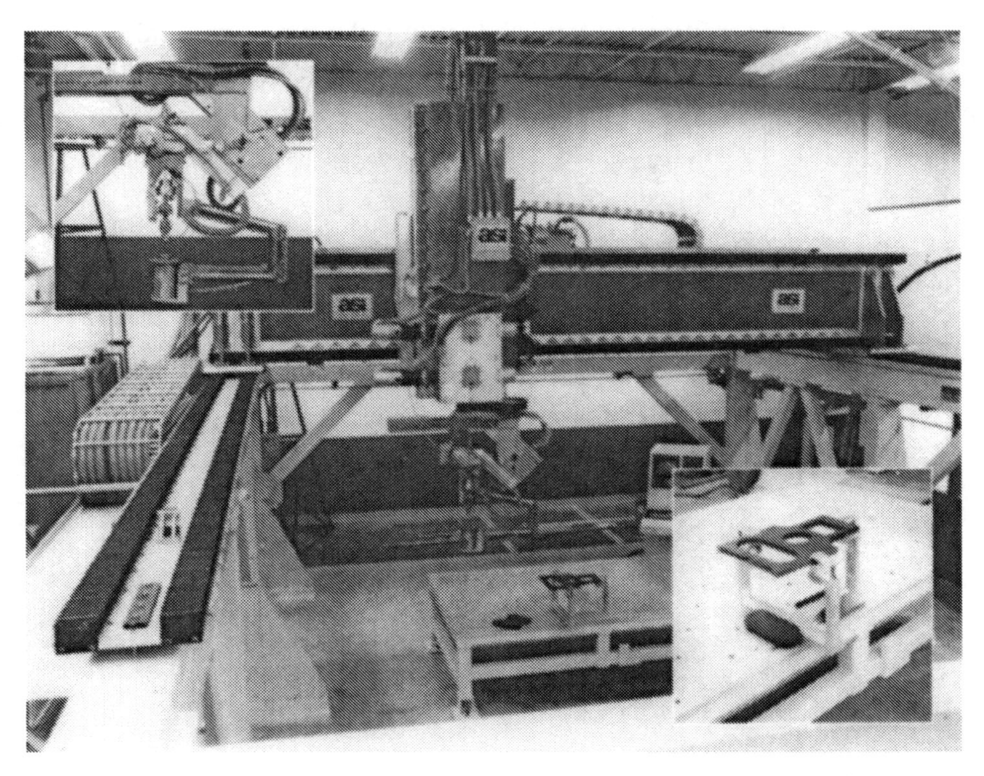

The Boeing Company

Fig. 4. Abrasive Water Jet Trimming Machine Tool

process rather than a true cutting process so there is very little force exerted on the part during trimming; therefore, only simple holding fixtures are required to support the part during cutting. In addition, no heat is generated during cutting negating the concern for possible matrix degradation. A cross section of a typical abrasive water jet head is shown in Fig. 5. Water pumped at low volume (1–2 gal min^{-1}) enters the top of the head and is then mixed with garnet grit that is expelled through a 0.040 inch diameter sapphire nozzle at 40,000–45,000 psi.[4] In general, higher grit size numbers (smaller grain diameters) produce better surface finishes, with a typical grit size being #80. Once the abrasive slurry has penetrated the composite laminate, there is a catcher filled with steel balls that spin to dissipate the flow. Other than the expense of these tools, the other main disadvantage is the noise level generated during the process. It is not unusual for trimming operations to exceed 100 dB; therefore, ear protection is required and many units are isolated within their own sound proof room.

The feed rate for abrasive water jet cutting should be kept fairly slow, especially as the laminate gets thicker. Too fast a feed rate will cause

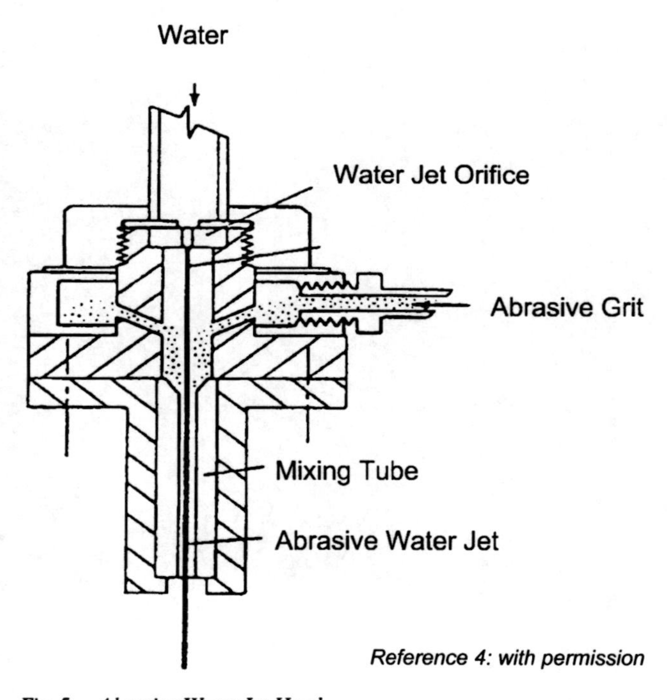

Fig. 5. Abrasive Water Jet Head

"trailback" or "weeping" as illustrated in Fig. 6, resulting in a tapered edge.[5] A typical feed rate for C/E up to 0.50 inches thick is 15 inch min^{-1} with slower values used for thicker materials. Surface roughness increases with increasing depths of cut. Standoff distance from the nozzle to the work piece is another consideration. The external diameter of the jet increases as the standoff distance increases due to jet disintegration caused by aeration and separation of the grit medium (i.e., the jet spreads like a Christmas tree). Therefore, high jet pressures and a low standoff distances results in cuts with less taper.[6] Delaminations can occur if the abrasive water jet "spits," which occurs when the nozzle gets plugged and then releases a sudden blast into the cutting area. This is usually a signal that either the nozzle is becoming worn or the grit is contaminated.

If edge sanding is required, die grinders at speeds of 4,000–20,000 rpm can be used along with 80 grit aluminum oxide paper for roughing and 240–320 grit silicon carbide paper for finishing.[5]

12.2 General Assembly Considerations

Mechanical fastener material selection for composites is important to prevent potential corrosion problems. Aluminum and cadmium coated steel fasteners will galvanically corrode when in contact with carbon fibers. Titanium (Ti–6Al–4V) is usually the best fastener material for carbon fiber composites based on its high strength-to-weight ratio and its corrosion resistance. When higher strength is required, cold worked A286 stainless steel or the nickel base alloy Inconel 718 can be used. If extremely high

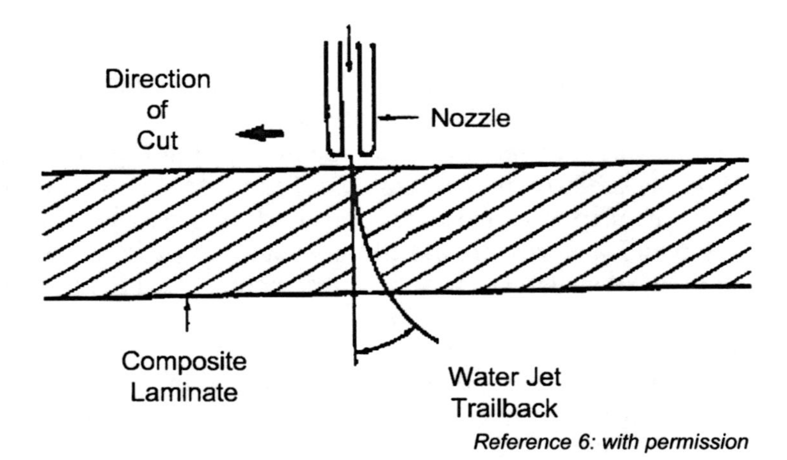

Fig. 6. Abrasive Water Trailback

strengths are required for very highly loaded joints, the nickel–cobalt–chromium multi-phase alloys MP35N and MP159 are available. It should be noted that glass and aramid fibers, being non-conductive, do not cause galvanic corrosion with metallic fasteners.

When a hole is placed in a composite laminate, it creates a stress concentration and the overall load bearing capability of the laminate is severely reduced. Even a properly designed mechanically fastened joint exhibits only 20–50% of the basic laminate tensile strength.[7] The various failure modes for composite joints are shown in Fig. 7. The only acceptable failure mode is when the joint fails in bearing since the joined members do not separate catastrophically. Bearing failures are characterized by localized damage such as delaminations and matrix crazing around the hole. Potential causes for the other failure modes shown include:

- *Shearout.* Insufficient edge distance or too many plies oriented in the load direction.
- *Tension.* Insufficient width or too few plies oriented in the load direction.
- *Cleavage-tension.* Insufficient edge distance and width or not enough cross-plies (e.g. +45° and –45°).
- *Fastener pull-through.* Countersink too deep or use of a shear head fastener.
- *Fastener failure.* Fastener too small for laminate thickness, unshimmed gaps or excessive shimmed gaps in joint, or insufficient fastener clamp-up.

Prior to starting hole drilling and fastener installation, it is important to check all joints for the presence of gaps. Since composites are more brittle and less forgiving than metals, excessive gaps can result in delaminations when they are pulled out during fastener installation. The composite is put in bending due to the force exerted by the fastener drawing the parts together and develops matrix cracks and/or delaminations around the holes. Cracks and delaminations usually occur on multiple layers through the thickness and can adversely affect the joint strength.[8] Gaps can also trap metal chips and contribute to backside hole splintering. If the skin is composite and the substructure is metal, the composite skin will often crack and delaminate if an appreciable gap is present during fastener installation. If both the skin and substructure are composite, then cracks can develop in either the skin or substructure, or both. Substructure cracking often occurs at the radius between the top of the stiffener and the web. To prevent cracking and delaminations, it is important to measure all gaps and then shim any gaps greater than 0.005 inches. Liquid shim, which is a filled thixotropic adhesive, can be used to shim gaps between 0.005 and 0.030 inches. If the gap exceeds 0.030 inches, then a solid shim is normally used but engineering approval is often required for a gap this large. Solid shims

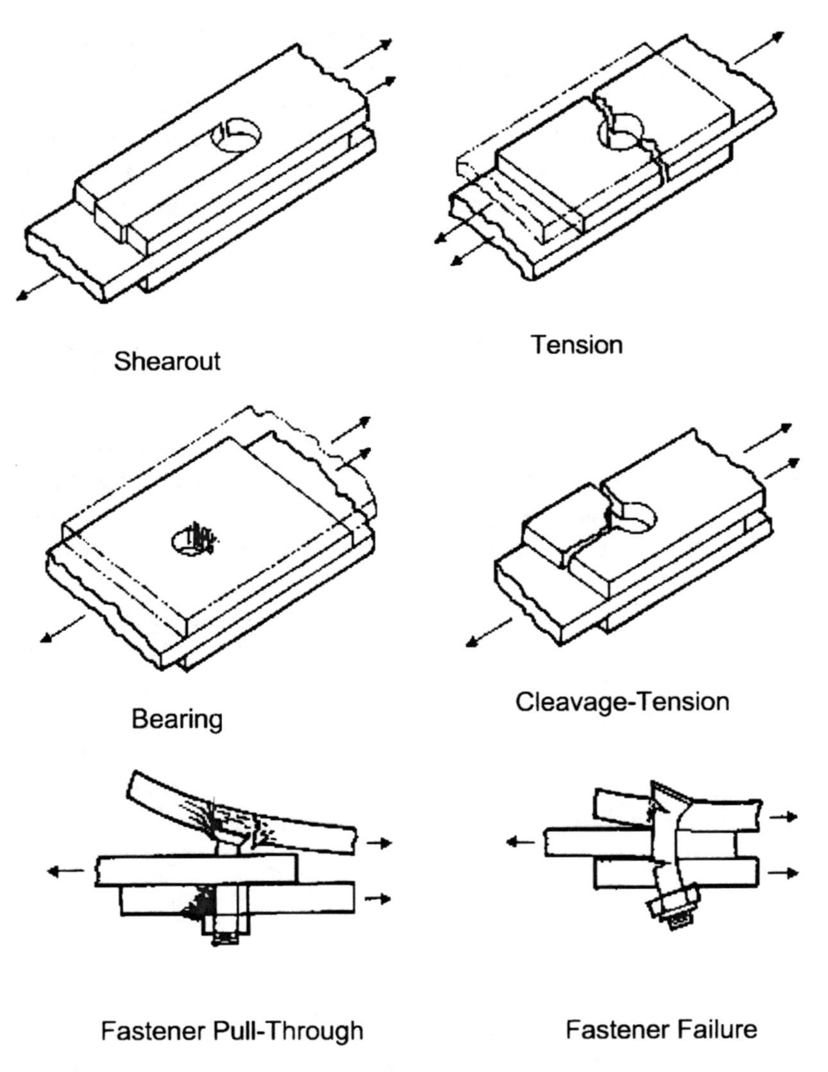

Fig. 7. Composite Joint Failure Modes

can be made from solid metal, laminated metal that can be peeled to the correct thickness or composite. When selecting a solid shim material, make sure there is no potential for galvanic corrosion within the joint.[9]

Liquid shimming can be accomplished by first drilling a series of undersize holes in the two mating surfaces for installing temporary fasteners to provide a light clamp-up during the shimming process. The liquid shim is usually bonded to one of the two surfaces. The surface that will be bonded should be clean and scuff sanded to provide adhesion. The other surface is covered with release tape or film. After the liquid shim is mixed, it is buttered onto one surface and the other surface is located and then clamped up with mold released temporary fasteners. The excess or squeeze-out is removed prior to gellation, which usually occurs within an hour of mixing. After the shim material is cured, typically for about 16 h, the part is disassembled and any voids or holes in the shim are repaired.

12.3 Hole Drilling

Hole drilling of composites is more difficult than in metals, again due to their relatively low sensitivity to heat damage and their weakness in the through-the-thickness direction. Hole diameters for aerospace structures nominally range from 0.164–0.375 inches with the predominate fastener size being 3/16 inches.[10] Composites are very susceptible to surface splintering (Fig. 8), particularly if unidirectional material is present on the surface. Note that splintering can occur at both the drill entrance and exit sides of the hole. As shown in Fig. 9, when the drill enters the top surface, it creates peeling forces on the matrix as it grabs the top plies, and when it exits the hole, it induces punching forces that again creates peel forces on the bottom surface plies. If top surface splintering is encountered, it is usually a sign that the feed rate is too fast, while exit surface splintering indicates that the feed force is too high.[9] It is common practice to cure a layer of fabric on both surfaces of composite parts, which will largely eliminate the hole splintering problem.

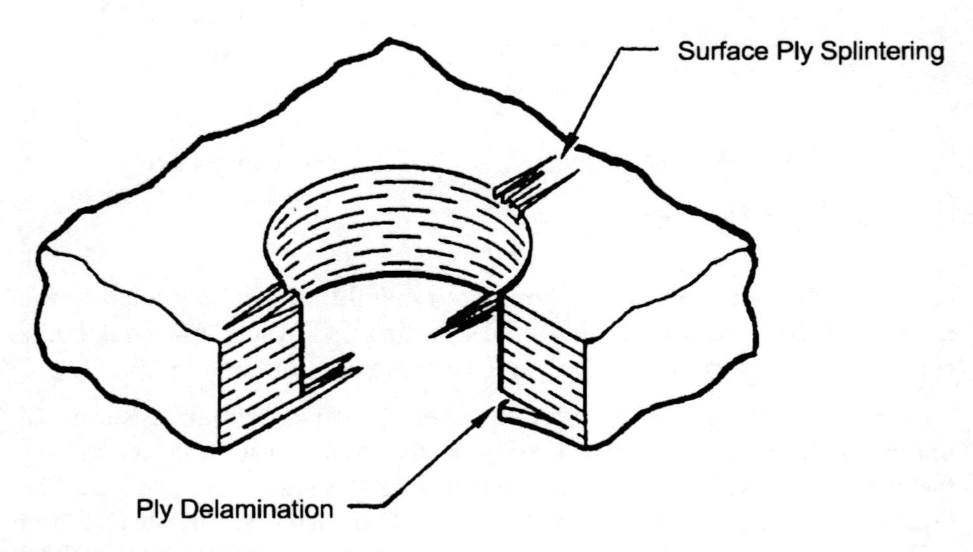

Fig. 8. Composite Hole Splintering

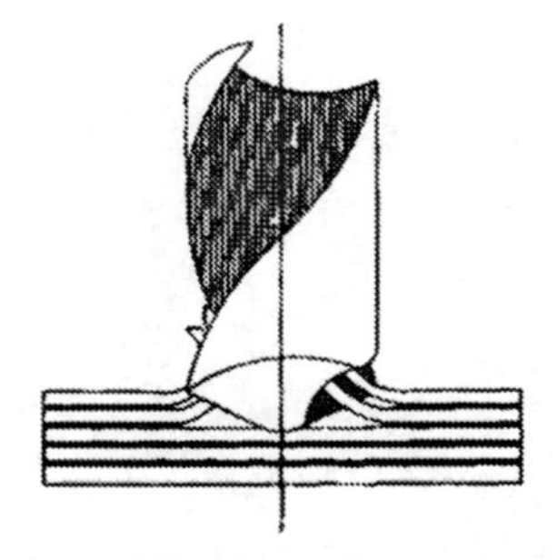

Drill Induces Peeling Forces on Top Plies During Entry

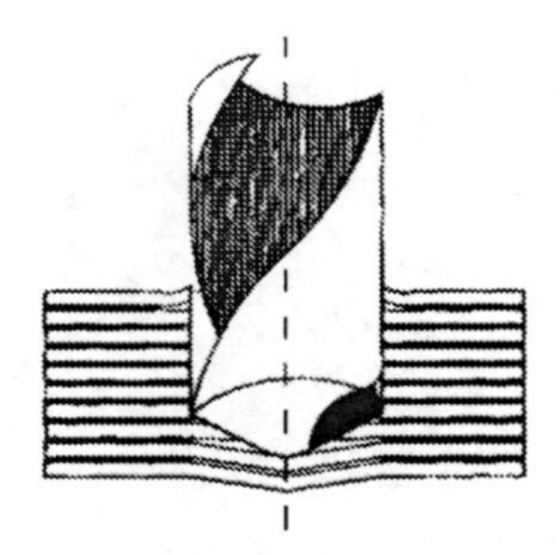

Drill Induces Punching Forces on Bottom Plies During Exit

Reference 2: with permission

Fig. 9. Drilling Forces on Composite Laminate

Since epoxy resin composites will start to degrade if heated above 400 °F, it is important that heat generation be minimized during the drilling operation. Typical drilling parameters are 2,000–3,000 rpm at feed rates of 0.002–0.004 inch per revolution (ipr), although this will vary depending on the drill geometry and the type of equipment used. Thermocouples and heat sensitive paints are often used during drilling parameter development tests to monitor the heat generated. Drilling parameters for composite-to-metal stack-ups are often controlled more by the metal than the composite. For example, when drilling C/E-to-aluminum, a speed 2,000–3,000 rpm with a feed rate of 0.001–0.002 ipr might be used, while a stack-up of C/E-to-titanium would require a slower speed (e.g., 300–400 rpm) and a higher feed rate (0.004–0.005 ipr). Titanium (Ti–6Al–4V) is very sensitive to heat buildup (hence the lower speed) and tends to rapidly work harden if light cuts are used (hence the higher feed rate).

Two other defects commonly encountered in composite drilling are shown in Fig. 10. Fiber pull-out occurs when selective pieces of the plies inside of the holes are pulled-out leaving an uneven hole finish. This condition is usually controlled by the selection of drill geometry and machining parameters. When drilling into composite-to-metal stack-ups, a phenomena called back counterboring can occur. As the metal chips (i.e., aluminum or titanium) travel up the flutes, they tend to erode the softer liquid shim and composite matrix material causing oversize and eroded holes. Back counterboring can be minimized by: (1) eliminating all gaps, (2) using a drill geometry that produces small chips, (3) changing speeds and feeds, (4) providing better clamp-up, (5) reaming the hole to final diameter after drilling, or (6) peck drilling.[9] Peck drilling (Fig. 11) is a process in which the drill bit is periodically withdrawn to clear the chips

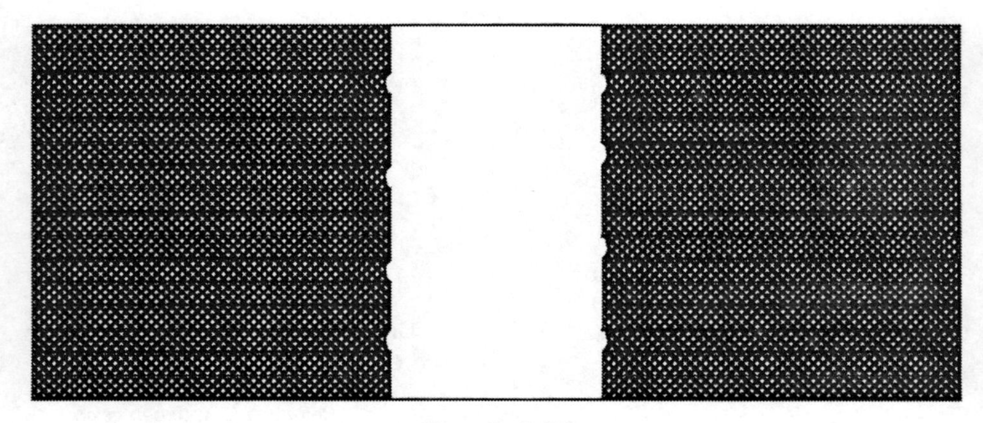

Fig. 10. Fiber Pull-Out and Back Counterboring

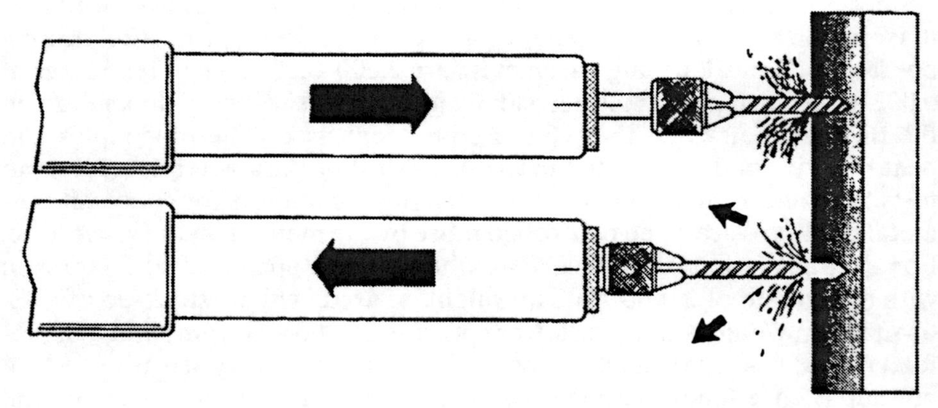

Fig. 11. Peck Drilling

from the flutes. Peck drilling is used almost exclusively when drilling composite-to-titanium stack-ups due to the back counterboring potential of the hard titanium chips, and the process greatly reduces the heat buildup that can rapidly occur when drilling titanium. Another method of reducing back counterboring and improving hole quality is to use through-the-drill cooling as shown in Fig. 12. The coolant helps to keep the chips flushed out of the hole. Note that in the setup shown, the drill and countersink are designed into the same body so that the countersinking operation is done during drilling.

There are many types of drill motors and units that can be used to drill composites, but they can be classified as either hand, power feed or

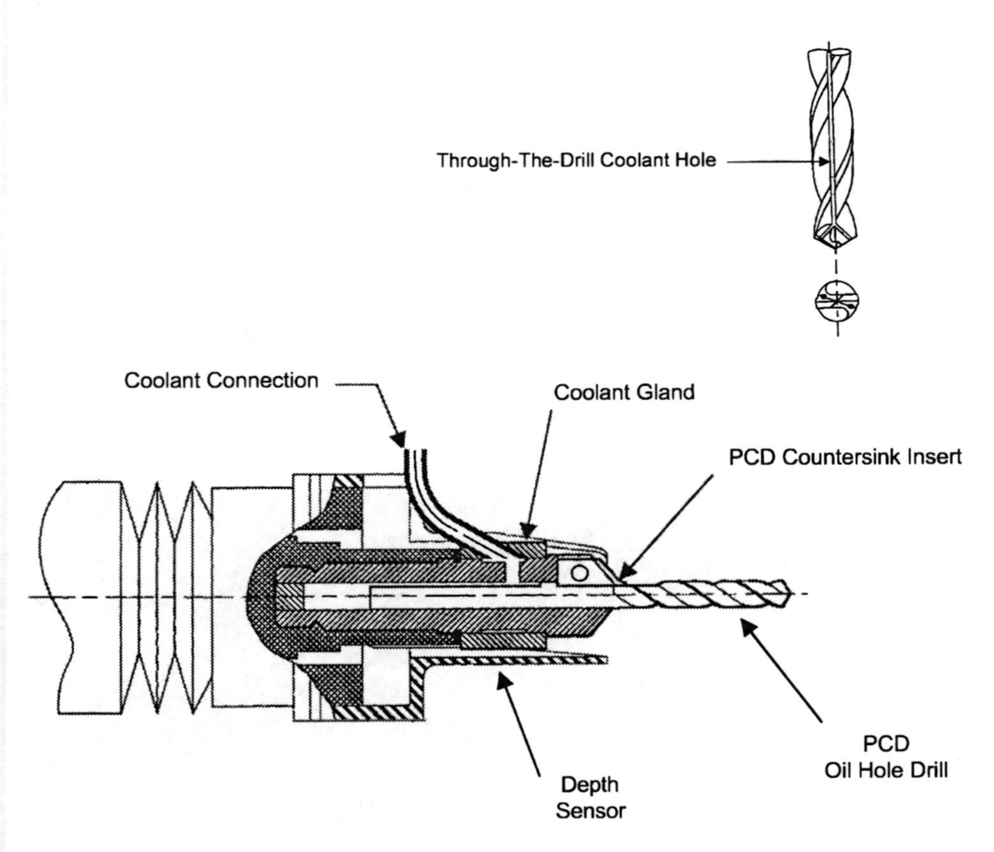

Fig. 12. Through-The-Drill Coolant

automated drilling units. A typical hand drill and power feed unit are shown in Fig. 13. Free hand drilling has the least chance of making a close tolerance hole (+0.003/–0.000 inches) in a composite or composite-to-metal stack-up. The only real control is the drill rpm. It is up to the operator to: (1) make sure the drill is located in the proper location, (2) is perpendicular to the surface, and (3) is fed with enough pressure to generate the hole but not too much pressure to damage the hole. A back-up material, such as aluminum or composite, clamped to the backside will frequently help to prevent backside hole splintering. Coolant is normally not used for carbon/epoxy laminates that are 0.250 inches thick or thinner. Although free hand drilling is obviously not the best method, it is frequently used because it requires no investment in tooling (i.e., drill templates) and in many applications where access is limited, it may be the only viable method. For tight access areas, right-angle drill motors are available. If free hand drilling is used, it is recommended that the operators use a drill bushing or tri-pod support to insure normality, and they be provided with detailed

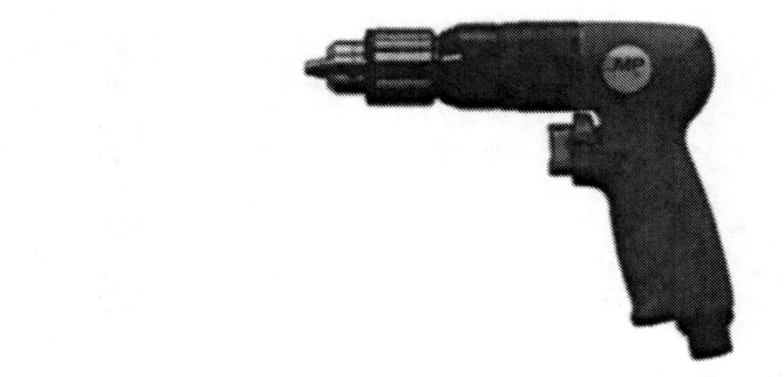

Free-Hand Drill Motor

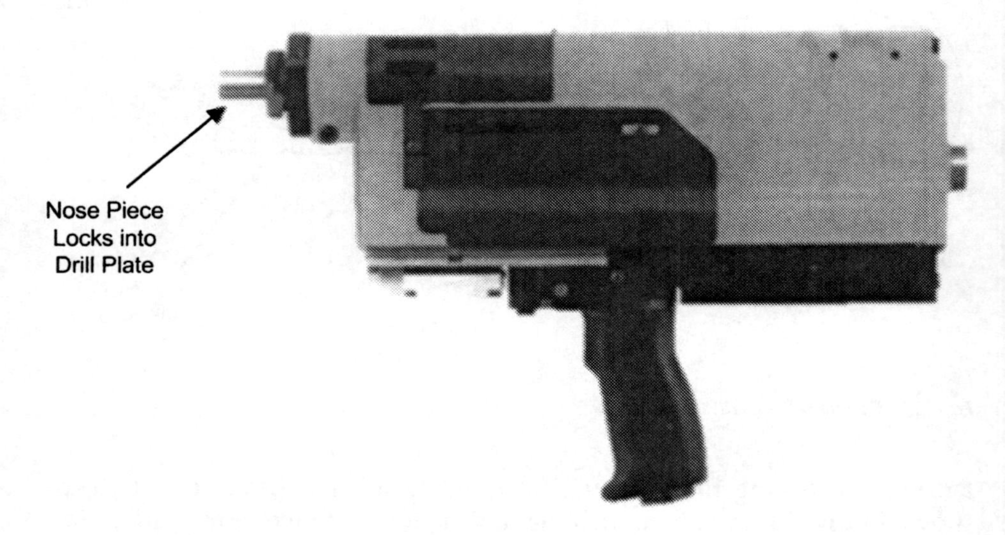

Power Feed Peck Drill

Cooper Power Tools

Fig. 13. Typical Drill Motors

written instructions for hole generation and inspection. As with hand routing, the operators should be provided with vacuum capability to suck up the dust and should always wear eye protection and a respirator.

Power feed drilling is much preferred to hand drilling. In power feed drilling, the drill unit is locked into a drill template that establishes both hole location and maintains drill normality. In addition, once the drilling operation starts, the unit is programmed to drill at a given speed and feed. Some units, such as the one shown in Fig. 13, can be programmed for different peck cycles. All of these controls lead to much better and more consistent hole quality, particularly when drilling composite-to-metal stack-ups. A typical peck drilling cycle for a 3/16 inch diameter hole through C/E-to-titanium would be a speed of 550 rpm with a feed rate of 0.002–0.004 ipr with 30–60 pecks per inch of thickness.[11]

For high volume hole generation, automated drilling equipment can be designed and built for specific applications.[12] Being large and sophisticated machine tools, these units are expensive so the number of holes drilled and the number of units produced needs to be large enough to justify the equipment investment. An example of one of these units is shown in Fig. 14. These machines are extremely rigid and allow for accurate hole location and normality. They are NC controlled so there is no need for drill templates. The one shown has a vision system that can scan the substructure and software that will then adjust the hole location to match where the substructure is actually located versus design nominal. All drilling parameters are automatically controlled with the capability to change speeds and feeds when drilling through different materials. Due to the thick stack-ups that must be drilled in a wing, a water soluble flood coolant is usually used during the hole drilling operations. All drilling data is automatically recorded and stored for quality control purposes. The drill

The Boeing Company

Fig. 14. Automated Wing Drilling System

holders contain bar codes that must match the drilling program to make sure the correct drills are used for the correct holes. These machines can also install temporary fasteners to clamp the skins to the substructure during drilling and frequently use integral drill-countersink cutters that drill the hole and then continue to countersink it during the same operation. The current trend in industry is to replace these large installations with smaller more flexible units.[13,14]

A number of unique drill geometries have been developed for composites, several of which are shown in Fig. 15. The design of the drill and the drilling procedures are very dependent on the materials being drilled. For example, carbon and aramid fibers exhibit different machining behavior and therefore require different drill geometries and procedures. In addition, composite-to-metal stack-ups will also require different cutters and procedures. The flat two-flute and four-flute dagger drills were developed specifically for drilling carbon/epoxy only stack-ups. The two-flute variety is normally run at 2,000–3,000 rpm, while the four-flute is run at 18,000–20,000 rpm.[9] When drilling through composite-to-metal stack-ups, the drill geometry is somewhat controlled by the metal and standard twist drills are often used. All aerospace companies have their own specifications for these drills with specific geometric details that are considered proprietary. Due to their low compressive strengths, aramid fibers have a tendency to recede back into the matrix rather than being cleanly cut resulting in fuzzing and fraying during drilling; therefore, the aramid drill contains a "C" type cutting edge that grabs the fibers on the

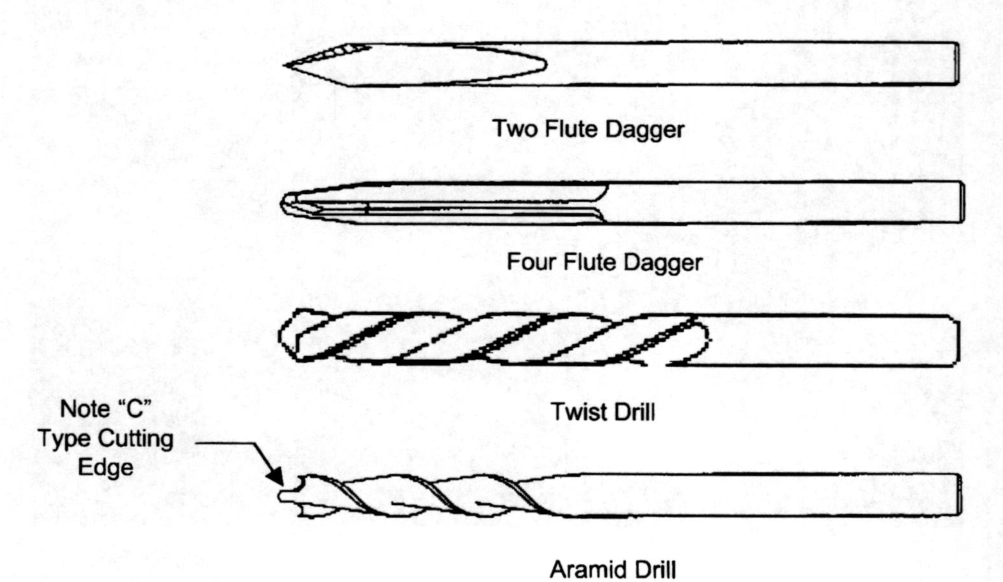

Fig. 15. Composite Drill Configurations

outside of the hole and keeps the fibers in tension during the cutting process. Typical drilling parameters for aramid fiber composites would be 5,000 rpm and a feed rate of 0.001 ipr.[11]

While standard high speed steel (HSS) drills work well in glass and aramid composites, the extremely abrasive nature of carbon fibers requires carbide drills to obtain an adequate drill life. For example, a HSS drill may only be capable of drilling one or two acceptable holes in carbon/epoxy, while a carbide drill of the same geometry can easily generate 50 or more acceptable holes. For drilling carbon/epoxy in rigid automated drilling equipment, polycrystalline diamond (PCD) drills (Fig. 16) have exhibited

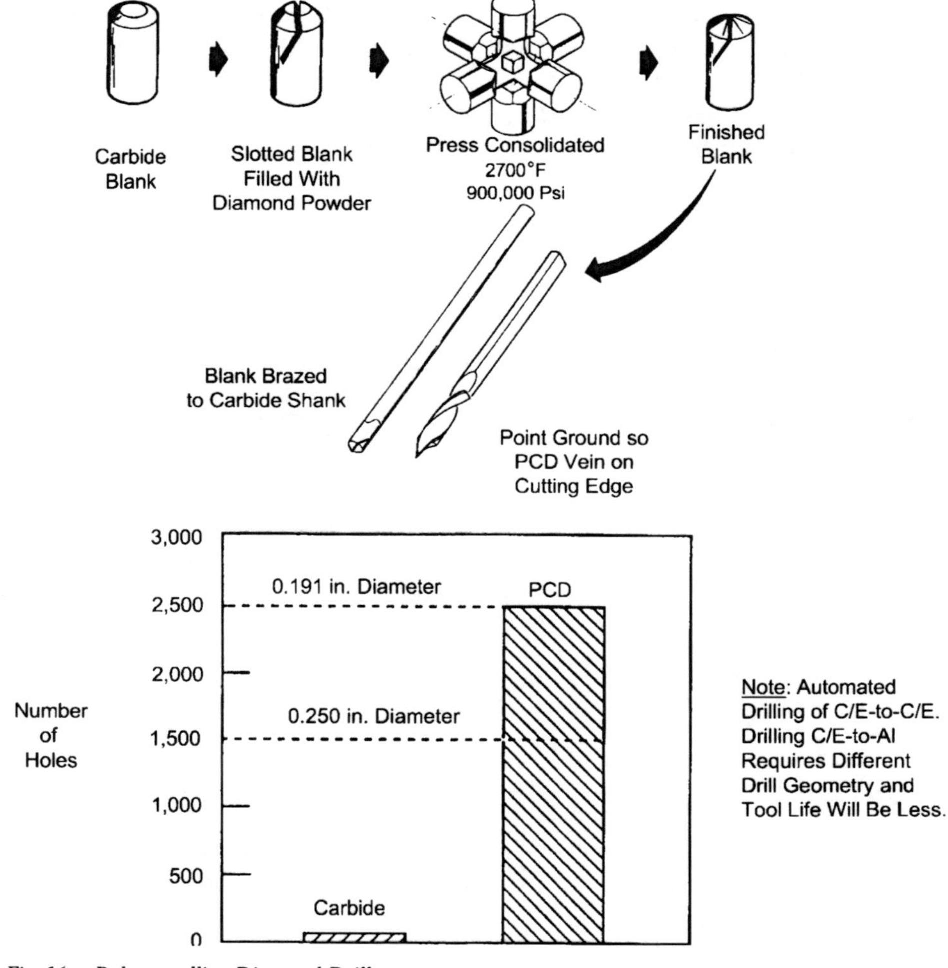

Fig. 16. Polycrystalline Diamond Drills

outstanding productivity improvements. While PCD drills are very expensive, the number of holes obtained per drill and the fewer changes required make them cost effective. It should be noted that PCD drills cannot be used with free hand or non-rigid setups; the point will immediately chip and break if any vibration or chatter is present during drilling.

Contersinking of composite structures is similar to that done in metals with two cautions: (1) the countersink depth has to be closely controlled to prevent a knife edge in thin structure (Fig. 17), and (2) the area where the countersink transitions into the hole must have the same radius as the fastener head-to-shank radius. Again, due to the low interlaminar shear strength of composites, either of these conditions can result in cracks and delaminations under the force of fastener installation. Countersinking cutters for composites are normally made of either solid carbide, steel bodies with carbide inserts, or steel bodies with PCD inserts. Piloted countersinks are helpful in centering the countersink tool in the hole and depth control can be obtained with microstop cages.

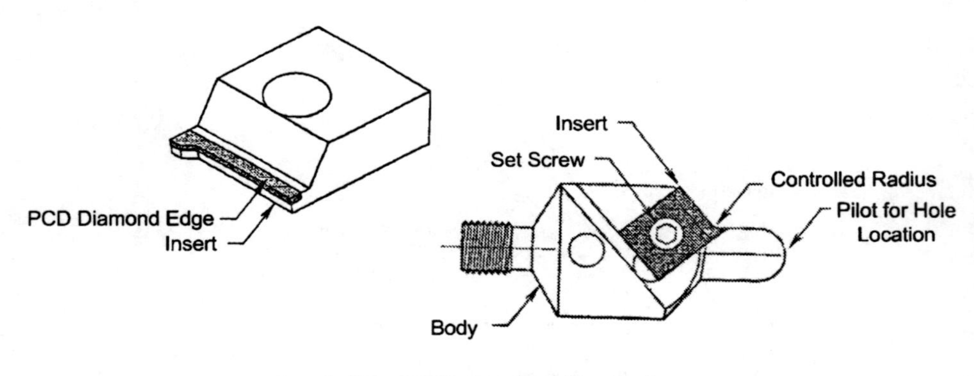

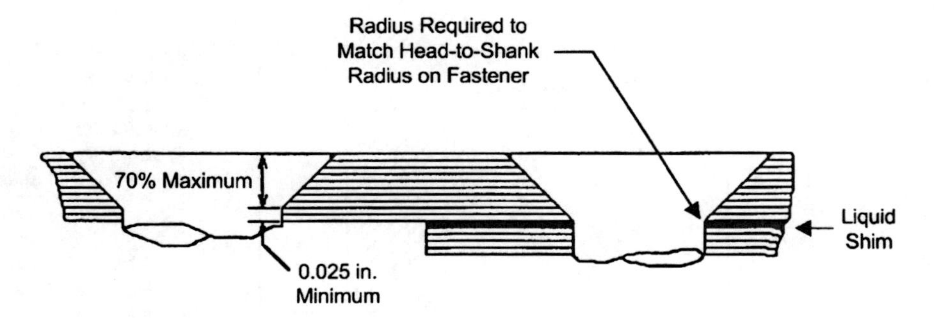

Fig. 17. Countersinking

Although it is desirable to drill the final hole size in one pass, it is sometimes necessary to ream the hole to final diameter. Reaming is done with carbide reamers at about one half the drilling speed (e.g., 500–1,000 rpm). In some composite-to-metal structures, fasteners are installed clearance fit in the composite and interference fit in the metallic structure for fatigue life enhancement. In this situation, the final hole diameter would be drilled in the composite-to-metal stack-up; the composite skin would then be taken down and the holes reamed to provide a clearance fit for the fastener. When the stack-up is reassembled, the fasteners would be installed clearance fit in the composite and interference fit in the metal.

12.4 Fastener Installation

Prior to installing fasteners in composite structure, it is important to measure the grip length of the fastener to be installed. There are commercial gages available that can be placed through the hole to measure the correct grip length. The correct grip length will insure that no threads in the fastener are placed in bearing when the structure is loaded.

Like hole drilling, fastener installation in composites is more difficult and damage prone than for metallic structure. Some of the potential problems with fastener installation are shown in Fig. 18. As previously discussed,

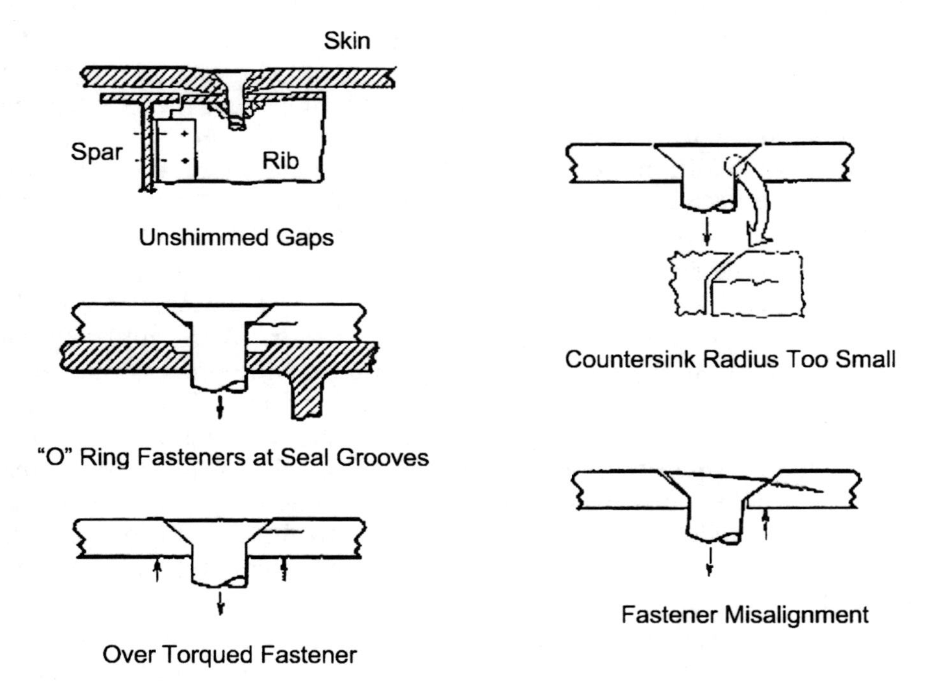

Fig. 18. Fastener Installation Defects

unshimmed gaps can cause cracking of either the composite skin or the composite substructure (or both) as the fastener is being installed and pulls the two pieces together. In fuel tanks, channel seal grooves are often used to help prevent fuel leakage. In addition, fasteners with O-ring seals can be used to further prevent leakage. It has been found through experience that this is a potential area for interlaminar cracking. While good clamp-up of the fastener is certainly desired, over torqueing fasteners can also result in cracking. If the countersink radius is too small and does not match that of the fastener head-to-shank radius, the fastener can apply a concentrated point load and cause matrix cracking. Likewise, fastener misalignment, where the hole and the countersink are not properly aligned, can result in point loading and cracking. In addition, fastener cocking during loading (Fig. 19) can result in point loading and lead to progressive damage during fatigue cycling.

In any mechanically fastened joint, high clamp-up forces are beneficial to both static and fatigue strength. High clamp-up produces friction in the joint, delays fastener cocking and reduces joint movement or ratcheting during fatigue loading. Most holes eventually fail in bearing caused by fastener cocking and localized high bearing stresses.[15] To allow the maximum clamp-up in composites without locally crushing the surface, fasteners have been designed specifically for composites that have large footprints (large heads and nut areas that bear against the composite) to help spread the fastener clamp-up loads over as large an area as possible. Washers are also frequently used under the nut or collar to help spread the clamp-up loads. In general, the larger the bearing area, the greater the clamp-up that can be applied to the composite, resulting in improved joint strength.[16] In addition, tension head rather shear head fasteners are normally used because they are not as susceptible to bolt bending during fatigue or fastener pull-through during installation in thin structure.

Rivets are extensively used for lightly loaded sheet metal structure but are rarely used in composites for two reasons: (1) aluminum rivets will galvanically corrode when in contact with carbon fibers, and (2) the vibration and expansion of the rivet during the driving process can cause delaminations. If rivets are used, they are usually a bimetallic rivet consisting of a Ti–6Al–4V pin with a softer titanium-columbium tail that are installed by squeezing rather than vibration driving. In addition, the head that is upset must be against metallic structure and not composite. There are also hollow end solid rivets designed to allow flaring of the ends without damaging expansion when used in double countersunk holes in composites.[16]

Various pin and collar fasteners are used for permanent installations and screws with nut plates or gang channels are used when the fasteners must be removable. Pin and collar fasteners are the most commonly used

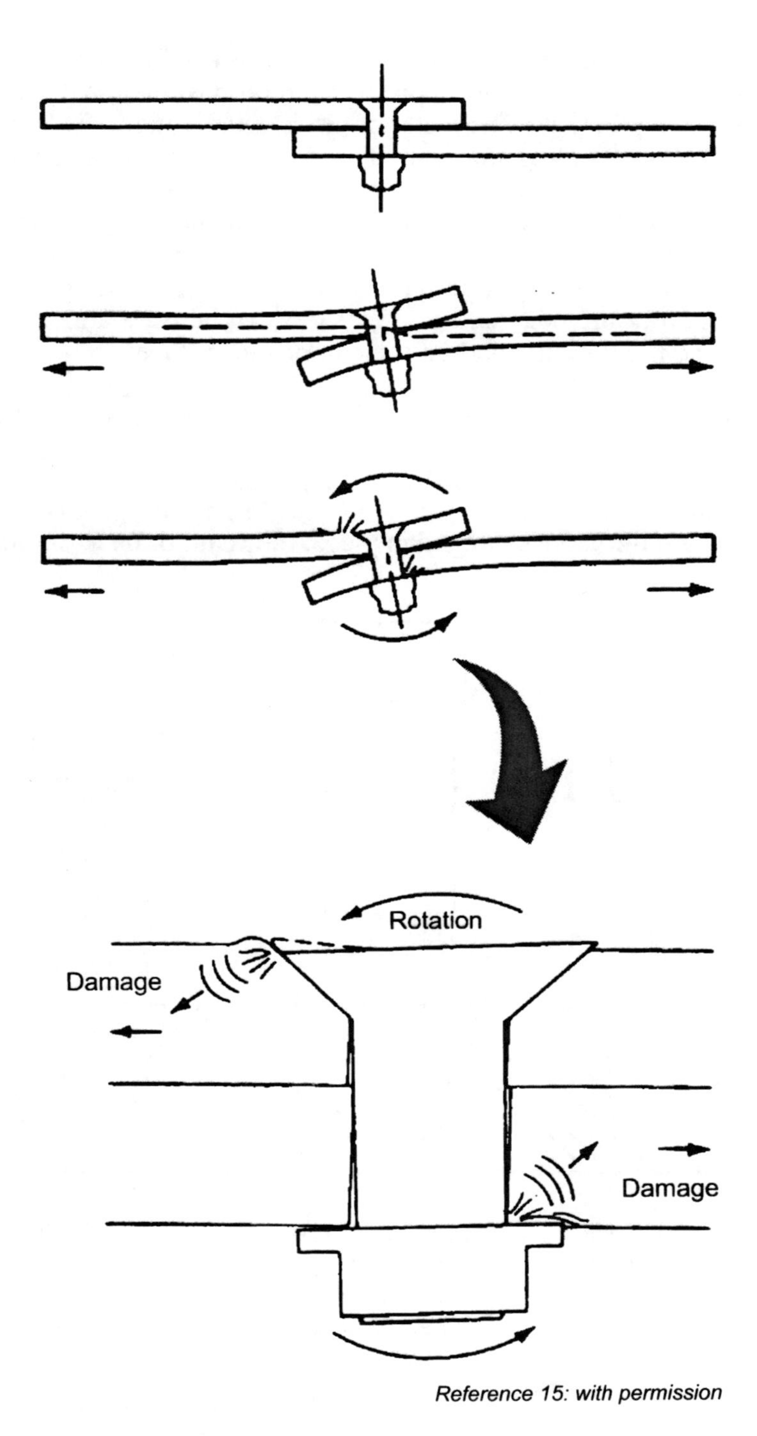

Fig. 19. Fastener Cocking in Single Lap Shear

fasteners for permanent installations where there is no requirement to remove the fastener. A typical pin and collar fastener is the Hi-Lok fastener shown in Fig. 20. The fastener pin is usually made from Ti–6Al–4V with an A286 stainless steel nut. Titanium nuts are occasionally used but the threads tend to gaul if they are not coated with an anti-gauling lubricant, which then adversely affects long term clamp-up. A hex key is inserted into the fastener stem to react the torque applied to the nut. The nut is tightened down until a predetermined torque level is achieved and the top portion of the nut fractures. Washers can be used under the head to help spread the bearing load on the composite surface.

Lockbolts are another common pin and collar fastener that can be installed by either pulling or swaging the collar from the backside. A typical pull type Lockbolt installation sequence is shown in Fig. 21. Lockbolts differ from Hi-Loks in that Hi-Loks have true threads that the nut is threaded onto, while Lockbolts have a series of annular grooves that the collar is swaged into. Once swaged in place, they cannot back off and have

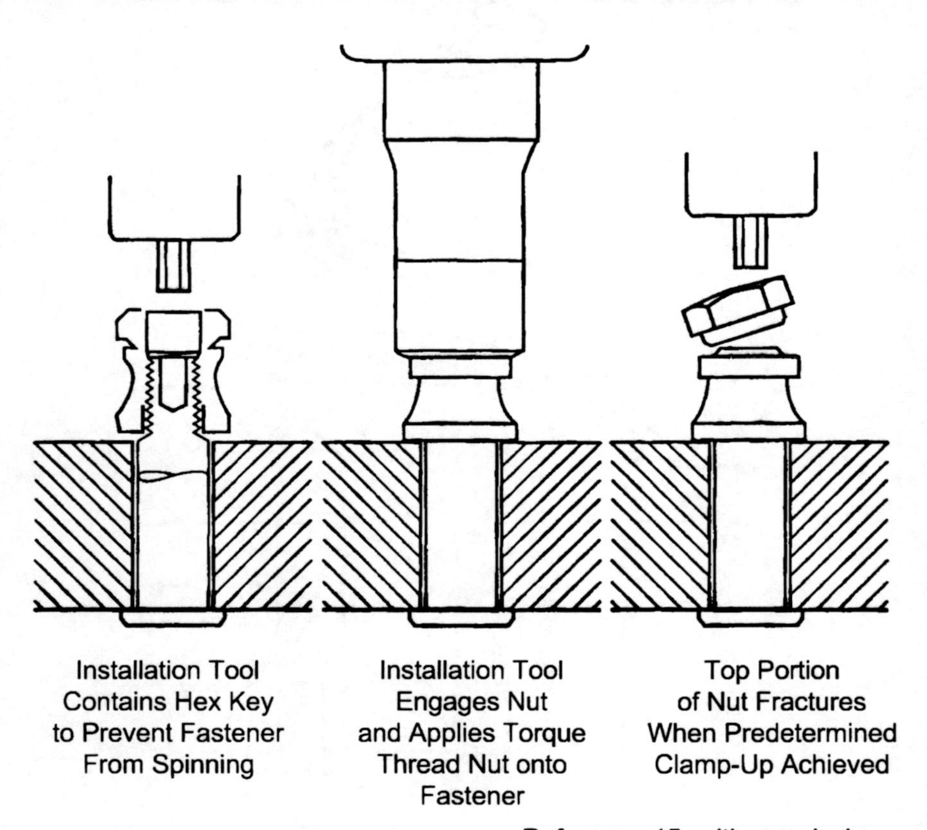

Fig. 20. Installation of Hi-Lok Fastener

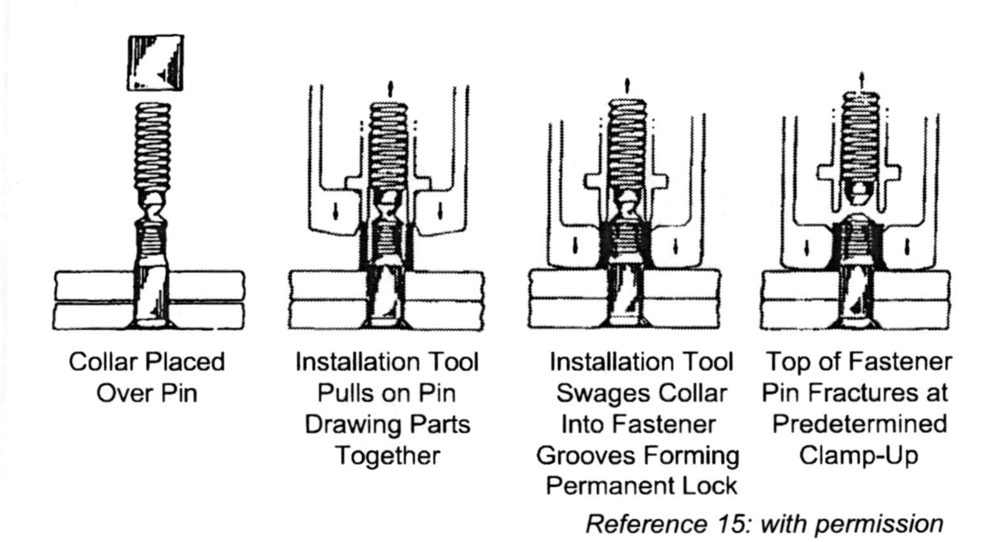

Fig. 21. Installation of Pull Type Lockbolt

superior vibration resistance.[16] There are two precautions that need to be followed when installing Lockbolts in composite structure: (1) pull type Lockbolts exert quite a bit of force on the composite due to the pulling action necessary to swage the collar onto the pin. If the composite is thin, fastener pull-through is a real possibility and if there are any unshimmed gaps, cracking and delaminations can occur when the fastener pin fractures; and (2) if backside Lockbolts (called stump Lockbolts) are installed in composites, they should be installed by a piece of automated equipment where careful control of the swaging operation can be exercised.

A third type of pin and collar fastener is the Eddie bolt shown in Fig. 22. As shown in the installation sequence, the collar initially threads onto the pin but then is swaged into flutes on the pin to provide a positive lock. The advantage of Eddie bolts is that they do provide a positive lock and will retain clamp-up loads better than Hi-Loks that rely on torque only. However, the fasteners and installation tools are expensive, the sockets are subject to wear and the installation procedure is more difficult. They are often specified in inlet duct areas where there is the potential for a fastener pin coming loose and flying into the engine blades and damaging the engine.

Blind fasteners are used in areas where there is limited or no access to the backside of the structure; however, the solid core pin and collar fasteners previously discussed are usually preferred because they are stronger and have better fatigue resistance. Two types of blind fasteners are the threaded core bolt type and the pull type shown in Fig. 23. The threaded core bolt (Fig. 24) relies on an internal screw mechanism to deform the

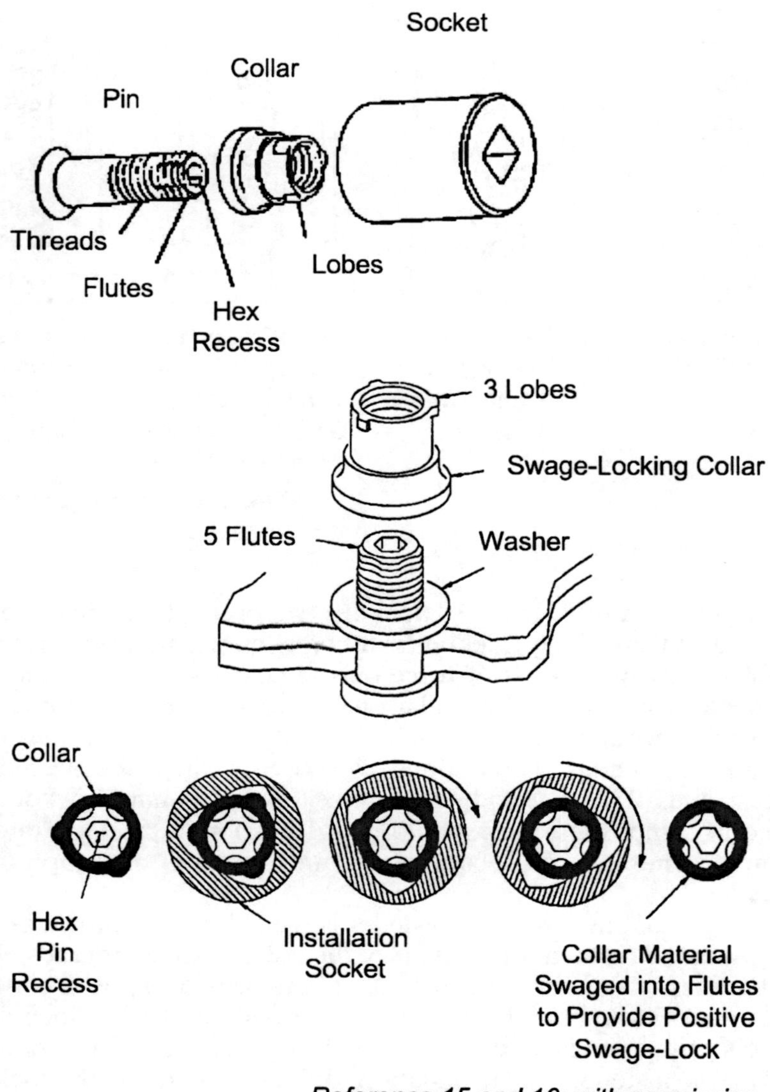

Fig. 22. Eddie Bolt Positive Lock Fastener

head and pull it up tight against the structure, while the pull type blind fastener uses a pure pulling action to form the backside head. Higher clamp-up forces and larger footprints are obtainable with the threaded core bolt leading to longer fatigue life; however, the pull types install quicker, are lighter and less expensive.

Interference fit fasteners are frequently used in metallic structure to improve the fatigue life. When an interference fit fastener is installed in

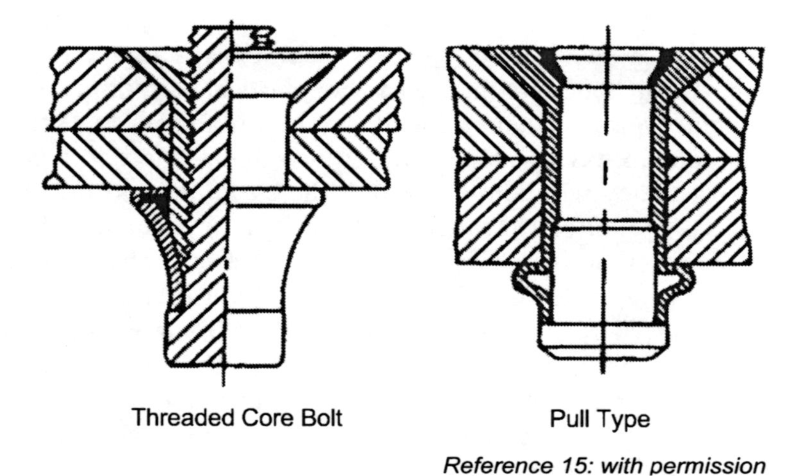

Fig. 23. Blind Fasteners

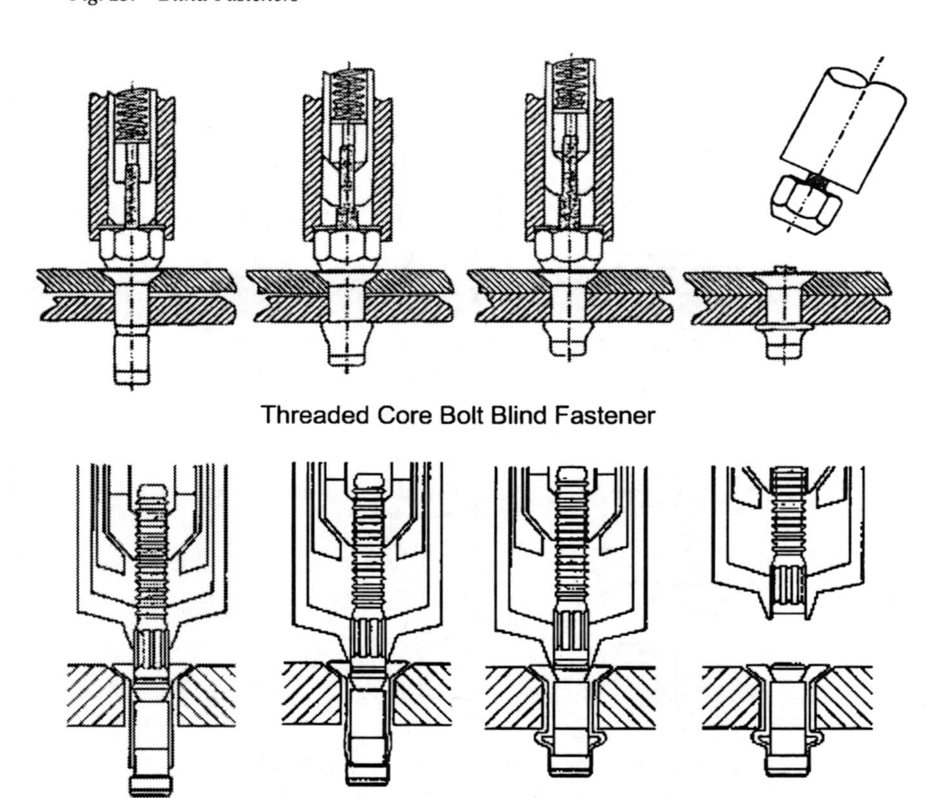

Fig. 24. Installation of Blind Fasteners

metal, it plastically deforms a small zone around the hole setting up a compressive stress field, which is beneficial when fatigue loading is primarily in tension. Since composites do not plastically deform, there is no fatigue life improvement when using interference fit fasteners. However, a potential benefit to having some interference fit fasteners in composite structure is that they will help "lock up" the structure and prevent any movement at the joint (called ratcheting) during fatigue loading. Previous work has shown that installing standard interference fit fasteners in composites with as little as 0.0007 inches of interference can lead to cracking and interlaminar delaminations.[15] To eliminate this problem, special sleeve-type interference fit fasteners (Fig. 25) have been designed so that the sleeve spreads the load evenly during installation to prevent the delamination problem. Interferences as high as 0.006 inches have been obtained without damaging the composite. Both pin and collar (Lockbolts) and threaded core blind bolt fasteners (Fig. 26) are available with sleeves.

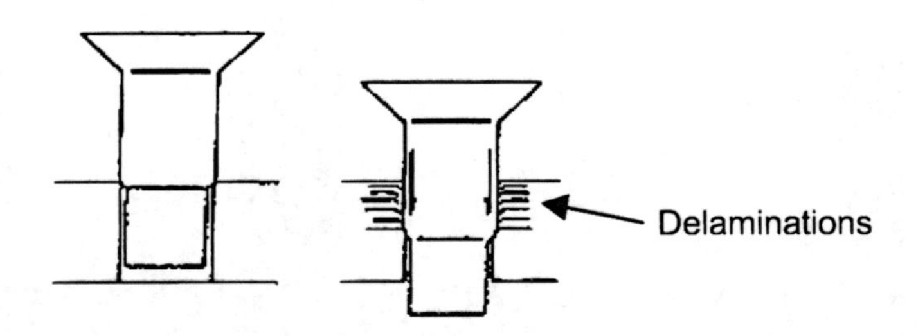

Installation of Conventional Interference Fit Fasteners
Can Cause Delaminations in Composite Structures

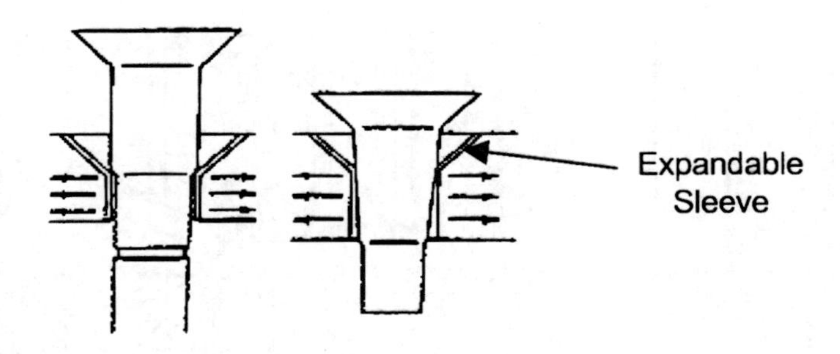

Installation of Interference Fit Fasteners With Sleeves
Spreads Forces to Eliminate Delaminations

Reference 15: with permission

Fig. 25. Interference Fit Fasteners in Composites

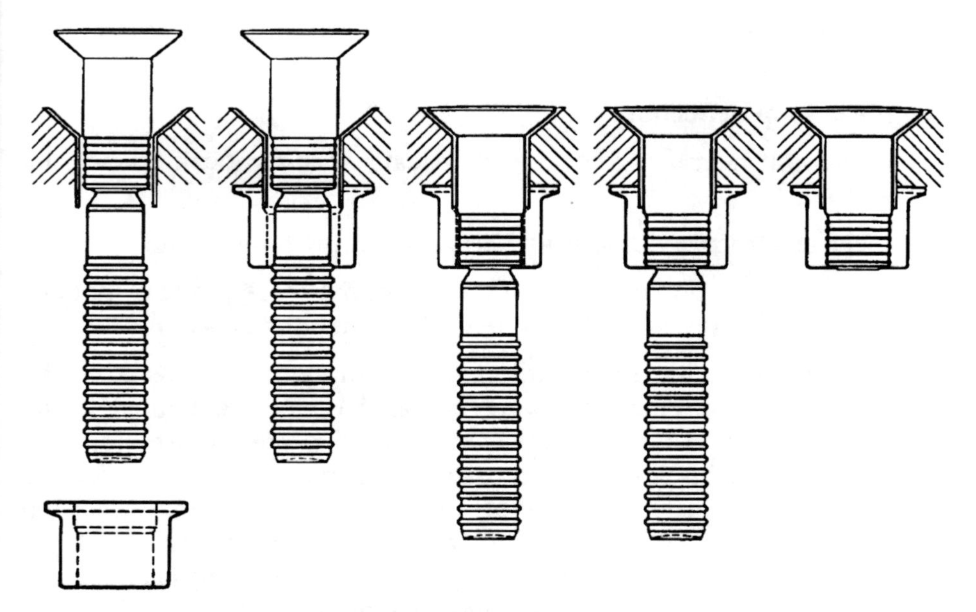

Interference Fit Sleeve Type Lockbolt

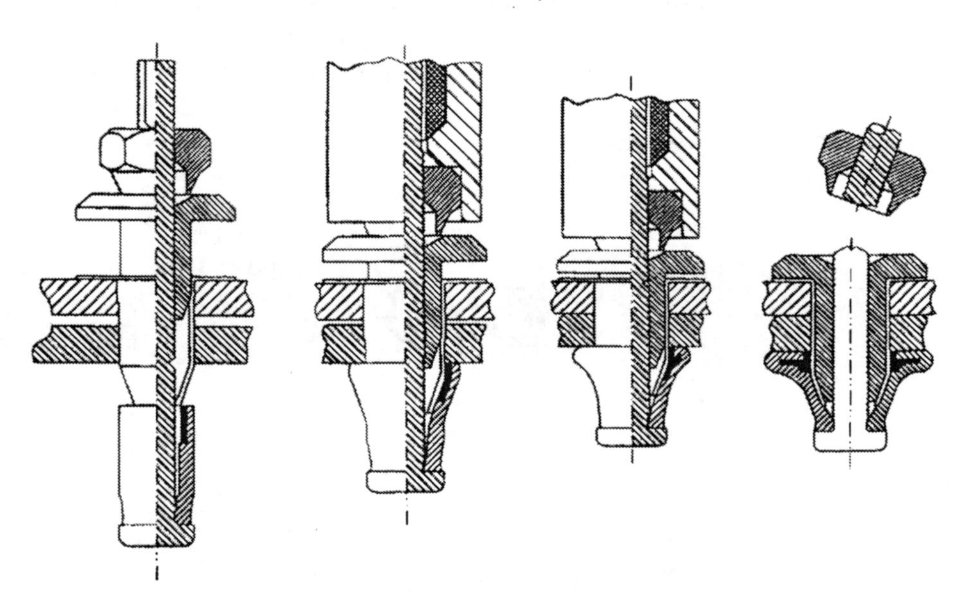

Interference Fit Threaded Corebolt Blind Fastener

Reference 15: with permission

Fig. 26. Installation of Sleeve Type Interference Fit Fasteners

There are several potential advantages to using interference fit fasteners in composite structure:[15]

- lower joint deflection;
- reduction in fastener cocking that leads to high localized bearing stresses;
- locks up structure to prevent ratcheting during fatigue; and
- reduced assembly costs when interference fits are required in metallic structure (no disassembly and ream operation for the composite).

High strength screws, along with plate nuts or gang channels, are used when there is a requirement for skin removal. A typical plate nut and gang channel are shown in Fig. 27. Note that three holes are required for each plate nut; two small holes for rivets to attach the plate nut to the structure and then the main fastener hole that is threaded to accept the threads in the screw. There are a wide variety of plate nut configurations available for different installations including self-aligning plate nuts. Gang channels are frequently used where there are long rows of fasteners to be installed because they do not require two rivet holes per fastener. They are attached at periodic points along the channel thus saving installation labor. Screws with plate nuts or gang channels do not perform as well as blind fasteners either in static or fatigue loading due to increased joint deflection and compliance of the fastening system.[15]

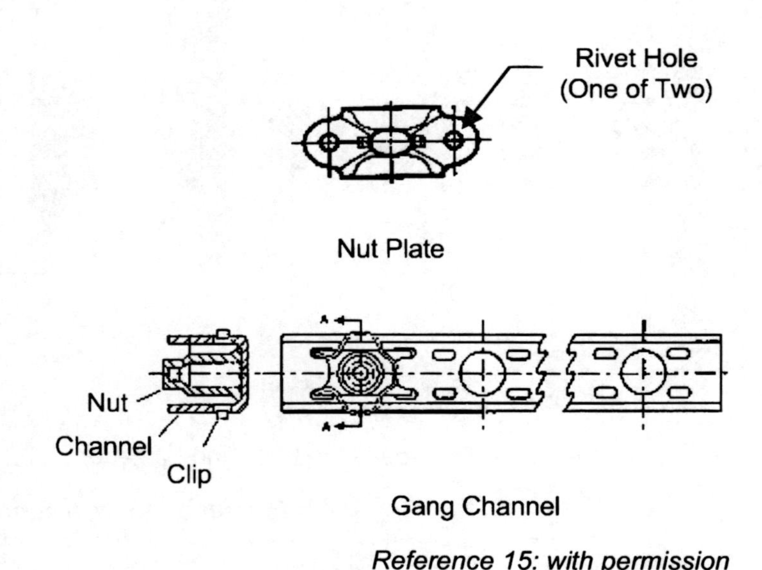

Fig. 27. Nut Plates and Gang Channels

12.5 Sealing

Many structures require sealing for (1) corrosion protection, (2) to keep water out of the structure or (3) to keep fuel in the structure. The typical wing fuel tank configuration shown in Fig. 28 will be used to explain the different sealing and corrosion protection methods. For joints between carbon/epoxy and aluminum parts, it is common practice to cocure or bond a thin layer of glass cloth to the surface of the carbon/epoxy part that acts as an electrical isolation barrier to prevent galvanic corrosion of the aluminum.

A good sealant must have good adhesion properties, high elongation, and be resistant to both temperature and chemicals. Sealing is usually accomplished using polysulfide sealants that are available in a variety of product forms with a range of viscosities and cure times. Polysulfide sealants are usable in the temperature range of –65 to 250 °F with short term capabilities to 350 °F. They contain leachable corrosion resistant compounds that aid in preventing the aluminum substructure from corroding. If higher temperatures are required, then silicone sealants can be used that have temperature capabilities as high as 500 °F.[17] Moldline fasteners are usually installed "wet" by applying sealant to the fastener before installation. The nuts are often over coated after installation.

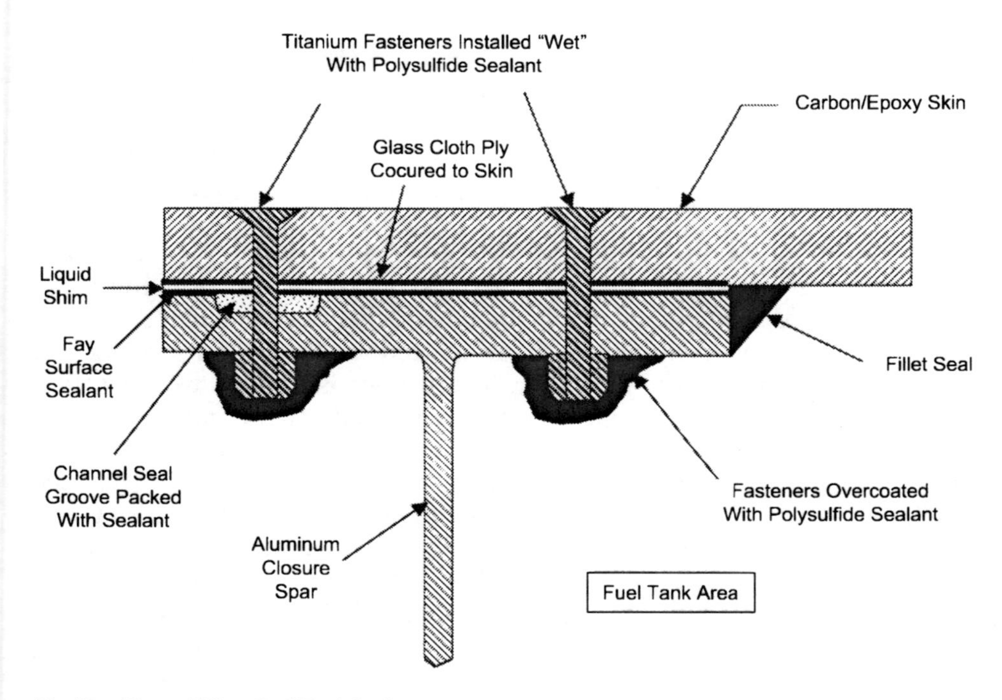

Fig. 28. Typical Wing Fuel Tank Sealing

During assembly, the faying surfaces are sealed and then fillet seals are placed around the periphery. The fay seal cannot be considered the primary seal because it is extremely thin and may be separated due to structural deflections. The fillet seal applied after assembly must be the primary seal. All potential leak paths must be fillet sealed. Fillet beads must be dressed with a filleting tool to work out air bubbles and voids and provide the finished fillet shape. Adequate fillet size is important in preventing leakage. Selecting the right sealant working life is important or the sealant may set up prior to finishing the job or it may take too long to cure affecting the production schedule.

Fuel tanks often have a channel groove that is packed with a fluorosilicone sealant that contains around 10% of small mirospheres that are graduated in diameter from 0.002 to 0.030 inches in diameter for more effective gap filling. They assist in keeping the compound in the groove and are effective in helping to seal gaps up to 0.010 inches wide. When fuel comes in contact with the sealant, it swells to help further seal the structure. Normally the channel is prepacked with sealant and then injected under pressure (up to 4,000 psig) after assembly. Injection points, usually spaced at 4–6 inches apart, can be at fastener holes or through specially designed fasteners that contain internal injection ports.

12.6 Painting

Paint adhesion to composite structure is not as difficult as it is with metallic structure. The surface should be clean of all dirt and grease. If the part contains a peel ply, it should be removed. Surface preparation can be accomplished by either scuff sanding with 150–180 grit sandpaper or by lightly grit blasting. For aerospace applications, the standard finishing system is an epoxy primer followed by a polyurethane topcoat. Epoxy primers are addition curing polyamides that contain: (1) strontium chromate that is an exceptional corrosion inhibitor for aluminum; (2) titanium dioxide for enhanced durability and chemical resistance; and (3) fillers such as silica to control viscosity and reduce cost. After sanding, the part should be primed within 36 h. The primer is applied to a dry film thickness of 0.0008–0.0014 inches and then cured for a minimum of 6 h. Polyurethane topcoats are aliphatic ester-based polyurethanes that exhibit good weathering, chemical resistance, durability and flexibility. They are applied to a dry film thickness of around 0.002 inches with an initial cure within 2–8 h and a full cure within 7–14 d.[18] Environmentally more friendly paint systems are being developed that are free of solvents or low in solvents, called low volatile organics compounds (VOCs) coatings. Also, toxic heavy metals (e.g., chromium) are being replaced with self-priming topcoats that are non-chromated high solids polyurethane coatings that replace both the epoxy primer and traditional polyurethane topcoat.[18]

12.7 Summary

Due to the labor intensity associated with assembly operations and the resultant high costs, composite structures should be designed to eliminate as much assembly as possible. However, the requirement to mechanically fasten composites to themselves and to metallic structure will not disappear in the foreseeable future.

Composite materials require more care when machining and drilling than comparable metals like aluminum. Their relative brittleness, low interlaminar shear and peel strengths, and low heat tolerance calls for special care in all machining and drilling operations.

Prior to hole drilling and fastener installation, it is critical to locate any gaps between structural members and shim them appropriately before starting the actual assembly. Failure to properly address gaps will lead to cracks and delaminations during fastener installation and clamp-up. Special drill geometries are available for composites to reduce the occurrence of defects such as hole splintering and fiber pull-out. Free hand drilling of composite structures should be avoided if possible. Power feed or automated drilling equipment gives much better and more consistent hole quality.

The first consideration in fastener selection is corrosion compatibility when joining carbon/epoxy. Fasteners with large footprints should be selected to spread the clamp-up loads across the composite surfaces. Fastener installation processes that induce vibration or sudden impact loads on the parts should never be used.

Sealing and painting of composites is very similar to the processes used for metallic structure. In fact, since composites themselves are not prone to corrosion, in some cases, the job is far less complex since special corrosion inhibiting compounds are not required. Likewise, paint adhesion to composites is as good or better than to metals provided that proper surface preparation techniques are followed.

References

[1] Taylor A., "RTM Material Developments for Improved Processability and Performance," *SAMPE Journal*, **36**(4), July/August 2000, pp. 1–24.

[2] Astrom B.T., *Manufacturing of Polymer Composites*, Chapman & Hall, 1997.

[3] Price T.L., Dalley G., McCullough P.C., Choquette L., "Handbook: Manufacturing Advanced Composite Components for Airframes," Report DOT/FAA/AR-96/75, Office of Aviation Research, April 1997.

[4] Kuberski L.F., "Machining, Trimming, and Routing of Polymer–Matrix Composites," in *ASM Handbook 21 Composites*, ASM International, 2001, pp. 616–619

[5] Strong A.B., *Fundamentals of Composites Manufacturing: Materials, Methods, and Applications*, SME, 1989.

[6] Ramulu M., Hashish M., Kunaporn S., Posinasetti P., "Abrasive Waterjet Machining of Aerospace Materials," 33rd International SAMPE Technical Conference, November 5–8, 2001.

[7] Niu M.C.Y., *Composite Airframe Structures: Practical Design Information and Data*, Conmilit Press, Hong Kong, 1992.

[8] Fraccihia C.A., Bohlmann R.E., "The Effects of Assembly Induced Delaminations at Fastener Holes on the Mechanical Behavior of Advanced Composite Materials," 39th International SAMPE Symposium, April 11–14, 1994, pp. 2665–2678.

[9] Paleen M.J., Kilwin J.J., "Hole Drilling in Polymer–Matrix Composites," in *ASM Handbook 21 Composites*, ASM International, 2001, pp. 646–650.

[10] Born G.C., "Single-pass Drilling of Composite/Metallic Stacks," 2001 Aerospace Congress, SAE Aerospace Manufacturing Technology Conference, September 10–14, 2001.

[11] Bolt J.A., Chanani J.P., "Solid Tool Machining and Drilling," in *Engineered Materials Handbook, Volume I Composites*, ASM International, 1987, pp. 667–672.

[12] Bohanan E.L., "F/A-18 Composite Wing Automated Drilling System," 30th National SAMPE Symposium, March 19–21, 1985, pp. 579–585.

[13] Jones J., Buhr M., "F/A-18 E/F Outer Wing Lean Production System," 2001 Aerospace Congress, SAE Aerospace Manufacturing Technology Conference, September 10–14, 2001.

[14] McGahey J.D., Schaut A.J., Chalupa E., Thompson P., Williams G., "An Investigation into the Use of Small, Flexible, Machine Tools to Support the Lean Manufacturing Environment," 2001 Aerospace Congress, SAE Aerospace Manufacturing Technology Conference, September 10–14, 2001.

[15] Parker R.T., "Mechanical Fastener Selection," in *ASM Handbook 21 Composites*, ASM International, 2001, pp. 651–658.

[16] Armstrong K.B., Barrett R.T., *Care and Repair of Advanced Composites*, SAE International, 1998.

[17] Hoeckelman L.A., "Environmental Protection and Sealing," in *ASM Handbook 21 Composites*, ASM International, 2001, pp. 659–665.

[18] Spadafora S.J., Eng A.T., Kovalseki K.J., Rice C.E., Pulley D.F., Dumsha D.A., "Aerospace Finishing Systems for Naval Aviation," 42nd International SAMPE Symposium, May 4–8, 1997, pp. 662–676.

Chapter 13

Nondestructive Inspection and Repair: Because Things Do Not Always Go As Planned

Unfortunately, flaws can occur at almost any stage of the composite manufacturing process. A summary of some potential flaws and damage are shown in Fig. 1. During ply collation, foreign objects are the most serious. These include the backing paper or plastic liner on the prepreg,

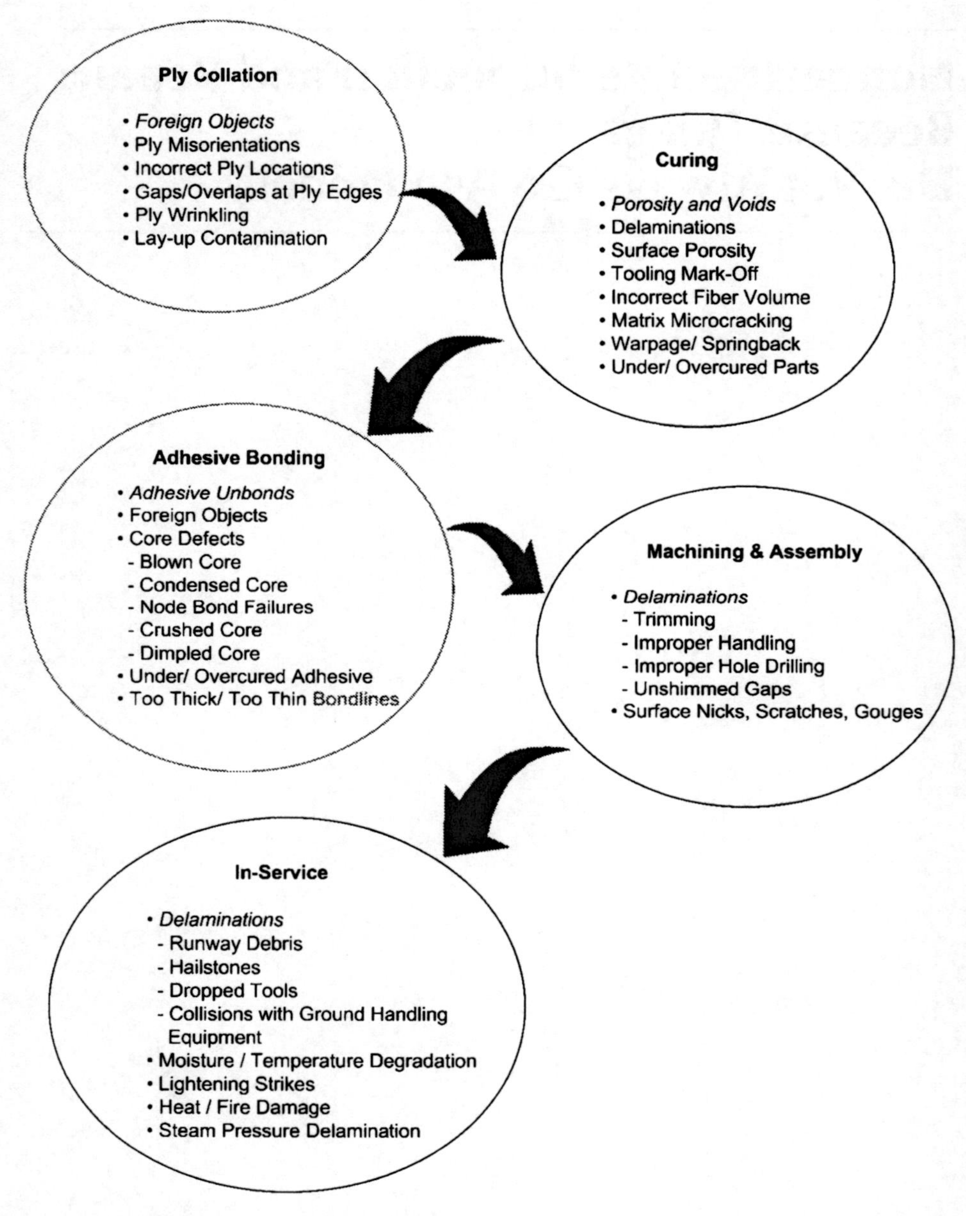

Fig. 1. Potential for Damage in Composite Parts

lay-up materials such as release films and tapes and tools such as knife blades. Since these can cause serious delaminations during curing, they normally result in either requiring a repair or scrapping the part if the foreign object is large enough or located in a highly stressed area. The cure process can also result in defects, the most serious being porosity and voids, which were discussed in Chapter 6 on Curing. If an adhesive bonding operation is involved, adhesive unbonds are some of the most serious flaws closely followed by a large number of potential honeycomb core defects. Delaminations are the most serious types of defects encountered during machining and assembly, resulting from improper trimming operations, part handling, hole drilling and installing fasteners with unshimmed gaps. During service, delaminations are again the most prevalent type of damage. Runway debris hitting control surfaces, hailstones from storms, tools dropped by maintenance mechanics and collisions with ground handling equipment, such as forklifts, can all cause varying degrees of damage. It is important to be able to reliably find these types of flaws and damage and be able to repair them if warranted.

Nondestructive inspection (NDI) methods are normally used to inspect completed parts and bonded assemblies to make sure that any flaws are not large enough or located in critical areas that they could cause the part to fail in service. Once a part is placed in service, NDI methods are also used to locate and evaluate the extent of damage. Based on the severity of the damage, a repair may or may not be required. In this chapter, the basics of NDI of composites are covered along with some of the common repair methods.

13.1 Nondestructive Inspection

Nondestructive inspection is an engineering discipline in its own right. All large aerospace companies and many part manufacturers have NDI groups that work full time at developing new and improved methods of inspecting parts. The literature on NDI is extensive;[1-5] therefore, in this chapter only the basics will be covered and how NDI can be used to detect flaws in composite parts. NDI methods range from simple visual inspections to very sophisticated automated systems with extensive data handling capabilities. It should be noted that composites are generally more difficult to inspect than metals due to their non-homogenous nature; i.e., laminated structures containing multiple ply orientations with numerous ply drop-offs. In addition, it is important that technicians conducting NDI be trained and certified in the method they are using.

13.2 Visual Inspection

Visual inspection is both a very valuable yet limited method for inspecting composite parts. Since only surface or edge defects are visible, it reveals nothing about the internal integrity of the part. Nevertheless, all parts should be periodically inspected for surface cracks, blisters, porosity, depressions or waviness, edge delaminations or paint discoloration. Proper lighting and low power magnifiers (5–10×) can help with visual inspection. Borescopes and mirrors are often used if the area to be inspected is hidden from direct view. Edge delaminations and tight surface cracks can be enhanced by taking a clean cotton cloth, dampening it with a solvent such as acetone, wiping the suspected area and then watching the solvent evaporate. If a crack or delamination is present, some residual solvent will continue to bleed out revealing its location and size. Normal dye penetrants should never be used, because they can contaminate the surfaces of the delamination making subsequent repair more difficult or impossible.

Although tap testing is normally classified as a sonic test method, actually a low-frequency vibration method, it will be discussed here since it is frequently used during visual inspection. In tap testing, a heavy coin, washer, or small hammer is used to *lightly* tap the surface. If a good region is tapped, the reflected sound will contain more high-frequency vibrations and will produce a ringing sound, while if a bad region (e.g., an unbond) is tapped, the reflected sound will contain less high-frequency vibrations and the sound will be much duller. Tap testing is capable of doing a credible job of finding unbonds or delaminations in honeycomb assemblies containing relatively thin skins (e.g., <0.040 inches or thinner).[6] For composite laminates, the tap test may be capable of detecting delaminations in the first few surface plies but cannot detect deeper defects. About the smallest defect that can be detected is 0.50 inches diameter at a depth up to about 0.25 inches.[1] Electronic tap testing equipment is available that will provide a more consistent test. Although tap testing can be used as a preliminary screening method, it is strongly recommended that instrumented ultrasonic equipment be used to make the final determination of defect size. For example, an impact delamination may result in only a small indication on the surface at the point of impact; however, the delamination often radiates from the impact point resulting in an extensive network of internal matrix cracks and delaminations.

13.3 Ultrasonic Inspection

Ultrasonic inspection is the most valuable technique for inspection of composite parts. The two most prevalent fabrication defects in solid laminates are porosity and foreign objects. Porosity is detectable because it contains solid-air interfaces that transmit very little and reflect large amounts of sound. Inclusions, or foreign objects, are detectable if the acoustic impedance of the foreign object is sufficiently different than that of the composite material.

Ultrasonics operates on the principle of transmitted and reflected sound waves. An ultrasonic wave traveling through a composite laminate that encounters a defect (e.g., porosity) will reflect some of the energy at the interface while the remainder of the energy passes through the porosity. The more severe the porosity, the greater the amount of reflected energy and the less that is transmitted through the defect. Ultrasonic waves are produced when an electrical signal generator sends a burst of electrical energy to a piezoelectric crystal in the transducer causing the crystal to vibrate and convert the electrical pulses into mechanical vibrations (sound waves). The piezoelectric crystal will also convert the returning sound waves back into electrical energy when the sound is received back from the part. A single crystal can be pulsed to send and receive sound waves, or two crystals can be used with one sending and the other receiving the pulse. Flaws are detectable since they alter the amount of sound returned to the receiver.

Ultrasonic inspection is defined as inspection conducted in the frequency range of 1–30 MHz, although most composite structure is usually tested at 1–5 MHz. High frequencies (short wavelengths) are more sensitive to small defects, while low frequencies (longer wavelengths) can penetrate to greater depths. As the ultrasonic beam passes through the composite, it is attenuated (i.e., lost) due to scattering, absorption and beam spreading. This loss or attenuation is usually expressed in decibels (dB). Thicker laminates will attenuate more sound than thinner laminates.

Through transmission (TT) ultrasonics is one of the two most common methods used to inspect fabricated composite laminates and assemblies. In the through transmission method, shown in Fig. 2, a transmitting

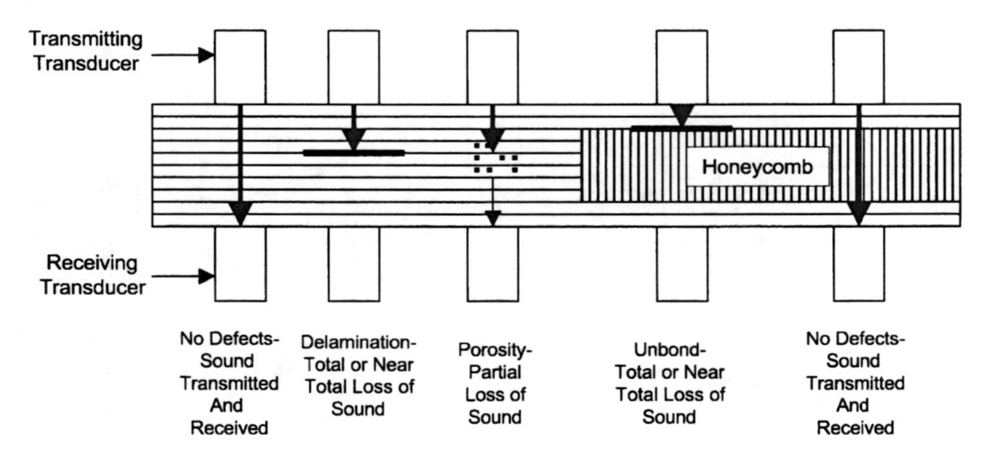

Fig. 2. Through Transmission Ultrasonics

transducer generates a longitudinal ultrasonic wave that travels through the laminate and is received by a receiving transducer placed on the opposite side of the part. If the part contains a defect, such as porosity or a delamination, some or (all) of the sound will be either absorbed or scattered so that some (or all) of the sound is not received by the receiving transducer. Through transmission is excellent at detecting porosity, unbonds, delaminations and some types of inclusions. However, this method cannot detect all types of foreign objects and it cannot detect the depth of the defects. Mylar film and nylon tapes are particularly difficult to detect with through transmission. Through transmission is usually conducted in a water tank or by using a water squirter method.

Since through transmission is not capable of detecting all types of foreign objects and the depth of defects, pulse echo (PE) ultrasonic inspection is frequently used in conjunction with through transmission ultrasonics to inspect parts. In the pulse echo method (Fig. 3), the sound is transmitted and received by the same transducer. Thus, it is an excellent method when there is access to only one side of the part. The amplitude of the echo received from the back surface is reduced by the presence of defects in the structure. Attenuation of the ultrasound is affected by internal defects and the time delay of the pulse is related to the depth of the defect. Pulse echo ultrasonics will detect just about all types of foreign objects but is not as capable of determining porosity levels as thoroughly as through transmission. If appropriate reference standards are available, pulse echo can be used to measure laminate thickness and the depth of defects. Pulse echo is more sensitive to transducer alignment than through transmission. For pulse echo inspection, the transducer needs to be within about 2° of

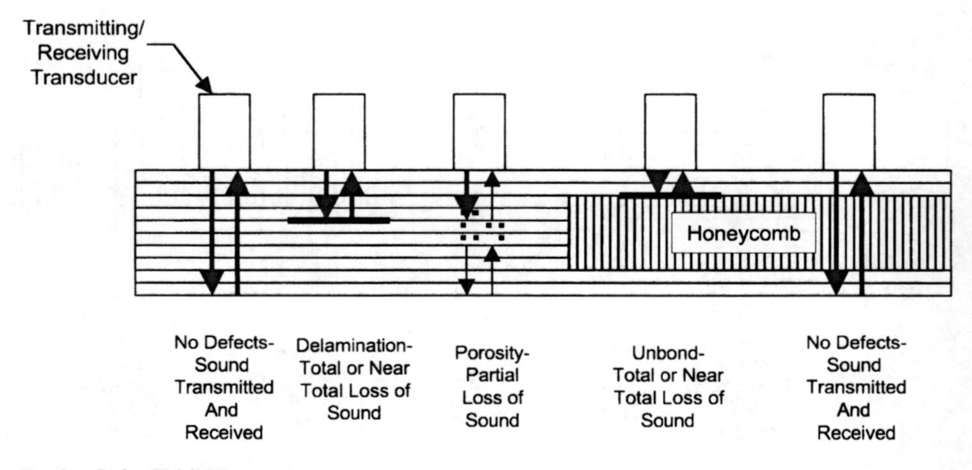

Fig. 3. Pulse Echo Ultrasonics

normal to the surface, while through transmission can tolerate misalignments up to about 10°.[1]

Since there is a large impedance mismatch between air and a solid interface, ultrasound does not propagate well through air; therefore, a couplant is used to more effectively transmit the sound from the transducer to the part. For hand inspection, glycerin compounds are frequently used while all automated systems use water. As shown in Fig. 4, automated systems can either be squirter systems or submerged reflector plate systems. Squirter systems, the most frequently used in production, are usually large gantry systems (Fig. 5) that are computer controlled to track the contour of the part and keep the transducers normal to the surface. They also index at the end of each scan pass. The ultrasonic energy is converted to digital data and stored in a file. Imaging software allows C-scan displays in either shades of gray or color. Modern units are capable of scan speeds of up to 40 inch s^{-1} and some units can record through transmission and pulse echo data simultaneously, eliminating the need to scan the part twice. There are also special units for cylindrical parts that contain turntables that rotate during the scanning operation.

The output from these automated units is displayed as a C-scan, which is a planar map of the part, where light (white) areas indicate less sound attenuation and are of higher quality than darker areas (gray to black) that indicate more sound attenuation and are of lower quality. Through transmission C-scans of both a good part and one with rejectable porosity are shown in Fig. 6. Lead reference standards that are placed on the part to identify location. As depicted in Fig. 7, the darker the area, the more severe the sound attenuation and the poorer the quality of the part. It should be noted that while through transmission is good at detecting porosity, it cannot tell the difference between scattered porosity (Fig. 8) and planar voids if the defect densities are similar. In addition, other defects, such as ply wrinkling, can often appear to be porosity. C-scan units can be programmed to print out the changes in sound levels as varying shades of gray or can be set in a go-no go mode where only rejectable areas are printed. Part manufacturers usually establish a baseline attenuation in decibels (dB) for each part. When the attenuation level exceeds the baseline by a predetermined dB, that area of the part is rejected. For example, if the baseline for a good laminate is 25 dB and the rejection threshold is 18 dB, then any indication over 43 dB (25 dB + 18 dB) would be rejected. Baselines and thresholds are determined by conducting effects-of-defects test programs in which known good laminates are compared with laminates of varying porosity levels. Both photomicrographs and mechanical property testing is used to establish the threshold levels. Part zoning can also be used to reduce cost. Areas that are

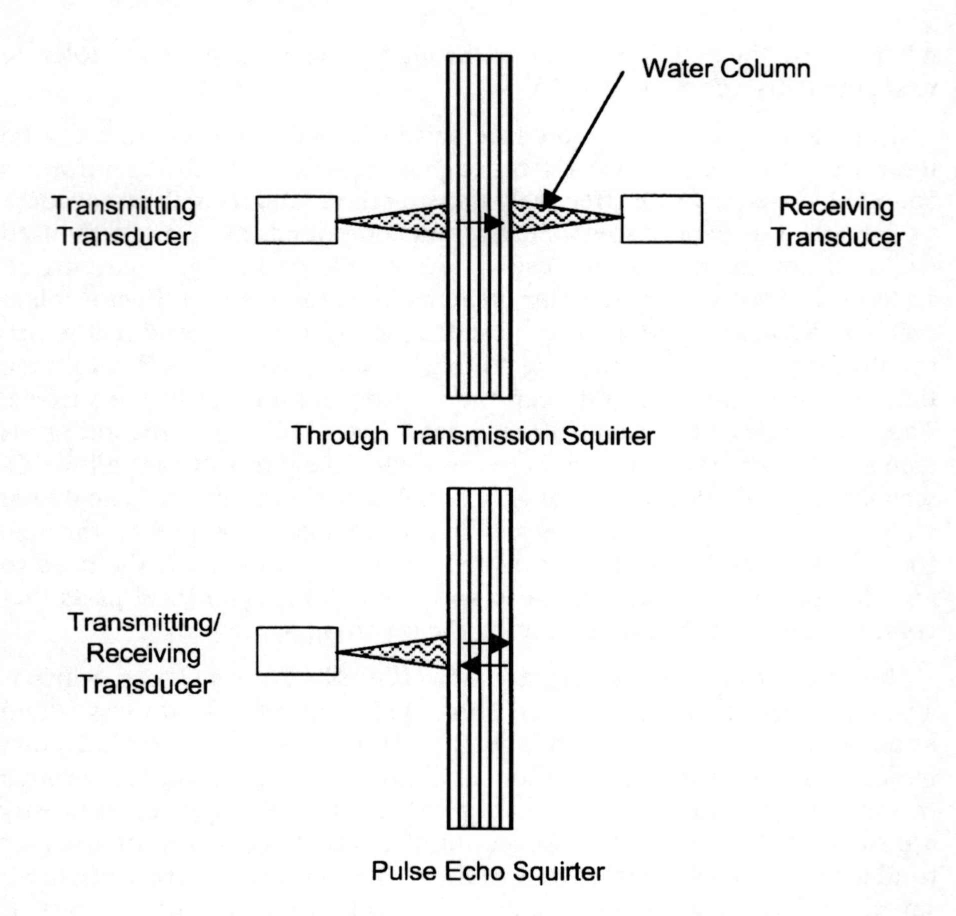

Fig. 4. Automated Ultrasonic Scanning Units

The Boeing Company

Fig. 5. Modern Ultrasonic Scanning Units

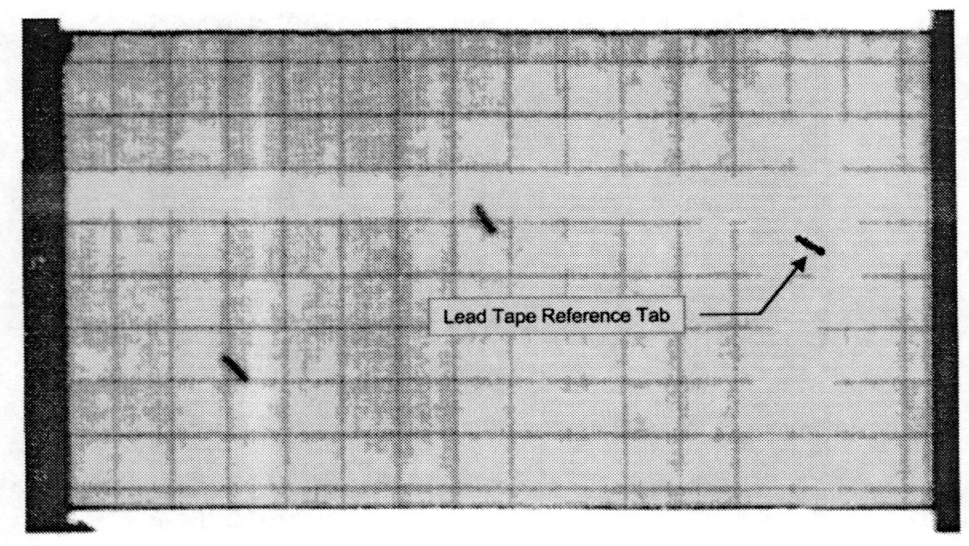

Acceptable Laminate

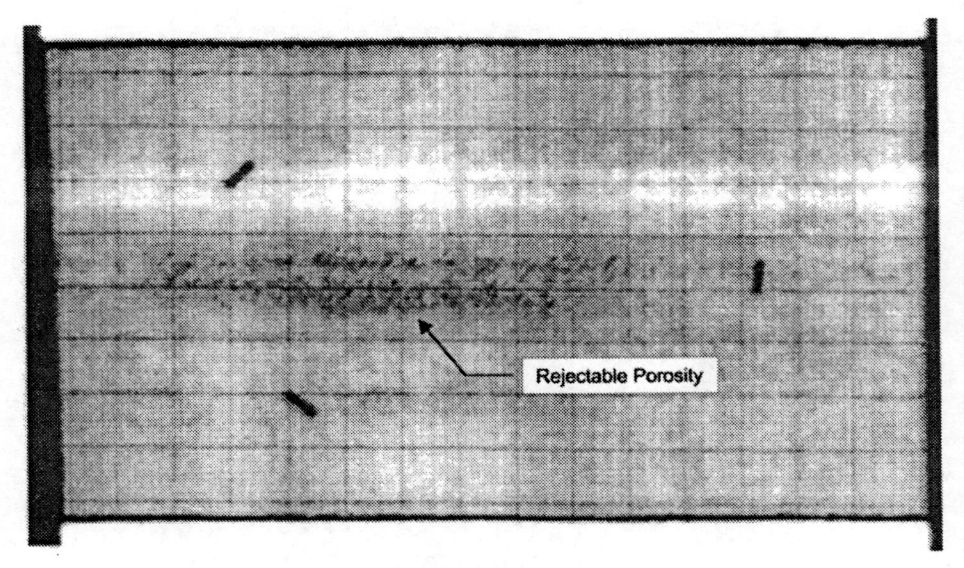

Rejectable Laminate

Fig. 6. Ultrasonic C-Scans of Composite Laminates

highly stressed would be zoned to lower threshold values than non-critical lower stressed areas.

Carbon/epoxy laminates are usually scanned at around 5 MHz while honeycomb assemblies require lower frequencies (1 or 2.25 MHz) to

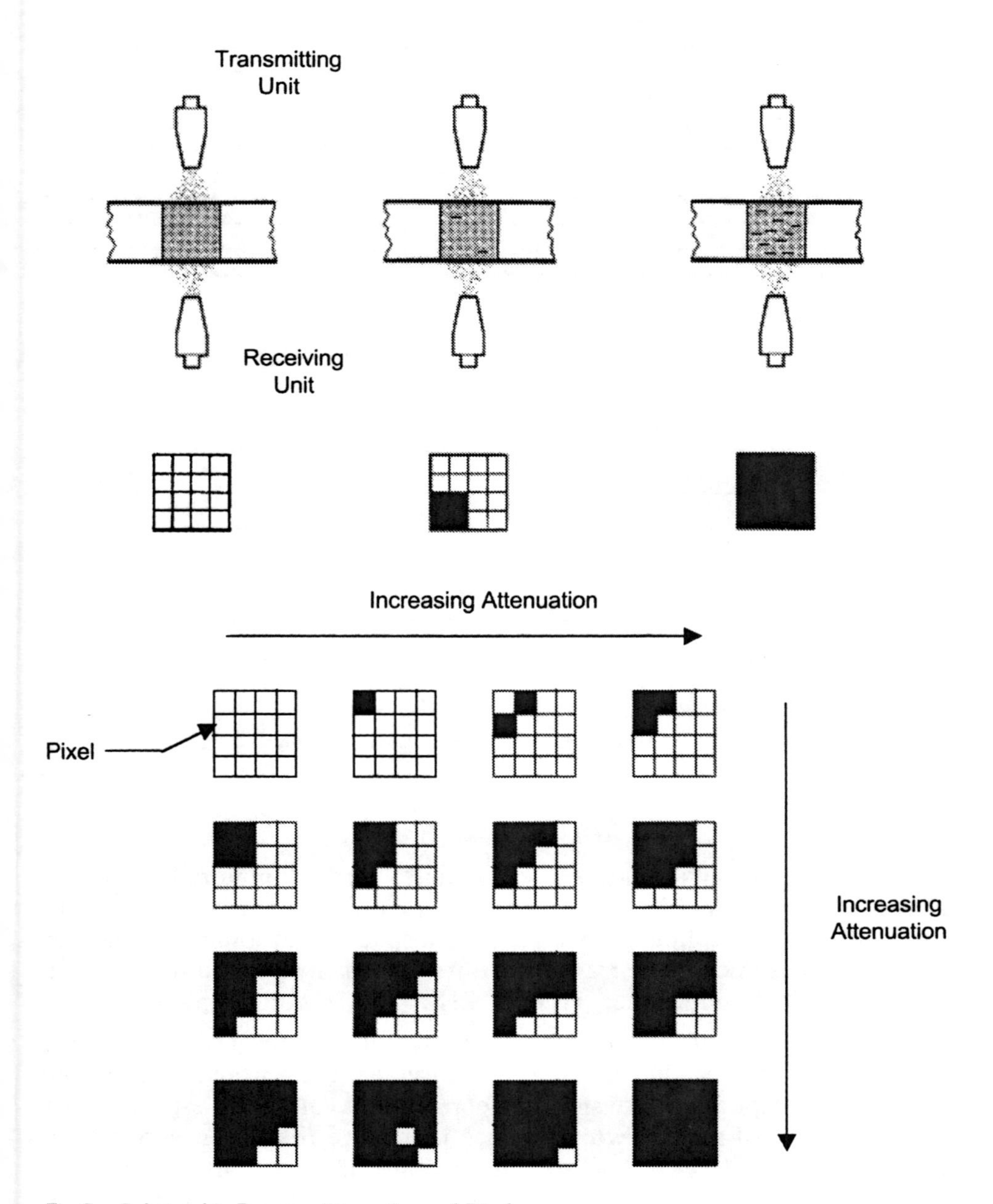

Fig. 7. Relationship Between Attenuation and Display

penetrate the thicker structure. Foam filled structures require even lower frequencies with 250 kHz, 500 kHz or 1 MHz being typical.[1] Since the ability to detect defects suffers at lower frequencies, parts are generally scanned with the highest frequency that can penetrate the part. This being

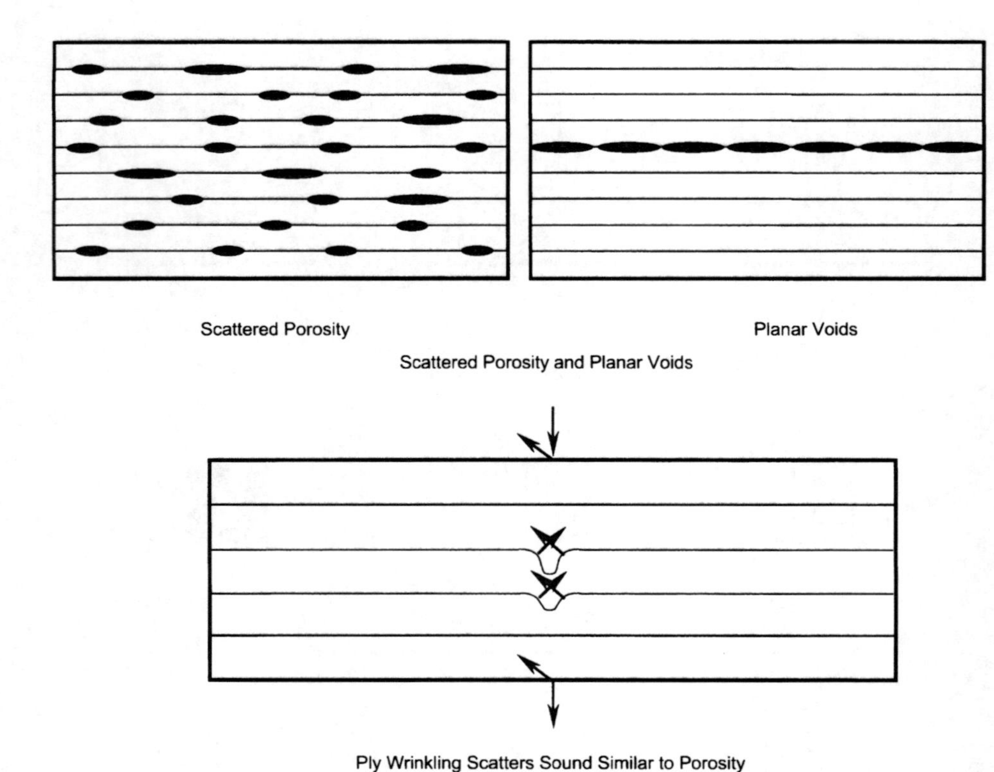

Fig. 8. Ultrasonic Indications

said, air coupled ultrasonics are occasionally used for materials with low acoustic impedance (i.e., lower density materials) such as honeycomb assemblies. Air coupling has been used to inspect honeycomb materials up to eight inches thick.[1] The transducers are placed close to the part surface (within an inch) and frequencies of 50 kHz to 5 MHz are employed.

A relatively new ultrasonic inspection technology is laser ultrasonics.[1] It provides essentially the same information as conventional ultrasonic inspection except that it is faster than conventional methods, especially for highly contoured parts. Two lasers are used. The first laser, generally a carbon dioxide laser, generates ultrasound in the part by causing thermoelastic expansion, while the second laser, normally a neodymium:yttrium-aluminum garnet laser, detects the sound signal as it returns to the top surface. Laser heating at the surface causes a temperature increase and a resultant local expansion of the material. If the laser pulses are short (10–100 ns), the expansion will create an ultrasonic wave in the 1–10 MHz range. The receiving laser detects light scattered off the surface that is analyzed by a Fabry-Perot interferometer to extract the ultrasonic signal. In this process, it is important to generate as much ultrasound as possible without causing heat damage to the composite surface. Surface temperatures are normally restricted to 150 °F or less. An additional benefit of laser ultrasonics is that the ultrasound propagates perpendicular to the surface somewhat independent of the laser angle of incidence. The transmitters and receivers can be off axis to normal as much as ±45° without loss of performance. However, since the part must have a thin layer of resin on the surface for effective sound generation, resin starved or machined surfaces may limit the success of this technique.

13.4 Portable Equipment

Both pulse echo and through-transmission equipment is available for field inspections. A typical setup, shown in Fig. 9, consists of a transducer connected to a console. Again, 1–5 MHz frequencies are normally used. Glycerin pastes are frequently used as couplants rather than water. Pulse echo is the most frequently used method because it requires only one transducer and can conduct inspections where access to the backside is not available. Note that the display is normally in the A-scan format in which the height of the amplitude signal is an indication of the severity of the defect and its location between the front and back surface gives an indication of its depth.

There are also a fairly large number of "bond testers" that generally operate below 1 MHz, some as low as the audio or near audio range (15–20 kHz). At the low end of the frequency range, a couplant is generally not required. A summary of some of these units is given in Table 1. Advanced portable units, such as the Mobile Automated Ultrasonic Scanning (MAUS) system shown in Fig. 10 can conduct multiple types of inspection, including ultrasonics (through transmission, pulse echo, shear wave), bond test methods (resonance, pitch/catch, mechanical impedance) and eddy current (single and dual frequency).

13.5 Radiographic Inspection

Radiographic inspection is normally used to look for microcracks in solid composite laminates and extensively to detect defects in honeycomb assemblies. Typical honeycomb defects detected include crushed core, core migration, blown core, dimpled core, node bond failures and the presence of water in the cells.

As shown in Fig. 11, the part is exposed to X-ray radiation that penetrates the part and produces an image on the film located under the part. Images produced on the film are a result of differential absorption of the X-rays due to changes in material makeup or construction. Since composites are nearly transparent to X-rays, low-energy X-rays are used with lower frequencies and longer wavelengths than high-energy X-rays. The

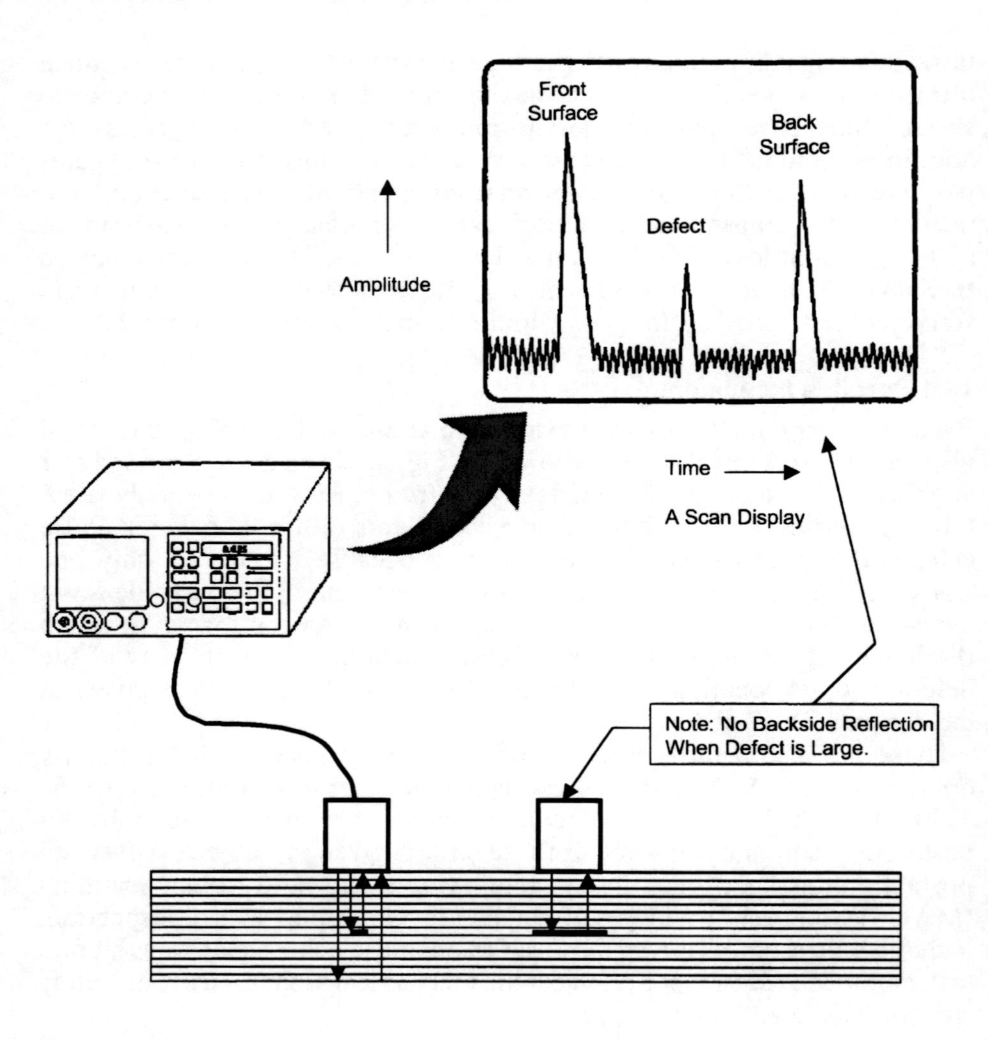

Fig. 9. Portable Pulse Echo Unit

sensitivity (radiographic contrast) to feature changes is improved by using X-rays with lower energy, usually less than 50 kV. Higher-energy X-rays are required to penetrate thicker or higher-density materials. An internal void or gap will decrease the amount of solid material through which the radiation passes increasing the intensity of radiation reaching the film, creating a darker area on the film. On the other hand, a metallic inclusion or metallic honeycomb core will increase the amount of solid material that the radiation passes through and decrease the intensity of radiation reaching the film, creating a lighter area on the film.

Table 13.1 Summary of Bond Testing Techniques

Resonance Method	Transducer is driven at resonance frequency (25-500 kHz) causing electrical impedance changes to detect unbonds or delaminations. Test requires a couplant and a variety of different transducer designs. Capable of detecting unbonds 0.50 in. or larger. Maximum thickness of 0.5 in. Delaminations in bottom one or two plies difficult to detect due to small change in material impedance detected by probe.
Pitch/Catch Sweep Method	Dual element transducer method in which one element transmits sound and the other element receives sound. Sound waves are transmitted across the part in a plate wave mode. Unbonds and deeper defects are detected as the sound loss. Transducer is swept in a circular manner to detect flaws. Sweep frequencies of 20-40 kHz or 30-50 kHz normally used. No couplant is required.
Pitch/Catch Impulse Method	Dual element transducer method in which one element transmits sound and the other element receives sound. Low frequency probes of 5-25 kHz used. Sound waves are transmitted in bursts into part. Unbonds are detected by differences in wave amplitude and/or phase changes. No couplant is required.
Eddy Current Sonic Tester	Transducer contains eddy current driver coil surrounding sonic receiver. Pulsed eddy currents causes unbonds to resonate at frequency detected by sonic receiver. Operates around 14 kHz and requires no couplant. Capable of detecting near and far side unbonds. Frequently used on aluminum honeycomb assemblies. Also known as a harmonic bond tester. Capable of detecting crushed and fractured core.
Mechanical Impedance Method	Dual element transducer in which transmitter element generates audible sound waves and receiver detects the effect of bond variations due to local stiffness changes in the part. The transmitter is swept through 2.5 kHz to 10 kHz during setup to establish test frequency. This method is capable of detecting unbonds, crushed core and defects in composite laminates. No couplant is required. Changes in surface contour will influence response.

Variations in density, thickness and part construction all cause variations in the radiographic image. Changes in density can be caused by changes in part thickness, cracks, porosity, crushed core or liquid water in honeycomb cells. The film used is a high contrast film with a small grain size. The source-to-film distance (SFD) should be as great as possible for the best resolution. As with ultrasonic inspection, reference standards are normally used. Reference standards can be fabricated, such as those shown in Fig. 12, but some of the best ones are pieces of scrapped assemblies that are cut up for dissection since they contain the same materials and thicknesses as the actual part being inspected. Flaw orientation is critical for detection reliability. The major dimension of the flaw should be parallel to the beam direction for maximum sensitivity. Some depth resolution can be obtained by titling either the X-ray source or the part.

There are several forms of radiography normally employed to inspect composite assemblies. Static X-ray units, like the one depicted in Fig. 11, do not move. The part is manually indexed under the X-ray source and

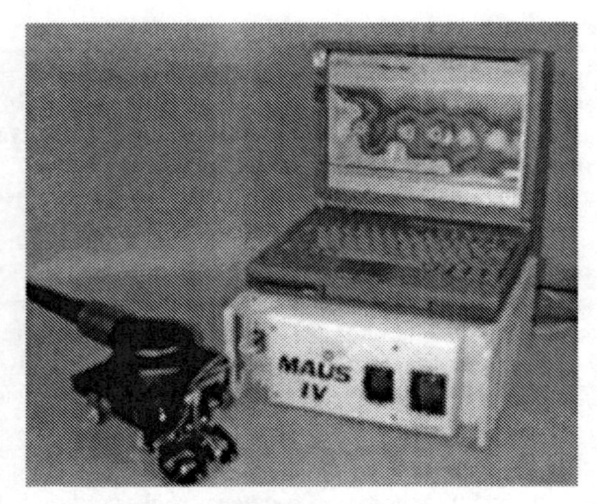

Mobile Automated Scanning System

Vacuum Attachment to Underside of Aircraft

The Boeing Company

Fig. 10. Mobile Automated Scanning System

multiple shots are taken to provide complete part coverage. Static X-ray units can be stationary or portable for on-aircraft inspection. Both methods require proper shielding of personnel. With in-motion X-ray, there are two robots (a source robot and a media robot) that are mounted on a gantry.

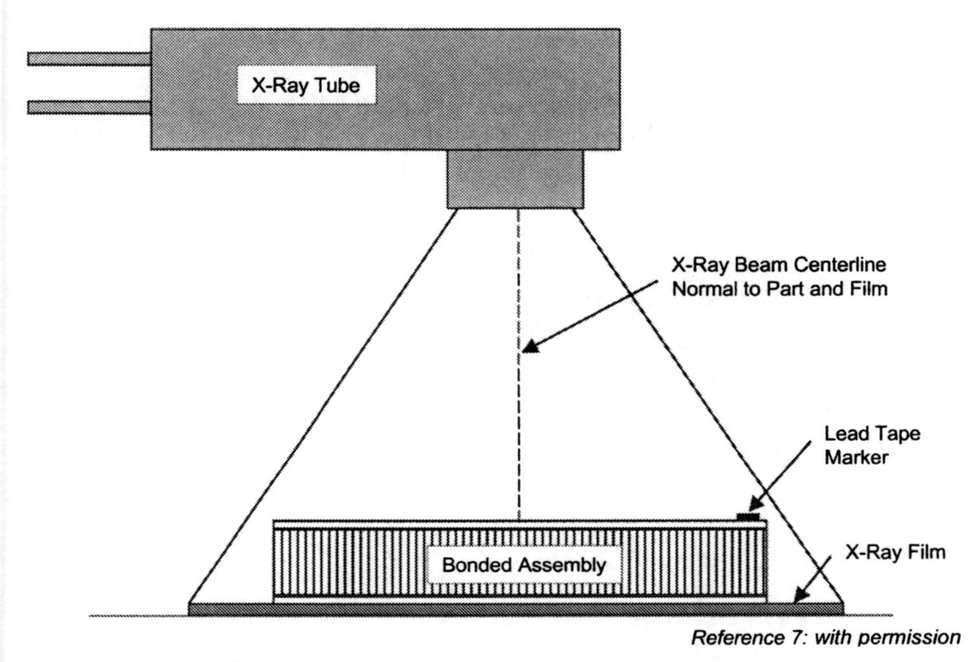

Reference 7: with permission

Fig. 11. Radiographic Inspection

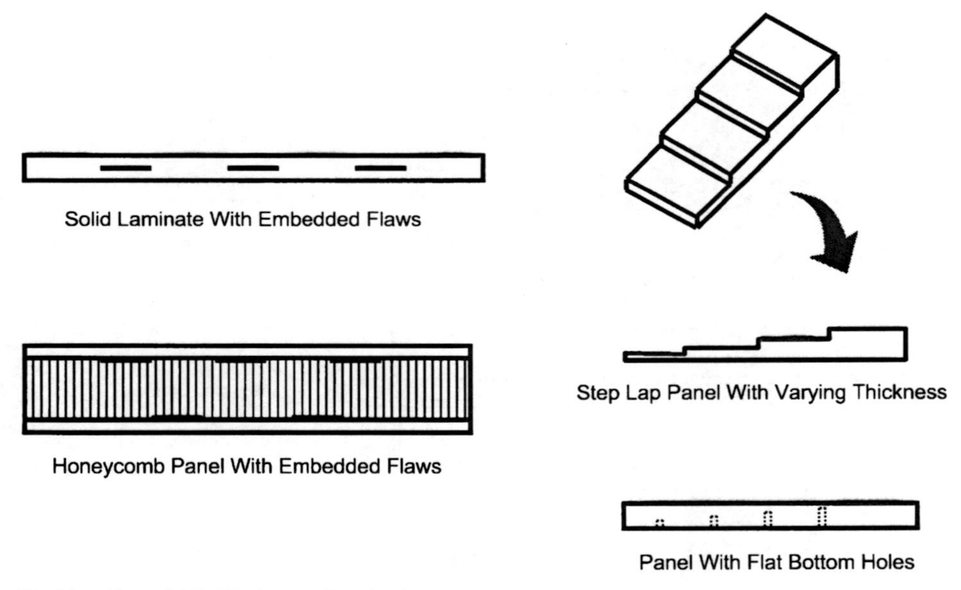

Fig. 12. Typical NDI Reference Standards

The system is computer-controlled to synchronize the movement of the two robots along the part as it is being X-rayed. Being computer controlled, the system can be programmed to vary the X-ray parameters along the length of the structure.

A summary of some of the defects detected by X-ray in honeycomb bonded assemblies is shown in Fig. 13.[8] Blown core is usually caused by a vacuum bag leak during bonding. Small leaks can result in small areas of

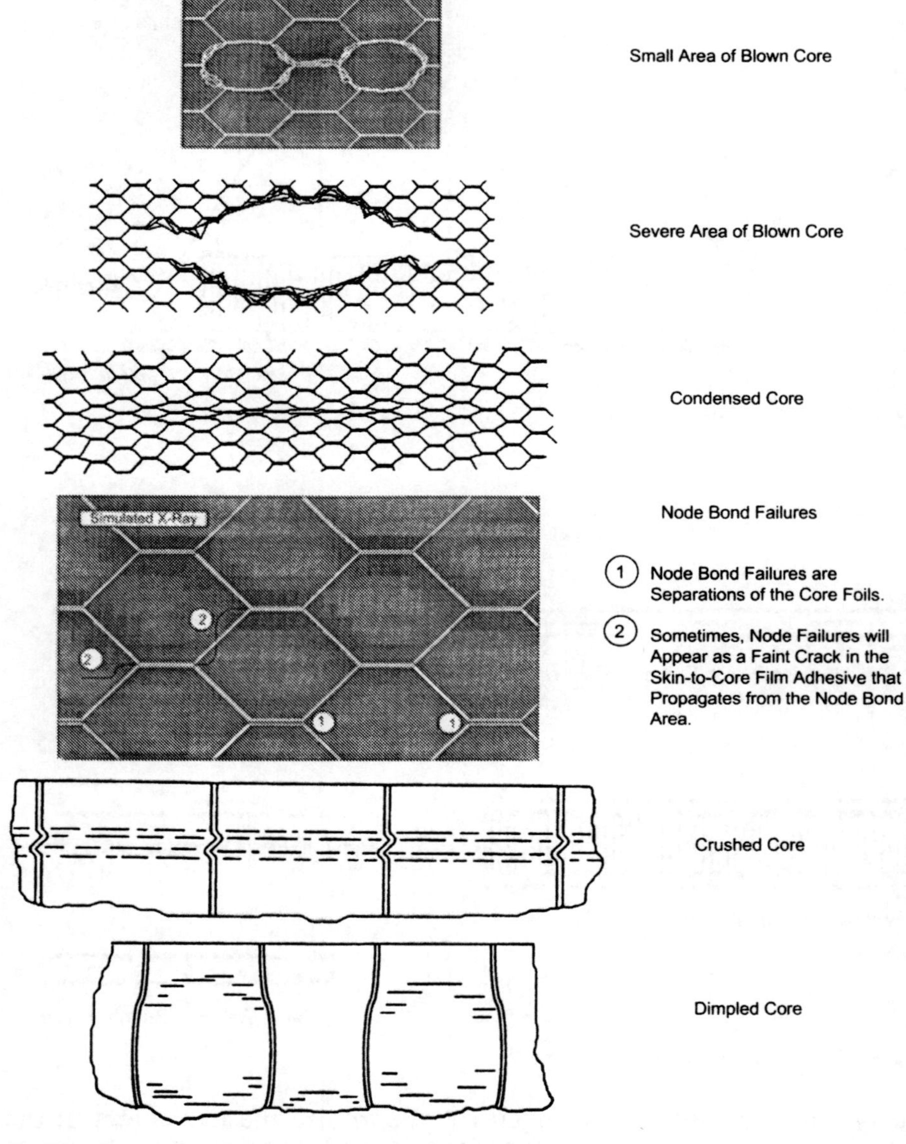

Fig. 13. Common Core Defects Detected By Radiography

blown core while large leaks (e.g., total loss of the vacuum bag) can result in massive areas of blown core, to the extent that the part will be unrepairable and have to be scrapped. Condensed core, or lateral compression of the cell walls, is usually a result of the core slipping or migrating before the adhesive gels during the bond cycle. It is more prevalent at edges of parts or at core ramp areas. Node bond failures, where the foil ribbons are separated at their connecting points or nodes, are usually a defect that occurs during the core manufacturing process but can also result from gas pressure differentials in the cells during bonding. Crushed core is usually associated with a dent in the skin or excessive pressure on thick sections of low-density core. Core dimpling, or waviness in the cell walls, is similar to crushed core but does not usually warrant repair. Both crushed core and dimpled core often require X-rays taken at a shallow angle for detection. Although not shown, water in the core is a serious problem with honeycomb assemblies. It can lead to corrosion of aluminum core, node bond failures and skin-to-core unbonds due to freeze-thaw cycles, and skin unbonds if the structure is heated above the boiling point of water. For water to be detectable in honeycomb core, the cell has to be 10% or more filled with liquid water. For thicker composite or aluminum skins, more water height is required for detection. Water will appear as dark gray areas in discrete cells or groups of cells. This can pose a problem if the part is in service because water in cells appears very similar to resin in cells, and it is difficult to distinguish between the two without the original radiograph that was taken when the part was manufactured.

Foaming adhesive bond lines can form voids (Fig. 14) during bonding that can be difficult to detect. If the foaming adhesive void is less than one half the height of the foam bond, detection will be questionable. If a void is suspected at a closure member, then pulse echo or a bond test method may be warranted.

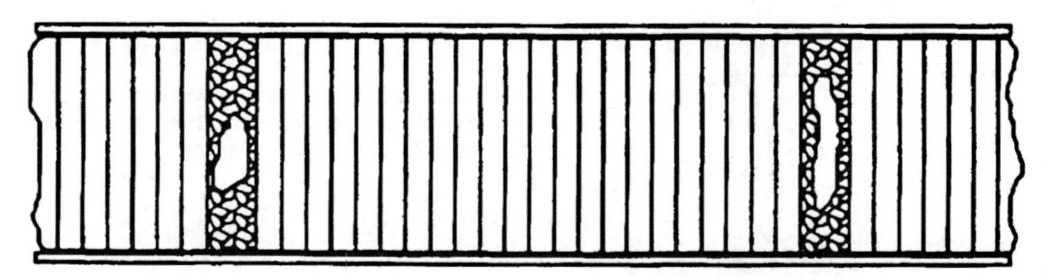

Fig. 14. Voids in Foaming Adhesive

Skin-to-core unbonds in sandwich structures (Fig. 15) are usually more detectable with ultrasonic methods than with radiography. These adhesive unbonds can be a result of mismatches between detail parts (skins, core, closure members), lack of locally applied pressure due to bridging, entrapped volatiles (air, water, residual solvents) or contaminated skins or core. Although adhesive unbonds are detectable, there are no NDI methods currently available that will determine the strength of an adhesive bond. For example, a high-strength bond (e.g., 5,000 psi shear strength) will appear the same as a low-strength bond (e.g., 500 psi shear strength). Therefore, mechanical property process control specimens are fabricated along with the production unit using the same surface preparation procedures and adhesives.

Fatigue damage in honeycomb core has been observed on in-service aircraft. Radiographically it appears to resemble crushed core. However, small pieces of the honeycomb cell wall break off and fall into the center of the cells. Static radiography, which projects the cell wall onto the film, is the most capable method of detection.

13.6 Thermographic Inspection

Although not as widely used as ultrasonics or x-ray, thermographic inspection is a relatively fast, non-contact, single sided process that has a wide coverage area. It can be used to detect delaminations, impact damage,

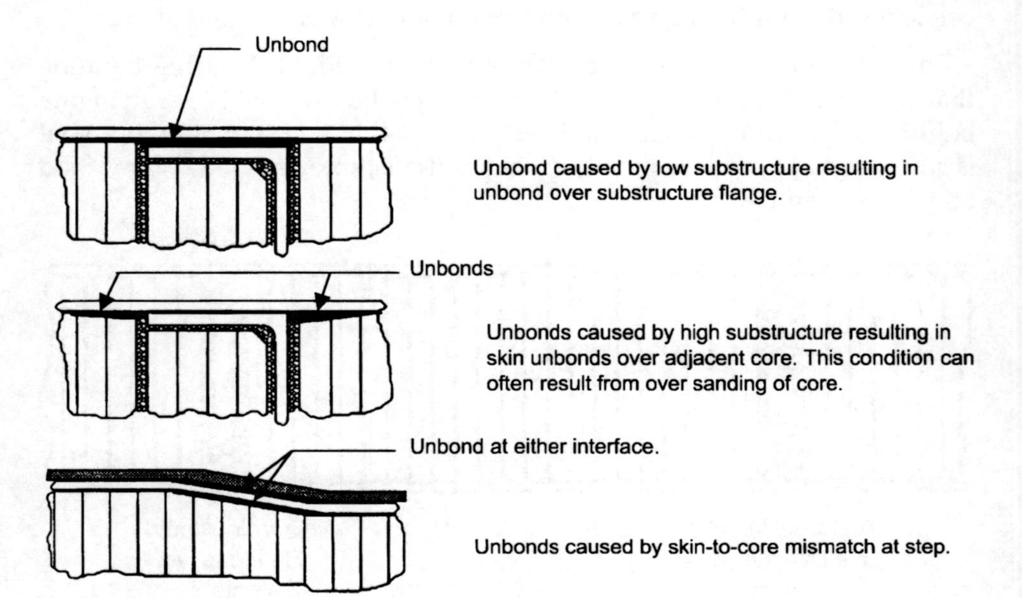

Fig. 15. Typical Skin-to-Core Unbonds

water ingression into honeycomb, inclusions and density variations. In thermography (Fig. 16), it is first necessary to uniformly heat the surface. This is normally accomplished by flash lamps that pulse for a few milliseconds to provide heat to the surface. High wattage tungsten halide lamps are frequently used that raise the surface temperature by 10–30 °F. As heat conducts into the part, the surface temperature falls. Since defects cause differences in heat conduction, the surface temperature cools at a different rate above a defect than in a defect free area. The surface of the part is monitored with an IR camera that collects the radiation from the surface. Imaging software is then used to examine the radiation received pixel by pixel and provide a map of the surface showing the defects. The output is often a colored map depicting temperature variations as contrasting colors. Aluminum transfers heat rapidly and aluminum core defect detection is difficult in aluminum bonded sandwich structures. When water is trapped in the core, heat is transferred to the to the water through the skin and core. Airlines often use thermographic inspection to check honeycomb bonded assemblies for the presence of liquid water. This

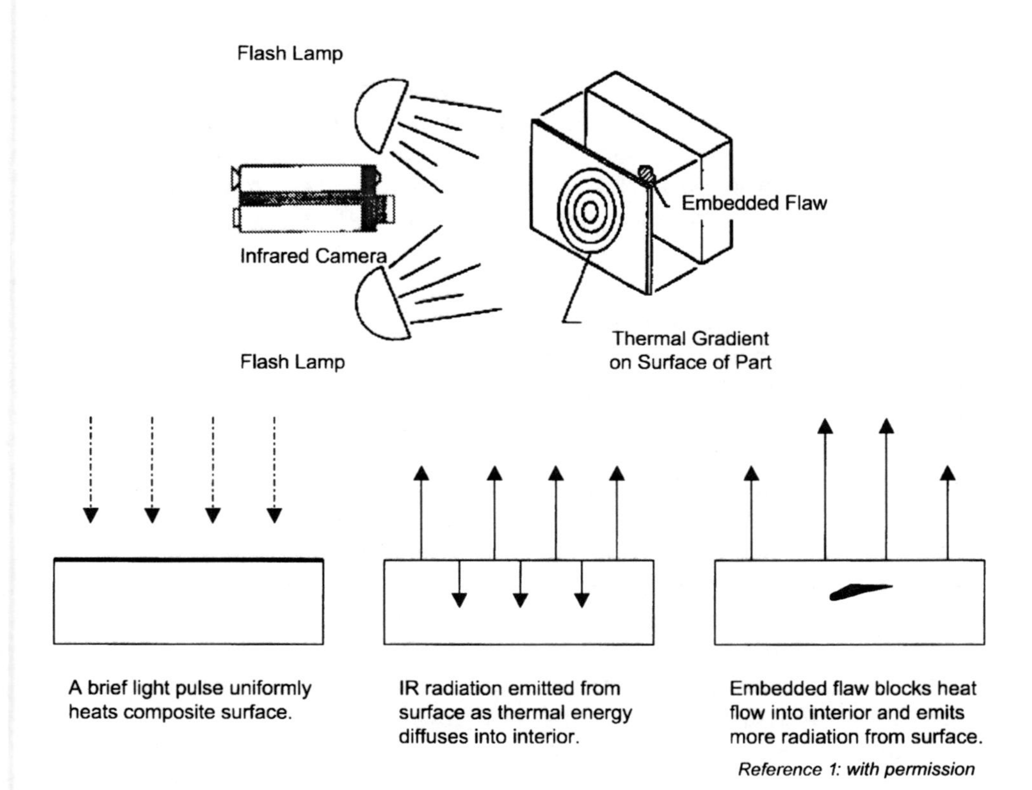

Fig. 16. Infrared Thermography

inspection is done immediately after the aircraft lands before the water (i.e., ice) melts and heats up to ambient temperature.[6]

13.7 Repair

All repairs of composite or bonded assemblies should be conducted per the specific instructions outlined in the Structural Repair Manual (SRM) or Technical Order (TO) for the aircraft. These manuals are prepared by the aircraft manufacturer and approved by the appropriate governing agency, such as the Federal Aviation Agency (FAA) for commercial aircraft or the Air Force/Navy/Army agency for military aircraft. If the damage exceeds the limits specified in the manual, it is imperative that a qualified stress engineer approves the repair procedure. All personnel conducting structural repairs should be trained and certified in the repair procedure. The instructions in the repair manual must be followed to the letter. A repair that is done incorrectly can often result in a second more extensive and complicated repair.

Repairs can be categorized as fill, injection, bolted or bonded repairs. Simple fill repairs (Fig. 17) are conducted with paste adhesives to repair non-structural damage such as minor scratches, gouges, nicks and dings. Injection repairs use low-viscosity adhesives that are injected into composite delaminations or adhesive unbonds. Bolted repairs are usually done on thick highly loaded composite laminates while bonded repairs are often required for thin skin honeycomb assemblies. Like NDI, the literature on composite repair is quite extensive. An excellent in-depth treatment of repair technology can be found in Ref. 6.

13.8 Fill Repairs

It should be emphasized that fill repairs are non-structural and therefore should be confined to only minor damage. Two part high-viscosity thixotropic epoxy adhesives are normally used for this type of repair. The surface to be repaired should be dry and free of any contamination that would inhibit the filler from adhering. Prior to filling, the surface should be lightly sanded with 180–240 grit silicon carbide paper. Once the adhesive is mixed and applied to the surface, most epoxy adhesives will cure sufficiently within 24 h at room temperature so they can be sanded flush with the surface. It normally takes 5–7 d at room temperature for them to develop their full strength. Heat lamps are often used to accelerate the cure by heating the adhesive to 180 °F for 1 h. There are also special fillers, called aerodynamic smoothing compounds, that are rubber toughened epoxy resins that will resist crazing and cracking in service. It is advisable not to heat these types of repairs to over 200 °F as moisture in the laminate could cause a steam pressure delamination problem resulting in a much more extensive repair.

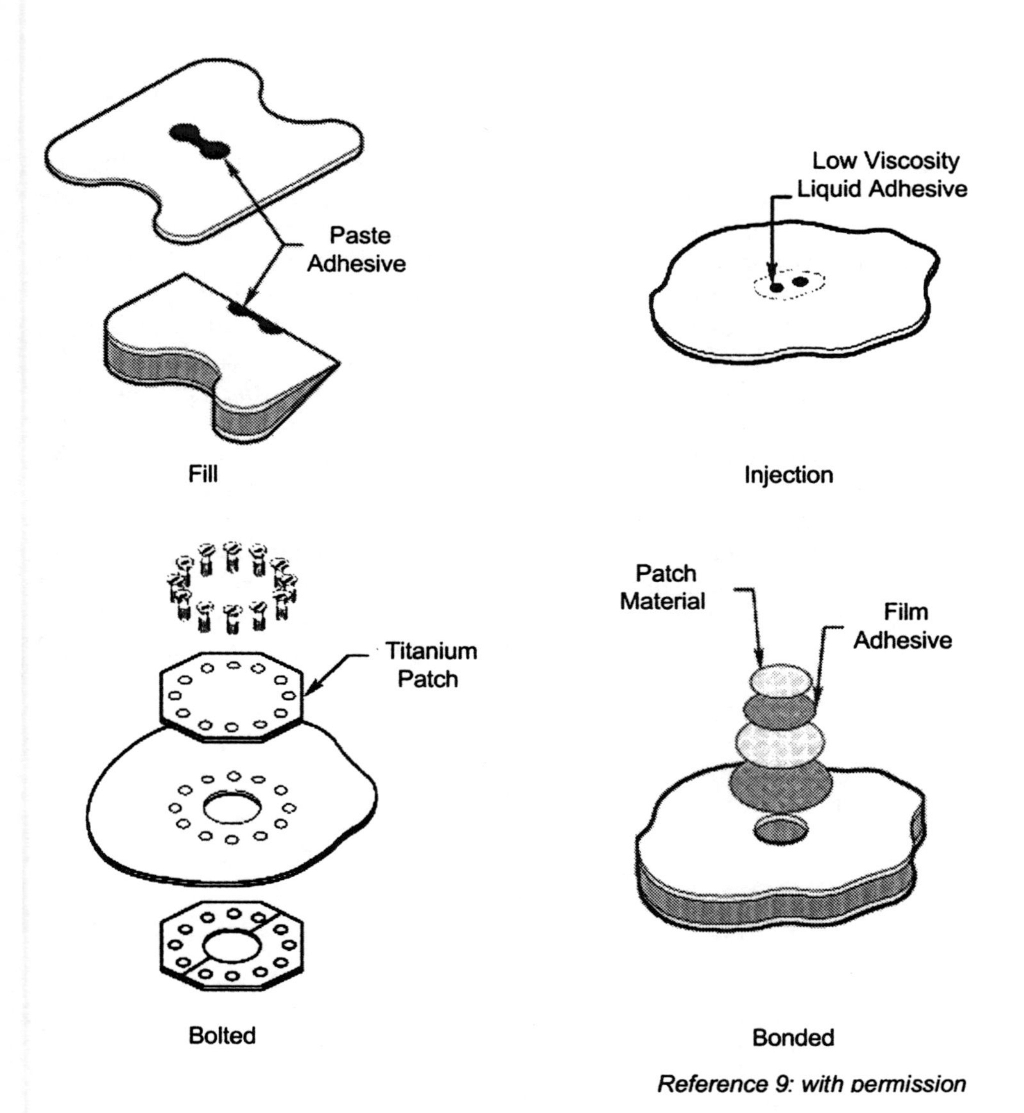

Fig. 17. Typical Composite Repairs

13.9 Injection Repairs

Injection repairs are somewhat controversial and can produce mixed results. If the repair is for an adhesive unbond or a delamination during cure, the internal surfaces of the unbond or delamination will normally contain a glossy oxidized surface that the injected adhesive may not adhere to structurally. On the other hand, if the delamination occurred due to an impact that broke the layers apart, the surfaces will be amendable to bonding with the injected adhesives. However, delaminations due to impacts are often in multiple layers and connected with tight microcracks that are difficult or impossible to fill.

Low-viscosity two part epoxy adhesives are injected under low or moderate pressures as shown in the two examples in Fig. 18. If the delamination does not extend to an edge, small diameter flat bottom holes (0.050 inches diameter) are drilled to the depth that is usually determined by pulse echo ultrasonics. Two or more holes are generally required, one for injection and one for venting. To help the resin flow into a tight delamination, it is helpful to preheat the delaminated area to 120–140 °F. This will reduce the resin viscosity helping it flow into the delamination. For tight delaminations, pressures as high as 40 psig can be used if lower pressures do not initially indicate filling by the adhesive flowing out of the vent hole. If honeycomb core will be exposed to the pressure, it should be limited to 20 psig to prevent blown core. Flow verification can be determined by injecting air pressure in one hole and monitoring air flow in the other hole by sealing a piece of rubber hose over the vent hole and emerging the other end in a cup of water. Air flow can then be monitored by looking for bubbles. Edge delaminations are somewhat easier to fill and

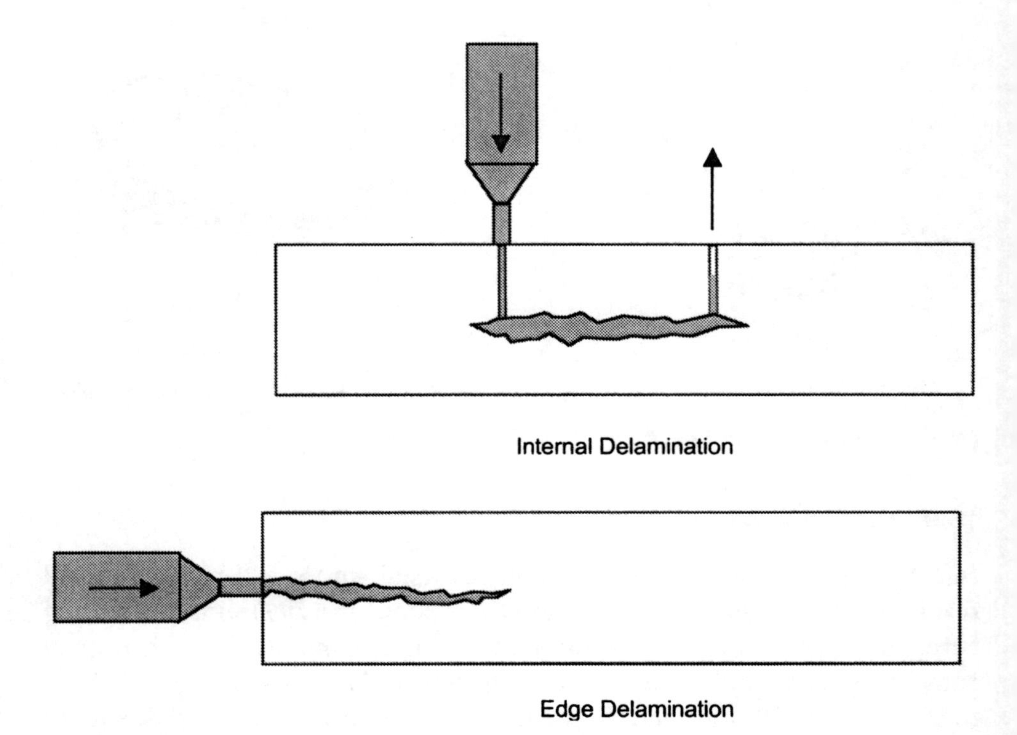

Fig. 18. Filling Delaminations With Low Viscosity Resins

do not normally require vent holes. In addition, after filling an edge delamination, C-clamps can be applied to provide pressure to spread the adhesive and push the plies back together.

The multiple layer nature of a delamination around a fastener hole that resulted from torquing a fastener down over an unshimmed gap (Refer back to Chapter 12 on Assembly) is shown in Fig. 19. A photomicrograph of a section through the hole (Fig. 20) reveals a complex network of delaminations and matrix cracks on the multiple layers through the laminate. Note that some of the delaminations do not even extend to the edge of the hole. An injection scheme for this type of delamination (Fig. 21) involves a series of injection and venting holes. To provide pressure during cure, a temporary mold released fastener with oversize washers can be placed in the hole and tightened. Again, ultrasonic pulse echo can be used to map the boundaries and depths of the delaminations.

If a skin-to-core adhesive unbond needs to be injected, the honeycomb assembly should be inverted so that the unbond is on the bottom surface. This will prevent the low-viscosity resin from running down the cell walls.

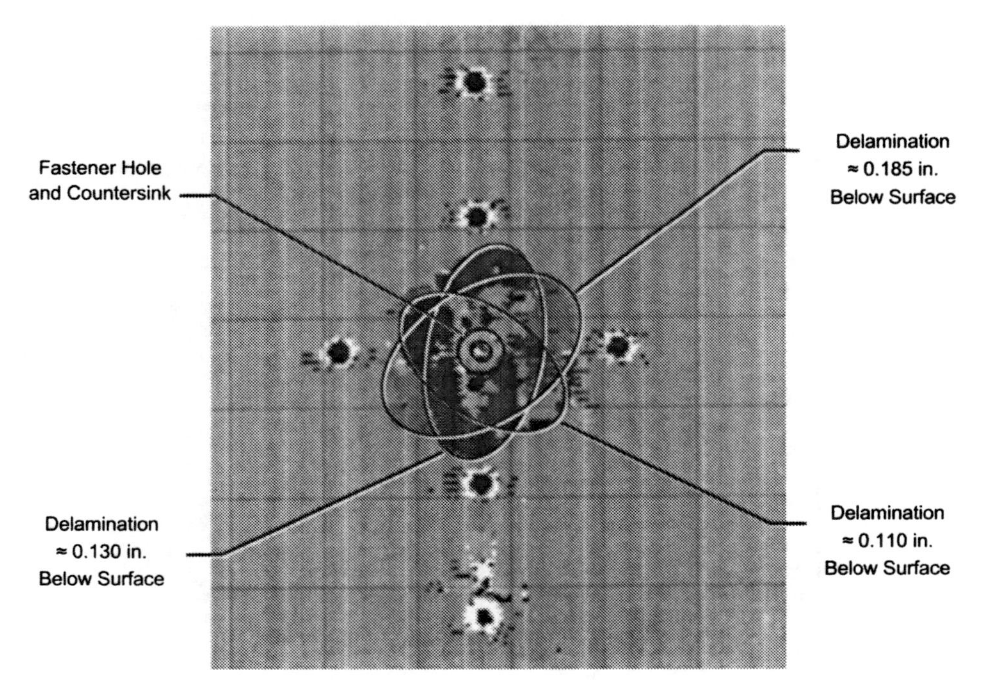

Fig. 19. Multiple Levels of Delaminations Around Fastener Hole

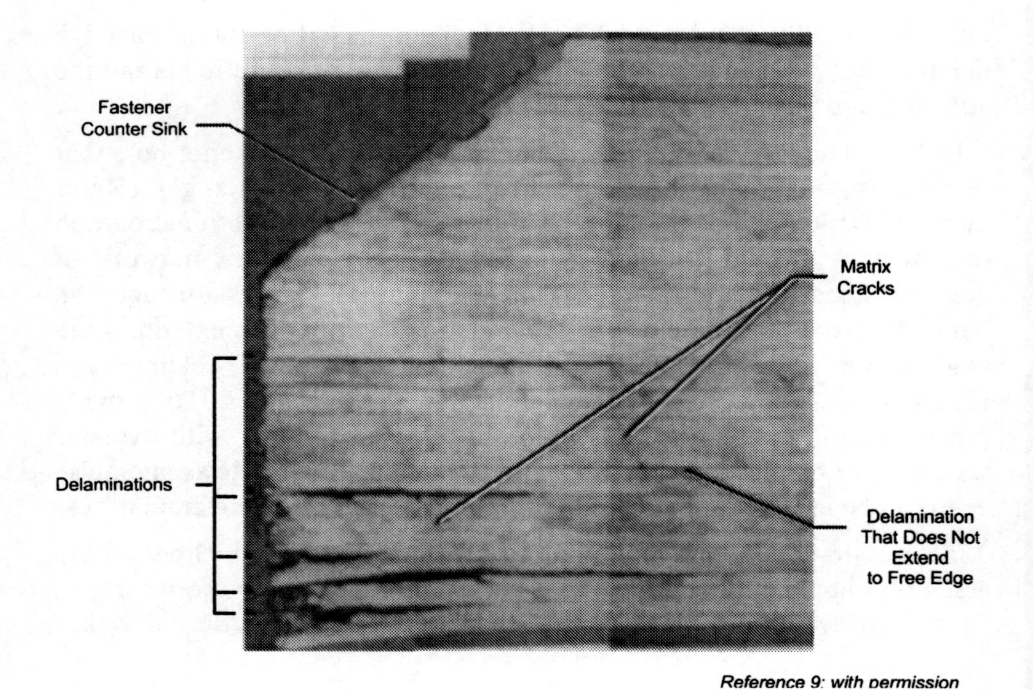

Fig. 20. Photomicrograph of Delaminations and Matrix Cracks

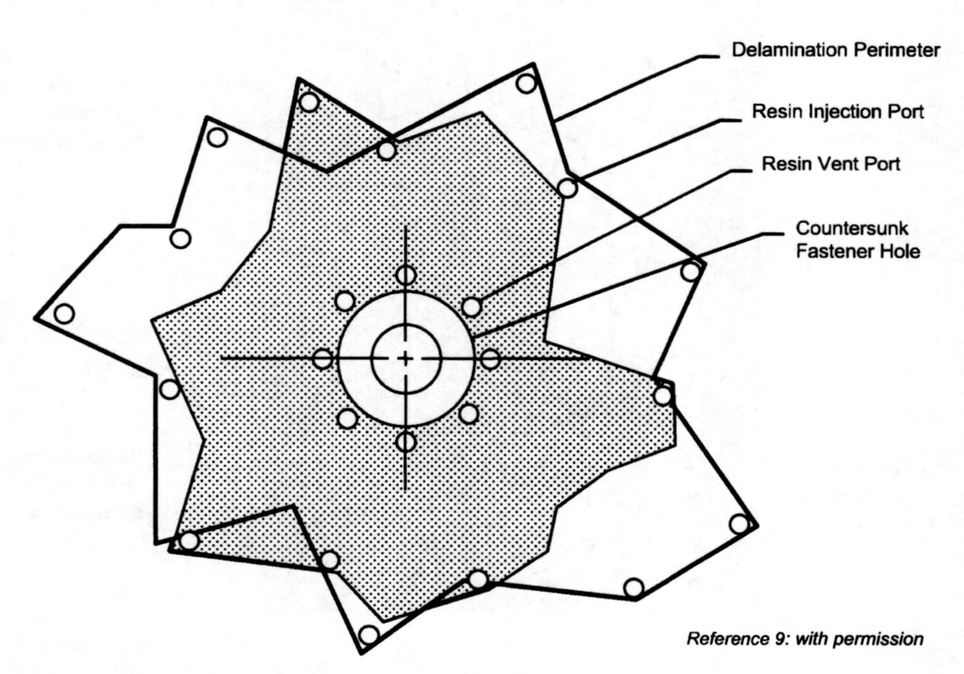

Fig. 21. Injection Scheme for Delaminated Fastener Hole

13.10 Bolted Repairs

Bolted repairs are usually preferred over bonded repairs because they are simpler and less prone to errors. A comparison of when to use bolted or bonded repairs is given in Table 2. The advantages of bolted repairs are

Table 13.2 Considerations When Selecting Structural Repair Method

Conditions		*Recommended*	
		Bolting	*Bonding*
Lightly Loaded, Thin (~<.10 in)		×	×
Highly Loaded, Thick (~>.10 in)		×	
Peeling Stresses High		×	
High Reliability Required		×	
Repairing Honeycomb Structure			×
Adherend Surfaces	Dry	×	×
	Wet	×	
	Clean	×	×
	Contaminated	×	
Sealing Required		×	×
Disassembly Required		×	
Restore "No Hole" Strength			×

Reference 9

obvious; however, the initial design of the structure must be able to accommodate the fastener bearing loads. In other words, the original design strains of the structure must be low enough that a bolted repair can be designed to carry ultimate design loads with a positive margin of safety.

Bolted repair patches (Fig. 22) can be applied to the external surface, the internal surface or both surfaces. A typical bolted repair (Fig. 23) consists of an external titanium patch, a center plug in the damaged area and a two-piece internal patch. The internal patch is split into two pieces so that it can be inserted inside the skin. Normal composite drilling and fastener installation procedures are used. Both protruding patches (Fig. 24) and flush patches (Fig. 25) can be designed, although protruding patches are easier to install.

If a one sided patch is being installed, the area to be repaired should be routed to form a circle or oval with no sharp edges or other stress concentrations. Undersize predrilled pilot holes should be drilled in the patch. To locate the patch, the skin should be marked with intersecting lines establishing the 0° and 90° directions. The patch is then located on the skin and two opposing end hole locations are marked on the skin. The patch is removed and two pilot holes are drilled in the skin. The patch can then be repositioned on the skin and held in place by using temporary

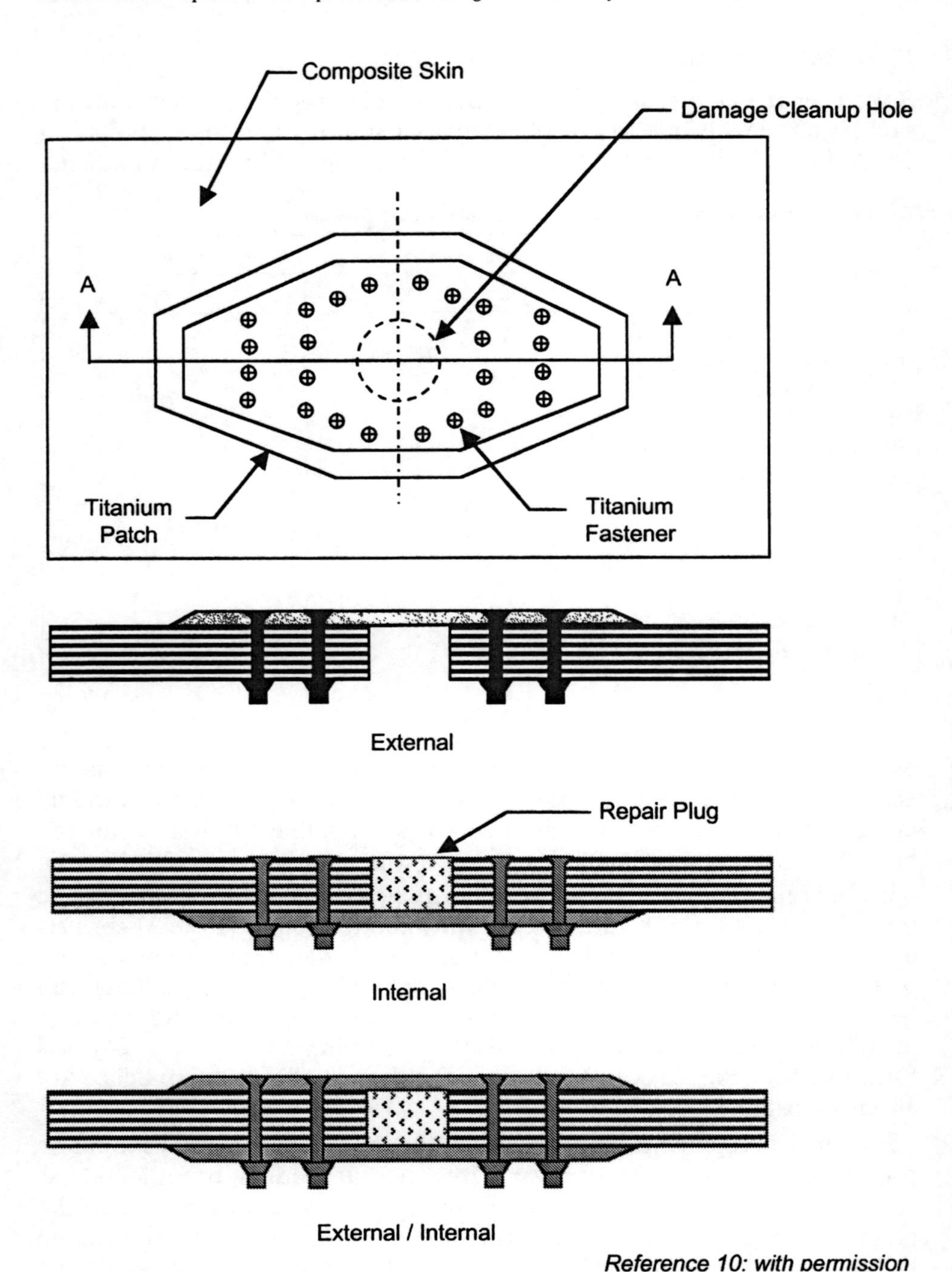

Fig. 22. Basic Bolted Repair Joints

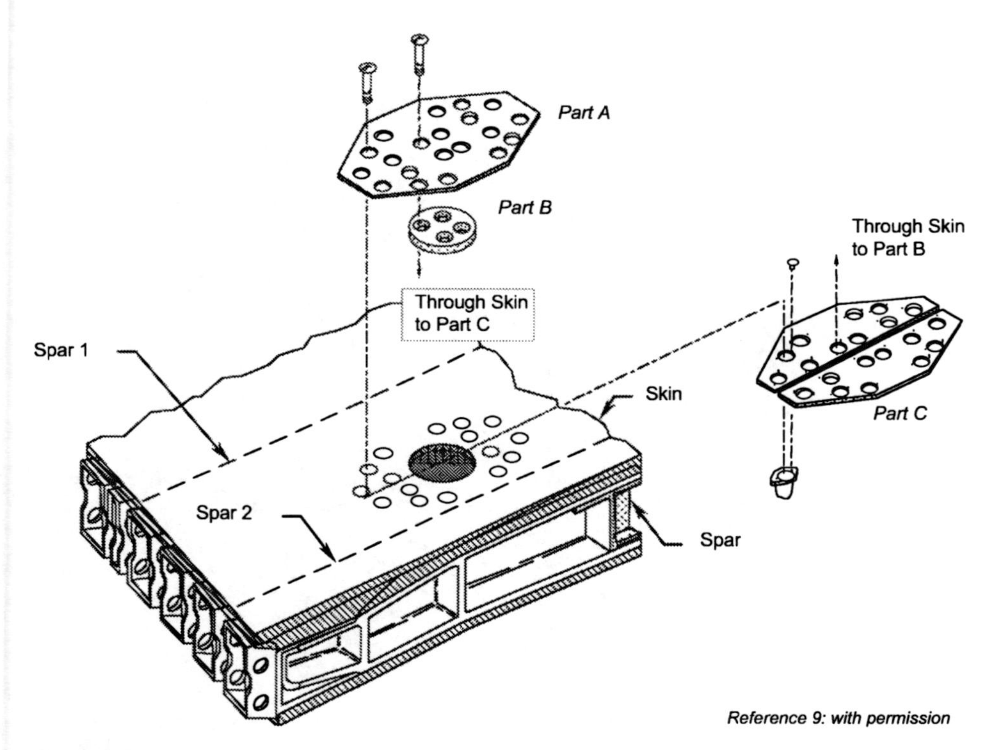

Fig. 23. Bolted Repair Concept

fasteners through the patch into the skin. A number of other pilot holes are then transferred from the patch to the skin. Additional temporary fasteners are installed as the pilot holes are drilled to provide clamp-up between the patch and skin. After all pilot holes are drilled and clamped in place, they can be enlarged to final diameter by first drilling and then reaming to final diameter, usually with a tolerance of +0.003 to –0.000 inches. Again, temporary fasteners are used to provide clamp-up between the patch and fastener. After all final hole sizes have been drilled and reamed, the patch is removed and deburred. A layer of woven glass cloth is impregnated with sealant to provide both sealing and corrosion protection. The full size holes in the patch are contersunk and the patch is installed with either one-sided blind fasteners if there is access to only one side or pin and collar fasteners if two-sided access is possible. All fasteners should be installed wet with sealant and the edges of the patch should be fillet sealed. One sided patches are the easiest to install but provide only single shear and asymmetry in the load path.

Two-sided patches provide double shear and more balanced load paths but are more difficult to install, particularly if only single sided access is

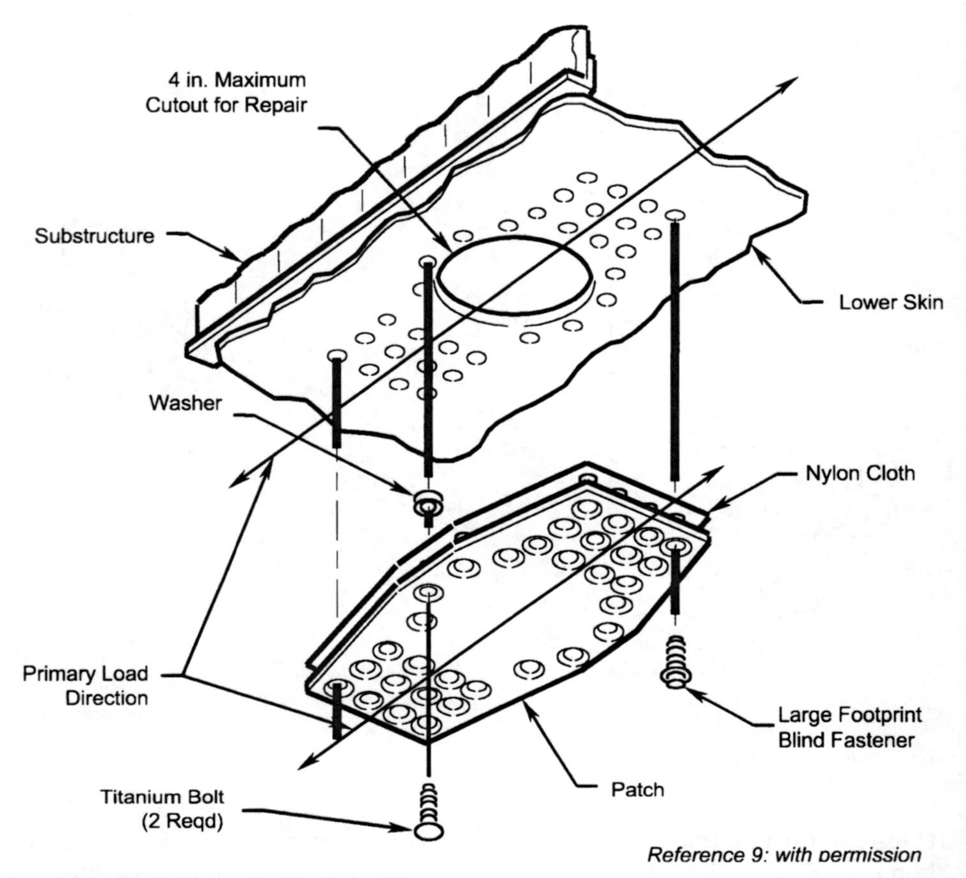

Fig. 24. Bolted Protruding Patch Repair

possible. Pilot holes, at least one through each piece, should be mate drilled through the outside and inside patch pieces prior to the start of installation. The interior patch pieces are often split so they can be slipped through the hole in the skin. Again, the outer patch is used to establish pilot holes in the skin. The interior and exterior patches are located and held in place with temporary fasteners inserted through the pilot holes located at each end. Using the existing pilot holes in the exterior patch and skin, pilot holes are drilled through the interior patch sections. Temporary fasteners are installed in the pilot holes as they are drilled. After all pilot holes have been drilled, they are brought up to full size by first drilling and then reaming to final size. The patch sections are then removed, countersunk on the exterior surface, deburred, fay surface sealed and reinstalled with temporary fasteners. If access to only one side is possible, then either blind fasteners can be used or plate nuts can be installed on the interior patch

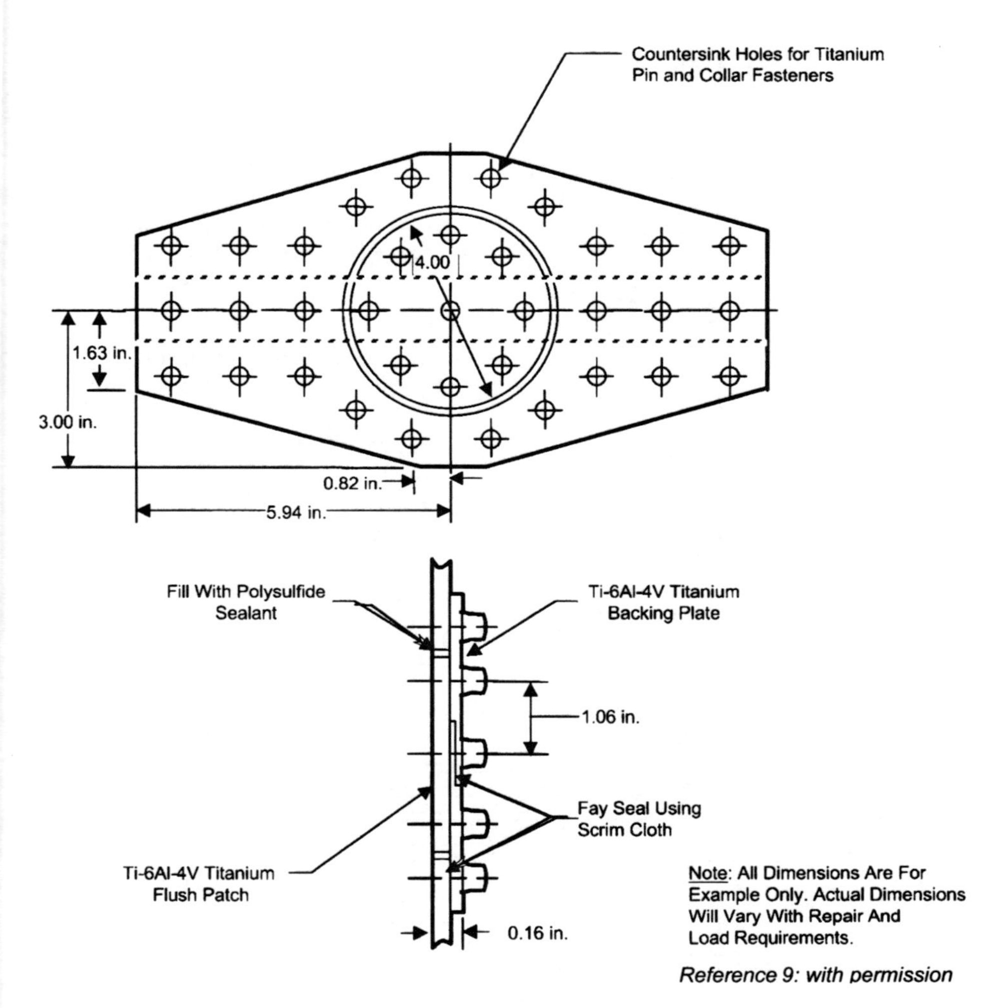

Fig. 25. Bolted Flush Patch Repair

sections for removable screws. Pin and collar fasteners can be used if access to both sides is available.

13.11 Bonded Repairs

Bonded repairs are the most difficult and error prone repairs to conduct. There are many variations of this type of repair and only a few examples will be given. Three typical bonded repairs (Fig. 26) are the bonded patch, the scarf repair and the stepped lap repair. The bonded patch repair can be accomplished using prepreg, wet lay-up or titanium sheets bonded together with layers of adhesive. For structure requiring higher load-bearing

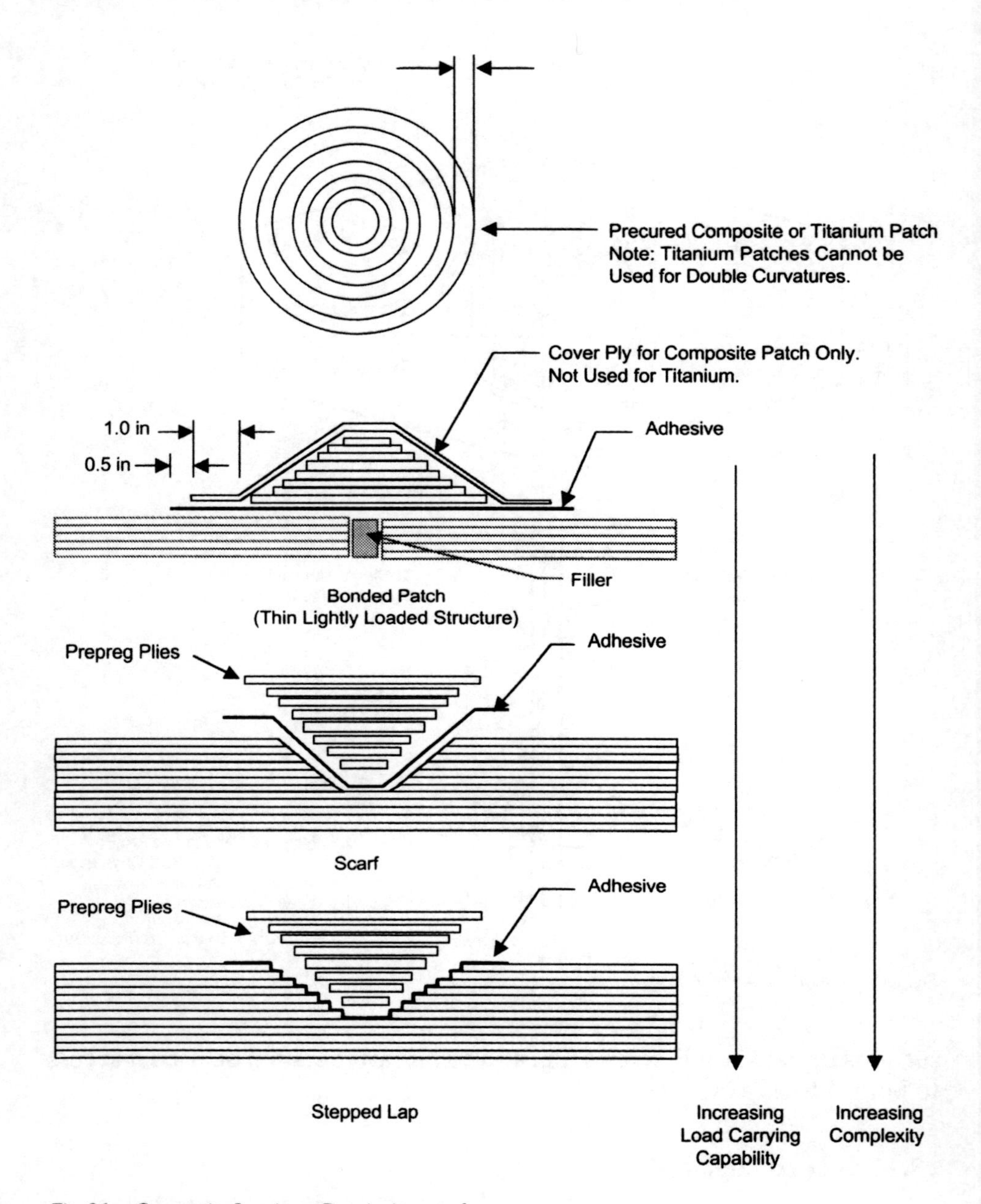

Fig. 26. Composite Laminate Repair Approaches

capability, the scarf or stepped lap repair is normally required; however, both of these methods require precise machining operations for effective load transfer. Bonded repairs also require significant areas of material to be removed (Fig. 27) to prevent high peeling forces on the adhesive.

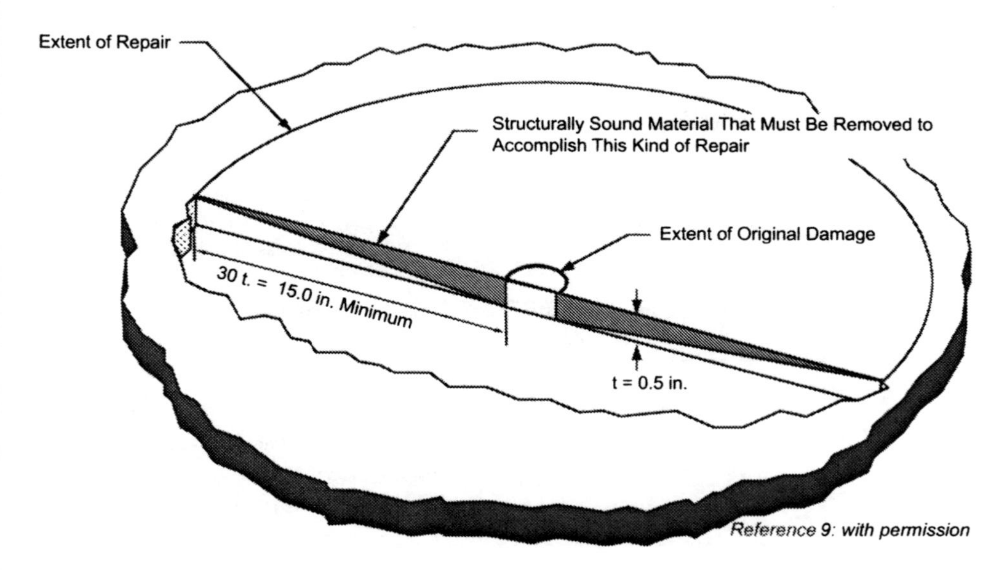

Fig. 27. Taper Angle Required for Effective Load Transfer for Bonded Repair

A typical hot bonded field repair procedure could consist of the following steps:

- Map out the damaged area with pulse echo ultrasonics and check to see if the repair falls within the limits of the structural repair manual.
- Carefully remove the damaged plies using high speed routers with depth control. If the stepped lap configuration is required, this will probably end up being a tedious job of hand cutting the steps ply-by-ply.
- If the repair is to be cured at a temperature of 200 °F or higher, dry the repair area at 200 °F for a minimum of 4 h.
- Bag the repair area with a heat blanket as shown in Fig. 28. Pull a full vacuum (22 inches of Hg vacuum minimum) on the repair area and slowly heat to the cure temperature (e.g., 250 or 350 °F). Hold under full vacuum at the cure temperature for the required period of time (e.g., 2–4 h). Maintain vacuum pressure on the cured patch until it cools down to 150 °F.
- Unbag the cured patch and clean-up the area. Inspect the quality of the repair with pulse echo ultrasonics.

For field repairs, there are commercial units (Fig. 29) that provide electrical power, thermocouple readouts, vacuum sources and can be programmed to provide uniform heat-up, hold and cool-down temperatures.

Bonded repair patches are usually made of prepreg, wet lay-ups, precured composites or thin titanium sheets adhesively bonded together. Since field repairs are conducted with only vacuum bag pressure (≤14.7 psia

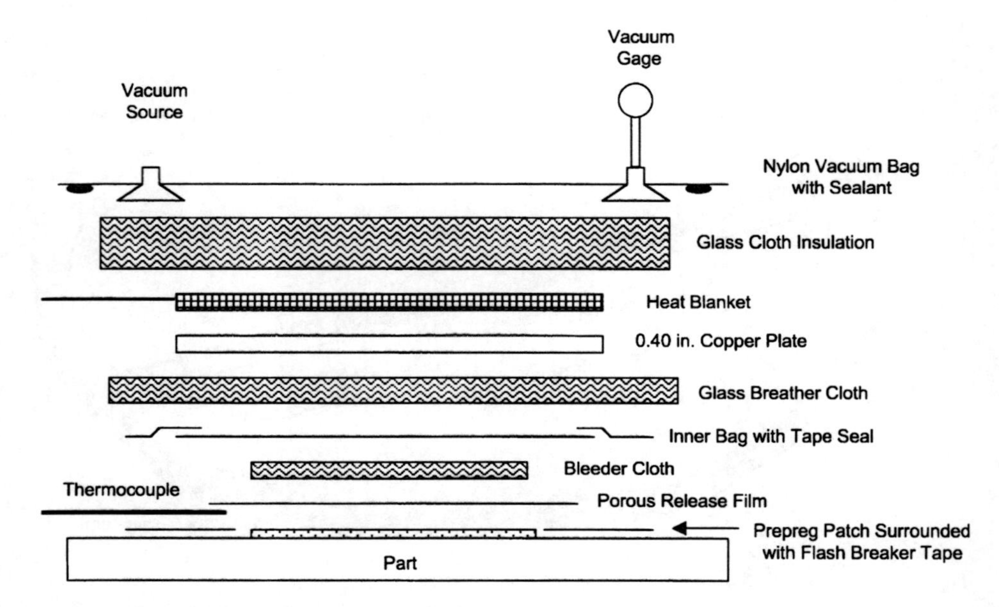

Fig. 28. Typical Bagging Sequence for Field Repair

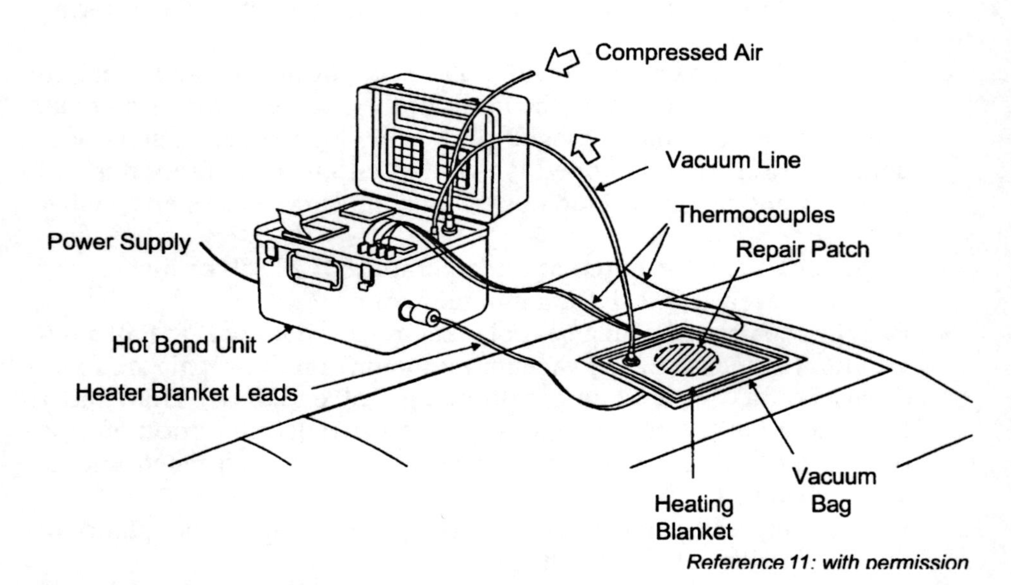

Fig. 29. Setup For Bonded Field Repairs

or less), the quality is not going to be as high as the original laminate that was cured in an autoclave at 100 psig. Prepreg patches produce higher-quality patches than wet lay-up patches. However, prepreg must be stored in a 0 °F freezer and has a shelf life of a year or less. Some resins used for wet lay-up patches can be stored at room temperature for 6–12 months. An additional concern with wet lay-up patches is that it may be difficult or impossible to match the strength and stiffness of the original structure if the structure is thin and made from unidirectional tape since the wet lay-up material will be woven cloth. Both prepreg and wet lay-up patches have the advantage they can be easily formed to the contour required for the repair.

Precured patches and patches constructed from adhesively bonding multiple layers of 0.010–0.020 inches thick titanium sheets bonded together can produce repairs with little or no voids. The bagging operations are also simplified since there is no bleeder required. The biggest disadvantages of these patches are they have very limited formability to contour and are only useful if there is a repetitive repair that has to be made. If titanium patch material is used, it must also be precleaned and primed and stored in moisture proof bags.

All damaged material should be removed and the hole or bottom of the repair should be round or an oval with no sharp edges. Machining of a scarf or stepped lap repair is an operation that requires a lot of care and skill. Templates and a depth control fixtures should be used with high-speed sanders or routers. On step lap repairs, it is important not to damage the plies under the step during the machining operation.

It is important that the area to be repaired is thoroughly dry, particularly if the repair area is going to be heated to above 200 °F. Absorbed moisture in skins or liquid water in honeycomb can turn to steam resulting in laminate delaminations and blistering, and water in core has been known to blow skins or cause additional core damage by blowing core cells. To help prevent moisture damage, the entire repair area should be thoroughly dried by slowly heating (e.g., 1 °F min^{-1}) to 200 °F and holding for 4 h for thin skins. Thick skins may take a considerably longer time. The drying operation can be combined with a trial run of the repair process by substituting dry glass cloth for the prepreg plies and bagging in the same manner that will be done for the actual patch cure. This trial run will reveal potential hot or cold spots in the repair area that can be corrected before actually committing to the actual repair cure cycle. Extra insulation or multiple heat blankets with zone control may be required to obtain uniform temperatures.

The type of resin and adhesive selected for the repair is also important. For example, if the original structure was bonded using a 250 °F curing adhesive, then it would certainly not be appropriate to use a 350 °F curing adhesive for a repair. In general, the lower the cure temperature for the composite patch plies and adhesive, the better the chances are for a successful repair that does not result in additional damage to the structure. If the repair is done by the Original Equipment Manufacturer (OEM) or at a depot with an autoclave, then the repair can be conducted using the same materials as used for original part manufacture since the part can be bagged and the repair cured at the same temperature and pressure as used during original fabrication. For example, if 350 °F curing systems were originally used, the autoclave pressure will hold everything together during a repair conducted at 350 °F. On the other hand, if only a local vacuum bag is used and the part is heated to 350 °F, the part might delaminate or unbond due to the adhesive bondlines becoming extremely weak at the 350 °F curing temperature. In addition, any moisture in either the laminate or honeycomb will reach much higher vapor pressures at 350 °F than at lower temperatures such as 250 °F or lower. Therefore, for field repairs, where only vacuum pressure is normally available, it is important to use a repair materials that cure at temperatures significantly lower (50–100 °F) than the original cure temperatures. Again, it is important to make sure the assembly is dry if placed in an autoclave or the repair area is dry if a vacuum bag is used for a field repair. In general, complicated repairs involving precision machining and complex lay-ups, such as bonded scarf and step lapped repairs should be conducted at either the OEM or a depot where more extensive tooling and autoclaves are available.

Epoxy film adhesives and prepregs are available that cure at 250–350 °F for 1–4 h, while wet-lay resins and adhesives will generally cure at room temperature in 5–7 d or can be heat cured at 160–180 °F in 1–2 h. While these materials require 5–7 d to obtain full strength, they generally cure sufficiently at room temperature in 24 hrs. so that the pressure can be removed and clean-up conducted, such as hand sanding and refinishing. While lower-temperature curing materials are attractive from a repair standpoint, the lower the cure temperature, the lower the glass transition temperature (T_g) and ultimate service temperature. If a wet lay-up or precured composite patch is cured with a paste adhesive, then a light weight layer of glass cloth should be embedded in the bondline to provide bondline thickness control and corrosion protection if a carbon patch is cured against aluminum honeycomb core.

Pressure for repair cures can be provided by an autoclave, mechanical (e.g., C-clamps) or a vacuum bag. An autoclave is certainly best because it can provide full pressure (100 psig) and the pressure application is isostatic; however, it does require drying of the entire assembly, usually in an oven, and either envelope bagging or placing on the original tool and vacuum bagging. Mechanical pressure can be very effective if the repair area is small and is located in an area that are accessible to clamps. Two concerns with clamps are that it is possible to apply too much pressure and locally crush the part, and clamps can serve as heat sinks making it difficult to obtain uniform heating. If clamps are used, pads should be used to spread the localized clamp loading across the surface. Vacuum bags are frequently used for field repairs since the repair can often done without having to remove the part from the aircraft. There are several disadvantages to vacuum bags: (1) the maximum pressure obtainable is between 10-15 psig; (2) the pressure is applied only at the repair area leaving adjacent areas without pressure if they get hot during heating; and (3) curing under vacuum tends to draw volatiles out of the matrix and adhesive during heating resulting in voids, porosity and frothy bondlines. To improve the quality of vacuum bag cures, the patch (prepreg or wet lay-up) can be staged prior to cure using the double bag technique shown in Fig. 30. In this concept, the freshly layed-up patch is bagged and a full vacuum is applied to the inner nylon vacuum bag. Then a vacuum is pulled on the outer hardback chamber and the patch is heated to the staging temperature. The staging temperature depends on the resin system used, but should be high enough to allow volatiles to escape from between the plies since there is no pressure pushing them together but low enough that the resin does not gel. For example, the staging temperature for a 350 °F curing adhesive might be 240–260 °F for 30 min. After the staging cycle is completed, the outer vacuum on the hardback is released allowing the inner vacuum bag to consolidate the plies. The patch is then cooled to below 150 °F before the pressure is released. After staging, the patch can be formed to the required contour by reheating with a hot air gun to soften the resin. The patch is then cured in the normal manner using a vacuum bag. This procedure greatly helps in reducing cured patch porosity and voids.[10]

Heat for curing may be supplied by an autoclave, oven, heat blankets or heat lamps depending on where and how the repair is conducted. Autoclaves and ovens provide the best temperature control. As previously discussed, if a heat blanket is used it is advisable to conduct a trial run to make sure there is sufficient power and identify any hot or cold spots. Heat blankets are available in a large range of sizes and coil designs that allow multiple blankets to be used if necessary to obtain uniform heating. Heat lamps are often used for simple repairs, such as fill repairs or to preheat localized areas for injection repairs. In all cases, ample thermocouples should be strategically placed so that the hottest and coldest areas are monitored. The cure temperature should be restricted by the hottest thermocouple while the cure time should be controlled by the lagging or coldest thermocouple.

After the patch has cured and allowed to cure below 150 °F under pressure, it can be unbagged, cleaned-up, inspected, sealed and refinished. Clean-up consists of removing all resin flash and sanding smooth. Flashbreaker release tape may be placed around the patch area before cure to minimize clean-up, but should not be located so close to the top patch ply that it prevents filleting between the patch and the skin. During lay-up the adhesive film can be allowed to extend beyond the composite patch plies to ensure a good fillet. All bonded repair patches should be inspected

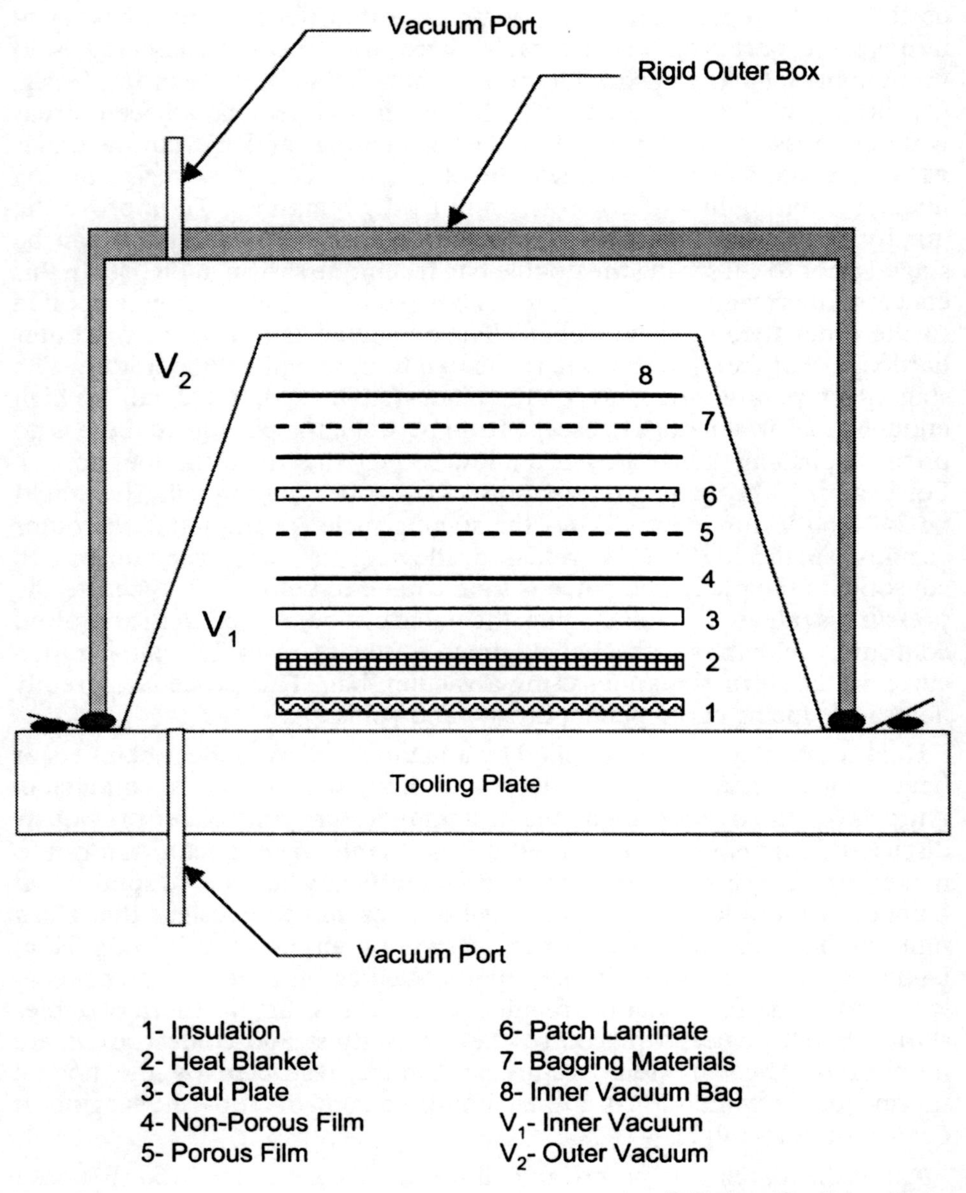

Fig. 30. Double Bag Method For Debulking Prepreg and Wet Lay-Up Patches

after curing using ultrasonic pulse echo to determine patch quality. If necessary the patch can be sealed by painting with a low-viscosity resin and then sanding when cured. The original finish system (primer and top coat) can then be applied.

Repairs of bonded honeycomb assemblies are similar to laminate repairs except that the honeycomb itself is often damaged and has to be replaced. Drying honeycomb assemblies is problematic. The drying cycle recommend for thin laminates (200 °F for 4 h) may not be sufficient for honeycomb assemblies if there is water in the core. If the part has been in service, the unit should be X-rayed before drying to determine if there is liquid water present. If it is found, then dry it at 200 °F for 4 hrs. and X-ray again. Repeat this process until all evidence of water disappears, then give it one final cycle at 200 °F for 4 h before bonding. While this process seems arduous, moisture is probably the main reason a lot of bonded repairs are unsuccessful.

A simple repair on a small area (Fig. 31) can often be accomplished by removing the damaged section of core and replacing it with a mixture of paste adhesive and milled glass fibers. After the filler cures, the surface should be scuff sanded and then a repair patch can be bonded. If the damaged section is larger, it is necessary to remove the damaged core and replace it with a new piece of the same density core. A typical single and double sided repair is shown in Fig. 32. Core can often be removed by slicing the cell walls with a sharpened putty knife and pulling pieces out

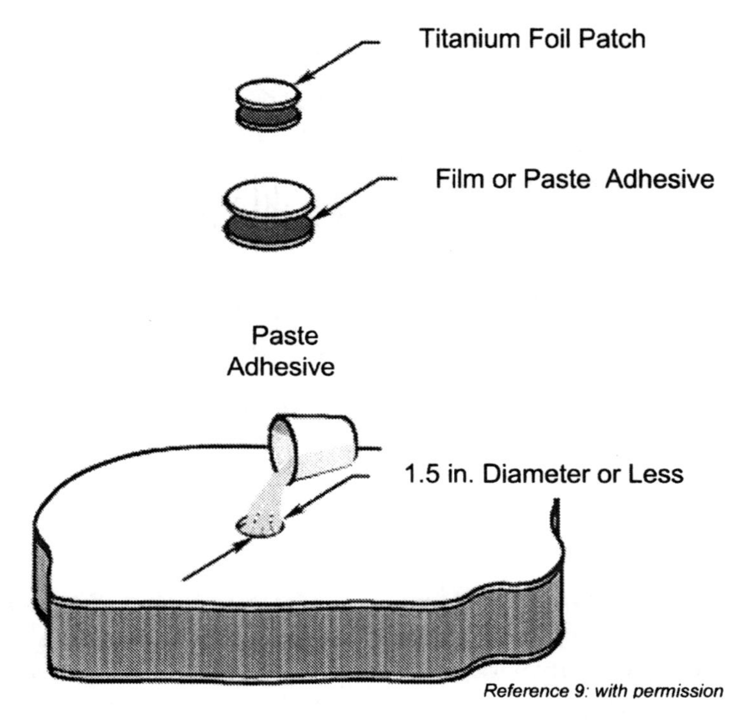

Fig. 31. Composite Honeycomb Minor Repair

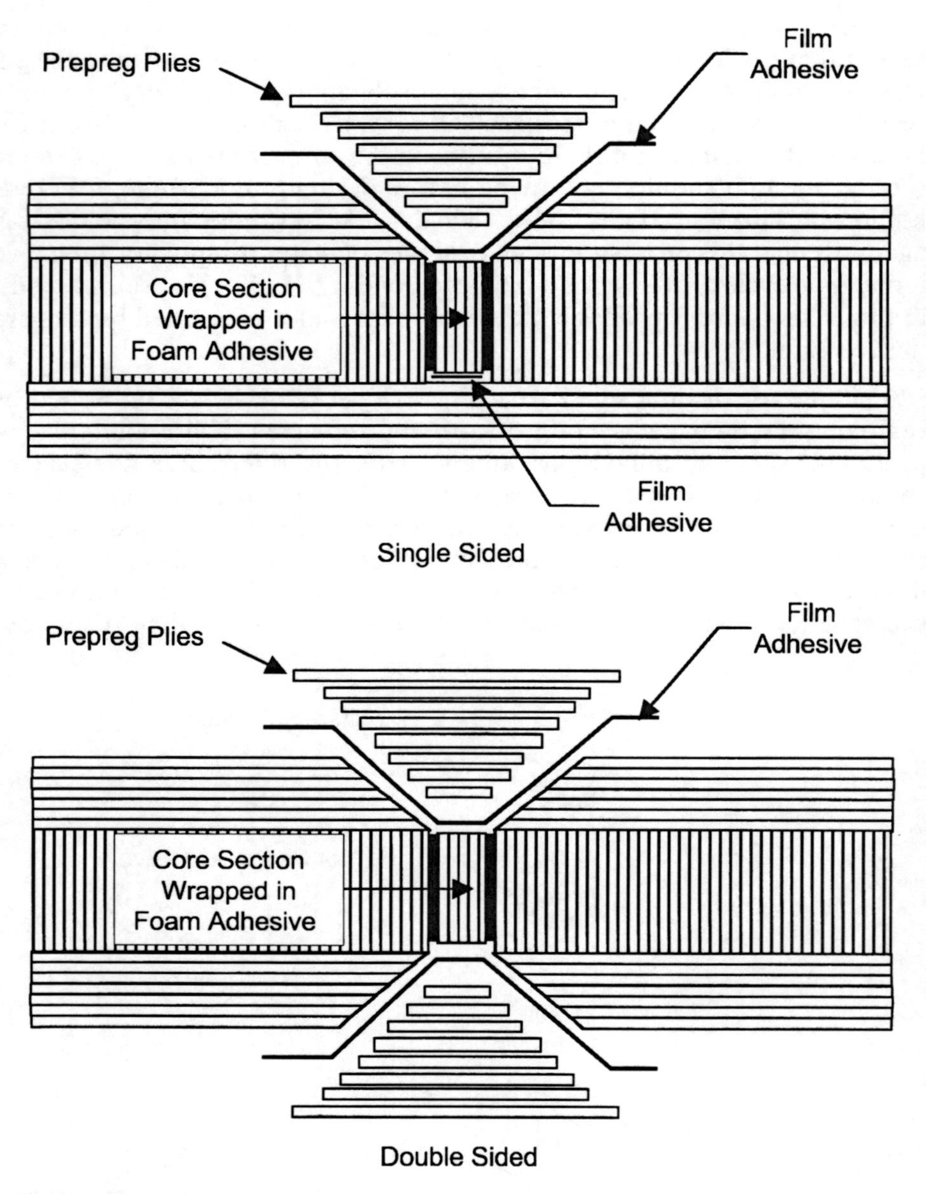

Fig. 32. Composite Honeycomb Bonded Repairs

with needle nose pliers. The remainder can then be ground down to the film adhesive on the undamaged skin. Replacement core is usually bonded into the main core body with either a foaming or paste adhesive. Foaming adhesive is normally used at the OEM and depot level where ovens are available since foaming adhesives will froth excessively under vacuum, while paste adhesives are usually used for field repairs. The core ribbon direction for the replacement section should match that of the remainder of the structure.

13.12 Summary

Damage can occur at almost any point during the composite manufacturing process and during service. The most serious composite manufacturing defects are porosity, voids, adhesive unbonds and delaminations. Once the part is placed in service, delaminations are the most prevalent and serious defects. Liquid water in honeycomb core remains a serious problem for bonded assemblies and makes repair difficult.

NDI is used to evaluate part quality after initial fabrication and periodically during its service life. Ultrasonics and radiography are the two most useful methods of inspection. Composite laminates are normally inspected by both through transmission for porosity and pulse echo for foreign objects. Radiography is occasionally used for composite parts with radii that may be subject to microcracking. Bonded honeycomb assemblies are inspected with both ultrasonics and radiography. Radiography is capable of finding many types of defects in honeycomb core that would go undetected with ultrasonics alone.

Repairs can be classified as fill, injection, bolted or bonded repairs. Fill repairs are primarily cosmetic in nature and do not restore any load-carrying capability. Depending on the nature of the delamination or unbond, injection repairs can be beneficial in restoring some portion of the original strength. Bolted repairs are normally used in solid composite laminates where high load-carrying capability is required; however, the original design must be capable of sustaining the bearing loads before a bolted repair can be considered. Bonded repairs are frequently required for thin skinned honeycomb assemblies.

References

[1] "Nondestructive Testing," in *ASM Handbook Volume 21 Composites*, ASM International, pp. 699–725, 2001

[2] Krautkramer J., Krautkramer H., *Ultrasonic Testing of Materials*, 4th edition, Springer, 1990.

[3] *Nondestructive Testing Handbook*, 2nd edition, American Society of Nondestructive Testing, 1991.

[4] *Nondestructive Evaluation and Quality Control*, ASM Handbook Volume 17, ASM International, 1989.

[5] Maldague X.P.V., *Nondestructive Evaluation of Materials by Infrared Thermography*, Springer, 1993.

[6] Armstrong K.B., Barrett R.T., "Care and Repair of Advanced Composites," Society of Automotive Engineers, 1998.

[7] Price T.L., Dalley G., McCullough P.C., Choquette L., "Handbook: Manufacturing Advanced Composite Components for Airframes," Report DOT/FAA/AR-96/75, Office of Aviation Research, April 1997.
[8] Hagemaier D.J., "Adhesive-bonded Joints," in *ASM Handbook Volume 17 Nondestructive Evaluation and Quality Control*, ASM International, 1989, pp. 610–640.
[9] Bohlmann R., Renieri M., Renieri G., Miller "Advanced Materials and Design for Integrated Topside Structures," training course given to Thales in The Netherlands, April 15-19, 2002.
[10] Chapter 8, "Supportability," in *MIL-HDBK-17-1F, Volume 3, Materials Usage, Design, and Analysis*, Department of Defense, December 12, 2001, pp. 8-1–8-60.
[11] "Composite Repair," Hexcel Composites, April 1999.

Appendix-A

Metric Conversions

To Convert From	To	Multiply By
Area		
in.^{2}	mm.^{2}	6.451 600E+02
in.^{2}	cm^2	6.451 600E+00
in.^{2}	m.^{2}	6.451 600E-04
ft^2	m^2	9.290 304E-02
Force		
lbf	N	4.448 222E+01
kip (1000 lbf)	N	4.448 222E+03
Length		
mil	μm	2.540 000E+01
in.	mm	2.540 000E+01
in.	cm	2.540 000E+00
ft	m	3.048 000E-01
yd	m	9.144 000E-01
Mass		
oz	kg	2.834 952E-02
yd	kg	4.535 924E-01
Mass Per Unit Area (Areal Weight)		
oz/in.^{2}	kg/m^2	4.395 000E+01
oz/ft.^{2}	kg/m^2	3.051 517E-01
oz/yd^2	kg/m^2	3.390 575E-02
lb/ft^2	kg/m^2	4.882 428E+00
Mass Per Unit Volume (Density)		
lb/in.^{3}	g/cm^3	2.767 990E+01
lb/in.^{3}	kg/m^3	2.767 990E+04
lb/ft^3	g/cm^3	1.601 846E-02
lb/ft^3	kg/m^3	1.601 846E+01
Pressure (Fluid)		
lbf/in.^{2} (psi)	Pa	6.894 757E+03
in. of Hg (60°F)	Pa	3.376 850E+03
atm (Standard)	Pa	1.013 250E+05

To Convert From	To	Multiply By
Stress (Force Per Unit Area)		
lbf/in.^{2} (psi)	MPa	6.894 757E-03
ksi (1,000 psi)	MPa	6.894 757E+00
msi (1,000,000 psi)	MPa	6.894 757E+03
Temperature		
°F	°C	5/9 (°F-32)
°K	°C	°K-273.15
Thermal Conductivity		
Btu/ft-h-°F	W/m-°K	1.730 735E+00
Thermal Expansion		
in./in.-°F	m/m-°K	1.800 000E+00
in./in.-°C	m/m-°K	1.000 000E+00
Velocity		
in./s	m/s	2.540 000E-02
ft/s	m/s	3.048 000E-01
ft/min	m/s	5.080 000E-03
ft/h	m/s	8.466 667E-05
Viscosity		
poise	Pa-s	1.000 000E-01
Volume		
in.^{3}	m^3	1.638 706E-05
ft^3	m^3	2.831 685E-02

Index

CPSIA information can be obtained
at www.ICGtesting.com
Printed in the USA
LVOW10*2105291117
558089LV00009B/35/P